Energy Efficient Technologies for Food Safety, Quality, and Security

Ahmed Alhussaini Hamad
Benha University, Egypt

Pallavi Singh
Poornima University, India

Dipali Saxena
Shri Vaishnav Vidyapeeth Vishwavidyalaya, India

A.K. Haghi
University of Coimbra, Portugal

IGI Global
Scientific Publishing
Publishing Tomorrow's Research Today

Vice President of Editorial	Melissa Wagner
Managing Editor of Acquisitions	Mikaela Felty
Managing Editor of Book Development	Jocelynn Hessler
Production Manager	Mike Brehm
Cover Design	Phillip Shickler

Published in the United States of America by
IGI Global Scientific Publishing
701 East Chocolate Avenue
Hershey, PA, 17033, USA
Tel: 717-533-8845
Fax: 717-533-7115
Website: https://www.igi-global.com E-mail: cust@igi-global.com

Library of Congress Cataloging-in-Publication Data

LCCN: 2024054398 (CIP Data Pending)
ISBN13: 9798369383421
EISBN13: 9798369383445

British Cataloguing in Publication Data
A Cataloguing in Publication record for this book is available from the British Library.

Table of Contents

Detailed Table of Contents

The food industry is one of the most energy-intensive sectors globally, encompassing processes from food production and processing to storage and distribution. As energy consumption rises, the need for effective energy efficiency measures becomes critical not only to reduce operational costs but also to mitigate environmental impacts. Reports have suggested that maximum inefficiency of energy in the food industry is witnessed in conventional technologies of processing and preservation right from farm to fork. Novel technologies such as novel thermodynamic cycles and non-thermal and novel heating processes hold the potential to reduce energy consumption and production costs and improve the sustainability of food production. This chapter investigates the different opportunities for energy efficiency in the food industry. Initially, it provides an overview of the various food industries and related energy consumption. Then, fundamentals of energy efficiency, and the potential of renewable energy resources.

The chapter "Energy Management System Applications in the Food Sector" in Energy Efficient Technologies for Food Safety, Quality, and Security examines how energy management systems (EMS) are utilized to optimize energy use and boost efficiency in the food industry. It explores the foundational principles of EMS, including energy monitoring, control, and optimization, and details the deployment of advanced sensors and automated controls across various food production processes. Through case studies, the chapter illustrates successful EMS applications that lead to significant energy savings, reduced costs, and enhanced sustainability. It also highlights the role of data analytics, machine learning, and emerging technologies like AI and IoT in advancing EMS capabilities, ultimately supporting goals of improved food safety, quality, and environmental stewardship.

A vital component of the global economy, the food industry produces, processes, and distributes food to support people all over the world. The need for food production has grown along with the world's population, which has caused food factories in both developed and developing nations to expand quickly. This chapter looks at how the food industry uses energy management systems (EMS) to maximize energy use and increase productivity. It examines the fundamental ideas of EMS, such as energy monitoring, control, and optimization, and describes how sophisticated sensors and automated controls are being implemented in a variety of food production processes. However, the quick growth has resulted in serious environmental problems, such as pollution, emissions, and excessive energy and water use. This chapter provides examples of effective EMS applications that result in notable cost savings, increased sustainability, and energy savings.

The rapid depletion of conventional fossil fuels is a result of factors such as urbanisation, population growth, and increased food consumption. The adoption of renewable energy sources (RES) presents a feasible way to address the growing energy needs while lowering the threats to the world's food supply. Thus, the use of various renewable energy sources (RES) for food processing—such as solar, wind, geothermal, biomass, etc.—is covered in this chapter. One of the most researched uses of solar energy is drying food products for longer shelf lives. Many different types of solar dryers have been created. In addition to drying food, heat produced by solar and geothermal energy can be used to heat, evaporate, distil, sterilise, and pasteurise beverages. Additional broad-spectrum uses of clean energy include milling and grinding, solar ovens, stoves, and cookers. To determine potential routes for lowering the reliance of the global food processing industry on traditional fossil fuels, a variety of well implemented renewable energy-based alternatives have been examined.

of high energy price volatility. The sector is energy-intensive, particularly in processing activities like drying, freezing, and canning, which cost about $800 million annually in electricity and fuel. This chapter examines energy consumption patterns and technologies that improve efficiency at different stages of fruit and vegetable production to enhance food security and minimize environmental impacts.

The integration of energy-efficient technologies in the food sector presents a transformative approach to addressing global challenges related to food safety, quality, and security. These technologies, which include solar-powered cold storage, LED lighting, and precision agriculture, reduce energy consumption and enhance food preservation, production efficiency, and resource management. This research explores the impact of these innovations on food safety and environmental sustainability, with particular emphasis on their ability to support Sustainable Development Goals (SDGs) like Zero Hunger and Climate Action. By analyzing case studies, the study demonstrates how energy-efficient solutions can mitigate environmental degradation while improving the resilience of food systems.

In particular, the food industry is steadily dependent on fossil fuels, and the food demand is look ahead to grow significantly in the coming years. Therefore, developing energy efficiency strategies for this particular industrial sector is very pivotal. In this chapter study about the different opportunities for energy efficiency in the food industry. It provides a brief overview of the various food processing and related food safety issues. Then, the different options for food improvement energy efficiency in the thermal and electric sector are discussed. Latest possibilities and trends, the food industry's sustainable growth depends on energy saving. In recent decades to enhance the food processing sustainability and food industry has also focused on recovering waste heat and enhancing energy efficiency. Reducing energy consumption, cutting production costs, and enhancing food production sustainability can also be achieved by substituting new technologies, like non-thermal and novel heating methods, for existing energy-intensive food processes, such as novel. In the food sector, new food processing technologies have been developed to replace conventional energy-intensive unit operations for chilling, freezing, evaporating, dehydrating, pasteurization and sterilizing. The majority of energy-saving technology are easily adaptable from other industrial industries to the food processing industry.

This chapter explores sustainable packaging solutions, highlighting the importance of reducing environmental impact through the adoption of renewable materials, energy-efficient processes, and cutting-edge technologies. It emphasizes the principles of reducing, reusing, and recycling, with a focus on biodegradable and compostable materials that mitigate the adverse effects of conventional packaging. The role of smart packaging in enhancing food safety and quality is also examined, presenting new opportunities in the industry. The stakeholders—manufacturers, retailers, policymakers, and consumers can collaborate to drive the adoption of eco-friendly packaging, overcome challenges, and ensure scalability and cost-effectiveness. Looking to the future, it envisions a path of continuous innovation, where advancements in biopolymers, intelligent systems, and digital technologies pave the way for packaging solutions that are both environmentally and economically sustainable.

This article delves into the recent advancements in food processing, exploring their impact on food safety, quality, and sustainability. Alongside safety and quality, the food industry also strongly emphasises sustainability and environmental impact. The adoption of renewable energy sources, the reduction of waste, and the implementation of circular economy principles in the food supply chain are becoming increasingly important. Furthermore, the integration of technologies with other preservation methods, such as antimicrobial agents or modified atmospheres, has shown promising

Preface

The food industry today faces an extraordinary challenge. With the global population rising and natural resources becoming scarcer, the need to produce safe, high-quality, and secure food has never been greater. At the same time, there's growing urgency to reduce environmental impacts and embrace more sustainable ways of working. That's where energy efficiency comes in—it's not just a buzzword but a critical piece of the puzzle for creating a more sustainable food system. This book, *Energy Efficient Technologies for Food Safety, Quality, and Security*, dives into the exciting ways energy-efficient solutions are reshaping the food industry. From the farm to the factory floor, it examines how new technologies, and smarter practices can help us use energy more wisely, minimize waste, and improve the safety and quality of the food we produce. Each chapter takes a closer look at a specific part of the food industry—whether it's livestock, dairy, fruits and vegetables, or post-harvest processing and packaging—and explores how innovative, energy-efficient approaches can make a difference. We also look at the bigger picture, including the role of renewable energy, the importance of sustainability, and even the legal and intellectual property issues tied to these innovations. This book is for anyone who cares about the future of food—researchers, industry professionals, policymakers, and even curious readers who want to learn more about how energy efficiency can transform the way we eat and live. It's packed with ideas, strategies, and success stories that show what's possible when we prioritize sustainability in every step of the food chain. Ultimately, *Energy Efficient Technologies for Food Safety, Quality, and Security* is about building a food system that works for everyone—one that meets today's demands while protecting the planet for future generations. We hope this book inspires you to think differently about energy, food, and the role you can play in creating a more sustainable world.

Ahmed Hamad
Pallavi Singh
Dipali Saxena

A. K. Haghi

Chapter 1

Introduction to Energy Efficiency in the Food industry

Garima Sengar
https://orcid.org/0000-0002-8177-319X
Sant Longowal Institute of Engineering and Technology, India

Dipali Saxena
https://orcid.org/0000-0002-8177-319X
Shri Vaishnav Vidyapeeth Vishwavidyalaya, India

ABSTRACT

The food industry is one of the most energy-intensive sectors globally, encompassing processes from food production and processing to storage and distribution. As energy consumption rises, the need for effective energy efficiency measures becomes critical not only to reduce operational costs but also to mitigate environmental impacts. Reports have suggested that maximum inefficiency of energy in the food industry is witnessed in conventional technologies of processing and preservation right from farm to fork. Novel technologies such as novel thermodynamic cycles and non-thermal and novel heating processes hold the potential to reduce energy consumption and production costs and improve the sustainability of food production. This chapter investigates the different opportunities for energy efficiency in the food industry. Initially, it provides an overview of the various food industries and related energy consumption. Then, fundamentals of energy efficiency, and the potential of renewable energy resources.

DOI: 10.4018/979-8-3693-8342-1.ch001

1. INTRODUCTION

The food industry plays a crucial role in global economies, providing sustenance to billions while also contributing significantly to energy consumption. From agriculture to food processing, packaging, storage, and transportation, the sector involves a range of energy-intensive processes. As global awareness of climate change and resource depletion increases, the food industry is under growing pressure to adopt energy-efficient practices to reduce its carbon footprint, optimize operational costs, and contribute to environmental sustainability.

Energy efficiency in the food industry not only improves profitability by lowering energy bills but also promotes more sustainable use of resources, enhancing the industry's long-term viability. This chapter delves into the fundamental concepts of energy efficiency within the food sector, exploring the technologies, strategies, and innovations that can lead to substantial energy savings. It addresses the energy demands of different stages of food production, identifies key opportunities for energy conservation, and highlights the benefits of integrating energy-efficient technologies across the supply chain.

Additionally, this chapter examines the broader environmental and economic implications of energy efficiency in the food industry, emphasizing the need for a comprehensive approach that incorporates energy management, regulatory frameworks, and consumer awareness. By exploring successful case studies and best practices, this chapter provides a practical roadmap for food industry stakeholders seeking to implement energy-saving measures and adopt a more sustainable approach to production and consumption.

1.1 Status of Energy Consumption in Food Industry Globally

Globally the consumption of energy in processing and distribution of food is recorded as 45% (FAO 2017;EIA 2017). India's food processing sector which is one of the largest in the world is expected to reach an output of $ 535 Bn by 2025-26. It lies at the heart of the Government of India's Make in India initiative and resonates with Prime Minister Modi's call for 'vocal for local' (Bhatia et al., 2020). All segments of food production are energy intensive and are heavily dependent on the use of fossil fuel. The two most energy intensive activities are found within industries are agriculture production and processing.

Energy consumption of the Indian food processing industry (including the production of meat, fish, fruit, vegetables and oil, beverages, products by grain mills and dairies, and other foods such as bakery products, convenience products, cocoa products or salty snacks) requires 5,300 kilotons of oil equivalent per year. Electrical energy consumption accounts for about found 50% of total energy consumption, especially for preservation and processing. In further segments such as beverages and other foods, a high proportion of the required energy is needed for heating processes – which to a large extent is delivered by petroleum and biomass (Ministry of New and Renewable Energy, Gov of India)

Energy employed in manufacturing and distribution of foods- within the UK and globally- within the food industry was collected from literature and clustered by product, processing technique and transportation method. Researchers marked instant coffee, milk powder, French fries, crisps and bread among the most energy intensive food products. The thermal processes of these products were found to consume large proportions of the total processing energy. In the meat and dairy processing sectors, energy and water use were increased due to a rise in hygienic standards and cleaning requirements. Additionally, meat products are processed- and sometime over processed- to a higher degree for consumer convenience, all this increasing the associated energy usage for manufacture (Ladha-Sabura et al., 2019)

Food and Drink Federation documented that, the food processing industries in the U.K is the largest single manufacturing sector, with an annual turnover of £97.3b (Food and Drink Federation, 2018). It is also the fourth largest industrial energy user: 117 petajoules (PJ) consumed in 2017 (Department for Business, Energy and Industrial Strategy, 2018).

In the EU-27 the overall energy consumption of the food industry has been found stable since 1990. The annual consumption of energy consumed by this sector was around 30 Mtoe. The use of natural gas in the food sector started to fall over the period 1990–2003, which was mainly due to a sudden increase in gas price, which doubled between 1996 and 2006. However, the abandoning of oil started much earlier with a constant increase in consumption of electricity (Kaminski and Leduc 2010).

A literature survey was carried out to collect energy consumption for the food manufacturing sector for the time period 1980 to 2015. Specific energy consumption (SEC) data for products, processes and food distribution was collected based on product-based energy intensity (PEI) metrics (Briam et al., 2015). The collected data informed about location of the study, product description, processes involved, energy sources, and bibliographic reference.

1.2 Energy Efficiency Conceptualization

The practice of utilizing lesser energy to accomplish the same amount of task and provide useful output from inventory (such as heating water, lighting, or cooling a fridge) of any small- or large-scale production unit, is Energy efficiency. It is measured as the ratio between the useful output of the end-use service and the associated energy input. It can also be characterized as the relationship between how much energy is needed to power a technology (for example, a light bulb, boiler, or motor) and the end-use service (for example, lighting, space heating, or motor power) that the technology provides.

Every year a lot of energy is wasted because of heat loss and inefficient technology. This leads to increased carbon pollution and costs consumers and businesses money. Energy efficiency is "one of the easiest and most cost-effective ways to combat climate change, clean the air we breathe, improve the competitiveness of our businesses and reduce energy costs for consumers" (US Department of energy).

Role of energy efficiency in economic development, and in reducing greenhouse gas (GHG) emissions was studied by Ayres et al., 2007. The study also addressed the deficiency of neoclassical economic growth theory. Raymond and Leung 2021 generated a model frame work where they specify a production equation in neoclassical economic growth framework, with renewable energy consumption alongside the usual capital and labour inputs to analyse the effect that renewable energy may have on real GDP:

$$Y_{it} = f(K_{it}, L_{it}, RE_{it})$$

Where Y_{it} is real GDP of country *I* in time period *t*.

K is capital input; L is labour input and RE is renewable energy input.

The whole expression can help to find out the effect of renewable energy resources in a particular sector on economic growth of a country.

2. ENERGY CONSUMPTION PATTERNS IN FOOD PROCESSING INDUSTRIES

The energy consumption and energy type used for the processing of a certain quantity of goods depends heavily on their nature (Clairand et al., 2020). In a food processing industry, the energy consumption mostly occurs in the areas of heat processing, refrigeration, ventilation and air conditioning (HVAC systems) (Wang 2008). Food processing systems can be categorized based on their size and range from small family consumption to large commercial consumption (Sims 2014). High energy consuming systems must be marked and taken in to consideration for reducing the energy input.

3.1 Energy Consuming Systems

i. **Subsistence:** Families engaged in the most basic forms of small-scale farming and fishing, and producing food for their own use solely are subsistence producers and they use very low inputs of energy, usually from human and animal power.
ii. **Small farms:** Small family farming units can engage in different activities, including the development of small gardens or rice fields, organic vegetables, orchards, cattle rearing, private fishing boats and dairy herds. They work slightly more and hire more people than small farms. These companies can reduce their fossil fuel dependence by improving energy efficiency and generating renewable energy on-farm
iii. **Large farms:** Corporate food systems depend on high direct external supply chain energy inputs, and they include fish trawler fleets, feed farms, sugar companies and palm oil farms.

3.2 Energy Consuming Technologies

i. **Drying:** This process consists of artificially drying cereals after harvesting and before storage and transport. This step could be one of the most energy-intensive operations, especially for developing countries.
ii. **Storage:** This process accounts for maintaining food at the proper temperature conditions in order to avoid degradation of quality of the product and provide both safe and high-quality foods. The typical machines used for storage in the food industry are energy consuming and include refrigerators and freezers.
iii. **Food and beverage processing:** The process involves the transformation of agricultural raw products into finished food and requires energy for heating, cooling and refrigeration.
iv. **Food cooking:** The process involves heat application to food in order to obtain a safe and nutritionally rich finished product. The time and temperature for cooking is pre-decided based on type of raw material and overall microbial load.
v. **Evaporation:** this process involves partially removing water from liquid food via boiling.
vi. **Dehydration:** this process involves reducing moisture in food to low levels for improved shelf life by adding one or more forms of energy to the food.
vii. **Filtration:** this process involves separating solids from a suspension in a liquid via a porous medium, screen or filter cloth, which retains the solids and allows the liquid to pass through.

Table 1. Technology based energy usage

Technology	Type of energy used	Unit of energy used
Drying	Electricity Natural gas Liquefied petroleum gas (LPG)	0.5-0.75MJ/Kg
Storage	Electricity	1-3MJ/Kg
Processing (Food and Beverage)	Electricity (for heating and cooling)	50-100MJ/Kg
Food Cooking	Heat through electricity/fuel	5-7MJ/Kg
Evaporation	Heat through electricity/fuel	2.5-2.7MJ/Kg

Source: Clairand et al 2020

3. ENERGY EFFICIENCY TECHNOLOGIES

The pursuit of energy efficiency in the food industry is driven by the need to reduce operational costs, lower environmental impact, and enhance sustainability. A wide array of technologies has been developed and implemented to address energy consumption across various stages of food production, processing, storage, and distribution. This section highlights some of the most promising energy efficiency technologies that are transforming the food industry by improving energy use, reducing waste, and optimizing processes.

3.1 Heat Recovery Systems

Heat recovery is one of the most effective ways to increase energy efficiency in food processing plants. Many food production processes, such as cooking, drying, and pasteurization, generate substantial amounts of waste heat. By capturing and reusing this waste heat, companies can significantly reduce their energy needs. Technologies such as heat exchangers, evaporative cooling systems, and regenerative heat exchangers are commonly used to recover heat and reuse it in other parts of the production process, reducing the need for additional energy input (Zhou et al., 2023).

3.2 High-Efficiency Motors and Drives

Electric motors are critical components in the food industry, powering various equipment such as pumps, mixers, conveyors, and refrigeration units. The use of energy-efficient motors and variable speed drives (VSDs) can lead to significant energy savings by optimizing motor performance based on the actual demand of the process. VSDs allow for better control of motor speed and power consumption, ensuring that motors run at the most energy-efficient levels for different tasks (Singh & Kumar, 2022).

3.3 Advanced Refrigeration Systems

Refrigeration is essential in food preservation, but it is also one of the most energy-consuming operations in the food industry. Advances in refrigeration technology, such as the adoption of natural refrigerants (e.g., CO2, ammonia) and more efficient compressor systems, have the potential to drastically reduce energy consumption. Additionally, innovations like thermoelectric cooling systems and improved insulation materials can further enhance the efficiency of refrigeration units, particularly in large-scale food storage and processing facilities (Mohammed et al., 2021).

3.4 Lighting and Building Energy Management Systems

Lighting systems in food processing and storage facilities are often a significant contributor to energy use. Implementing energy-efficient lighting technologies, such as LED lighting, and incorporating automated control systems can reduce energy consumption by adjusting lighting levels based on occupancy and time of day. Additionally, building energy management systems (BEMS) enable the monitoring and control of various energy-consuming systems within a facility, including heating, ventilation, and air conditioning (HVAC), lighting, and equipment, optimizing energy usage in real-time (Patel et al., 2023).

3.5 Cogeneration (Combined Heat and Power - CHP)

Cogeneration systems, or combined heat and power (CHP) systems, generate both electricity and useful heat from a single energy source. This technology is particularly effective for food industries that require both electricity and thermal energy for their operations, such as in cooking, drying, and sterilization. By using a single fuel source to generate both forms of energy, cogeneration systems can achieve efficiencies of up to 80% or more, significantly reducing energy costs and improving overall plant efficiency (Jia et al., 2022).

3.6 Smart Sensors and Internet of Things (IoT) Integration

The integration of smart sensors and IoT technologies allows food manufacturers to monitor and optimize energy use in real-time. Sensors can track temperature, pressure, flow rates, and other process parameters, providing valuable data to identify inefficiencies and potential areas for energy savings. IoT systems can also enable predictive maintenance, which helps reduce downtime and improve equipment longevity by ensuring machines operate at peak efficiency (Liu et al., 2021).

3.7 Automation and Process Optimization

Automation plays a key role in improving energy efficiency by ensuring processes are consistent and well-controlled. Through advanced control systems and process optimization algorithms, food manufacturers can minimize energy wastage, adjust processes based on real-time conditions, and optimize the use of energy-intensive machinery. For example, automation in mixing, heating, and cooling processes can reduce the amount of energy consumed by ensuring that equipment is used only when necessary and at optimal conditions (Ghosh et al., 2023).

3.8 Solar and Renewable Energy Integration

Incorporating renewable energy sources like solar, wind, and biomass into food industry operations offers an opportunity to reduce dependence on conventional energy sources, decrease operational costs, and reduce the carbon footprint. Solar photovoltaic (PV) systems can be integrated into food processing plants to generate electricity, while biomass systems can be used to generate heat or power from organic waste. Renewable energy integration, along with energy storage systems, can enhance energy security and contribute to more sustainable food production practices (Chen et al., 2022).

3.9 Application of Non-Thermal Processes

Table 2. Nonthermal processes and their medium of application

Technology	Medium Used	Out come	Reference
Food irradiation	High energy electrons in a wave length rages from 10^{-7} to 10^{-12m}	has lower costs compared to the conventional pasteurization process	Bouzarjomehri et al., 2020
Pulsed Electric Field	Application of electric field to biological cells	The temperature of the treated food or beverage is not increased. At high temperature the pulsed electric fields energy input might be reduced to the amount of 20 kJ/kg	Toepfl et al., 2006
High Pressure processing	High pressure to liquids causes alteration in lipids and proteins of biological cells, 52 kJ/kg energy is generally required to reach a pressure of 600 MPa.	Several health benefits of this technique are found out as compared to conventional pasteurization	Huang et al., 2020
Membrane processing	Separation of liquids are performed by membrane filtration	Saves up to 50% energy as compared to conventional processes	Nazir et al., 2019
Infrared, microwave and radio frequency heating	Electromagnetic waves are used to sterilize food	Radio frequency heat is more controlled technique with more penetration power and less energy requirement	Guo et al., 2019
Ohmic heating	Electric current is directly applied to the food to generate heat	no thermal damage and nutrition loss because of high conductivity of the material	Bhattacharjee et al., 2019

5. ENERGY PERFORMANCE INDICATORS (ENPI)

For energy planning and implementation of energy efficiency measures (EEMs), comprehensive knowledge of energy use at the process level is of utmost importance (Soederstrom, 1996). EEU (energy end use) categorization according to a standardized taxonomy would be an effective way to improve the EE (energy efficiency) potential (Andersson et al 2017). To get a better idea of energy end use energy performance indicators have to be established and monitored.

ISO 50006 defines an EnPI as a "quantitative value or measure of energy performance defined by the organization". EnPI is a gauge and it is very important to define the boundary of EnPIs. Several researchers have studied EnPIs for organizational levels and system level like Schmidt et al. 2016. The researcher proposed EnPIs for factory level to the process level, LJ Energy Pte Ltd in year 2016 proposed EnPIs for organizational level and system level for food manufacturing plants. Kanchiralla et al. (2020) proposed several EnPIs for organizational level, system level, and process levels. For the same levels they also found out End energy use (EEU) which is described in table 3. The organizational level, system level, and support process levels are generalizable for all industrial sectors, whereas the production processes vary widely between industrial sectors (Kanchiralla et al. 2020). Since the production processes are industry specific, EnPIs should be developed specifically for production processes in the food industry.

Table 3. Specific levels and the energy implied to operate

Levels	Type of Operation	Definition
1	Energy Carriers	Supply side of energy flow
2	Major Processes	Role of energy in value addition
3	Unit processes	The level is characterized on the basis of purpose of energy flow to that unit
4	Sub-unit processes	Energy end use of different technologies implied in the unit
5	Equipment/Tool	Energy used to operate individual machinery working under a subunit.

Source: **Kanchiralla et al. (2020)**

4. ISO CONTRIBUTION TO ENERGY PLANNING

This process involves optimizing how much energy is required in functional areas and developing an action plan according to the energy requirements (Ottosson et al 2016). Energy planning covers activities such as energy reviews, identification of energy objectives, EnPIs, EnB, and action plans (ISO 2011). ISO has established various requirements for the energy planning step; the steps for energy planning are very unspecific and involve a large number of concepts (Beisheim et al 2020). The conceptual diagram for the energy planning process provided by ISO 50001 is shown in Fig. 1(ISO 2011). The figure shows the three steps, inputs for this process, and outputs from the energy planning. Ottosson et al. 2016 developed a conceptual framework of energy management there by developing a comprehensive approach for effectively tapping the EE potential. The so developed framework included an insight in to developing energy policies, an energy strategy and action plan with risk management. Binasova et al. 2015 modified the energy planning mentioned in ISO 50001 to "the advanced approach of energy-efficient manufacturing". The study added more inputs and outputs to the energy planning and added a fourth step to already identified frame work by ISO, which is identifying advanced technology for improving EE.

Energy planning is also widely addressed in terms of national or regional levels, but this is different from the energy planning defined by energy management at the industry level. To the authors' knowledge, there is a scarcity of relevant studies aimed at formulating guidelines on the planning process for industries. As shown in Fig. 1, the energy review is central to energy planning, and identifying the opportunities for improvement is the final step of an energy review.

Figure 1. Energy planning

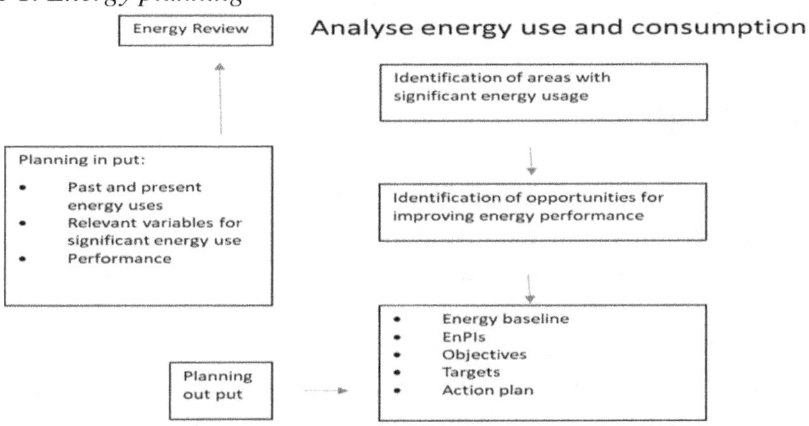

Source: ISO 2011

ISO 2011 also established a phenomenon of "benchmarking" where the process is identified, analysed, and compared on the basis of energy performance data in order to identify the opportunities for any improvement possible.

Benchmark helps to compares a facility's actual energy use against itself in earlier times, while the company-wide benchmark deals with comparing several facilities or processes in the same company Ottosson et al 2016.

5. BENEFITS OF ENERGY EFFICIENCY

i. Implementing energy-efficient practices in food production processes often leads to significant cost savings. Reduced energy consumption translates directly to lower utility bills, and the adoption of renewable energy sources can reduce long-term reliance on external energy suppliers. Additionally, the use of energy-efficient technologies can increase operational efficiency, reducing waste and downtime.

ii. Energy efficiency is also a key strategy in reducing the environmental impact of the food industry. By adopting energy-efficient technologies, food manufacturers can lower their carbon footprint, reduce greenhouse gas emissions, and contribute to global sustainability goals. The transition to renewable energy sources further enhances the industry's environmental profile.

iii. Adopting energy-efficient practices not only reduces costs but can also enhance a company's market position by differentiating it as a sustainable, environmentally-conscious brand. This can attract environmentally-aware consumers and create new business opportunities in markets where sustainability is a priority.

6. CONCLUSION

In future the food demand will not be proportional to the growth in energy capacity. To accomplish this gap significant initiatives and technologies are required to be developed. Energy efficiency is a critical concern in the food industry, offering both economic and environmental benefits. As energy costs rise and environmental regulations become stricter, food manufacturers must adopt innovative technologies and strategies to reduce energy consumption. From upgrading refrigeration systems to integrating renewable energy sources, there are many opportunities for the food industry to improve energy efficiency. Moreover, incentives must be provided to purchase technologies associated with smart food processing that are profitable over the long-term. Green house gas emission must be significantly decreased to a safe level and in order to achieve this, investments in renewable energy usage must be established by growing food industries.

REFERENCES

Ayres, R., Turton, H., & Casten, T. (2007). Energy efficiency, sustainability and economic growth. *Energy*, *32*(5), 634–648. DOI: 10.1016/j.energy.2006.06.005

Binasova, V., Micieta, B. and Kubinec, L., (2015). *Information and Communication Technology. Enabled Energy Efficiency*. International Scientific Book; 2015.

Brennan, J. G., O'Callaghan, P., & Ghosh, A. (2022). Advancements in food drying technologies: Energy efficiency and sustainability. *Journal of Food Engineering*, *323*(10), 24–36.

Briam, R., Walker, M., & Masanet, E. (2015). A comparison of product-based energy in tensity metrics for cheese and whey processing. *Journal of Food Engineering*, *151*, 25–33. DOI: 10.1016/j.jfoodeng.2014.11.011

Chen, J., Li, X., & Zhang, W. (2022). Integrating renewable energy sources into the food industry: A review of technologies and applications. *Renewable & Sustainable Energy Reviews*, *150*, 111488.

Clairand, J, M., Marco, B, L., Guillermo, E, E. and Pantaleo, A, M., (2020). *Review of Energy Efficiency Technologies in the Food Industry: Trends, Barriers, and Opportunities.* IEEE Access journals, 4, 1-15.

Department for Business, Energy and Industrial Strategy. (2018). Digest of United Kingdom Energy Statistics (DUKES) July 2018. London, UK. https://assets .publishing.service. gov.uk/government/uploads/system/uploads/attachment_data/file/736148/ DUKES_2018.pdf, Accessed date: 16 January 2025.

Energy Information Administration. EIA (2017). International Energy Outlook 2017. https://www.eia.gov/outlooks/archive/ieo17/pdf/exec_summ.pdf, Accessed date 23 January 2015 2018.

FAO. (2017). The future of food and agriculture– Trends and challenges. Rome, Italy: Food and Agriculture Organization of the United Nations. http://www.fao.org/3/a-i6583e. pdf, Accessed date: 20 January 20.

FDF. (2018). Facts and Stats. Stats at a glance. London, UK: Food and Drink Federation. www.fdf.org.uk/statsataglance.aspx, Accessed date: 15 January 2025.

Ghosh, A., Srivastava, A., & Bansal, S. (2023). Energy optimization and automation in food manufacturing processes. *International Journal of Energy Research*, *47*(6), 245–263.

International Organization for Standardization, "ISO 50001. (2011). Swedish Standards Institute.

ISO. (2023). *ISO 50001:2018 - Energy management systems.* International Organization for Standardization.

Kaminski, J., & Leduc, G. (2010). *Energy efficiency improvement options for the EU food industry.* POLITYKA ENERGETYCZNA Tom13 Zeszyt 1 PL ISSN:1429-6675.

Kanchiralla, F. M., Jalo, N., Thollander, P., Andersson, M., & Johnsson, S. (2020). Energy end-use categorization and performance indicators for energy management in the engineering industry. *Energies*, *13*(2), 369. DOI: 10.3390/en13020369

Ladha-Sabur, A., Bakalis, S., Fryer, P. J., & Lopez-Quiroga, E.Ladha-Sabura. (2019). Review Mapping energy consumption in food manufacturing. *Trends in Food Science & Technology*, *86*, 270–280. DOI: 10.1016/j.tifs.2019.02.034

Ministry of new and Renewable energy, Government of India. https://eerem.delhi .gov.in/sites/default/files/eerem/generic_multiple_files/com_solar_energy.pdf. Accessed date 14 January 2025. Identification of Industrial Sectors Promising for Commercialisation of Solar Energy Commercialisation of Solar Energy in Urban and Industrial Areas – ComSolar

Mohammed, M. S., Yousif, F., & Rehman, S. (2021). Energy-efficient refrigeration systems in the food sector: Recent developments and future perspectives. *Journal of Refrigeration Science*, *54*(2), 85–99.

Ottosson, M., Thollander, P., Schulze, M., & Nehler, H. (2016). Energy management in industry - A systematic review of previous findings and an integrative conceptual framework. *Journal of Cleaner Production*, *112*(3), 692–708.

Patel, S., Kumar, N., & Ahuja, A. (2023). Building energy management systems for energy efficiency in food production facilities. *Energy Efficiency*, *16*(1), 1–19.

Raymond, L., & Leung, G. C. K. (2021). The relationship between energy prices, economic growth and renewable energy consumption. *Energy Reports*, *7*, 712–1719.

Sims, R. E. H. (2014) *Global energy resources, supply and demand, energy security and on-farm energy efficiency, in Sustainable Energy Solutions in Agriculture.* J. Bundschuh and G. Chen, Eds, 1, 1952

Singh, R., & Kumar, P. (2022). Energy-efficient motors and drives in the food industry: Trends and challenges. *Industrial Electronics*, *68*(11), 9200–9214.

U.S department of energy About the Office of Energy Efficiency and Renewable Energy | Department of Energy (access on 16 January 2025).

Wang, L. (2008). *Energy Efficiency and Management in Food Processing Facili ties*. CRC Press. DOI: 10.1201/9781420063394

Zhou, X., Wu, Q., & Chen, Z. (2023). Advancements in heat recovery technologies for energy efficiency in food processing. *Journal of Food Science and Technology*, *60*(5), 1327–1339.

Chapter 2
Energy Management System Applications in the Food Sector

Dipali Saxena
https://orcid.org/0000-0002-8177-319X
Shri Vaishnav Vidyapeeth Vishwavidyalya, India

Ahmed Hamad
https://orcid.org/0000-0001-5037-9379
Benha University, Egypt

ABSTRACT

The chapter "Energy Management System Applications in the Food Sector" in Energy Efficient Technologies for Food Safety, Quality, and Security examines how energy management systems (EMS) are utilized to optimize energy use and boost efficiency in the food industry. It explores the foundational principles of EMS, including energy monitoring, control, and optimization, and details the deployment of advanced sensors and automated controls across various food production processes. Through case studies, the chapter illustrates successful EMS applications that lead to significant energy savings, reduced costs, and enhanced sustainability. It also highlights the role of data analytics, machine learning, and emerging technologies like AI and IoT in advancing EMS capabilities, ultimately supporting goals of improved food safety, quality, and environmental stewardship.

DOI: 10.4018/979-8-3693-8342-1.ch002

1. INTRODUCTION

The global food industry is a major consumer of energy, and as energy prices rise and environmental concerns intensify, it becomes imperative for food manufacturers to optimize their energy usage. Energy Management Systems (EMS) have emerged as a critical tool in achieving these objectives, enabling food producers to reduce energy consumption, improve efficiency, and enhance sustainability. EMS provide a structured approach to monitoring, controlling, and optimizing energy use across food processing operations. This chapter explores the application of EMS in the food sector, focusing on its benefits, technologies, and successful case studies from around the world.

The food sector's diverse operations—ranging from raw material handling to production, storage, and distribution—are all energy-intensive, thus making energy optimization a significant challenge. EMS can help address these challenges by integrating smart technologies, such as sensors, machine learning algorithms, and IoT devices, to improve energy efficiency. The primary aim of this chapter is to analyze how EMS applications can be implemented to achieve energy savings, reduce operational costs, and enhance food safety, quality, and security.

2. FOUNDATIONS OF ENERGY MANAGEMENT SYSTEMS (EMS)

2.1 What is an Energy Management System (EMS)?

An Energy Management System (EMS) refers to a set of technologies, processes, and tools designed to monitor, control, and optimize energy consumption. EMS typically involves the following core functions:

- Energy Monitoring: The continuous tracking of energy use throughout production processes to identify inefficiencies and areas for improvement.
- Energy Control: Automated systems that adjust energy consumption in real-time to minimize waste, such as adjusting machine speeds or regulating temperatures.
- Energy Optimization: Advanced analytics and modeling techniques that provide insights into energy use patterns and suggest strategies for improving efficiency.

In the food sector, EMS focuses on optimizing energy use in various aspects, including refrigeration, heating, cooking, and packaging. This optimization is achieved by installing smart sensors, integrating control systems, and leveraging software platforms that help in decision-making and process management.

2.2 Components of EMS in the Food Industry

The major components of an EMS in the food sector include:

1. Sensors: These devices capture real-time data on energy consumption, temperature, humidity, and equipment status, providing the necessary inputs for optimizing processes.
2. Control Systems: These systems regulate the operation of machines and equipment based on data from sensors. They adjust energy usage dynamically to meet demand and avoid waste.
3. Data Analytics and Artificial Intelligence (AI): Data analytics and AI algorithms analyze the data collected by sensors, identifying inefficiencies, predicting future energy needs, and recommending corrective actions.

3. APPLICATIONS OF EMS IN THE FOOD INDUSTRY

3.1 Energy Monitoring in Food Production Processes

Energy monitoring is the first step in deploying an EMS. In the food sector, this involves tracking energy consumption at various stages of food production—from raw material handling to processing, packaging, and storage.

For instance, in a meat processing facility, energy consumption typically spikes during cooking and refrigeration stages. By using smart meters and real-time monitoring systems, energy usage in each of these stages can be accurately tracked, and any excessive consumption or inefficiency can be identified.

Table 1. Typical energy consumption in food processing facilities

Process Area	Energy Consumption (%)	Energy Saving Potential (%)
Raw Material Handling	15%	10%
Cooking and Baking	25%	20%
Cooling and Refrigeration	20%	25%
Packaging	10%	15%

continued on following page

Table 1. Continued

Process Area	Energy Consumption (%)	Energy Saving Potential (%)
HVAC and Lighting	25%	15%
Equipment Standby	5%	5%

3.2 Control Systems for Energy Optimization

Control systems within EMS are designed to automate processes to optimize energy consumption. These systems can dynamically adjust energy use based on real-time data inputs. For example, in refrigeration, the system can adjust temperature settings based on the load, thereby avoiding overcooling and unnecessary energy use. Similarly, in cooking or baking processes, control systems can adjust the temperature based on batch size or product type.

A notable example is the implementation of a demand-controlled ventilation system in a bakery, where the ventilation system operates at varying speeds based on the number of ovens in operation, leading to reduced energy use when demand is low.

Table 2. EMS control systems and their energy saving impact

System	Purpose	Energy Saving (%)	Example Application
HVAC Systems	Optimize heating and cooling	15%	Climate control in production areas
Refrigeration Systems	Optimize refrigeration temperature based on load	20%	Dairy processing
Lighting Systems	Control lighting based on occupancy or time of day	10%	Lighting in production and storage areas

3.3 Advanced Technologies Enhancing EMS

With the advent of advanced technologies like Artificial Intelligence (AI), Internet of Things (IoT), and machine learning, EMS has evolved from basic energy monitoring tools to more sophisticated systems capable of real-time optimization and predictive analysis.

- AI and Machine Learning: AI algorithms can process large sets of operational data to predict energy demands, optimize machine settings, and schedule maintenance activities. For example, predictive algorithms in an industrial bakery can predict energy consumption based on weather patterns, machine performance, and production schedules, optimizing oven usage.

- IoT Devices: Smart sensors and connected devices communicate in real time with EMS, providing continuous feedback to operators. These devices can identify inefficiencies, such as underutilized equipment or incorrect machine settings, and automatically make adjustments to improve energy use.

4. CASE STUDIES ON EMS APPLICATIONS IN THE FOOD SECTOR

4.1 Case Study 1: Energy Savings in a Meat Processing Plant

A large meat processing plant implemented an EMS that integrated energy monitoring sensors across its production line. By optimizing the operation of cooking and refrigeration systems and using predictive algorithms to reduce energy consumption during non-peak hours, the facility achieved significant savings.

Table 3. Energy savings in meat processing facility

Process Area	Pre-EMS Energy Consumption (kWh)	Post-EMS Energy Consumption (kWh)	Energy Saved (%)
Meat Cooking	50,000	45,000	10%
Refrigeration	40,000	32,000	20%
Packaging	20,000	18,000	10%

4.2 Case Study 2: EMS Implementation in a Bakery

A bakery implemented an EMS to optimize its oven operation and energy-intensive cooling systems. By using AI-driven predictive algorithms, the bakery was able to reduce energy use in the ovens by 13% and its cooling system energy usage by 20%. This resulted in a total energy cost savings of 12%.

Table 4. Energy savings in bakery

Process Area	Pre-EMS Energy Consumption (kWh)	Post-EMS Energy Consumption (kWh)	Energy Saved (%)
Oven Operation	30,000	26,000	13%
Cooling Systems	10,000	8,000	20%
Lighting	5,000	3,500	30%

5. BENEFITS OF EMS IN THE FOOD SECTOR

5.1 Energy Efficiency

By continuously monitoring and optimizing energy consumption, EMS can help food producers improve energy efficiency. This leads to reduced energy waste and better resource utilization, which can significantly lower operational costs.

5.2 Cost Reduction

Reducing energy consumption directly impacts operational expenses. Food producers can save substantial amounts on energy bills by minimizing inefficiencies, thus improving profitability.

5.3 Sustainability and Environmental Impact

EMS plays a pivotal role in reducing the food industry's carbon footprint. By optimizing energy use, reducing waste, and promoting the use of renewable energy sources, food producers can achieve their sustainability goals and meet environmental regulations more effectively.

6. THE FUTURE OF EMS IN THE FOOD SECTOR

6.1 Emerging Technologies

The integration of advanced technologies such as blockchain for energy auditing, further advancements in IoT devices for real-time monitoring, and AI for predictive maintenance will continue to evolve EMS in the food industry.

6.2 Expansion of Smart Manufacturing

With the growing trend of Industry 4.0, which combines smart manufacturing techniques and automation, EMS will play an even more significant role in shaping the future of the food sector by enabling real-time, dynamic energy management.

7. CONCLUSION

Energy Management Systems are increasingly becoming an indispensable part of the food industry's operations. From energy monitoring and control to predictive analytics and optimization, EMS provide food manufacturers with the tools needed to reduce energy consumption, lower costs, and contribute to sustainability. The case studies highlighted in this chapter demonstrate the tangible benefits of EMS, with significant energy savings achieved through technology integration. As the food sector continues to face challenges related to energy efficiency, EMS will be at the forefront of efforts to improve operational performance, ensure food safety, and reduce environmental impact.

REFERENCES

Aghbashlo, M., & Taherzadeh, M. J. (2019). The role of renewable energy in food processing: Potential and opportunities for integration with EMS. *Renewable & Sustainable Energy Reviews*, *103*, 311–324. DOI: 10.1016/j.rser.2018.12.046

Barker, M., & Williams, P. (2020). Implementing energy management systems in the food industry: Benefits, challenges, and best practices. *Energy Efficiency*, *13*(3), 555–574. DOI: 10.1007/s12053-020-09728-3

Buchanan, R. A., & Wright, S. P. (2019). Leveraging IoT technologies for energy optimization in the food industry: A case study of intelligent temperature control systems. *Energy Reports*, *5*, 563–570. DOI: 10.1016/j.egyr.2019.07.004

Castellanos, R., & Balaguer, J. (2017). Modeling energy efficiency in food production through machine learning and data analytics. *Energy Procedia*, *105*, 1704–1711. DOI: 10.1016/j.egypro.2017.03.680

Dufresne, A., & Charlebois, S. (2021). Leveraging big data for energy efficiency and sustainability in food production. *Trends in Food Science & Technology*, *107*, 30–40. DOI: 10.1016/j.tifs.2020.11.010

Gadekar, K., & Kshirsagar, S. (2020). Energy management practices in the food processing industry: Challenges and opportunities. *Sustainable Energy Technologies and Assessments*, *39*, 100677. DOI: 10.1016/j.seta.2020.100677

Güler, O., & Uyar, A. (2018). Dynamic optimization of energy consumption in the food industry: A review of strategies and solutions. *Energy Efficiency*, *11*(3), 729–743. DOI: 10.1007/s12053-017-9577-0

Gupta, A., & Dey, P. K. (2017). Energy management in food industry: A review. *Journal of Food Engineering*, *193*, 47–58. DOI: 10.1016/j.jfoodeng.2016.09.015

Hughes, M. L., & Garcia, L. A. (2019). Advances in energy-efficient technologies for food safety, quality, and sustainability. *Comprehensive Reviews in Food Science and Food Safety*, *18*(2), 445–463. DOI: 10.1111/1541-4337.12456

Jafari, A., & Kheradmand, A. (2021). Data-driven energy optimization in the food industry: Insights from industrial case studies. *Journal of Cleaner Production*, *285*, 125417. DOI: 10.1016/j.jclepro.2020.125417

Kaur, R., & Bhagat, D. (2020). Application of Internet of Things (IoT) in energy management systems for the food industry. *Energy Reports*, *6*, 1245–1256. DOI: 10.1016/j.egyr.2020.05.008

Khan, M. S., & Abbas, A. (2021). Exploring AI-based energy optimization in food processing industries: A systematic review. *Energy and Building, 245*, 111027. DOI: 10.1016/j.enbuild.2021.111027

Koch, S., & Müller, S. (2019). The role of energy management systems in achieving sustainability targets in food production. *Sustainability, 11*(11), 3050. DOI: 10.3390/su11113050

López-González, R., & Peña-García, A. (2021). Industrial energy efficiency and optimization in food manufacturing: A case study approach. *Journal of Industrial Engineering and Management, 14*(1), 47–60. DOI: 10.3926/jiem.3336

Martínez-Hernández, E., & Rodríguez, A. (2020). Energy-saving measures and energy management in food processing: A holistic approach. *Food Research International, 132*, 109040. DOI: 10.1016/j.foodres.2020.109040

Mehmood, M. U., & Akhtar, P. (2020). A review of energy management systems in food processing and manufacturing: Current trends and future prospects. *Food Control, 110*, 107036. DOI: 10.1016/j.foodcont.2020.107036

Pereira, R., & Silva, F. (2020). Real-time energy monitoring and control systems in food manufacturing: A critical review. *Food and Bioprocess Technology, 13*(7), 1251–1265. DOI: 10.1007/s11947-020-02463-z

Tunc, Y., & Dincer, I. (2021). Energy efficiency and sustainability in food industries: Strategies and case studies. *Journal of Cleaner Production, 277*, 124040. DOI: 10.1016/j.jclepro.2020.124040

Zhao, L., & Chien, S. (2018). Optimization of energy consumption in food processing industries using energy management systems. *Renewable & Sustainable Energy Reviews, 81*, 296–304. DOI: 10.1016/j.rser.2017.07.040

Chapter 3
The Role of Technology and Energy Management System in Sustainable Food Industry Practices

Surendra Prakash Gupta
https://orcid.org/0000-0002-4873-6346
Shri Vaishnav Vidhyapeeth Viswavidhayalaya, India

Arun Bhatt
G.B. Pant Engineering College, India

ABSTRACT

A vital component of the global economy, the food industry produces, processes, and distributes food to support people all over the world. The need for food production has grown along with the world's population, which has caused food factories in both developed and developing nations to expand quickly. This chapter looks at how the food industry uses energy management systems (EMS) to maximize energy use and increase productivity. It examines the fundamental ideas of EMS, such as energy monitoring, control, and optimization, and describes how sophisticated sensors and automated controls are being implemented in a variety of food production processes. However, the quick growth has resulted in serious environmental problems, such as pollution, emissions, and excessive energy and water use. This chapter provides examples of effective EMS applications that result in notable cost savings, increased sustainability, and energy savings.

DOI: 10.4018/979-8-3693-8342-1.ch003

1. INTRODUCTION

The concept of energy management involves "the strategic and effective application of energy to boost profits and improve competitive standing through organizational initiatives and the enhancement of energy efficiency in processes" (Capehart et al., 2008). In the food industry, energy management systems (EMS) are essential for increasing sustainability and efficiency. Significant economic and environmental benefits arise from the ability of these systems to monitor and optimize energy consumption. A specific energy consumption approach in the yogurt manufacturing industry resulted in a financial gain of 5 percent, or around DZD 6.5 million per year, showing that EMS can identify energy saving opportunities (Kebour and Kebour, 2023). With increased global food demand and the resulting environmental impacts, the food industry can minimize energy waste through optimized processes (Corigliano and Algieri, 2024). As demonstrated by a seafood canning company that experienced a 7.48 percent decline in annual sales, resulting in significant cost savings, implementing standardized frameworks such as ISO 50001:2018 helps companies manage energy in a methodical manner (Araya, 2020). Real-time data collection and optimization controls are features of a complete energy management system that improve operational effectiveness and decision making. Despite the obvious advantages of EMS, obstacles such as initial implementation costs and the need for ongoing training may prevent its widespread use in the food industry. In order for energy management systems to achieve their full potential, these obstacles must be removed.

2. FUNDAMENTALS OF ENERGY MANAGEMENT SYSTEM

In order to maximize energy consumption, improve efficiency, and advance sustainability, a comprehensive energy management system (EMS) in the food industry integrates multiple components. Ikeyama and his coworkers, (2011) presented a strategy for maximizing energy resources in the manufacturing facilities used to produce food. They have explained how efficiently process data can be collected and efficient management in production facilities. The main elements of this strategy are data collection, analysis and control systems that enable real-time management and monitoring of energy resources. Energy consumption data from multiple terminals across the company has captured and visualized by systems such as the Energy Information Management System. Real-time data from production lines is collected by data collection modules and provides instant insights into energy consumption. To make informed decisions, modules examine energy data to find trends and inefficiencies. Systems optimize energy consumption and minimize waste by implementing control strategies based on data analysis. Integration with cloud platforms improves flexibility and responsiveness to energy demand by enabling remote monitoring and management (Chaabouni and Khemakhem, 2017). These platforms enable improved coordination and operational efficiency by providing a comprehensive perspective on energy management (Hui, L, et al., 2017). Potential obstacles such as the initial investment costs and the need for ongoing maintenance and updating of EMS technology must be taken into account.

3. EMS TECHNOLOGIES AND TOOLS

3.1 Sensors and Metering Devices

Recent developments in sensor technology have enabled the development of innovative applications in a variety of fields, including food science. Sensors and measuring devices are essential components of a comprehensive energy management system (EMS) in the food industry as they monitor a variety of key performance indicators (KPIs). These KPIs enable companies to monitor energy consumption, identify inefficiencies, and put improvement plans into action. Recent advances in chemical detection, the introduction of small sample volumes, miniaturization of instruments, improvements in maintaining sample integrity without destruction, and optimized manufacturing processes through detection techniques have improved the use and application of these indicators in food analysis. These advances have made it easier for new opportunities to emerge. However, sensing technologies have not yet been widely adopted by the food industry due to a lack of training and understanding regarding their potential as an analytical method, as well as issues with their reliability, which is highly dependent on the food matrix being studied.

3.1.1 Energy Consumption Metrics

a. **Total Energy Usage**: Measures the overall energy consumed across production processes, enabling identification of high-consumption areas (Jagtap & Rahimifard, 2018).
b. **Energy Intensity**: Assesses energy consumption relative to production output, helping to evaluate efficiency levels (Gorp, 2005).

3.1.2 Operational Efficiency Indicators

a. **Real-Time Monitoring**: Utilizes IoT-enabled sensors to provide continuous data on energy usage, facilitating immediate adjustments to operations.
b. **Cost Savings**: Tracks financial savings achieved through energy efficiency measures, with case studies showing significant reductions, such as 163,000 kWh saved in a beverage factory (Jagtap et al., 2019).

3.1.3 Demand Response and Load Management

Evaluates the effectiveness of demand response strategies in managing peak loads and reducing energy costs (Khelifa and Jelassi, 2016). While these KPIs are essential for enhancing energy efficiency, it is important to recognize that the sheer volume of data generated can overwhelm managers, potentially leading to decision paralysis if not effectively managed (Gorp, 2005).

3.2 Advanced Sensors and Metering Technologies

In the food processing and packaging sector, various sensor technologies are used to ensure the freshness, quality and safety of food. These sensors are essential for tracking conditions from production to consumption along the food supply chain. The most commonly used sensor types in this industry are biosensors, chemical sensors and indicators; Each has a specific purpose in food monitoring. A sensor can be a single device or a group of devices that utilize physical and chemical principles. Food science is one of the many areas where recent developments in sensor technology have enabled new applications.

3.2.1 Biosensors

Food safety and security require rapid, accurate and selective identification of analytes present in the food matrix (e.g. pathogenic bacteria, presence of pesticides and toxins). The quantitative detection and screening of various microorganisms and molecules in foods has been made possible by advances in biosensor technology. The use of a single sensor platform to detect multiple pathogens/toxins is at the heart of pathogen detection trends in the agri-food sector. To enable portability, cost reduction, analysis time reduction and commercial viability of this type of sensors, biotechnology has recently moved to ever smaller systems (Cho et al., 2014:Warriner et al., 2014; Derkus, 2016).

3.2.2 Chemical Sensors

Numerous chemical sensor types, including optical, mass, enzymatic, electrochemical (potentiometric, voltammetric, amperometric, impedimetric, and conductimetric), and sensor arrays, have been used or developed as electronic read instruments. An electronic tongue is a sophisticated analytical tool that combines a variety of sensors with pattern recognition technologies to perform both qualitative and quantitative analysis. Numerous other sensors, including colorimetric, a wide range of chemical and electrochemical sensors, and smartphone devices, have been developed or investigated. It is recommended that the reader seek information from the numerous publications, industry and scientific reports, and patents available in this field, as a detailed description of the various sensors is beyond the scope of this chapter. Sensor networks will impact the development of novel sensing technologies that could completely transform food analysis. These systems (e.g. single sensors or in combination) enable a more comprehensive examination of food safety and quality and therefore the entire supply and value chain. By evaluating and detecting multiple analytes at the same time, the use of sensor technologies will also improve the performance and effectiveness of current analysis systems.

3.2.3 Indicators

Improved energy performance and its management can provide companies with a competitive advantage and a valuable strategic asset by enabling them to respond adaptively to changing conditions and development scenarios and to make optimal use of their energy resources and assets, resulting in lower energy costs and consumption (Semkov et al., 2014; Wojdalski et al., 2015;Gupta & Bhardwaj, 2024). The two definitions of energy efficiency are the production ratio for a particular system or production device under normal operating conditions, or the amount of energy that the system, plant or production device uses to produce its output. By showing the relationships between energy consumption and products, indicators can also be used to measure energy efficiency (May et al., 2013). Economic and technical factors must be taken into account through indicators. Energy efficiency, environmental policy, energy price fluctuations, international trade in energy-intensive intermediate or final products and structural effects due to economic cycles can be linked to consumption-related factors through the analysis of energy indicators (Gielen & Taylor, 2009). Instruments for increasing energy efficiency can also be developed from these contexts. Despite the fact that these sensors significantly improve food safety and quality, there are still barriers to their integration and widespread use in the food industry, particularly in the areas of cost and consumer acceptance.

The efficiency of energy management systems in food production can be assessed using a number of key performance indicators (KPIs), including operational performance, environmental impact and energy efficiency. These indicators provide a comprehensive framework for assessing sustainability and energy consumption in the food industry. Global food production and security is currently facing major challenges due to climate change (Arora, 2019), which has multiple adverse impacts on agricultural production (FAO, 2019; World Bank, 2022). Recent reports suggest that lower crop yields, lower livestock productivity and reduced nutritional quality of major cereal crops are having a negative impact on global food production. To improve sustainability and minimize environmental impacts in food production and management, the use of energy efficiency indicators is necessary. However, the most difficult task of the scientific community is to reach a consensus on practical and useful criteria that can effectively link resource consumption with the services provided. According to several studies, integrating these indicators into agricultural systems is crucial for maximizing resource utilization and increasing productivity. The main features of this application are described in the following sections.

3.2.3.1 Energy Efficiency Indicators

Energy sustainability indicators, one of 17 Sustainability Performance Indicators (SPIs) found in the study, are critical for assessing the effectiveness of agricultural systems. This approach evaluates how well food production processes work and provides information about how much energy is used and how waste is reduced in the food industry. In the coming years, sustainability assessment will become increasingly important for the food industry. To measure energy consumption during the production process, numerous energy management standards have been examined.

a. **Cumulative Energy Use**: Measures total energy consumed throughout the production process, helping identify areas for improvement (Ghisellini et al., 2023).
b. **Overall Equipment Energy Effectiveness (OEE)**: Calculated as the ratio of production energy to process energy, reflecting the efficiency of energy use in manufacturing (Bannister, 2024).
c. **ISO 50001 Compliance**: Evaluates adherence to international energy management standards, promoting systematic energy efficiency improvements (Pradella et al., 2019).

3.2.3.2 Environmental Performance Indicators

When assessing the impact of food production on the environment and directing decision-making towards more sustainable methods, environmental performance indicators are essential (Benedetto and Klemeš, 2009; Thevanes, 2020). The Environmental Performance Strategy Map, which takes into account costs as an additional dimension and environmental indicators, is a method for evaluating environmental performance (Benedetto and Klemeš, 2009). This graphical representation makes it possible to compare different options and determine the most economical and sustainable solution. Another approach is eco-efficiency indicators, which quantify the material and resource consumption per unit of production, sales or value creation. However, these methods fall short because they focus primarily on the economic and environmental performance of companies and industries. The environmental performance of the entire product system, including the effects of upstream and downstream processes, must also be taken into account.

Incorporating energy and environmental performance metrics into EU policies and the circular economy promotes sustainable practices from production to consumption (Ghisellini et al., 2023). The EU views the adoption of more environmentally friendly agri-food systems founded on a "circular economy" and the "food to fork strategy" ("FF strategy") as a potential means of balancing society's demands for a sufficient supply of high-quality food with the preservation of biological resources. According to Mason et al., the "FF strategy" aims to create equitable, healthful, and ecologically sustainable food systems with reduced environmental impacts throughout the food life cycle (Mason et al., 2020).

4. CASE STUDY: EFFICIENCY GAINS IN COOKING PROCESSES

In the food production sector, increasing the efficiency of cooking processes requires a multi-pronged strategy that includes operations management, energy optimization and technology. The following tactics have proven particularly successful.

4.1 Automated Cooking Systems

To increase cooking efficiency, automated cooking systems (ACS) use robotics, IoT and AI. According to Bautista et al. (2023) These systems can adapt to different cooking situations, increasing precision and consistency while reducing labor costs. ACS is a useful tool in food production and can improve nutritional value, increase food safety and save time (Bautista et al., 2023). By using non-thermal heating techniques and innovative thermodynamic cycles, energy consumption in food processing can be drastically reduced (Wang, 2014). Workflow optimization and proper housekeeping are two strategies that can reduce energy consumption and operating costs (Klemeš and Perry, 2007). Process automation, intelligent data analysis, and effective inventory management are some strategies that can improve product quality, reduce costs, and streamline operations (Geminarqi and Purnomo, 2023). Despite the many benefits of these approaches, obstacles such as high upfront costs and the need for skilled personnel to operate sophisticated systems can prevent their widespread adoption. The sector's progress continues to depend on finding a balance between innovation and practical application.

By integrating cutting-edge technologies and optimizing energy consumption, energy management systems (EMS) significantly increase the efficiency of cooking processes in the food production industry. Through improved energy resource allocation that these systems enable, food processing can become more sustainable and use less energy. The power of multiple cooking systems can be dynamically distributed through EMS, ensuring that each system receives the energy it needs at the best time (Singer et al., 2014). This controlled allocation makes cooking processes more efficient overall and reduces energy waste.

Food processing can significantly reduce its energy consumption by using energy efficient technologies such as heat recovery systems and innovative thermodynamic cycles (Wang, 2014). For example, energy efficient cooking appliances use stirring and continuous temperature control to maintain ideal cooking temperatures while using as little energy as possible (Qiu, 2015). Food production facilities can utilize the additional heat generated during cooking by installing waste heat recovery systems, further improving energy efficiency (Wang, 2008). This method not only saves energy but also reduces the carbon footprint associated with food production. EMS can save a lot of energy, but for some food production facilities, the initial cost of sophisticated technologies and systems may be prohibitive, which could hinder widespread adoption.

5. FUTURE TRENDS

5.1. Artificial Intelligence (Ai) in Ems

Artificial intelligence (AI) plays a crucial role in optimizing energy consumption in food production by identifying inefficiencies in energy management systems. AI has the potential to improve the sustainability and operational efficiency of agricultural practices through the use of sophisticated data analytics and predictive modeling. Machine learning and other AI technologies examine large data sets to identify trends in energy consumption, enabling precise control of the energy system

(Olatunde et al., 2024). By predicting energy demand, predictive maintenance and demand forecasting help reduce waste and ensure energy supply stability (Manoharan et al., 2024). AI can improve real-time monitoring and control by combining with cloud computing and IoT devices to create a comprehensive energy management framework (Goel, 2024).

Artificial intelligence (AI) systems optimize temperature and humidity in agriculture with controlled environmental conditions, resulting in a significant reduction in energy consumption while maintaining crop yields (You, 2024). AI systems adapt to changing circumstances, including occupancy and weather, leading to continued advances in energy saving techniques (Olatunde et al. 2024). Digital twins that optimize energy consumption in plant factories are enabled through the use of physically based deep learning techniques that result in significant reductions in resource consumption (You, 2024).

Although artificial intelligence (AI) offers great potential for food production energy optimization, obstacles like data privacy, integration complexity, and the requirement for specialized skills continue to be crucial factors for successful deployment (Olatunde et al., 2024; Kuzmenko et al., 2024). Given that food is being wasted and lost, that food prices are rising, and that widespread food poverty is growing, the problem is getting worse. Production, transportation, storage, processing, sale, and consumption are all phases of the food chain where losses and waste happen. As a result of the "Agenda 2030 for Sustainable Development," the international community has committed to addressing food scarcity and malnourishment (SDG 2) by reducing food loss and waste. The United Nations Food Waste Index Report states that high-income countries lose an average of 79 kg of food per person annually (Delgado, L. M. Schuster and M. Torero,2017).

By predicting and preventing energy waste, AI-driven energy management systems (EMS) promise to dramatically reduce greenhouse gas emissions in the food production industry. Energy efficiency, crop yields and resource consumption can be improved through these systems through the use of sophisticated algorithms and data analysis.

5.2 AI-Enabled Optimization in Agriculture

Artificial intelligence (AI) systems can precisely control environmental factors such as temperature and humidity by using physics-based deep learning to create digital twins of plant factories. This reduces energy consumption by 13-41% (You, 2024). Green AI projects focus on streamlining agricultural processes, which can lead to lower emissions through higher crop yields and lower energy consumption (Pandey, 2023).

5.3 Enhancing Food Processing

In order to preserve food quality while reducing energy consumption, artificial intelligence (AI) can identify the ideal drying temperatures for heat drying processes, which are a major source of carbon emissions in the food production industry (Yudhistira et al., 2024). By efficiently decreasing food waste and strengthening the circular economy, artificial intelligence (AI) technologies can monitor and optimize food production and supply chains, which lowers greenhouse gas emissions (Onyeaka et al., 2023). While AI-driven EMS has the potential to lower emissions, issues like the energy usage of AI systems themselves need to be resolved to guarantee a sustainable outcome (Pandey, 2023).

5.4 Internet of Things and Its Impact

The use of IoT technology in food production greatly improves energy consumption optimization and real-time monitoring, which reduces waste. In order to make better decisions and promote sustainability, farmers can collect vital information on resource consumption, environmental conditions, and operational effectiveness by using a variety of sensors and devices.

5.5 Real-Time Data Collection

IoT devices enable accurate monitoring of agricultural conditions by collecting data on variables such as temperature, humidity, and soil moisture (Tan et al., 2024; Sodhi and Jamwal, 2024). Using this data can save water and energy to enable timely interventions, such as changing irrigation schedules according to soil moisture (Sharma et al., 2023). IoT systems are capable of examining patterns of energy consumption in the food production process, identifying inefficiencies and suggesting improvements. For example, one study showed that data-driven modifications could reduce energy consumption of certain machines by 5-6% (Ömerali and Kaya, 2023).

5.6 Waste Reduction

IoT technology reduces food waste by optimizing resource use, promoting a more sustainable agricultural sector (Stefanuto et al., 2023). The use of automated systems and smart irrigation reduces wasteful water and energy consumption (Sodhi and Jamwal, 2024; Sharma et al., 2023). The impact of IoT on sustainability in food production, however, may be limited due to barriers such as high acquisition costs and the need for technical know-how, which could prevent widespread adoption. The introduction of IoT-based energy management systems in the food industry brings significant financial and environmental benefits. These systems improve resource management and save money by increasing productivity, reducing waste and promoting sustainability.

5.7 Information and Communication Technologies (ICT)

A number of important energy efficiency metrics have been developed as a result of the incorporation of information and communication technologies (ICT) into food production and management. In the agricultural industry, these metrics are crucial for improving sustainability and maximizing energy use. Continuous data collection on crop conditions and energy consumption is made possible by ICT tools like IoT sensors and remote monitoring systems (Jagtap and Rahimifard, 2018). By enabling prompt interventions, this real-time data lowers energy waste and boosts operational effectiveness. According to Greco et al., (2020) these include the total energy used per unit of production, which aids in comparing the energy efficiency of various food production processes. IoT devices can be used to monitor additional metrics, such as machine energy consumption rates and downtime, enabling real-time adjustments to increase operational efficiency. Using IOT-based machine learning to optimize energy efficiency in food production results in a 5–6% energy savings for a single machine and a significant decrease in the factory's overall energy expenditures (Ömerali & Kaya, 2023). The idea of the "Internet of Things" (IoT) has been studied in this context in an effort to raise awareness, transparency, and visibility of different resource usage.

Greco et al., 2020, in another review article, addressed the issues of reducing fossil energy consumption and measures to improve energy efficiency in the European agricultural and food system and presented a list of key performance indicators for the fruit and vegetable sector that are found useful are comparative analyzes to select the best energy management methods and technologies to improve energy efficiency. While these KPIs provide a structured approach to evaluating energy management systems, it is important to recognize that the food industry faces challenges in adopting these indicators consistently. Many organizations still struggle to understand and implement effective energy efficiency measures (Pradella et al., 2018).

6. FUTURE DIRECTIONS: ANTICIPATED DEVELOPMENTS IN EMS TECHNOLGY IN FOOD SECTOR

There is growing pressure on the global food system to improve its resilience, sustainability and efficiency. Thanks to technological developments in EMS, these problems could be solved at various points in the food supply chain, from farm to table. The way EMS is currently implemented in the food industry varies greatly by supply chain and region. Major food processing plants and distribution centers in developed countries, particularly in the USA and some parts of Europe, have implemented EMS to maximize energy consumption and reduce operating costs (Harshini P and Sangeetha C. 2024).

This often requires the use of building management systems, the optimization of processes and the use of energy-efficient devices. Due to issues such as limited technical expertise and financial constraints, the adoption rate is still lower in smaller farms, particularly in the agricultural sector. These systems optimize energy use for heating, cooling and lighting by monitoring environmental factors such as temperature, humidity and lighting levels. This leads to higher yields and lower energy costs. Likewise, increased energy efficiency in food processing plants through process optimization and the use of energy efficient equipment reduces environmental impact and saves money (Osato Itohan Oriekhoe et al., 2024). An important example of a successful application is the use of EMS to optimize energy consumption in greenhouses (Zhang, 2023). However, integrating EMS into the entire food supply chain is still very difficult and requires the collaboration of various stakeholders and the creation of standardized procedures.

Renewable energy sources such as solar and wind energy are increasingly being integrated into food production and processing facilities. Growing concerns about climate change and the demand for greener energy sources are the main drivers of this change. However, fluctuations and grid stability pose serious obstacles to the integration of renewable energy sources. For the food system to successfully integrate renewable energy, smart grids and state-of-the-art energy storage technologies become essential (Lei et al., 2022).

Machine learning and artificial intelligence (AI) also make a significant contribution to energy efficiency (Choi et al., 2023). To predict energy demand and optimize energy distribution, AI-powered systems can evaluate enormous amounts of data from multiple sources such as sensors, weather forecasts and historical patterns of energy consumption. This allows energy consumption to be regulated more precisely, reducing waste and increasing efficiency. Another important building block in this context is the Internet of Things (IoT) (Hasan et al., 2023). IoT sensors placed throughout the food supply chain provide real-time information on temperature, humidity, energy consumption and other relevant factors. Using AI and machine learning algorithms, this data can be analyzed to maximize energy consumption and increase overall efficiency. For perishable goods, where maintaining the ideal temperature is crucial to reduce food waste and spoilage, the combination of IoT and AI is particularly promising for improving cold chain logistics (El Jaouhari et al., 2024).

Additionally, blockchain technology holds promise for improving the transparency and traceability of the food supply chain. Blockchain improves accountability and transparency by storing all data and transactions related to food production, processing and distribution on a secure, immutable ledger. This can reduce food fraud and counterfeiting while increasing consumer and manufacturer confidence. IoT sensors and blockchain technology enable real-time tracking of food across the entire supply chain, providing comprehensive information about product origin, handling and storage conditions. Thanks to this detailed traceability, food safety is increased and contamination problems can be identified and resolved promptly (Mason et al. 2020; Jokonya and Mbadlisa, 2024).

Another major obstacle in the food supply chain is the complexity of integrating different systems. Food supply chains often involve multiple parties with different systems and technological capabilities, making them fragmented. Coordination and standardization of these systems is critical to their integration into a seamless EMS. Data protection and data security aspects are also crucial factors that need to be taken into account. There are concerns about the possibility of misuse or illegal access in the collection and use of energy consumption data. To protect sensitive data and maintain customer trust, strict data security measures are required. It is also legitimate to be concerned that different food system stakeholders may not have equal access to new technologies (Kazancoglu et al., 2022). It is possible that the benefits of advanced EMS technologies are not shared equally, with larger companies and farms in developed countries benefiting more than smaller companies and farmers in developing countries. Addressing these disparities requires programs and policies that support equitable access to resources and technology.

7. CONCLUSION

The food industry could see significant change due to expected advances in EMS technology. IoT, blockchain, AI and machine learning, and renewable energy sources can all be integrated to significantly increase the resilience, sustainability, and efficiency of food supply chains. However, to fully exploit these technologies, it is important to address the issues of cost, training, system integration, data security and equal access. To build a more resilient, efficient and sustainable food system and accelerate the adoption of EMS technologies, collaboration between researchers, legislators and industry stakeholders is crucial (Astuti and Hidayati, 2023; Rajput et al., 2023). More research and development is needed to overcome current obstacles and unlock the transformative potential of EMS for the global food system.

The long-term economic and environmental impacts of various EMS strategies require further investigation (Carruthers and Vanclay, 2007). This includes determining the cost-effectiveness of various technologies, measuring reductions in greenhouse gas emissions and other environmental impacts, and analyzing financial and social benefits for various participants in the food system. It is important to examine how technology adoption impacts different societal stakeholders. This includes assessing the impact on food security, income distribution and employment, and developing plans to ensure that everyone has fair access to the benefits of EMS technologies. To enable wider adoption, standardized procedures and guidelines for EMS implementation across the food supply chain need to be developed. This requires the creation of data exchange protocols, interoperable systems and training courses for various stakeholders.

REFERENCES

Arora, N. K. (2019). Impact of climate change on agriculture production and its sustainable solutions. *Environmental Sustainability*, *2*(2), 95–96. DOI: 10.1007/s42398-019-00078-w

Astuti, R., & Hidayati, L. (2023). How might blockchain technology be used in the food supply chain? A systematic literature review. Cogent Business &amp. *Cogent Business & Management*, *2*(2), 2246739. Advance online publication. DOI: 10.1080/23311975.2023.2246739

Bannister, K. E. (2024). Key Performance Indicators. In Practical Lubrication for Industrial Facilities (pp. 717–721). River Publishers. https://doi.org/DOI: 10.1201/9781032632360-43

Capehart, B. L., Turner, W. C., & Kennedy, W. J. (2008). *Guide to Energy Management*. The Fairmont Press, Inc.

Carruthers, G., & Vanclay, F. (2007). Enhancing the social content of environmental management systems in Australian agriculture. *International Journal of Agricultural Resources, Governance and Ecology*, *3*(3), 326. DOI: 10.1504/IJARGE.2007.012840

Chaabouni, T., & Khemakhem, M. (2017). Energy management strategy in cloud computing: A perspective study. *The Journal of Supercomputing*, *12*(12), 6569–6597. DOI: 10.1007/s11227-017-2154-z

Cho, I.-H., Radadia, A. D., Farrokhzad, K., Ximenes, E., Bae, E., Singh, A. K., Oliver, H., Ladisch, M., Bhunia, A., Applegate, B., Mauer, L., Bashir, R., & Irudayaraj, J. (2014). Nano/Micro and Spectroscopic Approaches to Food Pathogen Detection. *Annual Review of Analytical Chemistry (Palo Alto, Calif.)*, *1*(1), 65–88. DOI: 10.1146/annurev-anchem-071213-020249 PMID: 24896312

Choi, J., Byun, S.-W., Lim, T., Lee, H. I., Shin, H. S., Kim, G. W., Kwon, J.-K., & Kang, B.-C. (2023). Artificial Intelligence-Based Plant Breeding using Genotype and Phenotype Data: Methods and Future Work. *2023 IEEE International Conference on Big Data (BigData)*, 3458–3463. https://doi.org/DOI: 10.1109/BigData59044.2023.10386874

Corigliano, O., & Algieri, A. (2024). A comprehensive investigation on energy consumptions, impacts, and challenges of the food industry. *Energy Conversion and Management*, *X*, 100661. DOI: 10.1016/j.ecmx.2024.100661

De Benedetto, L., & Klemeš, J. (2009). The Environmental Performance Strategy Map: An integrated LCA approach to support the strategic decision-making process. *Journal of Cleaner Production*, *10*(10), 900–906. DOI: 10.1016/j.jclepro.2009.02.012

Derkus, B. (2016). Applying the miniaturization technologies for biosensor design. *Biosensors & Bioelectronics*, *79*, 901–913. DOI: 10.1016/j.bios.2016.01.033 PMID: 26800206

El Jaouhari, A., Arif, J., Jawab, F., Samadhiya, A., & Kumar, A. (2024). Unfolding the role of metaverse in agri-food supply chain security: Current scenario and future perspectives. International Journal of Food Science &amp. *International Journal of Food Science & Technology*, *5*(5), 3451–3460. DOI: 10.1111/ijfs.16972

Fadhel Khelifa, R., & Jelassi, K. (2016). An energy monitoring and management system based on key performance indicators. 2016 IEEE 21st International Conference on Emerging Technologies and Factory Automation (ETFA), 1–6. https://doi.org/DOI: 10.1109/ETFA.2016.7733542

FAO (Food and Agriculture Organization of the United Nations). (2019). FAO's Work on Climate Change, United Nations Climate Change Conference 2019.. Available online: https://www.fao.org/3/ca7126en/CA7126EN.pdf (accessed on 25 January 2023).

Ghisellini, P., Ncube, A., Rotolo, G., Vassillo, C., Kaiser, S., Passaro, R., & Ulgiati, S. (2023). Evaluating Environmental and Energy Performance Indicators of Food Systems, within Circular Economy and "Farm to Fork". *Energies*, *4*(4), 1671. DOI: 10.3390/en16041671

Gielen, D., & Taylor, P. (2009). Indicators for industrial energy efficiency in India. *Energy*, *8*(8), 962–969. DOI: 10.1016/j.energy.2008.11.008

Gong, M., & Wall, G. (2001). On exergy and sustainable development—Part 2: Indicators and methods. Exergy. *International Journal (Toronto, Ont.)*, *4*(4), 217–233. DOI: 10.1016/S1164-0235(01)00030-9

Greco, C., Campiotti, A., De Rossi, P., Febo, P., & Giagnacovo, G. (2020). Energy consumption and improvement of energy efficiency for the European agricultural-food system. *Rivista di Studi sulla Sostenibilità*, *1*(1), 91–103. DOI: 10.3280/RISS2020-001006

Gupta, S. P., & Bhardwaj, A. (2024). Technology of Sensors. In *Advances in Environmental Engineering and Green Technologies* (pp. 1–9). IGI Global., DOI: 10.4018/979-8-3693-1930-7.ch001

Hasan, I., Habib, M., & Mohamed, Z. (2023). Blockchain Database and IoT: A Technology driven Agri-Food Supply Chain. *International Supply Chain Technology Journal, 3*(3), 40–45. DOI: 10.20545/isctj.v09.i03.01

Hui, L., Gui-rong, W., Jian-ping, W., & Peiyong, D. (2017, May). Monitoring platform of energy management system for smart community. In 2017 29th Chinese Control And Decision Conference (CCDC) (pp. 1832-1836). IEEE. DOI: 10.1109/CCDC.2017.7978814

Jagtap, S., Rahimifard, S., & Duong, L. N. K. (2019a). Real-time data collection to improve energy efficiency: A case study of food manufacturer. *Journal of Food Processing and Preservation, 8*. Advance online publication. DOI: 10.1111/jfpp.14338

Jagtap, S., Rahimifard, S., & Duong, L. N. K. (2019b). Real-time data collection to improve energy efficiency: A case study of food manufacturer. *Journal of Food Processing and Preservation, 8*. Advance online publication. DOI: 10.1111/jfpp.14338

Jagtap, S., Rahimifard, S., & Duong, L. N. K. (2019c). Real-time data collection to improve energy efficiency: A case study of food manufacturer. *Journal of Food Processing and Preservation, 8*. Advance online publication. DOI: 10.1111/jfpp.14338

Kazancoglu, Y., Ozbiltekin-Pala, M., Sezer, M. D., Luthra, S., & Kumar, A. (2022). Resilient reverse logistics with blockchain technology in sustainable food supply chain management during COVID-19. *Business Strategy and the Environment, 4*(4), 2327–2340. DOI: 10.1002/bse.3251

Kebour, O., & Kebour, D. (2023). A practical approach for Energy Management, in the Agro-food sector industry in Algeria. *2023 International Conference on Advances in Electronics, Control and Communication Systems (ICAECCS)*, 1–6. https://doi.org/DOI: 10.1109/ICAECCS56710.2023.10105084

Lei, T., Min, Z., Gao, Q., Song, L., Zhang, X., & Zhang, X. (2022). The Architecture Optimization and Energy Management Technology of Aircraft Power Systems: A Review and Future Trends. *Energies, 11*(11), 4109. DOI: 10.3390/en15114109

Mason, R. E., White, A., Bucini, G., Anderzén, J., Méndez, V. E., & Merrill, S. C. (2020). The evolving landscape of agroecological research. *Agroecology and Sustainable Food Systems, 4*(4), 551–591. DOI: 10.1080/21683565.2020.1845275

May, G., Taisch, M., Prabhu, V. V., & Barletta, I. (2013). Energy Related Key Performance Indicators – State of the Art, Gaps and Industrial Needs. In IFIP Advances in Information and Communication Technology (pp. 257–267). Springer Berlin Heidelberg. https://doi.org/DOI: 10.1007/978-3-642-41266-0_32

Mbadlisa, G., & Jokonya, O. (2024). Factors affecting the adoption of blockchain technologies in the food supply chain. *Frontiers in Sustainable Food Systems*, *8*, 1497599. Advance online publication. DOI: 10.3389/fsufs.2024.1497599

Menon, S., & Jain, K. (2024). Blockchain Technology for Transparency in Agri-Food Supply Chain: Use Cases, Limitations, and Future Directions. *IEEE Transactions on Engineering Management*, *71*, 106–120. DOI: 10.1109/TEM.2021.3110903

Omerali, M., & Kaya, T. (2023). Energy Efficiency Optimization Application in Food Production Using IIOT Based Machine Learning. In Applied Innovation and Technology Management (pp. 185–204). Springer International Publishing. http://dx.doi.org/DOI: 10.1007/978-3-031-38387-8_11

Oriekhoe, O. I., Ashiwaju, B. I., Ihemereze, K. C., Ikwue, U., & Udeh, C. A. (2024). Review of technological advancements in food supply chain management: A comparative study between the US and Africa. *International Journal of Management & Entrepreneurship Research*, *6*(1), 132–149.

Pradella, A. M., & Loures, E. de F. R., Costa, S. E. G. da, & Lima, E. P. de. (2019). Energy Efficiency in the Food Industry: A Systematic Literature Review. Brazilian Archives of Biology and Technology, spe. https://doi.org/DOI: 10.1590/1678-4324-smart-2019190002

Rajput, S., Jadhav, A., Gadge, J., Tilani, D., & Dalgade, V. (2023). Agricultural Food supply chain Traceability using Blockchain. 2023 4th International Conference on Innovative Trends in Information Technology (ICITIIT), 1–6. https://doi.org/DOI: 10.1109/ICITIIT57246.2023.10068564

Semkov, K., Mooney, E., Connolly, M., & Adley, C. (2014). Efficiency improvement through waste heat reduction. *Applied Thermal Engineering*, *1*(1), 716–722. DOI: 10.1016/j.applthermaleng.2014.05.030

The World Bank. 2022, Climate Change. Available online: https://www.worldbank.org/en/topic/climatechange/overview (accessed on 2 December 2022).

Thevanes, N. (2020). Environmental Management Practices of an Apparel Manufacturer in Sri Lanka: A Critique. South Asian Journal of Social Studies and Economics, 1–10. https://doi.org/DOI: 10.9734/sajsse/2019/v5i330149

TI, M. S., Prakash, S., & Vikraman, K. (2024). Cooked Food Supply Chain ERP Model.

Van Gorp, J. C. (2005). Using Key Performance Indicators To Manage Energy Costs. *Strategic Planning for Energy and the Environment*, *2*(2), 9–25. DOI: 10.1080/10485230509509683

Vásquez Araya, G. (2020). Sistema de Gestión de la Energía para una planta de conservas. *Yulök Revista de Innovación Académica*, *1*(1). Advance online publication. DOI: 10.47633/yulk.v4i1.263

Warriner, K., Reddy, S. M., Namvar, A., & Neethirajan, S. (2014). Developments in nanoparticles for use in biosensors to assess food safety and quality. Trends in Food Science &amp. *Trends in Food Science & Technology*, *2*(2), 183–199. DOI: 10.1016/j.tifs.2014.07.008

Wojdalski, J., Grochowicz, J., Dróżdż, B., Bartoszewska, K., Zdanowska, P., Kupczyk, A., Ekielski, A., Florczak, I., Hasny, A., & Wójcik, G. (2015). Energy efficiency of a confectionery plant – Case study. *Journal of Food Engineering*, *146*, 182–191. DOI: 10.1016/j.jfoodeng.2014.08.019

Yoe, H. (2024). ICT-based Digital Agriculture. 2024 IEEE/ACIS 22nd International Conference on Software Engineering Research, Management and Applications (SERA), 1–1. https://doi.org/DOI: 10.1109/SERA61261.2024.10685615

Zhang, J. (2023). Energy Management System: The Engine for Sustainable Development and Resource Optimization. Highlights in Science, Engineering and Technology, 618–624. https://doi.org/DOI: 10.54097/cvfd9m83

Zisopoulos, F. K., Rossier-Miranda, F. J., van der Goot, A. J., & Boom, R. M. (2015). The use of exergetic indicators in the food industry – A review. *Critical Reviews in Food Science and Nutrition*, *1*(1), 197–211. DOI: 10.1080/10408398.2014.975335 PMID: 26066523

Chapter 4
Integrating Renewable Energy Sources in Agriculture:
A Holistic Approach to Sustainable Food Systems

Rolly Mehrotra
https://orcid.org/0000-0003-2460-7467
Gorakhpur University, India

ABSTRACT

The rapid depletion of conventional fossil fuels is a result of factors such as urbanisation, population growth, and increased food consumption. The adoption of renewable energy sources (RES) presents a feasible way to address the growing energy needs while lowering the threats to the world's food supply. Thus, the use of various renewable energy sources (RES) for food processing—such as solar, wind, geothermal, biomass, etc.—is covered in this chapter. One of the most researched uses of solar energy is drying food products for longer shelf lives. Many different types of solar dryers have been created. In addition to drying food, heat produced by solar and geothermal energy can be used to heat, evaporate, distil, sterilise, and pasteurise beverages. Additional broad-spectrum uses of clean energy include milling and grinding, solar ovens, stoves, and cookers. To determine potential routes for lowering the reliance of the global food processing industry on traditional fossil fuels, a variety of well implemented renewable energy-based alternatives have been examined.

DOI: 10.4018/979-8-3693-8342-1.ch004

1. INTRODUCTION

Renewable energy also referred to as green energy or sustainable energy is described as the energy derived from natural processes that are replenished at a faster rate than they are consumed these comes from a source which does not run out. They are self-replenishing and have low or zero carbon footprints. Renewable energy are through inexhaustible in duration but are limited in the amount of energy that is available per unit of time. Renewable energy sources include biomass, geothermal energy, wind energy, and solar energy. Hydropower is regarded as a component of the mix of renewable technologies due to this definition and the fact that water supplies are restored in the yearly hydrologic cycle (Frey and Linke, 2002).

1.1. Brief Overview of Renewable Energy Sources

Renewable resources for the production of energy are considered especially important for their potential to replace nonrenewable, or finite, resources. Additionally, renewable resources can offer cleaner energy solutions than those provided by non-renewable resources such as coal and fossil fuels. A renewable resource is a resource of which there is a seemingly endless supply because it can be replenished. For example, renewable resources such as the sun, the wind, and geothermal heat are considered inexhaustible. Water is also considered a renewable natural resource, as long as there is precipitation. Changing climate patterns have underscored the need for conservation efforts to protect water supplies.

Other natural resources are considered renewable even though some time and effort must go into their renewal. In addition, most precious metals are considered renewable because they're reusable. Since they are not destroyed during their extraction and use, they can be recycled. Renewable energy as said above is the most widely growing technology for sustainable future. Different types of renewable energy are paving way for a better and sustainable future. Direct energy from the Sun is known as solar energy and is the most abundant energy source on Earth. The photovoltaic cell converts sunlight to electricity and is the fastest-growing alternative energy source.

Fossil fuels have been used since the late 1880s to produce the energy we use. Renewable resources such as hydropower and wood have been used much longer. In fact, they were the two primary renewable energy resources up to the 1990s. In the years since, renewable energy production has come increasingly from biomass, geothermal, solar, water, and wind resources. Renewable resources that can replace fossil fuels in the production of energy are a major focus of nations around the world. The challenges for successful renewable energy production include reliability and expense.

A great deal of research is taking place to determine the viability and best implementation of renewable energy on a mass scale. According to their current state and future potential, renewable energy sources will surpass nuclear (17%), natural gas (17%), and oil (8%), to become the second-largest power source after coal (39%) (Omer, 2008). There will be a need for a great deal of renewable energy potential, environmental interest, economic consideration of the use of fossil fuels, and a strong focus on long-term sustainable development.

Table 1. Source: Renewable Energy Resources in Powering Agriculture - energypedia

Source	Converstion to	Most applied technologies and applications
Solar energy	Heat Mechanical energy Electricity	Photovoltaic (PV) driven pumps for irrigation Crops, drying of fruits / spices, ice making and cold storage (through absorption or heat driven refrigeration)
Wind energy	Mechanical energy Electricity	Direct use: grinder, mills, mechanical water pumps Electrical water pumps
Biomass energy	Heat Electricity Liquied Biofuels Biogas	Dryer (fruits, herbs,spices) Fermenter (tea) Combustion motor or electric motor (fuels like ethanol and biodiesel for transportation) Anaerobic digester: biogas for lighting, cooking and heating and industrial biogas for decentralized electricity

1.2. Importance of Renewable Energy in Food Production

The agricultural sector uses non-renewable energy sources for the running of machinery and equipment, for lighting on farms, for heating and cooling buildings, and indirectly for industrial equipment, fertilizers, and off-farm chemicals that harm the environment (Aydoğan & Vardar, 2020). These kinds of environmental harm can be lessened by using renewable resources. To enhance the sustainability of the agriculture industry, it is imperative to discover more effective methods for producing energy and distributing it to farmers. One way to do this is to replace fossil fuels in agriculture with low-carbon and clean energy technology (R. Sims, 2015). The use of technology that do not have a negative impact on environmental products and services, are cost-effective and efficient for farmers, and enhance food production (Pretty, 2008). Utilization of renewable energy in agriculture can efficiently help in enhancing access to energy, lowering the industry's reliance on fossil fuels, diversifying farms, and advancing sustainable development.

The use of renewable energy in agricultural systems has several benefits, such as producing electricity for various agricultural tasks including pumping water for irrigation, livestock, or domestic usage, lighting farm buildings, producing power, and other things. According to Ali et al. (2012), these forms of renewable energy include geothermal, biomass, solar, and wind energy. The next sections go into further depth on how various renewable energy sources are used in agriculture to boost food production.

2. SOLAR ENERGY IN FOOD PRODUCTION

2.1. Definition and Benefits of Using Solar Energy

Anywhere that receives sunshine can use solar energy; however, the amount of solar energy that can be used to generate electricity depends on the time of day, the location, and the weather. Solar energy can be used in two primary ways: photovoltaics (PV), which turns solar energy into electricity, and solar thermal, which turns solar energy into heat.

2.1.1 Solar Cookers

Solar cooker is a type of solar thermal collector. It "gathers" and traps the Sun's energy. Heat is produced when high frequency light is converted into low frequency infrared radiation. Ultraviolet and visible light easily pass through glass, however when they strike a darkened surface they are converted into long wave infrared radiation (heat). The glass (called glazing on a solar collector) traps these long waves. The glass allows the Sun's energy,traps the thermal energy, and the air inside your car becomes hot. As more light enters the car, the air gets even hotter, until we say that it feels like an oven inside.

2.1.2. Solar Dryers

A solar dryer is another technology to harness the solar energy that is used to dry fruits, vegetables, and crops for preservation. Solar dryers are of two types: direct and indirect. In direct solar dryers, the substance that is to be dehydrated is exposed to the sunlight in a vast field. Indirect solar dryers consist of an insulated box coated inside with a black absorption surface, an air inlet and an air outlet, and a single- or double-glazed glass. A solar dryer may be considered as it comprises of three main components — a drying chamber, a solar collector, and some type of airflow system, as illustrated in Fig. 8.6. In the drying chamber, drying takes place, and the material is spread on the chamber to get dehydrated, whereas the solar collector converts the solar radiation spectrum into heat.

2.1.3. Solar Water Heater

The solar water heater is the simplest application of solar thermal energy. The solar water heating system consists of the water storage tank, piping system, and solar collector. Solar water heaters are classified as active and solar water heaters and direct and indirect solar water heaters. Active and passive classification is based on the requirement of the pumping for the solar water heater. In passive solar water heaters, heat is transferred from the collector to the water tank located at the top of the collector by the process of natural circulation. In active solar water heaters, the water is circulated from the solar collector to the water storage by using a water pump.

Solar water heaters in which heat is transferred from the working fluid to the water with the help of the heat exchanger are called indirect solar water heaters and solar water heaters in which no heat exchanger is used for the transference of heat are called direct solar water heaters

2.1.4. Solar Photovoltaic Systems

Solar photovoltaic (pv) systems generate electricity from sunlight. Solar PV cells that capture sunlight are placed in panels, which are in turn placed in arrays, to deliver solar power to homes and businesses. Most PV cells have two layers of semi-conductor material, one positively charged and one negatively charged. When light shines on the semi-conductor the electric field across the junction between these two layers causes electricity to flow, generating direct current (DC). By placing metal contacts on the top and bottom of the PV cell, we can draw that current off for external use.

Without causing any harm to the environment, solar radiation that strikes a solar cell is immediately transformed into electricity in solar photovoltaic (SPV) technology. Systems for spy pumping are perfect for raising water for irrigation and consumption without affecting the environment. These pumps can be put in rivers, cisterns, tanks, and boreholes. High flow rates at low heads are intended use case for DC surface pumps. DC floating pumps work well in a variety of flow and head conditions.

2.2. Application of Solar Energy in Food Production

Solar energy is important for agriculture in many ways, such as lowering pollution, conserving resources, and boosting independence. Because solar energy uses less energy, expenses are also reduced (UCS, 2008). In agricultural, solar electrical devices are utilized to power electric fences, small engines, water pumps, solar greenhouses, underfloor heating, and battery charging (Chel & Kaushik, 2011). Aquifers or the surface of rivers can be used to supply water for irrigation. The energy required can change based on how far the water moves in both a vertical and horizontal direction.

Solar energy may be utilized to pump water for crops and cattle and has great potential for solar irrigation. Decentralized irrigation facilities will receive electricity from solar-powered systems that are economical, ecologically benign, and efficient (IRENA, 2016). Food can be processed for an extended period of time and kept free from corruption by drying it, which removes moisture from the product. Fruit and vegetables are extracted in accordance with regulations and hygienic conditions using solar-drying technology. In addition to saving time and energy, it also reduces space requirements, enhances process efficiency and product quality, and safeguards the environment (Torshizi et al., 2017). In frigid climates, greenhouses provide a way to produce food and horticulture products near to the consumer, saving money and fuel. For greenhouse heating, a variety of passive solar system configurations are employed (Gorjian et al., 2011). Farms can have hot water thanks to solar water heaters. Homes and fields will be heated by the sun's energy. PV solar panels are another tool for harnessing solar energy to create electricity (Torshizi et al., 2017).

As a nascent sector, solar energy installations are anticipated to rise in order to lower costs and boost solar cell efficiency (Chel & Kaushik, 2011; Fischer et al., 2006).

3. WIND ENERGY IN FOOD PRODUCTION

3.1. Definition and Benefits of Using Wind Energy

Wind power or wind energy is a form of renewable energy that harnesses the power of the wind to generate electricity. It involves using wind turbines to convert the turning motion of blades, pushed by moving air (kinetic energy) into electrical energy (electricity). This requires certain technologies, such as a generator that sits at the top of a tower, behind the blades, in the head (nacelle) of a wind turbine.The electricity created can either provide power to specific needs (like a wind turbine powering a streetlight or isolated farm) or contribute to the electric grid, which then powers homes, businesses, and schools with the help of transmission and distribution cables (or power lines).

By employing the kinetic energy of flowing air to rotate the turbine's blades, which are coupled to a generator, a wind turbine transforms the power of moving air into electricity (Laverdure, 2009). Wind turbines, both offshore and onshore, can produce electricity. A sizable amount of the farm's energy needs may be met by wind turbines, but they typically require at least an acre of land and must be located in strong wind locations in order to produce adequate electricity.

Wind energy is also a form of clean energy, meaning wind turbines do not produce greenhouse gas emissions, like carbon dioxide, which means less pollution going into the air, oceans, and environment that can cause health problems or harm the environment. Wind energy can be used in many places, including isolated or remote areas, like islands, that are not able to access the utility grid for power. Wind farms can be installed both on land and offshore, taking advantage of wind currents across the United States and along its coastlines.

There are also financial benefits and economic advantages to installing wind energy.

3.2. Application of Wind Energy in Food Production

Generators, air conditioners, and agricultural machinery can all be powered by wind turbines. Small wind systems can assist agriculture in conventional ways by employing mechanical energy to crush grain or pump water (Chel & Kaushik, 2011). Dried grains, such as wheat, barley, and maize, have been ground using millstones to make flour or meals. Grain milling and water pumping are the two ways by which wind power has been used to generate electricity over the years. It is anticipated that farmers would increase their usage of wind power to reduce electricity prices and increase their level of resource self-sufficiency, even as technological advancements continue to improve wind energy's economic performance (Fischer et al. 2006). Wind energy is transformed into electrical energy by wind turbines. In areas with suitable wind conditions, they are being used more and more in close proximity to food producing facilities.

Farmers and ranchers can use the wind to generate their own electricity. Small wind generators can be tailored to individual uses or can power an entire farm with capacities ranging from 400 watts to 40 kilowatts or more. For example, a lot of ranchers in the West and Texas utilize wind generators to pump water for their livestock. They are also more affordable and more convenient than diesel generators, and they might be less expensive than expanding electricity lines.

4. GEOTHERMAL ENERGY IN FOOD PRODUCTION

4.1. Definition and Benefits of Using Geothermal Energy

It's clean, offering energy that can be extracted without burning fossil fuels such as coal, gas, or oil. Using geothermal for electricity produces only about one-sixth of the carbon dioxide of a natural gas power plant, and little—if any—nitrous oxide or sulfur dioxide.

4.2. Application of Geothermal Energy in Food Production

Energy is supplied by geothermal resources in two ways. The first is the direct use of hot steam or water for industrial or household heating, and the second is the production of electricity (Bel, 2009). Thermal or electrical energy can be produced using geothermal technology. According to Fischer et al. (2006), geothermal power is perfect for both national grid applications and small rural electrical networks.

One can use geothermal energy's heat directly. Geothermal fluids can be used to pasteurize milk, heat homes, grow greenhouses, and heat water for fish farming (Fischer et al., 2006). Industrial-scale greenhouse farming, which grows crops, fruits, and flowers, is the main use of geothermal energy in agriculture. The waste heat recovered from a geothermal plant or low-temperature geothermal fluids are utilized as energy sources to heat the air for drying agricultural products. Applications requiring pre-heating and heating make effective use of geothermal resources (Nguyen et al., 2015).

Future production of a range of agricultural products with added value could be supported by geothermal power due to its technological and financial capabilities (Nguyen et al., 2015). According to Kinney et al. (2019), geothermal energy systems have the potential to provide the necessary heat for greenhouse applications in isolated northern regions. In the long run, these systems can offer a dependable and strong source of affordable, sustainable energy, which can also serve as a foundation for enhanced food security and community economic empowerment.

5. HYDROELECTRIC ENERGY IN FOOD PRODUCTION

5.1. Definition and Benefits of Using Hydroelectric Energy

The natural flow of water produces hydroelectric electricity. The turbine's blades spin in order to generate energy as water falls. The generator that transforms the energy into electricity is attached to the turbine. The quantity of water flowing through a turbine (the volume of water flow) and the height at which the water "falls" (the head) determine how much energy a system can generate. More electricity is generated the larger the flow and head. A clean, domestic, renewable energy source is hydropower. It emits minimal pollutants and offers cheap electricity. Hydropower produces electricity without destroying water, in contrast to fossil fuels.

The only renewable energy source that can meet expanding energy demands and replace the production of electricity from fossil fuels is hydropower. Systems that use hydroelectric power range in size and purpose. Hydropower systems function according to this fundamental idea: A generator is attached to a turbine that is turned by the force of flowing water. There are two primary categories of hydropower plants: run-of-river hydropower plants and reservoir hydropower plants. The river's current is used by run-of-river hydroelectric power stations to spin turbines and produce electricity. Large amounts of water are collected and stored by a dam in hydroelectric power facilities that use reservoir technology. A turbine attached to the generator is run by dammed water (IEA, 2012; Tissot, 2009). Systems using hydropower range in size and purpose.

5.2. Application of Hydroelectric Energy in Food Production

Hydropower is used in agriculture for the following purposes: direct irrigation from hydropower dams, use of dams for fish farming, and electricity on farms for pumping water for irrigation, processing, and storing agricultural produce as well as lighting farmhouses and surrounding areas (Clement et al., 2018). As a well-established technology, hydropower has numerous uses in the food production chain and on farms. Mini-hydro systems require less capital than other renewable energy sources. Mini-hydro systems, which may run off-river without the need for a dam or water storage, may become more common in the future (R. Sims, 2015).

Among hydroelectric systems, micro-hydroelectric plants are the smallest in size. With a power output of 1 kW to 1 MW, they are perfect for small-scale operations like farms, processing plants, and towns. A lot of electricity may be produced by huge hydroelectric systems.

6. BIOMASS ENERGY IN FOOD PRODUCTION

6.1. Definition and Benefits of Using Biomass Energy

Clean energy produced from materials obtained from biological sources is known as biomass-based energy. Large volumes of biomass are produced by a variety of agri-food industry food processing operations, forestry, and agriculture activities.

These products are then used as feedstock for biomaterials, bioenergy, and biofuels (R. Sims, 2015). Organic waste generated during meal preparation can be turned into biogas by anaerobic digestion. You can directly consume this biogas as a source of energy or use it to create electricity.

The term "bio-energy" describes the process of using biomass to produce power, heat, or transportation energy. Liquid fuels made from biomass are typically referred to as biofuels. Biodiesel, made from transesterification of feedstocks like sunflower, palm oil, or rape seed (canola), and bio-ethanol, made from fermentation of starches or sugars like corn or sugar cane, with a little amount of biogas, are the two main types of biofuels. Because of government subsidies and policies, the importance of these biofuels has increased during the past ten years. Some agro-industries, including sugar mills and banana plantations, are also producing excess electricity for the grid by turning waste into heat and electricity.

6.2. Application of Biomass Energy in Food Production

Small-scale agriculture can use biomass energy without requiring any additional processing (Babatunde et al., 2019). Most developing-nation farmhouses get their energy from conventional solid biomass sources like wood (R. Sims, 2015). Additionally, biomass is utilized to create biorefineries, which are widely employed in agriculture. According to Babatunde et al. (2019), a biorefinery is an industrial facility or technology that transforms biomass resources into energy and other useful products. Fuel, or biofuel, is created when biomass is converted into ethanol, gas, diesel, or other types of energy. With special gas engines, biofuels can be used directly for on-site power and heating. According to Clement et al. (2018) and R. Sims (2015), biogas is occasionally refined to natural gas purity and then fed into the gas system. Through agricultural production and processing activities, biogas can be transformed into alternative energy sources that can be replenished (Weiland, 2006).

An essential technological and financial resource for renewable energy is biomass. Over time, biomass systems have become much more efficient, and new developments are expected to further reduce their cost (EIP-AGRI, 2019).

7. FUTURE TRENDS, INNOVATIONS AND CHALLENGES IN RENEWABLE ENERGY FOR FOOD PRODUCTION

7.1. Emerging Renewable Energy Technologies

Different renewable energy sources exist, such as – Solar, Wind, Hydrothermal, Geothermal, Biomass, etc. However, most of the food industry players commonly use only three sources – Solar, Wind, and Biomass renewable energy sources.

Table 2. Renewable energy sources used by the Indian food industry (Source: Company annual reports):

Category	Indicative Players	Renewable sources of energy used
Beverages	Hindustan Coca-Cola Beverages	Solar, Wind, and Biomass
	PepsiCo	Solar, Wind, and Biogas
Snack foods	Haldiram	Solar and Biomass/biogas
	Bikaji	Solar
Dairy Products	Amul	Solar and Biogas
	Heritage India	Solar and Wind
Processed Foods	ITC	Solar, Wind, and Biomass
	Nestle	Solar, Wind, and Biomass
Bakery and Confectionary	Jubilant FoodWorks	Solar
	Britannia	Solar, Wind, and Biomass
Health and Wellness	Patanjali	Solar
	Dabur	Solar and Biomass
Seafood and meat processing	Avanti Feeds	Solar and Wind
	Venky's Foods	Solar

7.2. Challenges in Adopting Renewable Energy

When integrating renewable energy into their operations, food manufacturing enterprises face numerous challenges despite the numerous benefits. These issues require careful consideration and original solutions:

1. **Intermittency of Renewable Sources:** Weather-related factors determine the intermittent nature of renewable energy sources, such as the sun and wind. Inconsistent energy generation can cause disruptions to ongoing industrial activities. Energy storage systems can be used to lessen this.

2. **High Upfront Capital Costs:** Installing renewable energy technologies, such as wind or solar turbines, can incur significant up-front costs. The initial capital investment burden can be reduced with the help of favorable financing options, government subsidies, and incentives. The initial costs may eventually be balanced by long-term operational cost benefits.
3. **Technological Integration and Compatibility:** It could be challenging to adapt existing production facilities to accommodate renewable energy systems. Ensuring compatibility with existing equipment and procedures is essential. Careful planning, technological developments, and employee training can all help ensure that renewable energy systems are seamlessly integrated into the existing infrastructure.
4. **Scalability and Flexibility:** The food production industry has a range of production levels. Renewable energy systems' ability to efficiently scale up or down to satisfy changing energy demands is challenging to ensure. Hybrid energy systems can be used to increase scalability and flexibility in response to fluctuating energy needs. These systems combine many renewable sources with sophisticated control systems.
5. **Availability of appropriate sites:** Proper geographic locations are necessary for some renewable energy sources, such as wind. The best places for these technologies could not be accessible to certain food processing companies. By choosing renewable energy sources that are appropriate for the local climate and geography, energy production can be maximized. Sometimes, businesses need to look into alternative sources or focus on solutions that are more site-independent.
6. **Public Perception and Stakeholder Engagement:** Stakeholders and the general public may be against the adoption of renewable energy due to concerns about aesthetics, loudness, or perceived challenges associated to renewable energy infrastructure. It is difficult to persuade contractors, suppliers, and other businesses to use the same technologies in their operations. Responsible project planning, clear communication, and community involvement can help create a favorable public perception and increase support.
7. **Technology Risks and Reliability:** Technologies that rely on renewable energy sources could be less reliable and more prone to unforeseen issues or equipment failures. Regular maintenance, close monitoring, and investments in reliable and tested technologies can all help lower the risks associated with the operation of renewable energy systems.

A complete approach that takes into account the technological, social, legal, and financial aspects is required to overcome these problems. Successful integration of renewable energy into the food manufacturing industry requires collaboration among government agencies, commercial players, and the community.

8. CONCLUSION

A shift from manual labor to machinery has occurred in agriculture since the industrial revolution. The use of chemicals and machinery in agriculture has made energy one of its primary inputs. Due to the negative consequences of greenhouse gas emissions from fossil fuels, which are typically utilized as an energy source for numerous uses in agriculture, such as water heating, irrigation, and lightning, management of energy usage in agriculture is a global concern. The demand for agricultural products is growing exponentially with the global population. Thus, one of the sectors that stands to gain the most from the use of renewable energy is agriculture.

Given the wide variety of energy requirements and the multitude of ways in which they can be satisfied, energy plays a critical role in enhancing productivity and agricultural production systems. Utilizing renewable energy technology in agriculture will improve sustainable agriculture, guarantee food security, and solve issues with energy and water scarcity. Because of this, using renewable energy and related technologies has a lot of promise to support sustainable agricultural goals. They are now utilized on a number of farms, and there are plenty of chances to use them more in the future. As a foundational approach to sustainable agriculture, the promotion of renewable agriculture requires increased government assistance. The plans for agricultural planning will take this into account.

REFERENCES

Alharbi, S. S., Al Mamun, M., Boubaker, S., & Rizvi, S. K. A. (2023). Green finance and renewable energy: A worldwide evidence. *Energy Economics, 118*, 106499.

Ali, S., Dash, N., & Pradhan, A. (2012). Role of renewable energy on agriculture. *International Journal of Engineering Sciences Emerging Technologies, 4*(1), 51–57.

Aydoğan, B., & Vardar, G. (2020). Evaluating the role of renewable energy, economic growth and agriculture on CO2 emission in E7 countries. *International Journal of Sustainable Energy, 39*(4), 335–348. DOI: 10.1080/14786451.2019.1686380

Babatunde, O., Denwigwe, I., Adedoja, O., Babatunde, D., & Gbadamosi, S. (2019). Harnessing Renewable Energy for Sustainable Agricultural Applications. *International Journal of Energy Economics and Policy, 9*(5), 308–315. DOI: 10.32479/ijeep.7775

Chel, A., & Kaushik, G. (2011). Renewable energy for sustainable agriculture. *Agronomy for Sustainable Development, 31*, 91–118.

Clement, O., Akinyele, O., Oladimeji, S., & Oladipo, O. (2018). Renewable Energy Usage for Agricultural Practices: A review. 2018 International Conference of The Nigerian Institution of Agricultural Engineers, Lagos, Nigeria, September, 2018. EIP-AGRI. (2019). Enhancing production and use of renewable energy on the farm. Brussels, Belgium.

Fischer, J. R., Finnell, J., & Lavoie, B. (2006). *Renewable Energy in Agriculture: Back to the Future?* Business Choices. The Magazine of Food, Farm, and Resources Issues.

Gorjian, S., Hashjin, T., & Ghobadian, B. (2011). Solar Powered Greenhouses. *10th International Conference on Sustainable Energy Technologies on Sustainable Energy Technologies*

Hyder, M. J., Khan, M. J., Khan, M. A., & Saeed, S. (2023). The Design and Development of a Solar Dehydrator for Fruits. *Engineering Proceedings, 2023*(45), 48.

IRENA. (2016). *Solar pumping for irrigation: Improving livelihoods and sustainability*. The International Renewable Energy Agency.

Kinney, C., Dehghani-Sanij, A., Mahbaz, S., Dusseault, M. B., Nathwani, J. S., & Fraser, R. A. (2019). Geothermal Energy for Sustainable Food Production in Canada's Remote Northern Communities. *Energies, 12*(21), 4058. DOI: 10.3390/en12214058

Laverdure, R. B. (2009). *Wind Systems Technology*. Renewable Energies., DOI: 10.1002/9780470612002.ch5

Majeed, Y., Khan, M. U., Waseem, M., Zahid, Z., Mahmood, F., Majeed, F., Sultan, M., & Raza, A. (2023). Renewable energy as an alternative source for energy management in agriculture. *Energy Reports*, *10*, 344–359. DOI: 10.1016/j.egyr.2023.06.032

Muller, A., Jawtusch, J., & Gattinger, A. (2011). Mitigating greenhouse gases in agriculture.

Nguyen, M. V., Arason, S., Gissurarson, M., & Pálsson, P. G. (2015). *Uses of geothermal energy in food and agriculture: Opportunities for developing countries. Food and Agriculture Organization of the United Nations*. FAO.

Olabi, A. G., & Abdelkareem, M. A. (2022). Renewable energy and climate change. *Renewable & Sustainable Energy Reviews*, *158*, 112111.

Pretty, J. (2008). Agricultural sustainability: Concepts, principles and evidence. *Philosophical Transactions of the Royal Society of London. Series B, Biological Sciences*, *363*(1491), 447–465. DOI: 10.1098/rstb.2007.2163 PMID: 17652074

Sedighi, Md., & Zakariapour, M. (2014)... *Sustainable Energy.*, *2*(2), 44–51. DOI: 10.12691/rse-2-2-2

Sims, R., Puri, A. F. M., & Bracco, S. (2015). *Opportunities For Agri-Food Chains To Become Energy-Smart. Food and Agriculture Organization of the United Nations*. FAO.

Tadvi, S., Jain, V., & Thakkar, K. (2014). A Review: Solar Water Heating Systems. DOI: 10.13140/2.1.1910.5281

Tissot, R. N. (2009). Small Hydropower. In Renewable Energies (pp. 227-260). DOI: 10.1002/9780470612002.ch8

Torshizi, M. V., Mighani, A. (2017). The application of solar energy in agricultural systems. 3(2), 234-240.

Türkmen, B. A. (2020). Renewable Energy Applications for Sustainable Agricultural Systems. *International Journal of Innovative Approaches in Agricultural Research*, *4*(4), 497–504. DOI: 10.29329/ijiaar.2020.320.11

Twidell, J. (2021). *Renewable Energy Resources* (4th ed.). Routledge., DOI: 10.4324/9780429452161

Tze-Zhang, A., Salem, M., Kamarol, M., Das, H. S., Nazari, M. A., & Prabaharan, N. (2022) A comprehensive study of renewable energy sources: Classifications, challenges and suggestions, Energy Strategy Reviews, Volume 43, 100939, ISSN 2211-467X

UCS. (2008). Renewable Energy and Agriculture: A Natural Fit. Retrieved 20.07.2020 from https://www.ucsusa.org/resources/renewable-energy-and-agriculture?print=t # 27

Weiland, P. (2006). Biomass Digestion in Agriculture: A Successful Pathway for the Energy Production and Waste Treatment in Germany. Engineering in Life Sciences, 6, 302-309.p https://doi.org/doi:10.1002/elsc.200620128

Chapter 5

The Integration of Renewable Energy Sources in Food Production:

A Critical Strategy to Enhance Sustainable Agriculture

Surendra Prakash Gupta
https://orcid.org/0000-0002-4873-6346
Shri Vaishnav Vidhyapeeth Viswavidhayalaya, India

ABSTRACT

The integration of renewable energy sources in food production plays a crucial role in reducing greenhouse gas emissions and enhancing sustainability. Studies have shown that renewable energy sources such as solar and wind power can be effectively used in agriculture, resulting in energy autonomy, reduced grid dependency and overall system efficiency. In addition, the use of renewable energy in agriculture is associated with positive economic impacts and a greater impact on the green economy in certain countries, highlighting the importance of promoting energy production from renewable sources. However, the optimization of renewable energy sources through IoT technology has significantly improved the efficiency of energy production, equipment maintenance, and the integration of renewable energy into power grids, contributing to greater sustainability and environmental benefits. This chapter addresses the multifaceted challenge requires a holistic approach that integrates renewable energy sources into food production, for improving the resilience and efficiency of agricultural systems.

DOI: 10.4018/979-8-3693-8342-1.ch005

1. INTRODUCTION

In recent years, the global agricultural sector has shifted to renewable energy sources to meet its energy needs in a sustainable manner. The integration of renewable energy sources into food production has become a significant approach in the field of sustainable agriculture. This transition is primarily influenced by concerns about climate change, energy self-sufficiency and the rising costs of conventional energy. Renewable energies in food production not only reduce greenhouse gas emissions, but also increase the robustness and effectiveness of agricultural operations. Transitioning to renewable energy sources in agricultural practice brings a number of benefits, including reducing carbon emissions, improving financial profitability, increasing energy efficiency and increasing resilience to changing energy prices as suggested by Nazarov et al., 2024. This change further supports environmental protection, the creation of employment opportunities and the promotion of financial equality.. Furthermore, the adoption of renewable energy technologies like solar, wind, biomass, and geothermal energy can lead to a decrease in greenhouse gas emissions, minimize dependence on fossil fuels, and contribute to combating global warming and climate change (Kar et al., 2023). Studies show that farmers who embrace renewable energy technologies experience higher technical efficiency, with adopters showcasing a 10% increase in efficiency compared to nonadopters (Allahverdiyev, 2023). By utilizing renewable energy sources, agriculture can meet the growing demand for food and energy in an environmentally friendly and sustainable manner.

Switching to sustainable farming practices in developing countries can bring numerous economic and environmental benefits. Implementing practices such as organic farming, agroforestry and water management can lead to expanded market opportunities, lower production costs and long-term profitability. These strategies improve soil health, protect biodiversity and improve human well-being, thereby increasing crop yields and diversifying revenue sources (Rasyid and Mumpuni Ningsih, 2024). In addition, sustainable agriculture contributes to carbon sequestration, reduces greenhouse gas emissions, strengthens ecosystem resilience and thus promotes environmental sustainability. However, obstacles such as limited technology access, high costs, and absent policy support hinder the complete execution of sustainable agricultural approaches in developing countries ("Opportunities for Sustainable Food Production" 2023). Subsequently, encouraging stakeholder engagement, knowledge transfer, and farmer education is essential for fully realizing the full economic and environmental benefits of sustainable food production methods in these regions.

This chapter examines the various renewable energy innovations that are reshaping the agricultural sector in food production and highlights their strengths, weaknesses and predicted future trends.

2. TYPES OF RENEWABLE ENERGY SOURCES IN FOOD PRODUCTION

2.1 Solar Energy

One of the most available and widely used renewable energy sources in agriculture is solar energy. Using solar panels, which collect sunlight and convert it into heat or electricity, farms can power irrigation systems, lighting and other agricultural operations. Solar energy provides a consistent source of energy in remote areas and developing countries with intermittent grid access, thereby reducing long-term operating costs and dependence on fossil fuels. The increasing feasibility and affordability of solar technologies for a range of agricultural applications such as irrigation, crop drying and greenhouse heating are current trends in integrating solar energy into agricultural practices (Makinde and Obikoya, 2024).

Adoption of solar energy in agriculture promotes clean energy access, sustainable economic growth, and climate action, all of which help to achieve the Sustainable Development Goals (Watto et al., 2024). Several obstacles still need to be overcome, such as the initial high costs in comparison to traditional fuels, the technical expertise required for system installation and maintenance, and the unpredictability of solar resources that dictate appropriate system sizing and demand integration. Overcoming these obstacles and encouraging the broad use of solar energy in agriculture requires policy support in the form of subsidies, tax breaks, and financing plans. Key tactics for ensuring the long-term sustainability and accelerated adoption of solar power in agricultural practices include providing technical expertise to local stakeholders, harmonizing regulatory bodies, and leveraging public sector incentives (Panda et al., 2024).

2.2 Wind Energy

Wind energy has gained importance in agriculture due to its renewable nature and ability to provide clean electricity. Wind turbines use the kinetic energy of the wind to generate electricity, which can be used directly on farms or fed into the grid. Companies in windy regions can install turbines to generate electricity for their own needs or even feed excess electricity into the grid. The integration of wind energy plays a crucial role in influencing energy consumption patterns in the food industry by providing sustainable solutions. Studies emphasize the potential of wind energy turbines in agricultural fields to extract groundwater, thereby reducing energy costs for pumping systems (Khayatnezhad, Fataei, and Imani, 2023).

Research on European food and beverage companies highlights the importance of integrating renewable energies such as wind power for reducing carbon emissions and improving energy efficiency within the industry ("A Survey of the Energy Consumption Patterns and Potential Methods for Energy Efficiency and Renewable Energy Integration ") in the food and beverage industries of five European countries" 2015). Furthermore, the development of large-scale renewable energy systems, particularly wind energy, is identified as a key component of modern agricultural energy systems in the food sector, contributing to stable energy supply and forecast accuracy ("Front Matter" 2017). The high energy consumption of the agri-food chain highlights the need for a transition to renewable sources such as wind energy to reduce dependence on fossil fuels and mitigate environmental impacts, ultimately promoting sustainability in the food production industry (Rosin, Drovtar and Kilter, 2024). Wind power is particularly beneficial for powering large operations such as dairy farms or grain storage facilities where energy requirements are significant.

2.3 Biomass Energy

Biomass and biogas technologies offer renewable energy solutions that utilize organic materials such as crop residues, animal manure and food waste. Anaerobic digestion processes are highlighted in various literature and play a crucial role in the production of methane-rich biogas for electricity generation, heat production and as a transportation fuel ("Assessing Potential of Biogas: Harnessing Sustainable Energy from Biomass for Renewable Solutions (review)" 2024; Ofon et al. 2024; Alwahab et al. 2024).

The versatility of biogas applications is demonstrated by studies like this one, which show how liquid biogas waste can be used in microbial fuel cells to generate electricity. In addition, research in (Honcharuk, 2023) emphasizes the importance of zero-waste biofuel production technologies in the circular economy, indicating that biogas and biomethane can significantly contribute to reducing greenhouse gas emissions and promoting renewable energy sources (Abera, 2023). Overall, these results highlight the enormous potential of biomass and biogas technologies to promote sustainable development goals and renewable energy sources. This process creates digestate, a byproduct that closes the waste management cycle and improves soil health by acting as a nutrient-rich fertilizer.

A promising approach to producing renewable energy is biomass energy, which uses manure to produce biogas. Research has demonstrated the possibility of producing biogas economically through the anaerobic digestion of cow dung, with signs such as a high Internal Rate of Return (IRR) and Net Present Value (NPV) (Abera,

2023; Muharja et al., 2024). Furthermore, farming waste's methane emissions can be decreased by using manure in anaerobic digestion processes, which helps achieve climate change mitigation objectives (Fragoso et al., 2024). Additionally,, the ability to produce energy and manage waste in rural areas can be demonstrated by the conversion of woody biomass into biogas through anaerobic digestion, which is a sustainable solution despite the biomass's recalcitrant nature requiring pretreatment (Sakthi Vignesh et al., 2024).

Moreover, innovative approaches that integrate biomass properties, statistical methods and advanced reactor models aim to improve the efficiency and sustainability of biogas production systems and provide insights for renewable energy development (Agoud et al., 2024). Furthermore, optimization of anaerobic co-digestion processes using *canna indica* biomass and swine manure with calcium oxide pretreatment has resulted in significant improvements in biogas yield, highlighting the potential for optimized waste management and renewable energy production.

2.4 Hydropower

Small-scale hydropower has emerged as a promising solution for providing sustainable energy to rural farming communities, particularly in developing regions where access to the electrical grid is limited. These systems can harness the power of local waterways, such as streams and rivers, to generate electricity that can be used to power essential farming equipment, irrigation systems, and other agricultural needs. By leveraging the abundant water resources often found in rural areas, small-scale hydropower projects offer a reliable and environmentally-friendly alternative to traditional fossil fuel-based energy sources, helping to improve the overall sustainability and productivity of farming operations.

Although less common in direct agricultural applications, hydropower can be used for irrigation systems and small-scale electricity generation on farms near streams or rivers. Micro hydroelectric systems are environmentally friendly and can provide reliable and sustainable energy in remote areas where other renewable options may be less feasible. These systems use the energy of flowing water, often from small streams or rivers, to generate electricity on a small scale. Compared to larger hydropower projects, micro hydropower systems have a much lower environmental impact and are easier to deploy in harsh, isolated locations. By providing a distributed, renewable source of electricity, microhydro can play a critical role in improving energy access and supporting sustainable development in remote communities.

Integrating small-scale hydropower in farming for food production offers numerous benefits. It contributes to the water-energy-food nexus by enhancing sustainability and reducing greenhouse gas emissions (Shih et al., 2022). The utilization of hidden hydropower in water systems can lead to a significant reduction in energy consumption across various sectors, impacting the costs of food, water, and energy, as well as reducing CO2 emissions (McNabola, Mérida García, and Rodríguez Díaz, 2022). Additionally, combining traditional farming methods with modern approaches can enhance the efficiency of food production, ensuring food security and offering circular flux management potential in small-scale food production systems (Oyama et al., 2009). Implementing small hydropower systems in farming, such as those using fish farm effluent water or seawater, can generate power while managing agricultural waste, contributing to sustainable energy practices in food production (Teixeira Leite, 2022).

In sub-Saharan Africa, where access to electricity is limited and small water resources are abundant, micro hydropower holds great potential. Micro hydroelectric power plants are an attractive alternative for communities looking to reduce their environmental impact because, unlike fossil fuel-based power generation, they generate electricity without generating waste or air pollution. Micro hydro is also a reliable and flexible source of electricity as it is able to provide both base and peak load electricity in a timely manner.

2.4.1. Environmental Impacts

Food production systems are significantly influenced by wind energy, particularly offshore wind farms (OWPs). As a solution, Agri-PV systems (AVS) combine solar energy production with agricultural production, minimizing land use conflicts and achieving positive environmental impacts by reducing fossil fuel use, eutrophication and climate change (Wagner et al., 2023). However, wind farms can disrupt food chains by affecting marine ecosystems and leading to changes in phytoplankton, zooplankton, zoobenthos, fish and mammals, with both positive and negative consequences for primary production and trophic interactions (Wang et al., 2024). Furthermore, wind farms can change the dynamics of predator-prey relationships. This is demonstrated in the Western Ghats, where fewer predatory birds are present due to wind turbine construction. This reduction in bird abundance leads to an increase in lizard density, which in turn affects the ecosystem's behavior, physiology, and morphology (Thaker, Zambre, and Bhosale, 2018). In summary, although wind energy can have a beneficial effect on food production by means of novel systems such as AVS, its impact on terrestrial and marine ecosystems must be carefully evaluated in order to minimize the possibility of upsetting food chains and biodiversity.

2.5 Geothermal Energy

In regions with accessible geothermal resources, such as hot springs or geologically active areas, geothermal energy can be tapped to provide both heat and electricity for greenhouse operations, aquaculture, and food processing facilities. Geothermal heat pumps can also be used for space heating and cooling in farm buildings, enhancing energy efficiency.

Geothermal energy integration can significantly enhance the energy efficiency and profitability of aquaculture and food processing facilities by providing a sustainable and renewable energy source. By utilizing geothermal resources, facilities can reduce their reliance on fossil fuels, mitigating adverse environmental effects (Vega-Garzon et al., 2022). Geothermal energy can be harnessed for various uses in agriculture, including aquaculture and food processing, addressing the lack of affordable energy in developing countries and advancing their agricultural sectors (ŚwiĄtek, 2020). Furthermore, integrating geothermal energy into circular food production systems can create revenue streams to cover heat extraction costs while supporting viable businesses, such as combining aquaculture and hydroponics to produce vegetables, fish, fertilizers, algae, and biogas(Thorarinsdottir and Unnthorsson, 2018). This integration not only enhances energy efficiency but also contributes to sustainable practices and profitability in aquaculture and food processing industries.

Integrating geothermal energy into food processing facilities presents both technical and economic challenges. From a technical perspective, the existing oil infrastructure may not always be suitable for geothermal energy production, requiring detailed analysis of each well's potential impact (Thorarinsdottir and Unnthorsson 2018; Tuschl and Kurevija 2023). Additionally, the use of geothermal energy in food processing faces environmental impacts and the need for mitigation strategies. On the economic front, the food industry heavily relies on fossil fuels, causing adverse environmental effects, making the transition to geothermal energy crucial for sustainability(Soltani et al. 2021). Despite the potential benefits, challenges such as regulatory complexity, validation of technical aspects, and the lack of turnkey solutions can hinder the economic viability of integrating geothermal energy into food processing facilities (Thorarinsdottir and Unnthorsson 2018). Addressing these challenges is essential for maximizing the energy potential of geothermal resources in the food industry.

2.6 Hydroponics and Aquaponics

Hydroponics and aquaponics offer promising solutions for urban agriculture and sustainable food systems by addressing challenges in food security and environmental sustainability. These innovative techniques utilize water-based systems to efficiently produce food in urban areas, minimizing environmental impacts and maximizing resource efficiency (Kok et al. 2024; Sousa et al. 2024). Hydroponics, a soil-free method, allows for increased productivity with limited land and water requirements, making it ideal for urban farming. Aquaponics combines aquaculture and hydroponics, creating a symbiotic relationship between plants and aquatic life, enhancing nutrient recycling and reducing waste. By integrating technologies like IoT for real-time monitoring and control, these systems can optimize conditions such as lighting, climate, and nutrient concentration, leading to higher agricultural yields and reduced traditional farming limitations (Raman and Vasmatkar 2024; Jaiswal et al. 2024). Overall, hydroponics and aquaponics represent a sustainable and efficient path towards enhancing urban food production and ensuring food security in a rapidly evolving world.

Energy efficiency in indoor farming of hydroponics and aquaponics in urban agriculture and sustainable food systems can be enhanced through various techniques. Utilizing Internet of Things (IoT) technologies for real-time monitoring and automated control can significantly improve energy efficiency in these systems (Ng, Chin, and Mahkeswaran 2023; Raman and Vasmatkar 2024). Implementing energy-saving techniques such as water pump control, filtration, aeration, system design, and ventilation can reduce energy consumption and enhance sustainability in urban hydroponics and aquaponics systems. Additionally, integrating renewable energy sources like solar panels can provide a sustainable power solution for compact aquaponics systems, further improving energy efficiency in urban farming practices (Sousa et al. 2024; Gillani et al. 2022). By optimizing energy usage through these methods, indoor farming of hydroponics and aquaponics can contribute to more resource-efficient and environmentally friendly food production systems in urban areas, addressing food security and sustainability challenges.

The integration of renewable energy sources like solar power in indoor farming systems such as hydroponics and aquaponics plays a crucial role in promoting sustainability in urban agriculture and food systems (Muchtarom and Jamaluddin 2024). By utilizing solar panels to power compact aquaponic systems, these setups can operate efficiently and sustainably, addressing the challenges of limited land availability in urban areas. Additionally, the use of solar power in hydroponic systems, coupled with automatic monitoring features facilitated by microcontrollers and sensors, ensures optimal conditions are maintained with high accuracy rates, enhancing the overall performance and sustainability of these farming methods (Gaspar et al. 2023). This integration not only supports food production in urban settings but also contributes to reducing environmental impacts and promoting resilience in food systems, aligning with the goals of enhancing urban agriculture sustainability and food security.

3. INNOVATIVE PRACTICES AND TECHNOLOGIES

Innovative practices and technologies in food production are increasingly integrating renewable energy resources to enhance sustainability and efficiency. The agriculture sector heavily relies on fossil fuels, contributing significantly to global greenhouse gas emissions (Rempelos, Kabourakis, and Leifert 2023). Utilizing renewable energy sources like solar, wind, geothermal, and biomass energy presents a viable solution to meet the energy demands of food processing while reducing dependency on conventional fossil fuels (Rani et al. 2023). Implementing advanced controlled environment agriculture models with robust predictive control frameworks can maximize renewable energy usage efficiency and optimize crop production, showcasing the potential for sustainable development in agriculture (Hu and You 2022). Furthermore, the adoption of energy-efficient systems and machines powered by renewable sources can revolutionize mechanized agriculture, offering small farmers access to new technologies for improved productivity and income generation (Hu and You 2022; Gangil et al. 2022).

3.1 Energy-Efficient Farming Equipment

Energy-efficient farming equipment plays a crucial role in reducing the carbon footprint of food production when utilizing renewable energy sources. By adopting renewable technologies like electric tractors and agricultural robots, which are increasingly prevalent in modern agriculture (Gorjian, Fakhraei, et al. 2022), farmers can significantly decrease their reliance on fossil fuels, thus mitigating greenhouse gas emissions associated with traditional energy consumption. Additionally, the adoption of renewable energy technologies on farms has been shown to increase technical efficiency, with adopters achieving 10% higher efficiency levels compared to nonadopters (Gembicki 2016). Implementing energy-efficient equipment not only reduces energy consumption but also contributes to sustainable practices in agriculture, aligning with the goal of achieving environmentally friendly and cost-effective food production systems (J. Wang et al. 2023).

3.2 Smart Farming and IoT Solutions

IoT-based sensors and automation systems play a crucial role in enhancing crop yields and reducing waste in Smart Farming by providing real-time data and enabling precise control over farming operations. These systems, as highlighted in various research papers, monitor essential parameters like soil moisture, temperature, humidity, and plant-specific requirements (Wasatkar, Chavhan, and Kolekar 2023; Karuna et al. 2024). By leveraging this data, farmers can make informed decisions regarding irrigation, fertilization, pest control, and overall crop management, leading to increased productivity and sustainability. Automation systems integrated with sensors enable accurate and timely interventions, optimizing watering schedules and resource utilization. This data-driven approach not only improves crop quality and quantity but also minimizes water wastage, over-fertilization, and environmental damage, ultimately revolutionizing traditional farming practices towards a more efficient and sustainable model (Jadhav et al. 2023).

3.3 Integrating Renewable Energy in Supply Chains

Integrating renewable energy into supply chains offers significant economic and environmental benefits. From an economic standpoint, renewable energy procurement can lead to cost savings, improved brand reputation, and regulatory compliance. On the environmental front, it helps reduce carbon emissions, promotes clean energy technologies, and enhances sustainability through transparent decision-making and community investments (Reynolds, 2024; Fazlollahtabar, 2024). Measuring these benefits can be challenging due to varying policy incentives, unobserved counterfactuals, and differing measurement standards among stakeholders. However, techniques like TOPSIS and VIKOR can be utilized to evaluate the most optimized renewable energy supply chain based on criteria such as energy procurement, production costs, operations, maintenance costs, and social and environmental impact, providing a comprehensive assessment of the economic and environmental advantages of renewable energy integration (Dewi 2024).

4. ECONOMIC AND ENVIRONMENTAL BENEFITS

Synergies between economic development and environmental conservation are crucial for sustainable growth. The traditional model of development has led to environmental degradation, emphasizing the need for a shift towards conservation-for-development approaches that integrate local community welfare. This shift requires a comprehensive strategy involving various stakeholders, including local authorities, communities, and businesses, to ensure balanced environmental and economic sustainability in rural areas (Mishenin et al. 2023). Blockchain technology has been identified as a catalyst for sustainable development, enhancing transparency, traceability, and resource efficiency, thus promoting economic growth while supporting environmental stewardship (Thanasi Boçe and Hoxha 2024). Additionally, the identification of synergy in economic science involves understanding the patterns in production relations and the production of public goods using nature's productive forces, highlighting the importance of finding optimal development paths

(Ryspaeva et al. 2024). Studies on the synchronized advancement of ecological and economic benefits in China's provincial regions emphasize the need for synergistic development to achieve high-quality economic growth while preserving ecological and environmental benefits (Gatiso et al. 2022).

4.1 Cost Savings and Financial Incentives

Financial incentives play a crucial role in driving the adoption of renewable energy technologies in various sectors, including food production. Studies have shown that price-based incentives, such as feed-in premiums (FIP) and feed-in tariffs (FIT), are more effective in promoting the diffusion of non-hydroelectric renewable energy technologies compared to quantity-based incentives like renewable portfolio standards (RPS)(Abbruzzese, Infante, and Smirnova 2024). In the context of domestic-scale renewable energy technologies, schemes like the feed-in tariff (FiT) and renewable heat incentive (RHI) have been introduced to incentivize adoption, with the FiT demonstrating a higher stimulation of adoption rates (Cox 2016). Optimal incentive structures should focus on providing financial rewards based on the quantity of renewable energy produced, as seen in successful schemes like FIP, to encourage widespread adoption of green technologies in food production and other sectors

4.2 Reducing Carbon Footprint

Reducing the carbon footprint in food production is crucial for achieving sustainability. Studies emphasize the significance of carbon footprint analysis in the food industry, with a focus on flour production and composting initiatives as effective ways to mitigate emissions and optimize processes (Cox, 2016). Research highlights the impact of various factors like fertilizer use on carbon emissions, suggesting a shift towards compound fertilizers to reduce environmental impact. Implementing renewable energy sources, waste reduction strategies, and advanced waste management systems are essential for lowering carbon emissions throughout the food supply chain, promoting eco-friendly practices and contributing to global climate change mitigation efforts (Bhatia et al., 2023). By customizing calculation methodologies, adopting innovative tools, and investing in sustainable practices, the food industry can significantly reduce its carbon footprint and work towards a more sustainable future (Holka, Kowalska, and Jakubowska, 2022).

4.3 Enhancing Food Security and Resilience

Enhancing food security and resilience in food production is crucial for mitigating vulnerabilities in global agricultural markets. Strategies such as biofortification, which enhances the nutritional quality of food crops, can combat hidden hunger and improve public health sustainably (Sandström et al. 2024). Moreover, transitioning towards agroecological practices based on ecological principles can enhance the resilience of smallholder food security while providing ecosystem services, especially in the face of multiple shocks like the COVID-19 pandemic and climate-related disasters (Haji and Himpel 2024). These approaches, coupled with the utilization of domestic food system byproducts to decrease trade dependency, can contribute to building more local and circular livestock and aquaculture production systems that are resilient to various shocks, ultimately advancing global food system sustainability and security.

5. CHALLENGES AND BARRIERS

Small-scale farmers face various challenges in maintaining sustainable food production. These challenges include overgrazing, shortage of grazing land, poor grazing veld, poor condition of livestock, predation, infrastructure issues, high costs of feed and fertilizers, diseases, unpredictable weather patterns, power failures, high electricity costs, cattle theft, lack of machinery and equipment, and lack of support (Zenda and Malan 2024). Additionally, the impacts of drought, such as livestock deaths and crop failure, further exacerbate the challenges faced by small-scale farmers (Mugiya et al. 2017). To address these issues and enhance sustainable food production, it is crucial to provide training on disease management, implement prevention strategies for livestock theft, offer financial support, promote market-oriented farming, and develop policies tailored to the specific concerns of small-scale farmers.

5.1 Technological Limitations

Technological limitations in food production using renewable energy sources are evident in the challenges faced in implementing eco-innovations, such as renewable energy sources in the agricultural sector. While renewable energy sources like solar, biomass, geothermal, and wind energies show great potential in agriculture, the design of these energy sources must align with the agricultural structure and processes to be effective (Dybikowska and Graczyk 2019; Nonhebel, 2005). Additionally, the use of renewable energy sources for food drying, a critical process in food preservation, is highlighted as a key area for sustainable development, emphasizing the need for reduced energy consumption and environmental impacts in drying technologies. Furthermore, the environmental impacts of different power-to-food (PtF) approaches, which aim to produce microbial protein sustainably, reveal that the energy source used and the efficiency of the bioreactor significantly influence the sustainability of the PtF approach, indicating the importance of technological choices in mitigating environmental impacts in food production using renewable energy sources (Sillman et al. 2020).

5.3 Policy and Regulatory Frameworks

Policy and regulatory frameworks play a crucial role in promoting the use of renewable energy sources in food production. The legal regulation of renewable energy sources in agriculture emphasizes the coordination between energy and agricultural legislation to support sustainable development. Energy-smart agri-food systems (ESAFS) aim to provide self-generated sustainable energy for agricultural production, reducing dependence on fossil fuels and enhancing resilience against energy price fluctuations, highlighting the need for improved legal regulations in this area(Gradziuk, Klepacki, and Stolarski 2022; Vázquez Pérez 2016). Analyzing the historical development of regulatory frameworks in various countries reveals the importance of appropriate regulations and support measures for the successful integration of renewable energy sources like solar photovoltaic and wind energy in distributed generation systems. U.S. policymakers have also implemented incentives and programs to encourage agriculture-based renewable energy production, driven by concerns such as energy security, greenhouse gas emission reduction, and boosting domestic demand for farm products(Nasiri et al. 2022; Garcia, Alzate, and Barrera 2012).

6. CASE STUDIES

6.1 Successful Implementations Worldwide

Successful implementations worldwide have shown that solar energy, biomass energy, wind energy, and geothermal energy are the most promising renewable energy sources for food production. These sources have been utilized for tasks such as heating, drying, and powering agricultural operations, contributing to reduced greenhouse gas emissions and environmental impact. However, limitations exist, including the intermittency of some sources like solar and wind energy, which can affect consistent energy supply for food production processes. Additionally, the initial high costs of setting up renewable energy systems and the need for further research on economic and environmental aspects, especially in integrating different renewable sources in Hybrid Renewable Energy Systems (HRESs), pose challenges to widespread adoption in the agriculture sector. Addressing these limitations through technological advancements and policy support will be crucial for maximizing the benefits of renewable energy sources in sustainable food production (Rani et al. 2023; Gorjian, Kamrani, et al. 2022).

7. FUTURE TRENDS IN RENEWABLE ENERGY AND FOOD PRODUCTION

The future trends in renewable energy and food production are intertwined in addressing global challenges such as energy security, food security, and climate change. As the world transitions towards sustainable practices, renewable energy sources like solar, wind, and geothermal energy play a crucial role in reducing greenhouse gas emissions and ensuring energy efficiency in the agri-food sector. The shift towards green electricity, hydrogen, and biofuels is essential for meeting energy demands while mitigating the depletion of fossil fuels and combating climate change ("Trends in Renewable Energy: An Overview" 2022). However, the competition between bioenergy production and food resources highlights the need for strategic management to prevent food price increments and ensure food security. Efforts to make agriculture and the agri-food supply chain independent from fossil fuel use have the potential to enhance global food security and climate protection for the coming century(Hernandez-Escobedo et al. 2022; Mitani 2018). By embracing renewable energy technologies and sustainable agricultural practices, the future holds promise for a more resilient and environmentally conscious food production system.

7.1 Emerging Technologies

Emerging technologies in food production encompass a wide array of innovative tools and methods that are revolutionizing the way we grow, process, and consume food. From non-thermal technologies like pulsed electric fields and cold plasma that enhance food quality and safety (Paucar-Menacho, Moreno-Rojo, and Chuqui-Diestra 2024; Rønning, Pedersen, and Bjørnerud 2024), to the groundbreaking field of Cellular Agriculture (CellAg) that has the potential to transform agriculture fundamentally through ethical considerations and technological advancements, the landscape of food production is rapidly evolving. Additionally, the integration of smart digital tools such as 3D printing, sensors, and artificial intelligence is enhancing the quality, safety, and sustainability of plant-based products along the food supply chain (Pattabhiramaiah and Mallikarjunaiah 2023). These advancements not only optimize resource utilization and reduce environmental impact but also offer new possibilities for creating nutritious and sustainable food products that cater to modern consumer demands.

7.2 Policy Recommendations for Sustainable Development

Policy recommendations for sustainable development in renewable energy and food production include agricultural and energy policy reform, increased investment in clean technologies, strong international cooperation, public education and awareness. Substituting fossil fuels with alternative low-carbon energy sources, such as renewable energy, is crucial for reducing greenhouse gas emissions in the agriculture and food supply chain (Masriadi et al. 2024). Implementing federal energy policies like the Energy Policy Act of 2005 and the Food, Conservation, and Energy Act of 2008 can provide support for renewable energy initiatives and biofuels development. Local-oriented indicators and policy initiatives focusing on smart water management, precision agriculture, renewable energy sources, and ecosystem protection are essential for successful implementation of Sustainable Development Goals (SDGs) in agriculture, as highlighted in the Agenda 2030 and local contexts analysis (Gorjian, Fakhraei, et al. 2022; Papadopoulou, Papadopoulou, and Laspidou, 2022). Sustainability in agriculture and food production is increasingly emphasized for viable development.

8. CONCLUSION

Despite the advantages, challenges such as high initial costs, intermittent nature of some renewables, and regulatory hurdles remain barriers to widespread adoption. Overcoming these challenges requires continued technological advancements, supportive policies, and financial incentives for farmers to invest in renewable energy infrastructure. The most promising renewable energy sources for food production include biomass, solar energy, and agri-voltaics. Biomass derived from food waste offers significant advantages in terms of green energy production, socioeconomic benefits, and environmental sustainability. Solar energy plays a crucial role in supplying heat and electricity for agricultural tasks, contributing to sustainable food and agriculture systems.

The production of food as part of renewable energy sources presents a complex set of challenges. These include ensuring food security while meeting energy demands, mitigating the environmental impacts of biofuel cultivation, and establishing economically viable integrated food and energy systems. Addressing these challenges will require innovative solutions that carefully balance the use of limited natural resources to achieve a sustainable future for both food and energy production.

One of the primary concerns is the potential conflict between food and fuel production. The expansion of biofuel crops can compete with land and resources needed for food production, leading to concerns about food security and affordability. As the global population continues to grow, the demand for both food and energy will increase, necessitating careful planning to ensure that the production of one does not compromise the other. Furthermore, the environmental impacts of biofuel production must be carefully considered. The cultivation of bioenergy crops can lead to land-use changes, deforestation, and the loss of biodiversity, potentially offsetting the greenhouse gas reduction benefits of biofuels. Researchers are exploring strategies to develop "low indirect-impact biofuel systems" that utilize biomass residues, such as logging residues, to minimize the environmental footprint.

In conclusion, the adoption of renewable energy sources in food production represents a critical step towards building a sustainable agricultural future. By harnessing solar, wind, biomass, hydropower, and geothermal energy, farmers not only reduce their carbon footprint but also contribute to a more resilient and efficient food supply chain. As technologies evolve and awareness grows, the potential for renewable energy to transform agriculture into a truly sustainable industry continues to expand. However, despite their promise, renewable energy sources like wind and solar energy may not be as abundant or environmentally friendly as commonly believed, highlighting the need for further research and development to address their limitations.

REFERENCES

Abbruzzese, M., Infante, D., & Smirnova, J. (2024). The Diffusion of Renewable Energy Production in European Countries: The Role of Incentives. *Economics of Innovation and New Technology*, (January), 1–24.

Abera, Y. 2023. "Renewable Energy Integration with Sustainable Materials and Technologies." DOI: 10.22541/au.170013430.05838314/v1

Agoud, S., Benchrifa, M., Mabrouki, J., Moufti, A., & Slaoui, M. (2024). Study of Energy Production by the Anaerobic Digestion System of Waste Management. In *Technical and Technological Solutions Towards a Sustainable Society and Circular Economy* (pp. 67–76). Springer Nature Switzerland. DOI: 10.1007/978-3-031-56292-1_5

Allahverdiyev, E. (2023). Use of renewable energy resources in agriculture. *Voprosy Istorii (Moscow, Russia)*, *2*(2), 250–255.

Alwahab, A., Maharani, P. T., Nur K, W. O. M., Ahmad, L. O., Alimin, A., & Zaeni, A. (2024). Optimization of Biogas Liquid Waste from Livestock Manure as a Source of Renewable Energy through Microbial Fuel Cell (MFC) Technology. *Advance Sustainable Science, Engineering and Technology*, *6*(2), 0240202. DOI: 10.26877/asset.v6i2.17931

Ani, O. I., Aniokete, T., & Agbo, A. (2024). Assessing Potential of Biogas: Harnessing Sustainable Energy from Biomass for Renewable Solutions. Al-Rafidain Journal of Engineering Sciences, 330-349.

Bhatia, L., Jha, H., Sarkar, T., & Sarangi, P. K. (2023). Food Waste Utilization for Reducing Carbon Footprints towards Sustainable and Cleaner Environment: A Review. *International Journal of Environmental Research and Public Health*, *20*(3), 2318. DOI: 10.3390/ijerph20032318 PMID: 36767685

Cox, S. (2016). *"Financial Incentives to Enable Clean Energy Deployment: Policy Overview and Good Practices." Office of Scientific and Technical Information.* OSTI., DOI: 10.2172/1239887

de Sousa, R., Bragança, L., da Silva, M. V., & Oliveira, R. S. (2024). Challenges and Solutions for Sustainable Food Systems: The Potential of Home Hydroponics. *Sustainability (Basel)*, *16*(2), 817. DOI: 10.3390/su16020817

Dewi, L. 2024. "Exploring Stakeholder Perspectives on Supply Chain Sustainability in Renewable Energy Industries." *Research Square*. DOI: 10.21203/rs.3.rs-4301842/v1

Dybikowska, A., & Graczyk, M. (2019). Technological Eco-Innovations Related to Res – Opportunities and Barriers. *Nongye Jingji Wenti*, *1*(358), 120–139. DOI: 10.30858/zer/104699

Fazlollahtabar, H. (2024). Sustainable Renewable Energy Supply Chain with Current Technological Adaptation: Macro Energy Progress Policy in Iran. *Environmental Progress & Sustainable Energy*, *43*(June), e14433. Advance online publication. DOI: 10.1002/ep.14433

Garcia, A., Alzate, J. M., & Barrera, J. (2012). Regulatory Design and Incentives for Renewable Energy. *Journal of Regulatory Economics*, *41*(3), 315–336. DOI: 10.1007/s11149-012-9188-1

Gaspar, R. F., Juliano, J. V., Natanauan, J., Zapanta, D. J., Santos, A., & Gevaña, S. (2023). Low-Cost Solar Powered Automated Modular Aquaponic System. In *TENCON 2023 - 2023 IEEE Region 10 Conference (TENCON)*. IEEE., DOI: 10.1109/TENCON58879.2023.10322522

Gatiso, T. T., Kulik, L., Bachmann, M., Bonn, A., Bösch, L., Freytag, A., Heurich, M., Wesche, K., Winter, M., Ordaz-Németh, I., Sop, T., & Kühl, H. S. (2022). Sustainable Protected Areas: Synergies between Biodiversity Conservation and Socioeconomic Development. *People and Nature*, *4*(4), 893–903. DOI: 10.1002/pan3.10326

Gembicki, J. (2016). Energy efficiency in the agricultural and food industry illustrated with the example of the feed production plant. In *E3S Web of Conferences* (Vol. 10, p. 00138). EDP Sciences.

Gillani, S. A., Abbasi, R., Martinez, P., & Ahmad, R. (2022). Review on Energy Efficient Artificial Illumination in Aquaponics. *Cleaner and Circular Bioeconomy*, *2*(100015), 100015. DOI: 10.1016/j.clcb.2022.100015

Gorjian, S., Fakhraei, O., Gorjian, A., Sharafkhani, A., & Aziznejad, A. (2022). Sustainable Food and Agriculture: Employment of Renewable Energy Technologies. *Current Robotics Reports*, *3*(3), 153–163. DOI: 10.1007/s43154-022-00080-x PMID: 35600253

Gorjian, S., Kamrani, F., Ebadi, H., Samanta, S., & Savoldi, L. (2022). Applications of Renewable Energy Sources in Agriculture from a Complementarity Perspective. In *Complementarity of Variable Renewable Energy Sources* (pp. 615–647). Elsevier. DOI: 10.1016/B978-0-323-85527-3.00019-4

Gradziuk, P., Klepacki, B., & Stolarski, M. J. (Eds.). (2022). *Energy Supply within Sustainable Agricultural Production: Challenges, Policies and Mechanisms*. Mdpi AG. DOI: 10.3390/books978-3-0365-5176-0

Haji, M., & Himpel, F. (2024). Building Resilience in Food Security: Sustainable Strategies Post-COVID-19. *Sustainability (Basel), 16*(3), 995. DOI: 10.3390/su16030995

Hernandez-Escobedo, Q., Muñoz-Rodríguez, D., Vargas-Casillas, A., José, M. J. L., Aparicio-Martínez, P., Martínez-Jiménez, M. P., & Perea-Moreno, A.-J. (2022). Renewable Energies in the Agricultural Sector: A Perspective Analysis of the Last Three Years. *Energies*, *16*(1), 345. DOI: 10.3390/en16010345

Holka, M., Kowalska, J., & Jakubowska, M. (2022). Reducing Carbon Footprint of Agriculture—Can Organic Farming Help to Mitigate Climate Change? *Agriculture*, *12*(9), 1383. DOI: 10.3390/agriculture12091383

Honcharuk, I. (2023). European experience of biogas and biomethane production from waste according to the circular economy principle in agriculture. Economy, finances, management: topical issues of science and practical activity, 4(66), 7-19.

Hu, G., & You, F. (2022). Renewable Energy-Powered Semi-Closed Greenhouse for Sustainable Crop Production Using Model Predictive Control and Machine Learning for Energy Management. *Renewable & Sustainable Energy Reviews*, *168*(112790), 112790. DOI: 10.1016/j.rser.2022.112790

Jadhav, N., Premalatha, S., & Poongodi, P. (2024). Enhancing Crop Growth Efficiency through IoT-enabled Smart Farming System. EAI Endorsed Transactions on Internet of Things, 10.

Jaiswal, S., Rawat, G., Khadse, C., & Sharma, S. (2024). A Smart Hydroponics System for Sustainable Agriculture. In *Data Science for Agricultural Innovation and Productivity* (pp. 25–47). BENTHAM SCIENCE PUBLISHERS. DOI: 10.2174/9789815196177124010006

Kar, S. K., Pradhan, N. C., Swain, S. S., & Kar, S. (2023). Applicability of Various Renewable Energy Sources to Agricultural Applications. In *Handbook of Energy Management in Agriculture* (pp. 1–19). Springer Nature Singapore. DOI: 10.1007/978-981-19-7736-7_22-1

Karuna, G., Kumar, R. R., Sanjeeva, P., Deepthi, P., Saeed, H. Y., Asha, V., & Praveen, P. (2024). Crop recommendation system and crop monitoring using IoT. In *E3S Web of Conferences* (Vol. 507, p. 01063). EDP Sciences.

Khayatnezhad, M., Fataei, E., & Imani, A. (2023). Retracted: Integrated Modeling of Food–water–energy Nexus for Maximizing Water Productivity. *Water Science and Technology: Water Supply*, *23*(3), 1362–1374. DOI: 10.2166/ws.2023.038

Kok, C. L., Bagus Pradnya Kusuma, I. M., Yan Koh, Y., Tang, H., & Boon Lim, A. (2024). Smart Aquaponics: An Automated Water Quality Management System for Sustainable Urban Agriculture. *Electronics (Basel)*, *13*(5), 820. DOI: 10.3390/electronics13050820

Leite, T., & Vicente, A. (2022). Design of a Smart Microgrid with Small-Scale Hydro Generation: A Practical Case Study. *Revista Facultad de Ingenieria Universidad de Antioquia (Medellín)*, (May). Advance online publication. DOI: 10.17533/udea.redin.20220577

Lycourghiotis, S. (2022). Trends in renewable energy: An overview. *Global NEST: the International Journal*, *24*(3), 505–525.

Makinde, P., & Obikoya, E. (2024). Implementation of Solar System for Electricity Generation for Rural Farmers: A Review. *World Journal of Advanced Research and Reviews*, *22*(3), 458–471. DOI: 10.30574/wjarr.2024.22.3.1705

Masriadi, L. F., & Gusmailina, L. M. P. (2024). Addressing the Global Food Security Crisis and Energy Shortages: Innovative Solutions and Policy Interventions for Sustainable Development. *The International Journal of Science in Society*, *6*(1), 748–759. DOI: 10.54783/ijsoc.v6i1.1060

McNabola, A., García, A. M., & Juan, A. R. D. (2022). The Role of Micro-Hydropower Energy Recovery in the Water-Energy-Food Nexus. In *EWaS5*. MDPI., DOI: 10.3390/environsciproc2022021027

Mishenin, Y., Zinovchuk, N., Vysochanska, M., & Botsula, A. (2023). Environmental and economic sustainability and conservation of natural resources in the context of rural territories. *Ukrainian Journal of Applied Economics and Technology*, *8*(4), 305–312. DOI: 10.36887/2415-8453-2023-4-49

Mitani, T. (2018). Recent Progress on Microwave Processing of Biomass for Bioenergy Production. *Journal of the Japan Petroleum Institute*, *61*(2), 113–120. DOI: 10.1627/jpi.61.113

Muchtarom, A., & Jamaluddin, J. (2024). Seamless Solar-Powered Hydroponics Revolutionizes Urban Farming in Indonesia. *Indonesian Journal of Innovation Studies*, *25*(4). Advance online publication. DOI: 10.21070/ijins.v25i4.1192

Mugiya, D., & Hofisi, C. (2017). Climate change adaptation challenges confronting small-scale farmers. *Environment and Ecology*, (8, Iss. 1), 57–65.

Muharja, M., Khotimah, H., Fitria, D. R., Andika, P., Isnan, K. A., & Badril, A. (2024). Techno-Economic Analysis of Biogas Production from Cow Manure. *Jurnal Teknik Kimia Dan Lingkungan*, *8*(1), 1–13. DOI: 10.33795/jtkl.v8i1.3335

Nasiri, E., Rocha-Meneses, L., Inayat, A., & Kikas, T. (2022). Impact of Policy Instruments in the Implementation of Renewable Sources of Energy in Selected European Countries. *Sustainability (Basel)*, *14*(10), 6314. DOI: 10.3390/su14106314

Nazarov, D., Sulimin, V., Shvedov, V., & Larionova, N. (2024). Renewable energy sources for the agricultural sector. In *E3S Web of Conferences* (Vol. 541, p. 01002). EDP Sciences.

Ng, A. K., Jing, Z. C., & Mahkeswaran, R. (2023). "Energy Conservation Strategies for Increased Sustainability and Energy Efficiency in Urban Farming." In *2023 IEEE International Conference on Service Operations and Logistics, and Informatics (SOLI)*. IEEE. DOI: 10.1109/SOLI60636.2023.10425538

Nonhebel, S. (2005). Renewable Energy and Food Supply: Will There Be Enough Land? *Renewable & Sustainable Energy Reviews*, *9*(2), 191–201. DOI: 10.1016/j.rser.2004.02.003

Ofon, U. A., Ndubuisi-Nnaji, U. U., Udo, I. M., Udofia, E. S., Fatunla, O. K., & Shaibu, S. E. (2024). Biogas Production Potential from Anaerobic Co-Digestion of Food Waste and Animal Manure. *Journal of Chemical Society of Nigeria*, *49*(1). Advance online publication. DOI: 10.46602/jcsn.v49i1.949

Oyama, S., Kashihara, H., Kazari, M., Matsuda, Y., Shimizu, M., Yamada, K., & Watanabe, T. (2009). "D105 Development of an Efficient Small-Scale Distributed Power Generation System Using Gasification of Woody Biomass(biomass-2)." *The Proceedings of the International Conference on Power Engineering (ICOPE)* 2009.1 (0): _1–213_ – _1–218_. DOI: 10.1299/jsmeicope.2009.1._1-213_

Panda, S. N., Saikia, R., Sagar, , Swamy, G. N., Panotra, N., Yadav, K., Singh, B. V., Rathi, S., Singh, R., & Pandey, S. K. (2024). Solar Energy's Role in Achieving Sustainable Development Goals in Agriculture. *International Journal of Environment and Climate Change*, *14*(5), 10–31. DOI: 10.9734/ijecc/2024/v14i54167

Papadopoulou, C.-A., Papadopoulou, M. P., & Laspidou, C. (2022). Implementing Water-Energy-Land-Food-Climate Nexus Approach to Achieve the Sustainable Development Goals in Greece: Indicators and Policy Recommendations. *Sustainability (Basel)*, *14*(7), 4100. DOI: 10.3390/su14074100

Pattabhiramaiah, M., & Mallikarjunaiah, S. (2023). "Advances in Innovative and Emerging Technologies in Food Processing." In *Advances in Environmental Engineering and Green Technologies*, 1–32. IGI Global. DOI: 10.4018/978-1-6684-9094-5.ch001

Paucar-Menacho, L. M., Moreno-Rojo, C., & Chuqui-Diestra, S. R. (2024). Emerging Non-Thermal Technologies in the Food Industry: Advances and Potential Applications in Food Processing. *Scientia Agropecuaria, 15*(1), 65–83. DOI: 10.17268/sci.agropecu.2024.006

Pérez, A. V., Castillo, G. A. L., Alava, L. A. C., & Chilan, J. C. H. (2016). The regulatory framework for renewable energy sources. Journal of College and University. This is an open access article under the, 2395, 7492.

Raman, R., & Vasmatkar, A. 2024. "IoT-Enabled Aquaponics and Hydroponics for Efficient Indoor Farming Systems with Cloud Computing." In *2024 International Conference on Advances in Data Engineering and Intelligent Computing Systems (ADICS)*. IEEE. DOI: 10.1109/ADICS58448.2024.10533550

Rani, P., Dubey, A., Kumar, P., & Kumar, A. (2023). Sustainable Renewable Energy Sources for Food and Dairy Processing. In *Food Process Engineering and Technology* (pp. 65–95). Springer Nature Singapore. DOI: 10.1007/978-981-99-6831-2_4

Rasyid, H., & Ningsih, G. M. (2024). Application of Sustainable Agricultural Technology to Increase Productivity and Food Security. *Journal of Wood Science, 3*(4), 500–506.

Rempelos, L., Kabourakis, E., & Leifert, C. (2023). Innovative Organic and Regenerative Agricultural Production. *Agronomy (Basel), 13*(5), 1344. DOI: 10.3390/agronomy13051344

Reynolds, S. 2024. "Exploring the Influence of Corporate Social Responsibility on Supply Chain Sustainability in Renewable Energy." *Preprints*. DOI: 10.20944/preprints202405.1888.v1

Rønning, S. B., Pedersen, M. E., & Bjørnerud, E. (2024). Emerging Food Trends: Cellular Agriculture—novel Food Production Technology. [Elsevier.]. *Food Industries, 4*(0), 233–246. DOI: 10.1016/B978-0-443-15516-1.00011-6

Rosin, A., Drovtar, I., & Kilter, J. (2024). "Solutions and Active Measures for Wind Power Integration." In *Large Scale Grid Integration of Renewable Energy Sources: Solutions and Technologies*, 91–137. Institution of Engineering and Technology. DOI: 10.1049/PBPO222E_ch4

Ryspaeva, A., Temirov, N., Kasymbaeva, N., & Chinara, K. (2024). Sources of synergy in the financial and environmental regulation of the state. In *E3S Web of Conferences* (Vol. 535, p. 02010). EDP Sciences.

Sandström, V., Kastner, T., Schwarzmueller, F., & Kummu, M. (2024). The Potential to Increase Food System Resilience by Replacing Feed Imports with Domestic Food System Byproducts. [Web Site]. *Environmental Research Letters*, *19*(8), 084018. DOI: 10.1088/1748-9326/ad5ab3

Shih, J.-C., Lin, F.-Y., Hong, M.-D., Lin, H.-R., & Wen, J.-C. 2022. "Application on the Micro-Hydropower Generation Benefits of Agricultural Channels and the Water- Energy-Food Nexus." DOI: 10.5194/egusphere-egu22-2132

Sillman, J., Uusitalo, V., Ruuskanen, V., Ojala, L., Kahiluoto, H., Soukka, R., & Ahola, J. (2020). A Life Cycle Environmental Sustainability Analysis of Microbial Protein Production via Power-to-Food Approaches. *The International Journal of Life Cycle Assessment*, *25*(11), 2190–2203. DOI: 10.1007/s11367-020-01771-3

Soltani, M., Moradi Kashkooli, F., Souri, M., Rafiei, B., Jabarifar, M., Gharali, K., & Nathwani, J. S. (2021). Environmental, Economic, and Social Impacts of Geothermal Energy Systems. *Renewable & Sustainable Energy Reviews*, *140*(110750), 110750. DOI: 10.1016/j.rser.2021.110750

Sturm, B., Meyers, S., Jones, M. C., & Schmitt, B. (2015). A survey of the energy consumption patterns and potential methods for energy efficiency and renewable energy integration within the food and beverage industries of five European countries. In *2015 ASABE Annual International Meeting* (p. 1). American Society of Agricultural and Biological Engineers.

ŚwiĄtek, L.ŚwiĄtek. (2020). A Case Study of Geothermal Resources Use for the Innovative Aquaculture from Perspective of Syntropic Development Concept. *International Journal of Energy Production and Management*, *5*(1), 60–69. DOI: 10.2495/EQ-V5-N1-60-69

Thaker, M., Zambre, A., & Bhosale, H. (2018). Wind Farms Have Cascading Impacts on Ecosystems across Trophic Levels. *Nature Ecology & Evolution*, *2*(12), 1854–1858. DOI: 10.1038/s41559-018-0707-z PMID: 30397304

Thanasi Boçe, M., & Hoxha, J. (2024). Blockchain Technology as a Catalyst for Sustainable Development: Exploring Economic, Social, and Environmental Synergies. *Academic Journal of Interdisciplinary Studies*, *13*(2), 151. DOI: 10.36941/ajis-2024-0041

Thorarinsdottir, R., & Unnthorsson, R. (2018). Direct Use of Geothermal Resources for Circular Food Production. *Proceedings, 2*(8), 497.

Tuschl, M., & Kurevija, T. (2023). Revitalization Modelling of a Mature Oil Field with Bottom-Type Aquifer into Geothermal Resource—Reservoir Engineering and Techno-Economic Challenges. *Energies, 16*(18), 6599. DOI: 10.3390/en16186599

Vega-Garzon, L. P., Jeimy, A. P. R., Maria, P. G. S., Maria, A. R. G., & Liz, A. P. J. (2022). Use of Geothermal Energy in the Food Industry: A Review. *Revista Ingenierias Univsidad de Medellin, 21*(40), 67–86. DOI: 10.22395/rium.v21n40a5

Vignesh, N. S., Vimali, E., Rajkeerthana, S., Kaleeshwari, R., Kiruthika, M., Ashokkumar, B., & Rajendran, K. (2024). Utilization of woody biomass for biogas production. In *Sustainable Biorefining of Woody Biomass to Biofuels and Biochemicals* (pp. 103–123). Woodhead Publishing.

Wagner, M., Lask, J., Kiesel, A., Lewandowski, I., Weselek, A., Högy, P., Trommsdorff, M., Schnaiker, M.-A., & Bauerle, A. (2023). Agrivoltaics: The Environmental Impacts of Combining Food Crop Cultivation and Solar Energy Generation. *Agronomy (Basel), 13*(2), 299. DOI: 10.3390/agronomy13020299

Wang, J., Li, W., Haq, S., & Shahbaz, P. (2023). Adoption of Renewable Energy Technology on Farms for Sustainable and Efficient Production: Exploring the Role of Entrepreneurial Orientation, Farmer Perception and Government Policies. *Sustainability (Basel), 15*(7), 5611. DOI: 10.3390/su15075611

Wang, L., Wang, B., Cen, W., Xu, R., Huang, Y., Zhang, X., Han, Y., & Zhang, Y. (2024). Ecological Impacts of the Expansion of Offshore Wind Farms on Trophic Level Species of Marine Food Chain. *Journal of Environmental Sciences (China), 139*(May), 226–244. DOI: 10.1016/j.jes.2023.05.002 PMID: 38105050

Wasatkar, N., Namrata, P. G. C., & Kolekar, V. K. 2023. "IoT and AI-Based Smart Farm: Optimizing Crop Yield and Sustainability." In *Research Trends in Artificial Intelligence: Internet of Things*, 1–17. BENTHAM SCIENCE PUBLISHERS.

Watto, M. A.CSIRO Environment Business Unit,(2024). Solar-Powered Irrigation Systems: Recent Developments and Future Trends. In *Burleigh Dodds Series in Agricultural Science* (pp. 209–232). Burleigh Dodds Science Publishing.

Zenda, M., & Malan, P. J. (2024). Investigation of Challenges Faced by Small-Scale Sheep Farmers in the Northern Cape (Hantam Karoo), South Africa. [SAJAE]. *South African Journal of Agricultural Extension, 52*(1), 189–209. DOI: 10.17159/2413-3221/2024/v52n1a14663

Chapter 6
Energy Efficiency in the Livestock Sector

Neha Prasad
https://orcid.org/0009-0006-1857-0521
VLCC Health Care Limited, India

ABSTRACT

Livestock sector is one of the perishable food commodities among other food and dietary sources of protein and the requirement of energy efficient technologies in this sector is of greater importance. Livestock sector is a significant contributor to the global economy, providing nutritious food products to billions of people. There are several emerging technologies that help for enhancing energy efficiency, such as processing optimization, refrigeration management, energy-efficient equipment, water conservation, and renewable energy integration. This chapter deals with the principles related to livestock sector for providing energy efficient technologies for providing security, producing high quality safe and sustainable meat products while promoting animal welfare and with challenges faced in implementation and policy recommendations to support sustainable practices.

1. INTRODUCTION

The livestock sector is a cornerstone of global agriculture, providing essential food resources while contributing substantially to greenhouse gas emissions and energy consumption. As demand for animal products rises, enhancing energy efficiency is critical for mitigating environmental impacts and ensuring sustainable production. The livestock sector is a pillar of the global food system and a contributor to poverty reduction, food security and agricultural development (FAO,2020). According to the FAO, livestock contributes 40% of the global value of agricultural output and

DOI: 10.4018/979-8-3693-8342-1.ch006

supports the livelihoods and food and nutrition security of almost 1.3 billion people. At the same time, there is wide scope to improve livestock sector practices so that they are more sustainable, more equitable, and pose less risk to animal and human health. Excesses and insufficiencies are jeopardizing the stability of the global food system. While the proportion of undernourished people has been rising again in recent years, the problem of obesity and the associated health effects is becoming increasingly relevant. The livestock industry is essential for food security and rural livelihoods, but it is also one of the largest users of energy, contributing significantly to global greenhouse gas emissions. According to the Food and Agriculture Organization (FAO), livestock accounts for approximately 14.5% of all anthropogenic greenhouse gas emissions (FAO, 2013). As the world's population continues to grow, the demand for animal products increases, leading to heightened energy consumption and environmental impacts. Therefore, improving energy efficiency in livestock operations is critical for sustainable development (Miller et,al,2019).

Livestock refers to domesticated animals that are raised for agricultural purposes. These animals are primarily kept for the production of food, fiber, labor, or other products (FAO,2023). Common types of livestock include:

- **Cattle**: Raised for beef, milk, leather, and labor.
- **Sheep**: Raised for meat (lamb/mutton), wool, and milk.
- **Goats**: Raised for meat (chevon), milk, and fiber (such as cashmere).
- **Pigs**: Raised for pork and other by-products.
- **Poultry**: Includes chickens, ducks, turkeys, and geese, primarily raised for eggs and meat.
- **Horses**: While primarily used for riding and pulling, they can also serve as work animals.

As global demand for livestock products continues to rise, the livestock sector faces substantial pressures to enhance food safety, quality, and security while mitigating environmental impacts(USDA,2022). One of the significant challenges is improving energy efficiency throughout the production, processing, and distribution of livestock products. Energy-efficient technologies can play a pivotal role in addressing these challenges. This chapter explores various strategies, technologies, and practices that enhance energy efficiency in the livestock sector, discussing their implications for food safety, quality, and security.

2. ENERGY CONSUMPTION IN LIVESTOCK SECTOR

2.1 Overview of Energy Inputs

Energy inputs in livestock operations or production can be categorized into direct and indirect energy. Direct energy includes fuels used for machinery and transportation, while indirect energy encompasses feed production, processing, and transportation. A significant portion of energy use arises from feed production, which often involves extensive use of fertilizers, pesticides, and machinery(Rosen et.al,2023).

2.2 Energy Use by Livestock Type

Energy consumption varies by livestock type:

- **Dairy Cattle:** Energy-intensive for feed production, cooling systems in barns, and transportation.
- **Beef Cattle:** High energy input due to extensive feedlot practices and land use.
- **Poultry:** Requires energy for heating, cooling, lighting, and feed production, but generally more efficient than ruminants.
- **Swine:** Similar to poultry, with significant energy used in maintaining optimal environmental conditions.

3. THE IMPORTANCE OF ENERGY EFFICIENCY IN LIVESTOCK PRODUCTION

Energy efficiency is critical in livestock production for several reasons:

3.1 Environmental Impact

The livestock sector is responsible for a significant portion of global greenhouse gas emissions, contributing to climate change and environmental degradation. Improving energy efficiency can help reduce emissions from production activities, feed transportation, and processing (Steinfeld et al., 2006).

3.2 Economic Viability

Energy costs represent a substantial portion of total production costs in livestock farming (Cottle et al., 2011). By implementing energy-efficient practices, producers can lower expenses and improve profit margins.

3.3 Food Security

Enhancing energy efficiency can lead to more sustainable production methods, ensuring a stable supply of livestock products to meet growing consumer demand (Thornton, 2010).

4. ENERGY EFFICIENCY STRATEGIES

Recent advancements in technology present significant opportunities to enhance energy efficiency in livestock operations, addressing both productivity and environmental sustainability. One of the most impactful innovations is Precision Livestock Farming (PLF), which utilizes an array of sensors and data analytics tools to monitor individual animals and herds in real time. This technology enables farmers to optimize feed efficiency by tailoring dietary inputs to the specific needs of each animal based on their health and production status, thus reducing waste and improving overall animal welfare.

In addition, anaerobic digestion technologies have emerged as a viable solution for managing manure. By converting organic waste materials into biogas, this process not only provides a renewable energy source but also significantly lowers methane and other greenhouse gas emissions associated with traditional manure management practices.

Moreover, advancements in genetic breeding further contribute to energy efficiency by focusing on selecting animals with superior feed conversion ratios and desirable growth traits. By enhancing these traits through selective breeding, livestock can produce more meat or milk with less feed and energy input, thereby minimizing their environmental footprint. Collectively, these technological innovations represent a transformative shift in livestock management, fostering a more sustainable and energy-efficient industry while addressing the considerable challenges of climate change and resource scarcity.

5. ENERGY EFFICIENCY STRATEGIES IN LIVESTOCK PRODUCTION

5.1 Optimizing Feed Utilization

Feed represents the largest operational cost in livestock production and accounts for a significant amount of energy input in the form of fertilizer, transportation, and processing (Nassif et al., 2020).

Precision Feeding is an innovative approach that harnesses technology to deliver individualized feed rations tailored to the specific nutritional needs of animals, thus minimizing waste and enhancing feed conversion rates. By employing advanced data analytics, sensors, and automated feeding systems, farmers can closely monitor each animal's dietary requirements based on factors such as age, weight, health status, and production goals. This technology allows for precise adjustments in feed formulations, ensuring that animals receive the optimal nutrition they need for maximum growth and productivity, while simultaneously reducing excess feed consumption and minimizing costs (Van Milgen et al., 2008).

Complementing precision feeding is the strategy of utilizing alternative feed ingredients, which involves incorporating locally sourced and energy-efficient feed alternatives, such as by-products from food processing or agricultural operations. This practice not only lessens the reliance on traditional feed sources, which can be energy-intensive, but also promotes sustainability by giving a second life to materials that would otherwise contribute to waste. By incorporating these alternative ingredients, farmers can effectively reduce the overall energy-outlay of feed production while often achieving cost savings. Combining precision feeding with the use of alternative feed ingredients represents a powerful strategy for optimizing livestock production, as it improves resource efficiency, reduces environmental impact, enhances animal welfare, and supports a more sustainable agricultural system (Belanche et al., 2019). This integrated approach is crucial for addressing the increasing global demand for livestock products while mitigating the industry's carbon footprint and preserving natural resources.

5.2 Renewable Energy Integration

The integration of renewable energy sources in livestock farms can significantly enhance energy efficiency like the integration of Solar energy and Biogas production in agriculture is revolutionizing energy practices on farms, enhancing sustainability and reducing reliance on fossil fuels. Many farms now utilize solar panels to harness renewable energy for electricity generation, significantly decreasing operational costs and mitigating greenhouse gas emissions.

A study by Li et al. (2020) highlighted that solar energy systems could meet a substantial portion of the energy needs of large-scale livestock facilities, thereby promoting energy independence. In addition, Biogas Production through anaerobic digestion of manure enables farms to convert waste into valuable energy. This process not only manages livestock waste effectively but also produces biogas, which can be used for heating, electricity generation, or as vehicle fuel, as noted by Mora et al. (2022).

Together, these renewable energy solutions allow farmers to create a more sustainable agricultural system by transforming potential environmental liabilities into clean energy resources, ultimately leading to a significant reduction in their carbon footprint.

5.3 Efficient Manure Management

Effective manure management is pivotal not only for enhancing energy efficiency but also for promoting environmental sustainability in agricultural practices. Proper techniques such as **composting** and **aerobic storage** of manure have been shown to significantly reduce methane emissions, which are potent greenhouse gases, and also lower energy consumption associated with the handling and transport of manure. By managing manure responsibly, farms can mitigate their environmental impact while simultaneously producing high-quality compost that enriches soil health and fertility, as noted by Kumar et al. (2019).

Moreover, the integration of technologies designed for nutrient recovery plays a vital role in this context. Such technologies extract valuable nutrients like nitrogen and phosphorus from manure, thereby diminishing the need for synthetic fertilizers, which are not only costly but also energy-intensive to produce. The reduction in reliance on these synthetic inputs leads to lower overall energy consumption in agricultural operations, contributing to a more sustainable farming approach as highlighted by Nguyen et al. (2021).

Consequently, adopting advanced manure management practices not only bolsters energy efficiency but also fosters a more sustainable agricultural system by minimizing environmental impacts, promoting resource conservation, and ultimately enhancing the profitability and resilience of farming operations. This holistic approach to manure management highlights the interconnectedness of energy efficiency, nutrient cycling, and environmental stewardship in modern agriculture.

6. ENERGY-EFFICIENT TECHNOLOGIES IN LIVESTOCK PROCESSING

6.1 Advanced Processing Techniques

Enhancing energy efficiency in the processing phase of livestock products is crucial for promoting sustainability and reducing operational costs in the food production industry. One effective strategy for achieving this is the implementation of **heat recovery systems** within processing plants. These systems capture waste heat generated from cooking or rendering processes, which can be repurposed to meet heating and cooling needs elsewhere in the plant. By effectively utilizing this otherwise lost energy, processing facilities can significantly decrease the overall energy required for their operations, leading to substantial cost savings and a marked reduction in greenhouse gas emissions, as highlighted by Fellows (2017).

In addition to heat recovery systems, the adoption of **low-energy processing equipment** plays a vital role in minimizing energy consumption while ensuring high standards of food safety and quality. Energy-efficient machinery, such as vacuum packers and steam cookers, is specifically designed to streamline operations and reduce energy usage without compromising the safety and integrity of the products being processed. The use of such technologies not only lowers the operational expenses for processing plants but also contributes to the overall reduction of the carbon footprint associated with livestock processing, as stated by Morrison et al. (2019).

By integrating these advanced technologies, the livestock processing sector can make significant strides in energy efficiency, thus supporting the industry's transition towards more sustainable practices. This focus on enhancing energy efficiency not only improves the competitiveness of processing facilities but also plays a vital role in addressing the broader challenges of rising energy costs and environmental sustainability within the global food system.

6.2 Automation and Control Systems

Automation technologies can enhance energy efficiency in processing plants:

Smart Sensors and IoT: The incorporation of smart sensors can optimize energy usage by monitoring and controlling temperature, humidity, and other environmental factors in real time (Haddad et al., 2020).

Predictive Maintenance: Maintenance systems equipped with predictive analytics can identify potential equipment failures before they occur, thus avoiding downtime and improving overall efficiency (Mahmoud et al., 2021).

7. CHALLENGES TO IMPLEMENTING ENERGY EFFICIENCY IN LIVESTOCK SECTOR

The adoption of energy-efficient technologies in agriculture is significantly hindered by various challenges, including economic constraints, knowledge gaps, and institutional barriers. One major economic challenge is the high initial investment required for implementing these technologies, which can be particularly prohibitive for smallholder farms lacking sufficient financial resources. The uncertainty surrounding the return on investment further discourages innovation among farmers.

Additionally, a lack of understanding and training about energy efficiency methods impedes adoption; extension services often fail to prioritize energy management, leaving farmers without the necessary support to integrate these practices effectively. Compounding these issues are institutional barriers such as fragmented supply chains, which complicate the procurement of energy-efficient technologies, and limited access to financial resources, restricting farmers' ability to invest in needed improvements. Addressing these interconnected challenges is essential for creating a supportive environment that encourages the adoption of energy-efficient practices, helping farmers transition toward more sustainable and productive agricultural operations.

5. FUTURE DIRECTIONS IN ENERGY EFFICIENCY

Continued innovation and research are essential for advancing energy efficiency in the livestock sector. Future directions may include:

Data-Driven Approaches: Utilizing big data analytics and machine learning algorithms to optimize production processes and energy utilization, tailoring practices to the specific needs of the operation (Wang et al., 2022).
Consumer Engagement: Increasing consumer awareness and demand for sustainably produced livestock products can encourage operations to adopt more energy-efficient technologies (Zhang et al., 2020).

6. CONCLUSION

Improving energy efficiency in the livestock sector is essential to achieving sustainability goals, food security, and reducing environmental impacts. Through technological innovations, better management practices, and supportive policies, the sector can transition towards more efficient energy use. Addressing the challenges faced by producers will be imperative to unlock the sector's full potential and ensure a sustainable future for food production. Energy efficiency stands as a critical component in the effort to enhance food safety, quality, and security within the livestock sector. By adopting innovative technologies and practices that optimize feed utilization, integrate renewable energy, and streamline processing, the industry can significantly reduce its environmental footprint while maintaining economic viability. Continued research and collaboration between producers, researchers, and policy-makers will be essential to promote sustainable practices in the livestock sector.

REFERENCES

Belanche, A.. (2019). The role of alternative feed ingredients in the sustainability of livestock systems. *Animal Feed Science and Technology*, *252*, 38–47.

Cottle, D. J.. (2011). *Livestock and greenhouse gas emissions: Facts and figures*. University of Queensland.

FAO. (2013). *Tackling Climate Change Through Livestock: A Global Assessment of Emissions and Mitigation Opportunities*. Food and Agriculture Organization of the United Nations.

FAO. (2020). [Food and Agriculture Organization of the United Nations.]. *The State of Food and Agriculture*, •••, 2020.

FAO (Food and Agriculture Organization of the United Nations). (2023). *Livestock*. Retrieved from FAO – Livestock.

Fellows, P. (2017). *Food Processing Technology: Principles and Practice*. Woodhead Publishing.

Haddad, S. M.. (2020). The Internet of Things in Food Safety: Research Progress and Future Directions. *Comprehensive Reviews in Food Science and Food Safety*, *19*(3), 1192–1209.

Kumar, S.. (2019). Sustainable livestock production: Environmental issues and their management. *Journal of Animal Science and Technology*, *61*, 1–11.

Li, H.. (2020). Solar energy utilization in livestock farms: A review. *Renewable & Sustainable Energy Reviews*, *134*, 110124.

Mahmoud, A. R.. (2021). Predictive maintenance in food processing: A review. *Computers and Electronics in Agriculture*, *183*, 106052.

Miller, J. D., & Babcock, B. A. (2019). The Role of Livestock in Sustainable Agricultural Production. *Journal of Animal Science*, *97*(5), 1–12. PMID: 30873559

Mora, N.. (2022). Biogas as an energy resource for livestock operations: A review. *Renewable Energy*, *186*, 1299–1310.

Morrison, J. H.. (2019). The impact of energy-efficient technologies on meat quality: A comprehensive review. *Foods*, *8*(5), 155. PMID: 31067817

Nassif, Y.. (2020). Innovations in livestock feeding strategies to enhance energy efficiency in animal production. *Food and Energy Security*, *9*(1), e224.

Nguyen, M. L.. (2021). Technologies for nutrients recovery from livestock waste: A review. *Waste Management (New York, N.Y.), 120*, 67–78.

Rosen, M. K., & Meilke, K. (2023). Livestock Production Systems: An Overview. *Agricultural Systems Review, 196*, 107832.

Steinfeld, H.. (2006). *Livestock's Long Shadow: Environmental Issues and Options*. FAO.

Thornton, P. K. (2010). Livestock production: Recent trends, future prospects. *Philosophical Transactions of the Royal Society of London. Series B, Biological Sciences, 365*(1554), 2853–2867. DOI: 10.1098/rstb.2010.0134 PMID: 20713389

USDA (U.S. Department of Agriculture). (2022). *Livestock and Poultry: World Markets and Trade*. Retrieved from USDA Foreign Agricultural Service.

Van Milgen, J.. (2008). Precision nutrition in growing pigs: Potential and limitations. *Animal Feed Science and Technology, 143*, 205–238.

Wang, Y.. (2022). Utilizing big data in animal agriculture: A review. *Computers and Electronics in Agriculture, 186*, 106215.

Zhang, B.. (2020). Consumer awareness and demand for sustainable livestock products. *Sustainability, 12*(5), 2188.

Chapter 7
Energy Efficiency in the Meat and Meat Processing Sector:
Introduction to Energy Efficiency in the Meat Industry

Ruby Celsia Arul Selvaraj
https://orcid.org/0000-0002-3805-7868
K.S. Rangasamy College of Technology, India

Chandrakumar Elumalai
https://orcid.org/0009-0002-7788-2657
K.S. Rangasamy College of Technology, India

Dhanush Anand Sekar
https://orcid.org/0009-0004-7830-0596
K.S. Rangasamy College of Technology, India

S. Kavin
https://orcid.org/0009-0004-7830-0596
K.S. Rangasamy College of Technology, India

Madhumathi Angamuthu
https://orcid.org/0009-0006-9335-5071
K.S. Rangasamy College of Technology, India

ABSTRACT

This chapter explores the significance of energy efficiency in the meat and meat processing industry, discussing the obstacles and possibilities for decreasing energy

DOI: 10.4018/979-8-3693-8342-1.ch007

usage while upholding product quality and safety. The meat industry is highly energy-intensive, particularly in the food processing sector. Significant energy consumption takes place during several stages, such as slaughtering, refrigeration, processing, and packing. This chapter examines methods for enhancing energy efficiency, such as developments in refrigeration technology, optimized processing procedures, and the integration of renewable energy sources. The text also explores the significance of energy management systems and automation in overseeing and diminishing energy consumption across different phases of meat processing. The chapter continues by providing a perspective on upcoming trends and advancements that have the potential to further improve energy efficiency in the meat processing industry.

1. INTRODUCTION TO ENERGY EFFICIENCY IN THE MEAT INDUSTRY

The food sector is a fundamental and essential component of the global economy, providing employment to a significant number of people and generating huge financial income on a yearly basis. Furthermore, it has a substantial impact on public health, environmental sustainability, and social fairness. With the global population rising, there is a significant need for more food. This requires making large investments in agricultural productivity, infrastructure, and enhancing food delivery and accessibility. Simultaneously, it is imperative to tackle the environmental consequences and guarantee long-term sustainability. The undeniable connection between energy, water, and waste in food production highlights the necessity of minimizing loss and waste at every stage of the supply chain, improving energy efficiency in food processes, and maximizing resource use (Corigliano & Algieri, 2024).

The meat processing industry is dynamic and constantly changing, requiring large quantities of water and power, among other natural resources. Consequently, the meat processing sector is heavily to blame for the ever-increasing air, water, and soil pollution that is a direct result of their massive energy and water use, not to mention the garbage they produce. Because of the high organic content in both solid waste and wastewater, the sector in question can have an even greater negative effect on the environment due to the energy and raw materials needed for waste management and treatment. Despite this, the waste's nature presents numerous opportunities to treat, recycle, reuse, or otherwise process it into energy, materials, or a combination of these. Methods such as the membrane bioreactor, aeration treatment, chlorination, ultraviolet treatment, anaerobic digestion, hydrothermal carbonization, and composting are ideal for treating and reusing water and energy from production in the meat processing industry. The meat processing industry chose these methods due to their efficacy in cleaning wastewater and producing renewable energy from waste use. A membrane bioreactor's combination of biological processes and membrane filtration makes it a sophisticated approach to wastewater treatment. The system consists of a bioreactor tank that breaks down biomass and then uses membrane filtration to filter out microorganisms.To purify wastewater, aeration treatment involves adding air to the mixture, which speeds up organic matter decomposition. One effective method of disinfecting treated water is ultraviolet light therapy. Using UV light to disinfect water is an excellent way to get rid of germs and viruses. One tried and true method for purifying water is chlorination. It can effectively remove microbial hazards while oxidizing a wide variety of organic and inorganic compounds. One method of producing biogas from organic materials in wet solid wastes is anaerobic digestion, which is a kind of anaerobic fermentation (Boukouvalas et al., 2024).

The production of meat, which encompasses several agricultural and food sectors, has significant adverse global environmental effects. Crops and cattle utilize approximately 70% of the world's freshwater resources, primarily for agricultural purposes and meat production. The agriculture sector is associated with numerous detrimental environmental effects, particularly in terms of direct and localized contamination. Moreover, the meat industry specifically produces a greater amount of carbon emissions compared to the transportation sector. Projections indicate that the continuous growth in the adult population consuming meat will drive a minimum 30% rise in global meat production by 2050. This study seeks to examine the environmental impact of meat production and technological advancements in a dynamic manner. They analyzed the panel of 37 developing and developed nations from 1998 to 2018 using the two-step Generalized Method of Moments (GMM) approach. Four distinct measures were used to determine the extent of environmental protection: carbon dioxide (CO2) emissions from agricultural activities, CO2 emissions from transportation, CO2 emissions associated with the use of liquid fuels for energy, and CO2 emissions per person (Chatti & Majeed, 2024).

Law No. 18 of 2012 states that in order to meet the needs of Indonesia's expanding population, it is essential to ensure the availability of healthy food while also maintaining stringent rules for the food industry. Among the several sectors targeted by the 2015–2035 National Industrial Development Master Plan is the meat processing industry. A combination of factors, including rising incomes, changing eating habits, and the proliferation of modern retail locations, has driven up demand for processed beef products across the nation. However, there are obstacles that the company faces, like low meat consumption and heavy reliance on imported raw materials. This highlights the need for a holistic strategy to boost competitiveness, increase consumption, and solve nutritional issues. Focusing on the need for strategic adjustments to fulfill both domestic and global demands, this study analyzes the competitiveness of Indonesia's meat processing sector. The study highlights issues such as restricted export market expansion and dependence on imported raw materials. It zeroes in on the analytical hierarchy process, the Industrial Competitiveness Index, and SWOT analysis as instruments for strategic decision-making. Technology and innovation investments, supply chain sustainability, market diversification, talent development, and regulatory help are all on the list of recommended actions (Wattie et al., 2024).

The world's food systems are currently facing significant challenges, not only in terms of production and nutritional goals, but also in terms of their environmental effects, including energy use. The contemporary agri-food system encompasses a diverse array of energy consumers, beginning with farmers and extending to processing, packaging, transportation, retail, households, and waste management. Energy is required for the various stages of food production, processing, distribution, consumption, and disposal. Furthermore, the sustainability of agricultural and food systems, rural development, renewable energy, and the diminishing natural resources of fossil fuels are all significant. The food system is witnessing an increasing need for energy, primarily due to the surge in global food production that stems from growing consumption. While fossil fuels now provide the majority of the food system's fundamental energy requirements, there is an increasing demand for sustainable energy sources. The energy transition holds significant importance within the context of climate change and the ongoing conflict in Ukraine, as it directly impacts the cost and security of energy supply (Zuba-Ciszewska & Suchoń, 2024). Figure 1 describe about the Energy Efficiency in the meat industry.

Figure 1. Energy efficiency in the meat industry

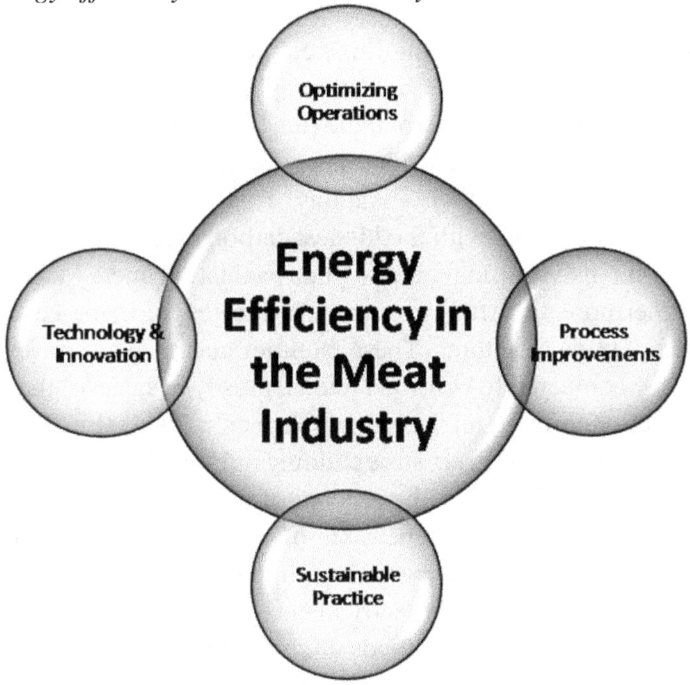

2. ENERGY CONSUMPTION IN MEAT PROCESSING

Currently, factors like climate change and the depletion of natural resources primarily influence European environmental policy. The European Green Deal aims to achieve net-zero greenhouse gas emissions in Europe by 2050 as a means of mitigating global warming. To attain climate neutrality and sustainability, it is necessary to shift towards energy efficiency and energy systems that rely on decarbonized energy sources. Thus, the shift towards green energy focuses on reducing carbon emissions by replacing carbon-intensive energy sources, such as fossil fuels, with environmentally friendly, renewable, and sustainable alternatives. In order to accomplish this transformation, the European Green Deal has established as one of its main principles the reduction of greenhouse gas emissions and the improvement of people's quality of life. We will achieve this by prioritizing energy efficiency, which entails improving building energy performance and fostering a power sector based on renewable sources.

Buildings are critical in achieving this goal because they have significant potential for cost-efficient emissions reduction. The building sector is responsible for approximately 40% of global carbon dioxide (CO2) emissions and 36% of ultimate energy consumption. In particular, there are concrete ways to reduce carbon emissions in commercial building operations. This paradigm classifies the retail industry as one of the sectors with the highest carbon emissions. Stores are one of the most energy-intensive building types when evaluating energy usage per unit of floor area. Furthermore, this study uses the terms stores, supermarkets, or retails to refer to food retail trade structures. These facilities consume a significantly higher amount of energy compared to non-food retails, mostly because of the requirement for food preservation through refrigeration (Gac et al., 2024) (Piselli et al., 2024).

Particularly important in grocery store buildings are three separate energy systems. Because these technologies affect the amount of energy used by these structures, they may help ease the switch to renewable power while simultaneously increasing efficiency and decreasing environmental impact. One system preserves food, another lights, and finally HVAC. Electrical energy accounts for a staggering 95% of all energy use. This component speeds up the energy system's shift to electrical power, an essential tactic for cutting down on carbon emissions. This method's premise is that fossil fuel thermal energy conversion should not be a prerequisite for producing electrical power. A major contributor to this energy is refrigeration, which includes both low and middle temperatures. Refrigeration uses an estimated 30% of all energy, with that number rising to 60% in extreme cases. About half of this energy goes into compressors. With less than 30% of overall energy usage. After buildings, HVAC systems utilize the most energy, making up about 15 to 20 percent of total consumption. Ancillary processes such as cooking, baking, and other household chores consume the residual energy (Zhang et al., 2024). Food stores account for nearly three percent of all electricity usage in the European Union, and thirty-five percent of all CO2-equivalent emissions (including refrigerant leakage) come from those stores' refrigeration equipment. The retail food sector relies on efficient refrigeration systems to keep perishable goods fresh for longer and reduce food waste.The use of technology to store and showcase frozen or chilled products for business transactions is known as business refrigeration, specifically in the context of food retail refrigeration. Specifically, the equipment utilized in these establishments can be categorized as mechanical refrigeration apparatus, display fridges, storage fridges, and processing fridges. Refrigerators can be classified as reach-in refrigerators, which have doors, or open refrigerators, which do not have doors. Other classifications focus on refrigerant type and temperature range (Xue et al., 2024) (Saitone et al., 2024).

Life cycle assessment, the emission factor method, the input-output method, and comparable techniques are frequently used to evaluate carbon emissions. The purpose of a life cycle assessment is to determine the whole or partial environmental impact of a product or process by tracking its usage through time. Determining goals and scope, making an inventory, executing an impact assessment, and interpreting the results are the usual phases in a life cycle assessment. This approach provides solutions for enhancing a product's performance while also helping to comprehend its ecological effects and resource use from its inception to its disposal. Reducing carbon emissions and promoting sustainable development can be achieved through the implementation of optimization methods (Han et al., 2024).

Management strategy known as “supply chain route optimization” seeks to increase logistical efficiency while decreasing costs through the deliberate planning of routes. Transportation planning optimization, scenario creation, and acceptable mode and route selection are some broad strategies for building efficient supply chain routes. In addition, companies can learn about their current GHG emissions and come up with suitable optimization tactics with the use of data provided by GHG assessment methodologies, which are crucial for improving supply chain routes. A producer's productivity efficiency can be defined as their output in relation to their maximum capacity.

In terms of maximal output, defined objectives, or an optimal situation, this ratio indicates the amount of achievement. It also shows how near we are to reaching or reaching the production possibility frontier. Based on their input-output ratios and cost-benefit analyses, producers frequently utilize it to evaluate performance levels. The primary premise of productivity efficiency, known as Pareto efficiency, is based on the idea of a completely competitive market. According to the theory, when all resources are being used to their full potential, a manufacturing unit is operating at peak efficiency if and only if cutting output of one product necessitates reducing production of another. By determining the present industry's production frontier—also called the Production Possibility Frontier (Nusri-Un et al., 2024) —academic studies usually identify and measure the efficiency of particular production entities.

It was in 1957 when Farrell first made the distinction between technical efficiency and technological development. According to his theory, there are two parts to a producer's efficiency: technical efficiency, which is the capacity to maximize output with a given set of inputs, and allocation efficiency, which is the capacity to gain advantages from given prices and technology by allocating production factor resources optimally. Economic efficiency, which includes all of the aforementioned traits, is the same as production efficiency. By highlighting the significance of allocation efficiency, Farrell introduced a novel method of analysis that investigates the effects of pricing and scale on production efficiency (Caputo et al., 2024). Figure 2 show the Energy Consumption in Meat Industry.

Figure 2. Energy consumption in meat industry

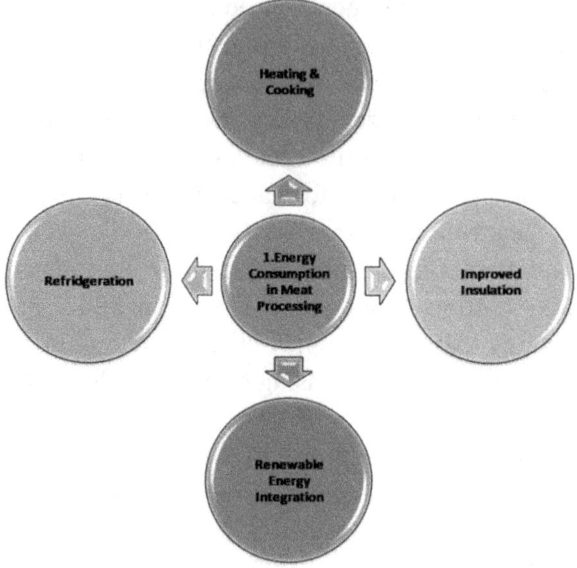

3. TECHNOLOGICAL INNOVATIONS FOR ENERGY EFFICIENCY

Grown meat is gaining recognition as a potential replacement for traditional meat due to its sustainable production methods and ability to provide healthier dietary options. Finding sustainable sources of protein has become a challenge for most countries, especially those with low rates of food self-sufficiency. In great detail, this study examines the stage of product development, technical research progress, and commercial market participation of farmed beef. In this analysis, we highlight the successes of industrialized cultured meat production and look forward to the challenges that may arise in the near future. Thanks to recent developments in cell cultivation, stem cell biology, tissue engineering, nutritional science, and biological process engineering, cultured meats (CMs) have recently attracted a lot of attention from researchers (Tuomisto & Ryynänen, 2024) (Casnici & Schulzmann, 2024). The National Academy of Sciences, Engineering, and Medicine recognized CMs as a promising biotech product in 2017. Compared to conventional meat processing methods, cultured meat production uses far less energy, produces substantially fewer greenhouse gas emissions, and uses less freshwater and arable land, according to life cycle assessment studies. Additional sustainability benefits provided by CMs included better health and safety, conservation of resources, and an increase in animal welfare. CMs can greatly lower their chances of getting traditional food-borne diseases like antibiotic residues and even lower their chances of getting animal diseases like the African Swine Fever virus by carefully watching the parts that are added to the culture medium and doing the first steps of production in a clean environment. Furthermore, CMs have instituted dietary supplementation. Biological methods, which increase carotenoid production and accumulation in bovine myocytes, produce beef with a higher carotenoid concentration than animal feeding (Azis et al., 2024) (Ji et al., 2024).

The significance of the cultured meat business has been steadily increasing, as it has the potential to replace animal-based production systems and help alleviate the impacts of climate change. Although there is concern about scalability and achieving price parity with traditional meat, as well as improving customer acceptance, we are observing a growing number of cultured meat companies entering the competition and extending and broadening their range of products. The emerging competitive ecosystems bolster their exponential increase in size and technological progress (Giménez et al., 2024).

A strategic initiative, the establishment of the super hybrid duck cultivation and breeding industry in Blitar Regency aims to enhance the productivity and economic prospects of local breeders. Implementing energy-efficient hatching machinery in Blitar Regency's super hybrid duck growing and breeding businesses is a pioneering measure in the livestock industry. The purpose of this technology is to enhance efficiency, productivity, and sustainability in the process of hatching duck eggs. The goal is to examine the potential obstacles and solutions to establishing a super hybrid duck farming and breeding enterprise in the area. Field surveys, interviews, and data analysis were the techniques used in the research. The program's findings indicate that Blitar Regency has significant potential for expanding a super hybrid duck farming enterprise. However, the enterprise faces several obstacles such as insufficient technical expertise, inadequate infrastructure, and restricted market entry (Zhang et al., 2024).

The adoption of technology in generating and raising super hybrid ducks involves the use of energy-efficient equipment, namely hatching machines. This technology offers benefits such as improved energy efficiency, lower operational expenses, better productivity, and enhanced duck seed quality. The adoption of technology can have significant repercussions, including reducing environmental problems and improving economic factors. Additionally, adopting this technology presents both challenges and opportunities (Wang et al., 2024). The proposed solutions, such as technical training and infrastructure improvements, are expected to provide guidance for all stakeholders, including breeders, government, and related groups, in supporting the growth of super hybrid duck cultivation and breeding businesses in Blitar Regency.Water retention technology is critical for improving the taste and quantity of meat products while also maximizing the economic benefits of modern meat processing procedures.

This technology improves the company's economic sustainability by enhancing the textural profile of meat products and maximizing yield. Water retention techniques mostly involve manipulating pH levels, inhibiting protein oxidation denaturation, and altering the micro-spatial structure of muscles. The several elements that contribute to water loss in meat and meat products include animal species and cuts, pH levels, processing techniques, storage and packaging conditions, and salt content. This research examined the use of water retention methods in meat processing and its resulting products. The main aim of this research is to provide theoretical rationale for minimizing juice loss, maintaining necessary water content, improving texture, and enhancing meat quality throughout processing and storage (Badawi et al., 2024). Figure 3 illustrate the Technological Innovations for Energy Efficiency.

Figure 3. Technological innovations for energy efficiency

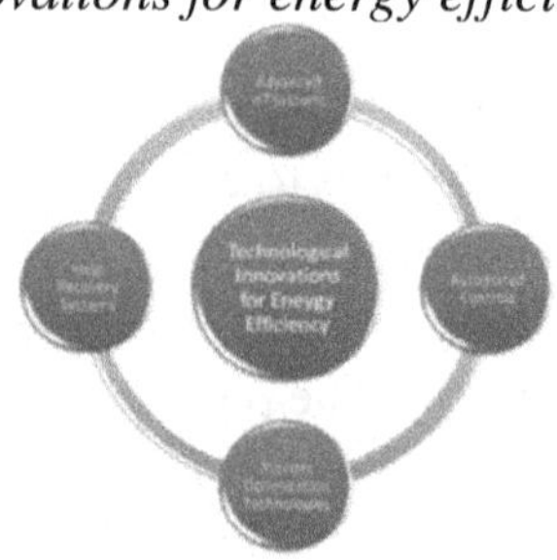

4. ENERGY MANAGEMENT PRACTICES

The meat manufacturing process in slaughterhouses results in a large amount of waste materials, which constitute around 37%-49% of the animal's live weight. The processing of cattle and poultry releases wastewater amounts ranging from 0.2 to 1.5 cubic meters. The food and drink sector consumes about a quarter of its freshwater from the processing of meat. The high protein and lipid content (up to 90.8% dry matter and 55.3% dry matter, respectively) in these wastewaters and byproducts makes them very spoilable (Kim et al., 2024). The release of foul odors and the contamination of wastewater and solid waste are only two of the many environmental challenges that these compounds are known to exacerbate. However, prompt and appropriate treatment can mitigate these issues. In an effort to create a meat industry that is both sustainable and kind to the environment, the Chinese government is working to standardize slaughterhouse operations while simultaneously improving meat quality and safety. The Chinese government has established regulations for disposing of by-products and effluent from slaughterhouses. This legislation mandates the use of burning, rendering, and deep burying as mandatory disposal processes in China. These technologies commonly require large amounts of energy and chemicals, along with a significant amount of land for farming (Patil et al., 2024).

The in vitro meat production system involves cultivating stem cells taken from farm animals inside a bioreactor utilizing advanced tissue engineering techniques, resulting in the creation of meat outside of traditional food animals. In addition to gaining the support of animal rights activists due to its ethical approach to meat production, the in vitro meat production system also avoids numerous problems associated with traditional meat production systems. These problems include the cruel slaughter of animals, diseases caused by poor nutrition, illnesses transmitted through food, excessive use of resources, the emergence of antibiotic-resistant pathogens, and the significant release of methane, which contributes to global warming.

Due to the controlled and manipulatable circumstances in an in vitro meat production system, it will be possible to manufacture customized, chemically safe, and disease-free meat in a sustainable manner. However, overcoming other obstacles is necessary before lab-grown meat can become economically viable. While the production cost and public acceptance are crucial, substantial funding is urgently needed for future research on the subject.The growing worldwide demand for sustainable and ethical animal protein production, combined with the need to reduce the adverse environmental effects of this industry, is a major challenge for animal agriculture. Livestock are responsible for utilizing 32.0% of the Earth's freshwater and 30.0% of the global arable land, while generating 14.5% of the world's greenhouse gas emissions. The cattle business is widely regarded as the least environmentally sustainable sector among livestock industries, responsible for approximately 61.7% of the total greenhouse gas emissions produced by the worldwide livestock sector (Aboagye et al., 2024).

In 2022, the Food and Agriculture Organizations will be present. The beef business in Canada is a significant component of the agricultural sector, contributing over CAD$10.2 billion in farm sales annually. Canada is the ninth-largest beef exporting nation, with exports reaching over 50 countries and totaling CAD$4.5 billion in 2021. The beef market is always changing, and the whole Canadian beef sector is striving to embrace environmentally sustainable and ethical techniques to satisfy the demands of both producers and consumers. Hence, it is imperative for the beef industry in Canada to enhance public confidence by actively involving multiple stakeholders and employing evidence-based methods to evaluate its advancements in tackling sustainability concerns.

Although the evaluation of greenhouse gas (GHG) emissions in the Canadian cattle value chain has advanced, research on the impact of beef production on ecosystem services like water and air quality has lagged. The United States has developed various comprehensive approaches to describe beef production systems and evaluate their environmental sustainability. Conducting thorough life cycle assessments is crucial for evaluating and ranking the most effective management techniques that can enhance the sustainability of Canadian beef production. To establish benchmarks, it is necessary to compare production methods and environmental impacts in the beef industry.The Canadian Roundtable for Sustainable Beef (CRSB) conducted a nationwide life cycle assessment (LCA) as part of the National Beef Sustainability Assessment (NBSA), utilizing data gathered in 2014. This effort offered the initial chance to assess the ecological, societal, and financial viability of the Canadian beef sector, as well as to set standards and pinpoint areas that need enhancement (Oke et al., 2024).

As consumer expectations grow, the importance of high-quality beef becomes increasingly apparent. The non-toxic, rapid, and energy-efficient characteristics of pulsed electric field (PEF), a recently invented processing technology, are attracting the interest of academics. By physically disrupting muscle fibers, PEF improves the meat's softness, which is a major benefit. This disturbance compromises collagen and connective tissue structures, breaking Z-disk and I-band connections. The release of calcium ions during membrane electroporation causes an increase in calpain activity. An increase in calpain activity may lead to the breakdown of troponin-T and desmin, potentially providing pain relief. In addition, PEF may improve the flavor and appeal of chicken that is pale, soft, and exudative (PSE), similar to how it looks. In addition, PEF may speed up the digestion of cooked meat, reduce the amount of salt used in items, and cut down on the time meat needs to cure. The meat processing industry has long utilized conventional hot-processing methods like smoking, grilling, roasting, cooking, and frying due to their versatility and ease of use. However, certain components, including proteins, vitamins, and minerals, are sensitive to high temperatures; thus, these procedures can damage meat quality. We can prevent the denaturation of proteins and oxidation of lipids by carefully controlling the heating temperatures and durations used during processing (Woiciechowski et al., 2024)(Guo et al., 2024)(Pudenz & Schulz, 2024)(Noble et al., 2024). Figure 4 show the Energy management practices.

Figure 4. Energy management practices

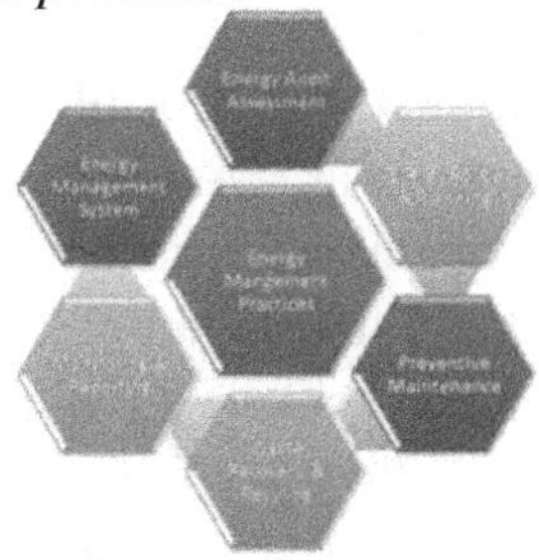

5. SUSTAINABLE PRACTICES IN MEAT PROCESSING

Today, the vast majority of U.S. beef packers openly practice multi-plant collaboration. Using a Salop Circular City model that incorporates both formula and negotiated pricing, we demonstrate that when the markdowns required for coordinating several plants outweigh the benefits, the disparity between the costs of upstream-fed cattle and downstream beef becomes substantially larger. A number of components work together to shed light on the reasons behind the constant huge price disparity between farm-raised beef and wholesale beef, including market concentration, geography, transportation costs, cow cycles, and the use of relevant packer capacity. When cattle numbers decline, our research reveals that a multi-plant coordinator will permanently close one plant before the other plants, operating as distinct profit centers, follow suit. This trend lines up with what packers have been doing recently (Dissanayake et al., 2024). In addition, we show that a newly-minted company's strategically placed packaging facility can help bring down the price differential across marketplaces. This paradigm has to do with ongoing policy concerns regarding the organization, conduct, and effectiveness of the cattle packing business. Solar water heating systems (SWHS) face obstacles in the agri-food industry due to the thermal demand's high degree of unpredictability. An in-depth examination of the effects of consumption patterns on the sector's SWHS's profitability and energy efficiency. In addition, the project hopes to find out whether other important factors impact SWHS practicality. To accomplish this, we took into account different purchasing patterns, economic variables, and geographic areas in our extensive examination of almost 1,300,000 cases (Benavente et al., 2024). Demand seasonality, weekly demand frequency, and energy prices significantly influence the feasibility of SWHS. The consistency of daily demand influences demand, location, and investment cost. As a result, most people can't stick to a weekly SWHS regimen. But with just three separate days, the chances of success are far higher, leading to paybacks in less than five years and, under ideal circumstances, energy savings of more than forty percent. Whether or not solar water heating systems (SWHS) are feasible for businesses depends heavily on how extreme the seasonal swings are. A staple of Irish food and cooking for generations, cured meats are a popular cut of meat in the Emerald Isle. Curing is a common processing step for several traditional meats, including ham, turkey, and bacon. Products like this often receive nitrite additions and have high salt content. We order to achieve variety, increase their shelf life, and improve their unique quality and safety features, we prepare these things in a number of ways. However, buyers are more cognizant of the processing involved in these products, and they may see some ingredients and compounds as harmful to their well-being. Researchers have been exploring methods to decrease the levels of salt, saturated fat, and chemical preservatives (like nitrites) associated with health issues in meat products (Molina et al., 2024). Because these compounds have such a profound techno-functional effect on the product's safety, quality, and identity, finishing the job is no simple feat. Despite these advancements, we still don't know enough about the industrial-level effects and sustainability of processing, reformulation, and packaging innovations on different meat products. The legislative aspect of this issue heavily influences research priorities and potential alternatives (Chandrasekaran & Rajasekaran, 2024). In an effort to enhance the nutritional profile of certain Irish meat products, researchers have undertaken multiple investigations into the possibility of reformulating these items. This includes studies that look at ways to lessen or replace ingredients like salt, saturated fat, chemical additives, and preservatives—elements that consumers are becoming more worried about. The development of such items presents a challenge for meat product producers because reformulations have the potential to change quality and safety standards. There is a lack of research on the correct use of nitrites and nitrates in Irish product manufacturing, as well as concerns about the production of nitrosamines and the microbiological safety of finished goods. When developing new items, it is essential to think about the target market as it is right now. Customers and the industry that is adopting the innovations are also important factors (Karkou et al., 2024)(Guven et al., 2024). The advent of the Internet of Things in the agricultural sector has drastically changed the ways in which farmers monitor and manage their crops. Several environmental variables, including soil moisture, crop growth, temperature, and humidity, may be immediately available with Internet of Things sensors. This data will assist us in making better decisions, which will increase crop yields. The massive amount of data generated by these sensors, however, poses a significant challenge to data processing and communication (Nsabiyeze et al., 2024). Clustering is a common tool for tackling this problem; it allows us to organize sensors into groups and select one to act as a communication hub with the gateway node. Critical factors influencing the operation of the IoT system include selecting an appropriate CH and the optimal path for data transmission. In this paper, we present a new approach to path selection and CH (Cluster Head) optimization. This approach combines an enhanced version of the whale optimization algorithm (WOA) and enhanced Crow swarm optimization (ECSO). Using fuzzy logic, the approach determines the best Cluster Head (CH) by considering important factors like energy, distance, overhead, trust, and node degree. Taking into account the chosen cluster head, we employ ECSO to ascertain the optimal data transmission path. We used simulated trials in a smart agriculture environment to assess the suggested methodology. The outcomes show that the proposed approach outperforms the state-of-the-art in terms of energy efficiency, latency, packet delivery ratio, and throughput (Du et al., 2024). Figure 5 show the Sustainable Practices in Meat Processing.

Figure 5. Sustainable practices in meat processing.

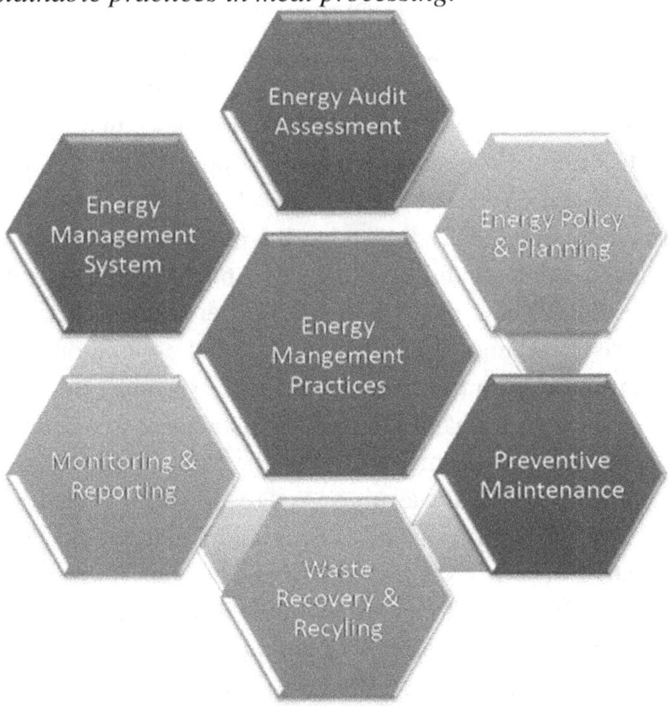

6. FUTURE TRENDS IN ENERGY EFFICIENCY

Constantly rising energy consumption is a result of both technological progress and the ever-increasing human population. Most of the energy that people use today comes from fossil fuels. Sustainable and renewable energy sources are necessary to meet human civilization's growing energy demands due to the diminishing supply of fossil fuels and the worsening problem of climate change. As a result of the depletion of fossil fuel supplies, the cost of energy is currently rising and fluctuating. As a result of rising energy costs, expenses rise. On the other hand, burning fossil fuels puts carbon dioxide into the air, which adds to the problem of climate change. Increased global temperatures pose serious threats to the biosphere because they drive climate change. As a result, 196 countries signed legally enforceable commitments in December 2015 to slow the rate of climate change and keep global warming below 2 degrees Celsius over pre-industrial levels. According to the 2015 UNFCCC statement, the Paris Climate Agreement suggests that reducing greenhouse gas (GHG) emissions by up to 43 percent could potentially limit global warming to 1.5 °C by 2030. The United Nations has established a goal of reaching net zero emissions by 2050 (Kheiralipour et al., 2024), in addition to the 2030 ambition. To achieve the UN's sustainable development goals, all equipment, including vehicles, must use renewable energy as a fuel source. The burning of fossil fuels emits carbon dioxide (CO2), which has the potential to exacerbate the warming trend. Because of this, studies looking at sustainable and renewable fuels as alternatives to fossil fuels have been ramping up. Energy consumption is increasing every day due to rising technology and a larger global population. Most of the energy that people use today comes from fossil fuels. However, in order to meet the rising energy demands of human civilization, sustainable and renewable sources are essential, since we face two major threats: the finite supply of fossil fuels and the worsening problem of global warming. As a result of the depletion of fossil fuel supplies, the cost of energy is currently rising and fluctuating. Excessive energy prices lead to a surge in prices. On the other hand, burning fossil fuels puts carbon dioxide into the air, which adds to the problem of climate change. The environment is under grave danger due to the effects of climate change, which include steadily increasing global temperatures. As a result, 196 countries signed legally enforceable commitments in December 2015 to slow the rate of climate change and keep global warming below 2 degrees Celsius over pre-industrial levels. The Paris Climate Agreement confirms, as stated by the UN Framework Convention on Climate Change in 2015, that we can limit the increase in global temperature to 1.5 °C by lowering emissions of greenhouse gases by as much as 43 percent by the year 2030. The United Nations has established a goal of reaching net zero emissions by 2050 (McCarthy, 2024)(Sleem et al., 2024)(Rezende et al., 2024)(Li et al., 2024), in addition to the 2030 ambition. To achieve the sustainable development goals set forth by the United Nations, we must power vehicles and other equipment with renewable and sustainable sources. The burning of fossil fuels emits carbon dioxide (CO2), which has the potential to exacerbate the warming trend. This has led to a surge in interest in studying sustainable and renewable fuels as possible alternatives to fossil fuels. There is growing scientific evidence that different identities affect energy consumption. But whether their impact differs in size across domains is still up for debate. This research adds to the literature by using identity theory to examine consumer habits related to energy efficiency. Furthermore, this study investigates the role of anticipated guilt as a mediator in the relationship between identity and energy efficiency (Umaraw et al., 2024). Since there is a dearth of literature on the subject of solar homes' ability to reduce energy consumption, this study uses multi-group analysis to contrast homes that have solar energy systems with those that do not. Using partial least squares structural equation modeling, we gathered and evaluated data from 607 respondents in Australia. Anticipated guilt influences energy-related behaviors, and the results show a link between different identities and these behaviors. Intriguingly, when looking at the mediated interactions, it turns out that being environmentally conscious is more effective than being frugal in getting consumers to buy energy-efficient equipment. Many are concerned about the effects of these changes on the environment and the amount of waste that will result (Zahran, 2024). This study uses material flow analysis to track nitrogen flows in Brazil from 2011 to 2021, during both cow slaughter and meat consumption. The analysis covers many co-production streams, including hair, tallow, viscera, and blood. By calculating the nitrogen cascade indicator and the nitrogen utilization efficiency, we were able to determine how well the manufacturing chain accumulated and used nitrogen. While beef output grew by 7.29%, nitrogen inputs into the system increased by 8.47%. In spite of a 2.84 percent increase in total consumption, population growth caused a 1.29 kilogram drop in beef consumption per capita in Brazil. The export of beef surged by 86.03%. Losses from human waste rose by 10.88%, while losses from food waste rose by 2.84%. In terms of NUE values, the killing phase stands out with a whopping 90%. The next three phases, processing, shipping, and storage, have values between 79% and 88%. With nitrogen utilization efficiencies (NUE) ranging from 29% to 34%, the consuming phase showed the weakest performance. The cumulative nitrogen effect, or NCI, varied from 77% to 82% throughout the experiment (Russo et al., 2024).It is evident that there is room for improvement in nitrogen use efficiency, particularly in the area of consumer food waste. This study focuses on nitrogen accumulation throughout Brazil's beef production process, which may cause environmental stress due to the nitrogen cascade effect. These problems, as well as the need to increase seafood's quality, longevity, and safety, are driving developments in culinary science and technology, which are creating new methods to address them. Researchers and practitioners alike place a premium on learning about and perfecting the many seafood processing techniques, which range from edible coatings to high-pressure processing. This study compiles the most recent findings from a number of emerging technologies used in the seafood industry and gives an evaluation of their efficacy, potential pitfalls, and future benefits (Sani et al., 2024). Included in this set of technologies are high-pressure plasma (HPP), ultrasonic waves, pulsed electric fields, plasma technologies, pulsed light, low-voltage electrostatic fields, ozone, vacuum cooking, microwave heating, purified condensed smoke, and edible coating. The seafood item's features and the desired result are the primary determinants of the technology's effective execution, while every method has its own set of advantages and disadvantages. There is a lot of promise in using microwave heating and high-pressure processing to extend shelf life and maintain quality. Edible coatings serve as a versatile method for preserving food and potentially enhancing its nutritional value. Despite the promising future of these technologies, it is imperative to address the SWOT analysis's highlighted concerns regarding cost-effectiveness, scalability, regulatory aspects, and customer acceptance. The seafood industry is about to undergo a technological revolution; thus, understanding these complexities is key to achieving long-term success (Lukose & Singh, 2024). Over the last half century, meat consumption has skyrocketed, and by 2020, we expect the worldwide market for fresh and processed meat to reach 277.5 million tons. Forecasts predict that the worldwide market for fresh and processed meat will reach 292.92 million tons by 2027. However, deteriorating and spoiling meat and meat products can lead to financial losses for producers and health risks for consumers. Manufacturers have responded to this concern by creating complex packaging materials that incorporate sensors and indicators such as radio frequency tags and barcodes, guaranteeing stringent safety and quality controls. Technological advancements have shaken up the packaging sector by reducing waste and increasing product longevity. In addition, smart packaging solutions provide customers with real-time product condition updates (Perry & Welsh, 2024)

(Kelliher et al., 2024). As a result, packaged perishable goods, like beef, are now safer and of higher quality. People all over the world enjoy meat and meat products, such as chicken, red meat, and fish, due to their high protein and nutritional value. Products with a short shelf life are susceptible to degradation and oxidation caused by microbes and metabolic/enzymatic processes, which shortens their useful life. A drop in product quality and safety, huge financial losses, higher costs, and an increased risk of widespread consumer poisoning are all possible outcomes. Consequently, while formulating regulations for these consumable goods, regulatory agencies may place constraints based on risk and safety evaluations (Neri et al., 2024). Therefore, it is critical to use new and existing methods to monitor food safety and quality throughout its whole lifecycle, from production to processing to final consumption. Twenty percent of the world's total meat production—337 million tons—comes from the cattle industry, which produces 67.4 million tons of beef annually. The many parts of the beef industry work together toward the same goal: satisfying customer demand. On the other hand, different parts of the system focus on different ways to improve efficiency. The primary goal of the cow-calf operator is to optimize production by ensuring that each cow produces a calf each year. From the moment of weaning until they are ready for slaughter, stockers and feeders tirelessly strive to raise calves efficiently. Offering the greatest quality product is the meatpacker's primary goal in order to maximize sales. Every segment's end goal is to make a product that consumers want, although different segments focus on different aspects of improvement to different degrees. Compared to growth or carcass quality, reproduction has a 5–10 times higher impact on a cow-calf operator's profitability. On the other hand, the meat and packing industries aim for lean, well-marbled beef with little surplus trim, whereas growth rates and feed-to-gain ratios primarily impact the profitability of the stocker/feeder. The livestock industry is a major source of greenhouse gas emissions, according to recent studies. Efficiently growing calves that reach slaughter weights at earlier ages benefits the industry and the environment, although the exact extent of the contribution is still up for debate. Incorporate energy-saving devices and equipment into your routine, and use them more often. We look at the newest innovations that aim to make beef preparation easier on the electrical grid. The Internet of Things is revolutionizing energy monitoring and control. Smart grid technologies can enhance the efficiency of energy distribution and consumption. Our meat processing firm is committed to achieving ambitious goals. To improve environmental activities, the corporation incorporates energy efficiency measures and they are also planning and projecting for the next decade. Figure 6 show the Future Trends in Energy Efficiency in meat industry.

Figure 6. Future trends in energy efficiency in meat industry

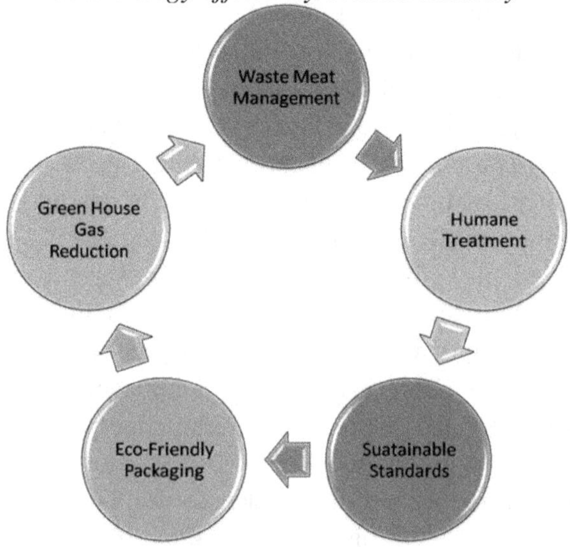

7. CONCLUSION

Technological advancements, regulatory regulations, and the growing demand for sustainable practices will significantly shape energy efficiency in the meat and meat processing industries. Automation, better energy management systems, and the Internet of Things (IoT) are examples of smart technologies that will be crucial in the future of the industry for maximizing energy utilization while minimizing environmental damage. In order to decrease the carbon footprint and reliance on fossil fuels, waste-to-energy technology and renewable energy sources will become increasingly important in meat processing facilities. To further enhance the industry's sustainability, the circular economy emphasizes reusing energy and minimizing waste. However, success is not without its challenges. Coordinated efforts from all stakeholders are required due to economic limits, technological difficulties, and the need for industry-wide collaboration. Regulatory frameworks will shape the environment for the foreseeable future, while new laws will propel the sector towards greater efficiency and sustainability.

The future of the meat and meat processing sector lies in innovation and proactive energy management, which are essential for meeting the demands of environmental responsibility while also ensuring economic viability. Sustainable and economically stable futures are within reach if the sector takes the lead in setting ambitious goals for energy savings and promoting collaboration throughout the supply chain. By taking the initiative, the sector will ensure compliance with new regulations and position itself as a leader in the global movement towards a more sustainable food system. Actions taken now will determine the meat processing industry's ability to take advantage of future possibilities and overcome future challenges, but the potential for energy efficiency is considerable.

REFERENCES

Aboagye, I. A., Valappil, G., Dutta, B., Imbeault-Tétreault, H., Ominski, K. H., & Marcos, R. C. (2024). Cordeiro, Roland Kröbel, Sarah J. Pogue, and Tim A. McAllister. "An assessment of the environmental sustainability of beef production in Canada.". *Canadian Journal of Animal Science*, *104*(2), 221–240. DOI: 10.1139/cjas-2023-0077

Akshay Patil, R., Bhavana, A., & Roopa Patil, B. (2024). Cultured Meat: The Upcoming Meat Production having Sustainable Benefits over Conventional Meat Production: A Review. *Agricultural Reviews (Karnal)*, *45*(1).

Azis, R., Priyanti, N. A., & Wahyuni, S. (2024). RadityaPratama, Muhammad Rizza Pahlevi, and PraptopYuliadi. "Development of a Super Hybrid Duck Cultivation and breeding business using energy efficient hatching machines in Blitar Regency.". *International Journal of Asia Pacific Community Service*, *1*(1), 24–30. DOI: 10.71131/jmm26n60

Badawi, A. K., Hassan, R., Farouk, M., Bakhoum, E. S., & Salama, R. S. (2024). Optimizing the coagulation/flocculation process for the treatment of slaughterhouse and meat processing wastewater: Experimental studies and pilot-scale proposal. *International Journal of Environmental Science and Technology*, *21*(13), 1–16. DOI: 10.1007/s13762-024-05591-y

Benavente, R. M., Perdigones, A., Baptista, F., García, J. L., & Mazarrón, F. R. (2024). Influence of hot water consumption patterns on the viability of low-temperature solar thermal systems in the food industry. *Results in Engineering*, *22*, 102140. DOI: 10.1016/j.rineng.2024.102140

Boukouvalas, C., Kekes, T., Oikonomopoulou, V., & Krokida, M. (2024). Life Cycle Assessment of Energy Production from Solid Waste Valorization and Wastewater Purification: A Case Study of Meat Processing Industry. *Energies*, *17*(2), 487. DOI: 10.3390/en17020487

Caputo, V., Sun, J., Staples, A. J., & Taylor, H. (2024). Market outlook for meat alternatives: Challenges, opportunities, and new developments. *Trends in Food Science & Technology*, *148*, 104474. DOI: 10.1016/j.tifs.2024.104474

Casnici, C. V. C., & Schulzmann, D. (2024). Future Feast: Mapping the Cultivated Meat Sector. In *Cultivated Meat: Technologies, Commercialization and Challenges* (pp. 407–423). Springer Nature Switzerland. DOI: 10.1007/978-3-031-55968-6_20

Chandrasekaran, S. K., & Rajasekaran, V. A. (2024). Energy-efficient cluster head using modified fuzzy logic with WOA and path selection using enhanced CSO in IoT-enabled smart agriculture systems. *The Journal of Supercomputing*, *80*(8), 11149–11190. DOI: 10.1007/s11227-023-05780-5

Chatti, W., & Majeed, M. T. (2024). Meat production, technological advances, and environmental protection: Evidence from a dynamic panel data model. *Environment, Development and Sustainability*, *26*(12), 1–26. DOI: 10.1007/s10668-023-04449-6

Corigliano, O., & Algieri, A. (2024). A comprehensive investigation on energy consumptions, impacts, and challenges of the food industry. *Energy Conversion and Management*, *X*, 100661. DOI: 10.1016/j.ecmx.2024.100661

Dissanayake, K., Rifky, M., Nurmukhamedov, K., Makhmayorov, J., Abdullayev, B., Farmanov, J., & Samadiy, M. (2024). A comparative analysis of traditional meat processing methods. In *E3S Web of Conferences* (Vol. 494, p. 04023). EDP Sciences.

Du, Y., Wang, H., Yuan, S., Yu, H., Xie, Y., Guo, Y., Cheng, Y., & Yao, W. (2024). Dielectric barrier discharge plasma pretreatment: A cleaner new way to improve energy efficiency and quality of wolfberry drying. *Journal of Cleaner Production*, *450*, 141951. DOI: 10.1016/j.jclepro.2024.141951

Gac, A., H. Chambaut, C. Perrot, A. Lorilloux, C. Mosnier, and J. B. Dollé. "28. Environmental footprint of milk and meat from the French cattle sector: improvements since 1990 and future trends until 2035." (2035).

Giménez, B., Zaritzky, N., & Graiver, N. (2024). Ozone treatment of meat and meat products: A review. *Frontiers in Food Science and Technology*, *4*, 1351801. DOI: 10.3389/frfst.2024.1351801

Guo, Y., Han, M., Chen, L., Zeng, X., Wang, P., Xu, X., Feng, X., & Lu, X. (2024). Pulsed electric field: A novel processing technology for meat quality enhancing. *Food Bioscience*, *58*, 103645. DOI: 10.1016/j.fbio.2024.103645

Guven, T. B., Degirmenci, H., Gunerhan, A., & Altuntas, O. (2024). HursitDegirmenci, Ali Gunerhan, and Onder Altuntas. "Energy efficiency and sustainability: Implementing circular economy principles for cabin waste management in aviation.". *Energy for Sustainable Development*, *81*, 101515. DOI: 10.1016/j.esd.2024.101515

Han, J., Dong, P., & Benjamin, W. B. (2024). Holman, Huixuan Yang, Xue Chen, Lixian Zhu, Xin Luo, Yanwei Mao, and Yimin Zhang. "Processing interventions for enhanced microbiological safety of beef carcasses and beef products: A review.". *Critical Reviews in Food Science and Nutrition*, *64*(8), 2105–2129. DOI: 10.1080/10408398.2022.2121258 PMID: 36148812

Ibatulin, I., Nedaskivsky, V., Golembivsky, S., Artimonova, I., & Prysiazhnyuk, N. (2024). Problems and prospects of the functioning of agricultural chains in the meat breeding of Ukraine. *Ekonomichnyyanaliz, 34*(2), 9–19. DOI: 10.35774/econa2024.02.009

Ji, L., Nie, Q., Zhou, Y., Ning, J., Wang, S., Zhou, C., Luo, Y., Yang, R., & Zhang, J. (2024). Advances and applications in water retention technology for meat and meat products: Status and future research directions. *International Journal of Food Science & Technology, 59*(5), 2823–2836. DOI: 10.1111/ijfs.16980

Karkou, E., Teo, C. J., Savvakis, N., Poinapen, J., & Arampatzis, G. "Industrial Circular Water Use Practices Through the Application of a Conceptual Water Efficiency Framework in the Process Industry." *Available atSSRN* 4790097. DOI: 10.1016/j.jenvman.2024.122596

Kelliher, M., Bogueva, D., & Marinova, D. (2024). Meta-Analysis and Ranking of the Most Effective Methane Reduction Strategies for Australia's Beef and Dairy Sector. *Climate (Basel), 12*(4), 50. DOI: 10.3390/cli12040050

Kheiralipour, K., Rafiee, S., Karimi, M., Nadimi, M., & Paliwal, J. (2024). The environmental impacts of commercial poultry production systems using life cycle assessment: A review. *World's Poultry Science Journal, 80*(1), 33–54. DOI: 10.1080/00439339.2023.2250326

Kim, Y.-J., Cha, J. Y., Kim, T.-K., Lee, J. H., Jung, S., & Choi, Y.-S. (2024). The Effect of Irradiation on Meat Products. *Food Science of Animal Resources, 44*(4), 779–789. DOI: 10.5851/kosfa.2024.e35 PMID: 38974724

Li, D., Gan, M., Liu, X., Hu, Q., & Liu, X. (2024). Energy efficiency and driving factors of railway cold chain transportation in China. *Journal of Environmental Planning andManagement, 67*(4), 852–869. DOI: 10.1080/09640568.2022.2133688

Lukose, S., & Singh, K. "Sustainability in Food Process Development: A Comprehensive Carbon Footprint Analysis Tool." In *Sustainable Food Systems (Volume I) SFS: Framework, Sustainable Diets, Traditional Food Culture & Food Production*, pp. 55-71. Cham: Springer Nature Switzerland, 2024. DOI: 10.1007/978-3-031-47122-3_4

McCarthy, B. (2024). Green, guilty, and frugal: Facilitators of energy saving and the choice of energy-efficient appliances in the Australian energy market. *Energy Strategy Reviews, 53*, 101400. DOI: 10.1016/j.esr.2024.101400

Molina, J., Frías-Celayeta, J., Bolton, D., & Botinestean, C. (2024). Molina, Jan Roland G., Jesús M. Frías-Celayeta, Declan J. Bolton, and Cristina Botinestean. "A Comprehensive Review of Cured Meat Products in the Irish Market: Opportunities for Reformulation and Processing.". *Foods, 13*(5), 746. DOI: 10.3390/foods13050746 PMID: 38472858

Neri, A., Hummel, F., Benalia, S., Zimbalatti, G., Gabauer, W., Mihajlovic, I., & Bernardi, B. (2024). Use of Continuous Stirred Tank Reactors for Anaerobic Co-Digestion of Dairy and Meat Industry By-Products for Biogas Production. *Sustainability (Basel), 16*(11), 4346. DOI: 10.3390/su16114346

Noble, K., Huatuco, L. H., Dyke, A., & Green, J. "The Environmental Impacts of Cultured Meat Production: A Systematic Literature Review." In *International Conference on Sustainable Design and Manufacturing*, pp. 91-101. Springer, Singapore, 2024. DOI: 10.1007/978-981-99-8159-5_8

Nsabiyeze, A., Ma, R., Li, J., Zhao, Q., & Zhang, M. (2024). Mitigating greenhouse gas emissions from sheep production system in China: An integrated approach of data envelopment analysis and life cycle assessment. *Resources, Conservation and Recycling, 207*, 107695. DOI: 10.1016/j.resconrec.2024.107695

Nusri-Un, J., Kabsuk, J., Binsulong, B., & Sommart, K. (2024). JirapornKabsuk, BhoowadolBinsulong, and KritaponSommart. "Effects of Cattle Breeds and Dietary Energy Density on Intake, Growth, Carcass, and Meat Quality under Thai Feedlot Management System.". *Animals (Basel), 14*(8), 1186. DOI: 10.3390/ani14081186 PMID: 38672334

Oke, O. E., Akosile, O. A., Uyanga, V. A., Oke, F. O., Oni, A. I., Tona, K., & Onagbesan, O. M. (2024). Climate change and broiler production. *Veterinary Medicine and Science, 10*(3), e1416. DOI: 10.1002/vms3.1416 PMID: 38504607

Perry, G. A., & Welsh, T. H.Jr. (2024). The importance of developmental programming in the beef industry. *Animal Reproduction Science, 265*, 107488. DOI: 10.1016/j.anireprosci.2024.107488 PMID: 38701640

Piselli, C., Balocco, C., Forastiere, S., Silei, A., Sciurpi, F., & Cotana, F. (2024). Energy efficiency in the commercial sector. Thermodynamics fundamentals for the energy transition. *Energy Reports, 11*, 4601–4621. DOI: 10.1016/j.egyr.2024.04.033

Pudenz, C. C., & Schulz, L. L. (2024). Multi-plant coordination in the US beef packing industry. *American Journal of Agricultural Economics, 106*(1), 382–415. DOI: 10.1111/ajae.12391

Rezende, V. T., Araújo Nacimento, R., Ali, S., Roberto Dias Rodrigues, G., Libório Romanelli, T., Noely dos Santos Gonçalves Cyrillo, J., Bonaudo, T., Lescoat, P., & Hauber Gameiro, A. (2024). Understanding nitrogen dynamics in the Brazilian beef industry: A comprehensive decadal analysis. *The Science of the Total Environment, 921*, 171045. DOI: 10.1016/j.scitotenv.2024.171045 PMID: 38402966

Russo, G. L., Langellotti, A. L., Torrieri, E., & Masi, P. (2024). Emerging technologies in seafood processing: An overview of innovations reshaping the aquatic food industry. *Comprehensive Reviews in Food Science and Food Safety*, *23*(1), e13281. DOI: 10.1111/1541-4337.13281 PMID: 38284572

Saitone, T. L., Schaefer, K. A., Scheitrum, D., Arita, S., Breneman, V., Boehm, R. N., & Maples, J. G. (2024). Consolidation and concentration in US meat processing: Updated measures using plant-level data. *Review of Industrial Organization*, *64*(1), 35–56. DOI: 10.1007/s11151-023-09923-z

Sani, M. A., Zhang, W., Abedini, A., Khezerlou, A., Shariatifar, N., Assadpour, E., Zhang, F., & Jafari, S. M. (2024). Intelligent packaging systems for the quality and safety monitoring of meat products: From lab scale to industrialization. *Food Control*, *160*, 110359. DOI: 10.1016/j.foodcont.2024.110359

Sleem, S. T., Salam, D. A., Ghaddar, N., Abou Ghali, K., Chehab, G., Daghir, N., Doughan, Y., & Haddad, N. (2024). NuhadDaghir, Youssef Doughan, and Nicolas Haddad. "Solar-assisted poultry production in small-scale farms: A case study in the Bekaa semi-arid region, Lebanon.". *Energy, Sustainability and Society*, *14*(1), 8. DOI: 10.1186/s13705-023-00437-w

Tuomisto, H. L., & Ryynänen, T. (2024). Environmental Impacts of Cultivated Meat. In *Cultivated Meat: Technologies, Commercialization and Challenges* (pp. 277–297). Springer Nature Switzerland. DOI: 10.1007/978-3-031-55968-6_14

Umaraw, P., & Verma, A. K. (2024). V. P. Singh, Pavan Kumar, N. Mehta, and Devendra Kumar. "Rabbit meat processing and safety.". In *Veterinary Care of Farm Rabbits: A Complete Practice Guide to Rabbit Medicine and Production* (pp. 317–335). Springer International Publishing. DOI: 10.1007/978-3-031-44542-2_14

Vazirov, R. A., Yu Sokovnin, S., Krivonogova, A. S., & Isaeva, A. G. (2024). Radiation surface antimicrobial processing of poultry meat and by-products using the nanosecond low-energy electron beam. *Radiation Physics and Chemistry*, *217*, 111528. DOI: 10.1016/j.radphyschem.2024.111528

Wang, S., Wei, Z., & Wang, L. (2024). Improving slaughterhouse byproducts utilization via anaerobic digestion, composting, and rendering. *Renewable & Sustainable Energy Reviews*, *189*, 113881. DOI: 10.1016/j.rser.2023.113881

Wattie, Y., Suparno, O., & Raharja, S. (2024, June). Enhancing competitiveness in the Indonesian meat processing industry: A literature review. [). IOP Publishing.]. *IOP Conference Series. Earth and Environmental Science, 1358*(1), 012042.

Xue, Y., Qi, Z., Yan, J., Li, D., Zhao, H., & Zheng, H. (2024). Technical Efficiency and Allocative Efficiency of Beef Cattle Fattening in the Content of Digital Economy: An Empirical Study Based on Survey in China. *Agriculture, 14*(7), 1007. DOI: 10.3390/agriculture14071007

Zahran, S. (2024). Optimizing Supply Chain Management of Fresh E-Commerce Agri-Consumer Products Using Energy-Efficient Vehicle Routing. *IEEE Transactions on Consumer Electronics, 70*(1), 1685–1693. DOI: 10.1109/TCE.2024.3370845

Zhang, L., Yin, G., Wei, Z., Li, W., Cui, C., Wang, M., Zhao, C., Zhao, H., & Xue, F. (2024). Potential and Pathways of Carbon Emission Reduction in China's Beef Production from the Supply Chain Perspective. *Agriculture, 14*(7), 1190. DOI: 10.3390/agriculture14071190

Zhang, X., Jiang, D., Li, J., Zhao, Q., & Zhang, M. (2024). Carbon emission oriented life cycle assessment and optimization strategy for meat supply chain. *Journal of Cleaner Production, 439*, 140727. DOI: 10.1016/j.jclepro.2024.140727

Zuba-Ciszewska, M., & Suchoń, A. (2024). The Role of State Aid in the Achievement of the Energy Efficiency Objective in the Food Industry—The Example of Poland. *Energies, 17*(12), 2895. DOI: 10.3390/en17122895

Chapter 8
Energy Efficiency in the Dairy Sector:
Innovations and Sustainable Practices

Neha Prasad
https://orcid.org/0009-0006-1857-0521
VLCC Health Care Limited, India

Dipali Saxena
https://orcid.org/0000-0002-8177-319X
Shri Vaishnav Vidyapeeth Vishwavidyalaya, India

ABSTRACT

The dairy sector is a cornerstone of global agriculture, contributing significantly to food security and economic growth. However, it is also a major consumer of energy, contributing to environmental degradation and climate change. This chapter explores the importance of energy efficiency in the dairy sector, presenting key innovations and sustainable practices that can reduce energy consumption and environmental impact while improving operational efficiency. Through a detailed review of technologies, practices, and case studies, this chapter highlights pathways for sustainable energy management in the dairy industry.

1. INTRODUCTION

The global dairy industry is one of the most energy-intensive sectors in agriculture. Energy consumption is a critical operational aspect, encompassing processes such as milk collection, storage, transportation, processing, and packaging. According

DOI: 10.4018/979-8-3693-8342-1.ch008

to the International Energy Agency (IEA), the dairy sector accounts for nearly 4% of global agricultural energy use, with an increasing trend driven by growing consumer demand for dairy products. Energy efficiency serves as a linchpin for achieving sustainability in the dairy sector. It involves optimizing energy usage to minimize waste and improve productivity. Implementing energy-efficient measures can significantly reduce operational costs and greenhouse gas (GHG) emissions. Additionally, it enables dairy producers to comply with environmental regulations and meet global sustainability targets, such as those outlined in the Paris Agreement.

Dairy processing consumes significant amounts of energy in activities such as pasteurization, cooling, drying, and evaporation. The pasteurization process, which is crucial to ensure milk safety and extend shelf life, requires heating milk to specific temperatures for precise time intervals, consuming a considerable amount of thermal energy (Zhao et al., 2020). Similarly, milk powder production, particularly the evaporation and drying stages, accounts for a large portion of the total energy consumption in dairy plants, with drying processes alone consuming up to 40% of the total energy used (Haas et al., 2017). The cooling of milk and products and the refrigeration of perishable items such as cheese and yogurt also add to the energy load, further emphasizing the need for energy-efficient solutions (Bielenberg et al., 2021).

Moreover, energy-efficient equipment such as high-efficiency motors and variable speed drives for pumps and compressors is gaining traction, providing dairy processors with the tools to reduce their energy consumption while maintaining or even improving production output (Almeida et al., 2022). By optimizing processes and integrating renewable energy sources, dairy producers can also benefit from enhanced energy security, mitigating the risks associated with fluctuating energy prices and the growing pressure to reduce their carbon footprint.

2. ENERGY USE IN DAIRY PROCESSING

Energy use in dairy processing is a significant factor influencing both the environmental sustainability and economic efficiency of the dairy industry. The dairy processing sector is highly energy-intensive, primarily due to the need for heating, cooling, and pumping operations involved in milk processing, pasteurization, drying, and packaging (González-González et al., 2019). Among the various processes, pasteurization and drying of milk are the most energy demanding. Pasteurization, required to kill harmful microorganisms, involves heating milk to temperatures between 63-85°C, which consumes substantial amounts of energy (Zhao et al., 2020). Similarly, milk powder production, which requires evaporating water from milk, is one of the most energy-intensive processes in dairy factories (Hough et al., 2015). It has been reported that energy consumption for milk powder production can account for up to 40% of the total energy usage in dairy processing (Haas et al., 2017).

To reduce energy consumption and enhance efficiency, many dairy plants have turned to energy recovery systems. These systems capture waste heat from processes like pasteurization or the refrigeration of products and reuse it for heating water or air in other parts of the factory (Lopez-Gonzalez et al., 2021). Additionally, combined heat and power (CHP) systems, which generate both electricity and useful heat from a single fuel source, have been increasingly adopted to improve overall energy efficiency (Marin et al., 2020). Furthermore, solar energy has been utilized in some dairy operations, especially in sunny regions, to reduce reliance on traditional energy sources for heating and hot water requirements.

In recent years, there has been a push for the adoption of energy-efficient technologies. For instance, the use of high-efficiency motors and variable speed drives in pumps and compressors can significantly reduce energy consumption (Bielenberg et al., 2021). Moreover, automation and smart technologies enable better control of energy use in dairy processing plants by continuously monitoring and optimizing the energy consumption of machinery based on real-time data. Greenhouse gas (GHG) emission reduction through energy efficiency improvements is another critical consideration. Studies show that energy-efficient practices in dairy processing could lead to a 20-30% reduction in GHG emissions from dairy plants (Haas et al., 2017).

3. INNOVATIONS IN ENERGY EFFICIENCY IN THE DAIRY SECTOR

The dairy sector is increasingly focused on adopting innovations that enhance energy efficiency and reduce environmental impact. With energy consumption being one of the largest operational costs in dairy production, advancements in technology and process optimization have paved the way for more sustainable practices. Innovations in waste heat recovery, combined heat and power (CHP) systems, energy-efficient equipment, and renewable energy integration are central to these improvements, helping the industry reduce both energy consumption and carbon emissions.

3.1 Waste Heat Recovery Systems

One of the most promising innovations in energy efficiency for the dairy sector is the implementation of waste heat recovery (WHR) systems. Dairy plants generate substantial heat during critical processes such as pasteurization, drying, and evaporation. Waste heat recovery systems capture this excess heat and repurpose it for other operational needs, like heating water or air. The recovered heat can be used in less energy-intensive processes, reducing the need for additional external energy sources. This approach can lead to energy savings of up to 30-40% and significantly lower operational costs (Lopez-Gonzalez et al., 2021).

3.2 Combined Heat and Power (CHP) Systems

Combined Heat and Power (CHP) systems are another key innovation in improving energy efficiency. These systems simultaneously produce electricity and useful heat from a single energy source, typically natural gas or biogas. A CHP unit consists of a turbine, generator, and heat recovery system, allowing the dairy plant to generate its own electricity while capturing waste heat for use in processes like pasteurization. By utilizing both heat and power, CHP systems achieve overall energy efficiencies of 75-80%, much higher than traditional separate generation methods. This can result in a significant reduction in energy costs and a decrease in greenhouse gas emissions (Haas et al., 2017).

3.3 Energy-Efficient Equipment and Automation

Upgrading existing equipment and integrating automation technologies have become essential steps for improving energy efficiency in the dairy sector. Key advancements include high-efficiency motors and variable speed drives (VSDs) that allow dairy plants to optimize the operation of motors in pumps, compressors, and fans. These upgrades ensure that energy is used only when required, resulting in significant savings. Additionally, the adoption of energy-efficient refrigeration systems, which utilize natural refrigerants and variable-speed compressors, help reduce energy consumption in milk storage and processing. The use of smart sensors and IoT-based automation systems has also revolutionized energy management. These technologies enable real-time monitoring of energy use, allowing for optimization of processes like cooling, pasteurization, and drying. Together, these innovations can reduce energy consumption by up to 25-30% in dairy operations (González-González et al., 2020).

3.4 Renewable Energy Integration

The integration of renewable energy sources is another critical innovation in improving energy efficiency and sustainability within the dairy sector. Solar energy can be harnessed both for electricity generation through photovoltaic systems and for thermal applications using solar thermal collectors to replace conventional heating methods. Biogas production, through anaerobic digestion of organic waste like manure, is another renewable energy source that can be used to meet a dairy plant's energy demands. Biogas can generate electricity and heat, reducing the plant's reliance on fossil fuels. Moreover, in regions with strong wind resources, wind turbines can contribute to the renewable energy mix, reducing both energy costs and the carbon footprint of dairy operations. The integration of these renewable energy sources can reduce energy costs by up to 40-50% in the long run, providing a more sustainable and resilient energy supply for dairy producers (Almeida et al., 2022).

3.5 Advanced Data Analytics and Energy Management Systems

The application of data analytics and energy management systems (EMS) is increasingly important for enhancing energy efficiency in dairy processing. EMS, equipped with real-time monitoring tools, provide insight into the energy usage patterns of various processes. By analyzing this data, dairy plants can identify areas of inefficiency, predict future energy demands, and optimize production schedules to ensure energy is used more effectively. The use of machine learning algorithms and AI further helps in predicting energy consumption and fine-tuning operations to minimize waste. By employing these technologies, dairy operations can achieve energy savings of 15-20% by better managing consumption and optimizing process flows (González-González et al., 2020).

3.6 Cold Chain Optimization

Efficient management of the cold chain is crucial in the dairy industry, as products need to be stored and transported at specific temperatures to ensure quality. Innovations in refrigeration technology have led to more energy-efficient systems for maintaining the required temperatures during processing and distribution. The use of variable-speed compressors, smart cold storage systems, and improved insulation helps reduce energy consumption associated with refrigeration. Cold chain optimization can lead to energy savings of up to 20-30%, improving overall sustainability in dairy storage and transportation (Zhao et al., 2020).

4. SUSTAINABLE PRACTICES IN DAIRY FARMS

Sustainable practices in dairy farming aim to reduce environmental impact while enhancing economic viability and ensuring animal welfare. One key practice is efficient feed management, where the use of precision agriculture and drought-resistant forage varieties helps optimize feed production while minimizing the reliance on chemical fertilizers and water (Herrero et al., 2021). Additionally, incorporating feed additives such as seaweed or tannins has been shown to reduce methane emissions from livestock (Hristov et al., 2022). Manure and waste management is another vital aspect, where technologies like anaerobic digesters capture methane from manure to generate renewable energy (Cheng et al., 2020). Efficient water conservation techniques, such as rainwater harvesting and drip irrigation systems, reduce water use, particularly in dry regions (Mekonnen & Hoekstra, 2016). Energy efficiency is improved by integrating renewable energy sources like solar panels and biogas systems, which lower dependence on fossil fuels (Bielenberg et al., 2022). Furthermore, pasture and land management through rotational grazing and agroforestry supports soil health, prevents overgrazing, and promotes biodiversity (Teague et al., 2016). Ensuring animal health and welfare through preventive healthcare and proper housing reduces the need for antibiotics and improves productivity (Haskell et al., 2020). Sustainable dairy farming also involves strategies to mitigate greenhouse gas emissions, such as carbon sequestration through agroforestry and soil management practices (Lemaire et al., 2021). Additionally, technological integration of precision dairy farming, automation, and blockchain for traceability enables more efficient operations and transparency, enhancing sustainability (Capper et al., 2014). Despite these advancements, challenges such as high initial investment costs, knowledge gaps, and climate variability still pose barriers to widespread adoption (Cui et al., 2020). However, with continued innovation and policy support, sustainable practices in dairy farming can play a significant role in reducing the environmental footprint of the sector.

5. CHALLENGES AND OPPORTUNITIES IN ENERGY EFFICIENCY IN THE DAIRY SECTOR

The dairy processing sector faces several challenges and opportunities in improving energy efficiency. These challenges stem from technical, economic, and regulatory aspects, while the opportunities arise from technological innovations, regulatory incentives, and market trends favoring sustainability.

5.1 Challenges

i. High Energy Demand in Processing Dairy processing is inherently energy-intensive. Key processes such as pasteurization, drying, evaporation, and cooling consume significant energy. For instance, milk powder production alone can account for up to 40% of total energy use in a dairy plant (Haas et al., 2017).

While energy-efficient technologies exist, the sheer volume of energy required remains a major challenge, especially for large-scale operations.

ii. High Initial Investment Costs - Many energy-efficient technologies, such as combined heat and power (CHP) systems, waste heat recovery systems, and high-efficiency equipment like motors and variable speed drives, come with high upfront costs (Almeida et al., 2022). Smaller dairy operations, in particular, may struggle to invest in these systems, making it difficult for them to adopt energy-efficient solutions. Although these systems lead to long-term savings, the initial financial burden can be a deterrent.

iii. Complexity of Integration into Existing Systems Implementing energy-efficient solutions often requires the retrofit of existing systems, which can be complicated and time-consuming. Existing dairy plants may have outdated infrastructure that is not designed for modern energy-efficient technologies, which increases the complexity and costs of integration (González-González et al., 2020). Additionally, integrating renewable energy sources, such as solar or wind power, can be challenging if the plant is not designed to accommodate such systems.

iv. Fluctuating Energy Prices -The dairy industry is sensitive to fluctuations in energy prices, especially in regions that rely heavily on fossil fuels for power generation. Rising energy costs can significantly impact the cost structure of dairy processing, making it more difficult to sustain energy efficiency initiatives (Zhao et al., 2020). In areas with volatile energy prices, managing energy use becomes even more critical to ensure financial stability.

v. Limited Skilled Workforce -Energy efficiency projects often require specialized knowledge to design, implement, and monitor. In many dairy processing plants, a lack of skilled workers with expertise in energy management or technical systems like automation and smart technologies can hinder the effective implementation of energy-saving measures (Bielenberg et al., 2021). The complexity of energy management systems requires continuous training and upskilling of employees.

vi. Regulatory and Policy Challenges-While some countries have strong incentives for energy efficiency, others have insufficient regulatory frameworks to encourage sustainability in dairy processing. Inconsistent policies or lack of financial support for green technologies in certain regions can slow the adoption of energy-efficient practices (González-González et al., 2019). Additionally, the complexity of navigating various environmental regulations can make it harder for companies to stay compliant while pursuing energy efficiency.

5.2 Opportunities

i. Technological Innovations- The development of energy-efficient technologies offers significant opportunities to improve energy use in dairy processing. Advanced automation systems, such as those that incorporate Internet of Things (IoT) devices, allow for real-time monitoring and optimization of energy consumption (González-González et al., 2020). High-efficiency motors, variable-speed drives, and smart sensors are also becoming increasingly affordable and effective at reducing energy consumption (Bielenberg et al., 2021).

ii. Waste Heat Recovery - One of the most promising opportunities in dairy processing is waste heat recovery. This involves capturing the heat generated during processes such as pasteurization, drying, and refrigeration, and repurposing it for other applications, such as heating water or air. Systems for waste heat recovery can significantly reduce the overall energy demand of a dairy plant, and many companies have already seen substantial savings from integrating these technologies (Lopez-Gonzalez et al., 2021).

iii. Renewable Energy Integration - The integration of renewable energy sources like solar power, biogas, and wind power can help dairy processors reduce reliance on fossil fuels and lower their carbon footprint. Dairy farms with anaerobic digesters can capture methane from manure, which can be converted into biogas and used for energy (Zhao et al., 2020). Additionally, solar thermal systems can be used to provide heat for pasteurization and other thermal processes, lowering energy consumption from conventional sources (Almeida et al., 2022).

iv. Government Incentives and Support - Many governments are introducing financial incentives and subsidies to encourage businesses to adopt energy-efficient technologies. Programs offering tax breaks, grants, or low-interest loans for energy efficiency investments can significantly reduce the financial barriers for dairy processors (Haas et al., 2017). In regions with stringent carbon reduction goals, regulatory frameworks increasingly support investments in clean energy technologies, providing both financial and regulatory incentives for energy-efficient dairy operations.

v. Consumer Demand for Sustainable Products - Growing consumer demand for sustainably produced products offers dairy processors a competitive advantage. Companies that adopt energy-efficient practices and reduce their environmental impact can market their products as more sustainable, potentially gaining access to new market segments or obtaining price premiums. In addition, increasing consumer awareness of climate change and sustainability has led to greater emphasis on eco-friendly production methods (Hough et al., 2015).

vi. Operational Cost Savings- Implementing energy-efficient measures can lead to significant cost savings in the long term. While the initial investment may be high, the reduction in energy bills over time can offset these costs. Moreover, improving energy efficiency helps reduce other operational expenses, such as maintenance costs, by utilizing more reliable and durable equipment (Haas et al., 2017). These savings can contribute to overall profitability, especially in the face of rising energy prices.

7. CONCLUSION

The dairy sector plays a vital role in global food production, providing a wide range of nutritional products that are central to human diets worldwide. However, the sector faces significant challenges in terms of energy consumption, environmental impact, and sustainability. The dairy processing industry is inherently energy-intensive, with high demand for energy in critical processes like pasteurization, cooling, drying, and packaging. This high energy consumption not only increases operational costs but also contributes to substantial greenhouse gas (GHG) emissions, making energy efficiency and sustainability crucial considerations for the future of the dairy industry.

Energy efficiency improvements present a clear pathway toward addressing these challenges. Technological innovations, such as the adoption of energy-efficient equipment, waste heat recovery systems, and the integration of renewable energy sources, have the potential to drastically reduce energy use and the carbon footprint of dairy production. Additionally, the implementation of smart technologies and energy management systems that monitor and optimize energy consumption in real time offers further opportunities for energy savings (González-González et al., 2020). These advancements can lead to both cost reductions and environmental benefits, providing a competitive edge for dairy processors in a rapidly changing market that increasingly prioritizes sustainability.

The dairy sector is at the forefront of an urgent transition towards more energy-efficient and sustainable practices. By leveraging innovative technologies, renewable energy sources, and smart systems, the dairy industry can significantly reduce its environmental impact while maintaining product quality and economic competitiveness. As policy frameworks evolve and consumer demand for sustainable products continues to rise, the dairy sector has the opportunity to lead the way in creating a more sustainable and resilient food system. However, addressing challenges such as high capital investment, regulatory complexities, and the need for skilled labor will be crucial for realizing these future prospects. With the right combination of technological innovation, policy support, and industry collaboration, the dairy sector can pave the way for a more sustainable and energy-efficient future.

REFERENCES

Almeida, R. M., et al. (2022). *The use of solar energy in dairy processing: A case study of energy savings*. Journal of Clean Energy Technologies.

Bielenberg, J.. (2021). Improving energy efficiency in dairy plants: The role of automation and high-efficiency motors. *Energy Reports*.

Capper, J. L.. (2014). The environmental sustainability of dairy production in the United States: A synthesis of greenhouse gas emissions, water use, and land use efficiency. *Journal of Dairy Science*.

Cheng, Y.. (2020). Biogas production from dairy farm waste: Efficiency, challenges, and future trends. *Waste Management & Research*.

Choudhary, M., & Singh, P. (2022). Renewable energy applications in dairy processing: A review. *Journal of Dairy Science, 105*(4), 1541–1550. DOI: 10.3168/jds.2021-20984

Cui, Y.. (2020). Barriers and opportunities for implementing sustainable dairy farming practices: A global perspective. *Environmental Sustainability*.

FAO. (2023). *Energy-smart food for people and climate: The dairy sector perspective*. FAO Publications.

González-González, A.. (2019). Energy use in the dairy industry: Overview and opportunities for improvements. *Dairy Science & Technology*.

González-González, A., et al. (2020). *Energy consumption optimization in dairy*

Haas, C.. (2017). Energy consumption and greenhouse gas emissions in milk powder production. *Renewable & Sustainable Energy Reviews*.

Haas, C.. (2017). Energy consumption and greenhouse gas emissions in milk powder production. *Renewable & Sustainable Energy Reviews*.

Haskell, M. J.. (2020). Animal welfare in dairy production: An overview of recent research and practices. *Animal Welfare (South Mimms, England)*.

Herrero, M.. (2021). Livestock and climate change: A review of the potential impacts of climate change on dairy production systems. *Environmental Research Letters*.

Hristov, A. N.. (2022). Mitigation of methane emissions from ruminants through dietary strategies: A review. *Journal of Animal Science*.

International Energy Agency (IEA). (2021). Tracking Industry 2021: Focus on dairy sector energy use.

Lemaire, G.. (2021). Regenerative agriculture practices for reducing greenhouse gas emissions in dairy farming. *Agricultural Systems*.

Lopez-Gonzalez, J.. (2021). Waste heat recovery in dairy processing plants: A review of opportunities and challenges. *Energy Conversion and Management*.

Mekonnen, M. M., & Hoekstra, A. Y. (2016). Water footprint assessment of dairy products: A global perspective. *Water Resources and Industry*.

Renewable Energy World. (2021). Solar thermal applications in the dairy industry.

Smith, L., & Jones, K. (2020). Advances in sustainable dairy farming. *Sustainability*, *12*(7), 2985. DOI: 10.3390/su12072985

Teague, W. R.. (2016). The role of rotational grazing and agroforestry in promoting soil health and biodiversity in dairy farming. *Land Use Policy*.

UNIDO. (2022). *Energy efficiency guide for dairy processing plants*. UNIDO Publications.

World Resources Institute (WRI). (2022). The role of precision agriculture in reducing dairy sector emissions.

Zhao, X.. (2020). Energy management in dairy processing: A comprehensive review. *Journal of Dairy Science*.

Chapter 9
Revolutionizing Post-Harvest Packaging:
Energy-Efficient Approach for Fresh Produce Preservation

Sapna Kundu
University Institute of Biotechnology, Chandigarh University, Mohali, India

Chhavi Sharma
https://orcid.org/0000-0003-3118-5100
University Centre for Research and Development, University Institute of Biotechnology, Chandigarh University, Mohali, India

Puneet Pathak
Agriliv Research Foundation, Sonipat, India

Juhi Saxena
Parul Institute of Technology, Parul University, India

Sneh Gautam
G.B. Pant University of Agriculture and Technology, Pantnagar, India

ABSTRACT

Post-harvest losses are a global challenge that impacts the quality and quantity of fruits and vegetables; thus, these losses should be mitigated. This chapter explores traditional preservation approaches like physical (regulating temperature, humidity, modified atmospheric conditions, plastic-based packaging, wax-based coatings, etc.) and chemical (chemical pesticides and plant hormones) that are not safe for humans and the environment. Thus, an alternative could be edible coatings/films as natural packaging systems. These are formulated from biopolymers, and their properties can

DOI: 10.4018/979-8-3693-8342-1.ch009

be improved by incorporating bioactive components (plant extracts, antimicrobial agents, etc.). They are safe, eco-friendly, and biodegradable. Nanotechnological approaches (nanosensors and nanoparticles) also play a key role in post-harvest management. This chapter will prove a step towards the achievement of the United Nations Sustainable Development Goals 2, 3, 7, and 12, which aim to end hunger, achieve food security, and promote sustainable consumption, production, and agriculture patterns.

INTRODUCTION

Agriculture affects the economy, environment, and health of consumers worldwide. Indian economy heavily depends on agriculture, around 42.39% of Indians earn their living from farming and associated sectors either directly or indirectly, contributing 18 percent to the total Gross Domestic Production (GDP) (Singh, 2022). The global population, currently around 7.5 billion, is projected to increase to approximately 9 billion by 2050 (UN-DESAPD, 2017). This anticipated population is expected to increase food demand, particularly in poor third-world countries. Food production is anticipated to rise by 100-110% to satisfy the needs of the population, with food demand predicted to double (Sjah & Zainuri, 2020).

Fruits and vegetables are highly prioritized by consumers due to their nutritional values and also play a key role in agriculture by significantly contributing to the economy, environment, and human health. Brazil, the United States, China, and India are the world's leading producers of fruits and vegetables and make major contributions to the world's supply (Springmann & Freund, 2022). The market for fruits and vegetables was estimated to be worth over $200 billion globally in 2020 (Oluwole et al., 2023). The demand for fresh produce in the market across the year is rising due to the growing consumption of natural and health-conscious foods (Dong et al., 2022). Fruit and vegetable production is 107.51 and 209.14 million metric tons in 2021-2022 with cultivation areas of 7.05 and 11.35 million hectares respectively. This production is estimated to reach 108.34 and 212.91 million tons in 2022-2023 (Joshi et al., 2024). India is the world's second-largest producer of fruits and vegetables, with exports worth Rs. 13,185.30 crores/ 1635.95 USD million, which comprised fresh fruits at Rs. 6219.46 crores/USD 770.70 million and vegetables at Rs. 6965.83 crores/ 865.24 USD million in 2022-2023 (apeda.gov.in). The demand for fresh fruits and vegetables will rise by 1.6 and 3.2% annually and is projected to reach 540 million tons by 2050, as reported by the Indian Institute of Horticultural Research (IIHR) (Bvenura & Sivakumar, 2017).

United States Food and Drug Administration (USFDA), Food and Agriculture Organization (FAO), and World Health Organization (WHO) have recommended more intake of fruits and vegetables due to several human benefits particularly reducing the risk of many diseases. They provide nutrients, energy, carbohydrates, minerals, organic acids, dietary fibers, carotenoids, vitamins (vitamins A, B6, B12, and C, etc.), amino acids (e.g., thiamine, riboflavin, niacin, etc.), antioxidants, water, phytochemicals, immunity, health-promoting substances, etc. (Arumugam et al., 2021). They prevent several chronic non-communicable diseases (NCDs) including cardiovascular diseases, diabetes, obesity, cancer, and respiratory disorders, as they contain key phytochemicals and have low fat, sugar, and sodium content (Juárez et al., 2023). It is frequently advised to consume a varied and vibrant diet to obtain a broad range of minerals, vitamins, and phytochemicals that promote overall health. The major nutritional components like water, protein, sugar, fat, starch, energy, dietary fibers, and micronutrients of fruits and vegetables are shown in **Figure1**. The recommended amount per capita is 120 g/day for fruits and 300 g/day for vegetables, but availability per capita is 104 g/day for fruits and 207 g/day for vegetables (Wallace et al., 2020; Piližota, 2023; Vicente et al., 2022).

Figure 1. Major nutritional features of fruits and vegetables

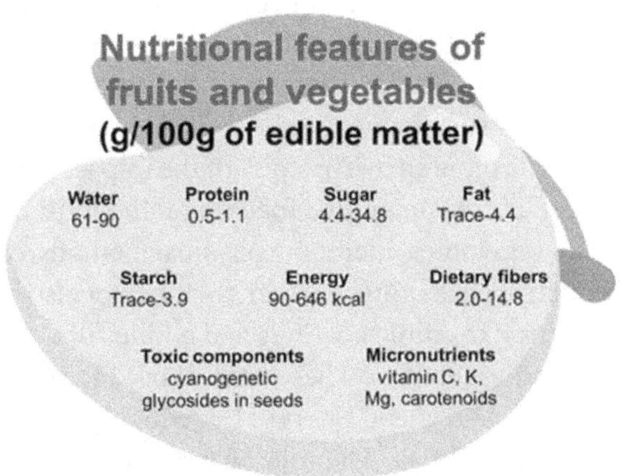

However, the availability is still much less than the advised amount due to the post-harvest losses. Approximately 40% of fresh produce is lost during the post-harvest stage (Gardas et al., 2018). This could result in an overabundance of waste output, which could endanger the ecosystem and limit availability during the off-season. Fruits and vegetables are the food group with the second-highest value of losses and waste (approximately 22%) at all stages in the food supply chain, due to their highly perishable nature (Palumbo et al., 2022). This waste leads to unequal food distribution in terms of location and season, financial loss for farmers, distributors, and consumers, and environmental harm. These losses should be mitigated to preserve the fruits and vegetables and to maintain their quality, quantity, and security (Wani et al., 2024).

Globally, hunger persists with nearly 1 in 10 of the world's population facing it in 2022, while 2.4 billion people experienced moderate to severe food insecurity. FAO estimates that 811 million people lack access to food and 132 million experience food insecurity (Braun et al. 2024). Despite this, 931 million tons of food are wasted annually, along with all the resources like soil, water, energy, and labor used in the production and dissemination phase (Nunes et al., 2023). Thus, there is a need to reduce hunger and food waste. This chapter is primarily associated with three Sustainable Development Goals (SDGs 2, 3, and 12) to tackle all these problems by mitigating food loss, attaining food security, resource consumption, and production patterns, and ensuring a healthy life. SDG 2 aims to end hunger, achieve food security and improved nutrition, and promote sustainable agriculture. Hence, to achieve zero hunger, more initiatives are needed to modify food systems such that they are resilient, equitable, and sustainable (Arora & Mishra, 2022). SDG 3 aims for good health and well-being by positively impacting the health of human beings. Similarly, SDG Target 12.3 aims to reduce global food waste at consumer and retailer levels to halve per capita and bring down the cumulative food losses resulting from poor production, transportation, storage, and harvesting practices by 2030 (Sharma et al., 2024). Hence, there is a need to preserve the produce and mitigate post-harvest losses sustainably.

This chapter provides a comprehensive study of the impact of post-harvest losses on fruits and vegetables. This chapter also focuses on the most common strategies for preserving fruits and vegetables, including physical methods, chemical methods, proper storage, and effective packaging. Further, this chapter also gives an idea about natural packaging systems like edible coatings and edible films for mitigating post-harvest losses in fruits and vegetables. These natural packaging systems contribute towards SDG 7 (Affordable and Clean Energy) by reducing energy consumption in climate-controlled storage facilities. The role of nanotechnology like nanosensors and nanoparticles in preserving produce is also discussed here. This chapter provides a study of food innovations with a specific focus on sustainable packaging methods in line with the objectives outlined in the United Nation's 2030 SDGs 2, 3, 7, and 12 and is also aligned with the recently approved "BioE3 Policy" by the cabinet.

POST-HARVEST LOSSES AND THEIR IMPACT ON FRUITS AND VEGETABLES

The term "post-harvest losses" describes the decline in the quantity and quality of fruits and vegetables that occur during harvest and either sale or intake (Jain et al., 2023). According to the Food Corporation of India (FCI), each year 21 and 12 million tons of losses occur for vegetables and fruits respectively totaling 240 billion rupees (Suresh et al., 2023). Food security and nutrition may be harmed by post-harvest losses which can impact the key components of food security: the presence of food, the ability to obtain it, its utilization as well as stability (Wijewardane, 2022). The main causes of these losses are inadequate supply chain actors, managerial ability, logistical challenges, technological limitations, insufficient skills and knowledge, and lack of markets (Mesterházy et al., 2020). Inadequate infrastructure, technology, and information access are major causes of post-harvest losses in developing nations (Debebe, 2022). Post-harvest losses and wastes of fresh fruits and vegetables occur at various stages: harvesting (20%), processing and packaging (10%), distribution (10%), post-harvest handling and storage (8%), and consumption (5%) as shown in **Figure2**. The highest contribution to the economic loss was by livestock produce-milk, meat, fish, and eggs (21.70%), followed by fruits (19.34%) and vegetables (17.97%) (Mohan et al., 2023).

Figure 2. Post-harvest loss percentage of fruits and vegetables at different stages

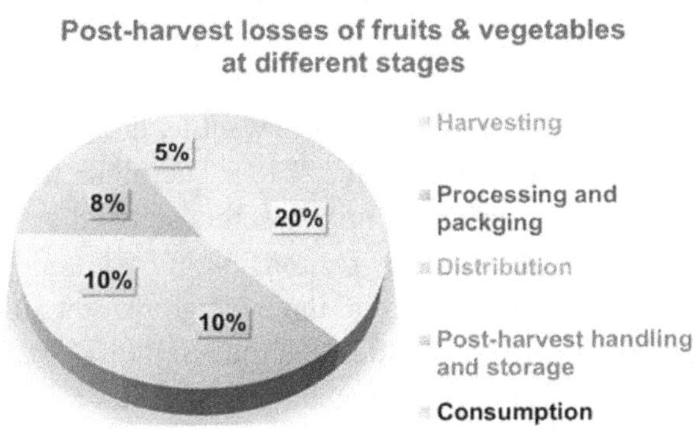

In India, around 30-80% of post-harvest fruit and vegetable losses occur due to insufficient market accessibility, inadequate infrastructure, and inadequate agricultural method expertise (Gardas et al., 2017). Improper post-harvest handling causes a 30 to 60% loss of the total world's yield of fruits and vegetables (Workineh & Enyew, 2021). After harvest fruits and vegetables lose 10-15% of their weight, compared to

20-40% in undeveloped countries. It means that roughly one-fourth of the output is never consumed (Ferdousi et al., 2024). On average, the total weight of vegetables (25%) and fruits (12%) contributes to total food waste (Joy et al., 2022).

Several variables that cause post-harvest losses include physiological, mechanical, environmental, and biological (pest and disease) damages. Biological factors cause more than 40% of the overall losses in fruits and vegetables (Yahia et al., 2019). The factors affecting post-harvest losses of fruits and vegetables can be classified into two major categories, primary and secondary factors as shown in **Figure 3**. The primary factor influencing how long horticulture perishables last after harvest is temperature, which includes both extremes. Because of this, the temperature of horticulture produce during harvesting, handling, transporting, and marketing is significantly greater than what is advised for the produce quality maintenance due to elevated respiration rate and additional associated metabolic reactions in the produce (Ariwaodo, 2022). The majority of physiological conditions are closely associated with water loss. Post-harvest water loss can affect marketable weight loss, shriveling, softening, other quality, and consumer satisfaction issues (Liguori et al., 2024; Ramjan & Ansari, 2018; Sharma & Kumari, 2020).

Similarly, mechanical injuries also cause degradation in the quality and quantity of fresh produce. The primary issue is how to distribute fresh fruits and vegetables. Fresh produce has a short shelf life, high water content, and can degrade quickly, which contributes to a longer transit and transaction time (Rennie & Sunjka, 2017). As a result, they need extremely safe transportation and effective infrastructure. However, as a result of mechanical damage during the distribution and sales process, there is a significant loss of quality because of underdeveloped distribution facilities and equipment (Gardas et al., 2018). Statistics show that fruits and vegetables can lose up to 25-35% of their freshness, whereas in wealthy nations, the corresponding loss rates are only 5% (Ishangulyyev et al., 2019). The other study report in 2006 from supermarket fresh logistics distribution centers states that shipping, sorting, storing, and transportation are primarily responsible for fruit and vegetable losses (Kirci et al., 2022). For example, in India, the average market value losses for tomatoes, cucurbits, okra, mangoes, and litchis at the retail level were 1%, 52%, 31%, 20, and 18%, respectively, due to perceived quality declines (Yahia et al., 2019).

Figure 3. Factors affecting post-harvest losses in fruits and vegetables

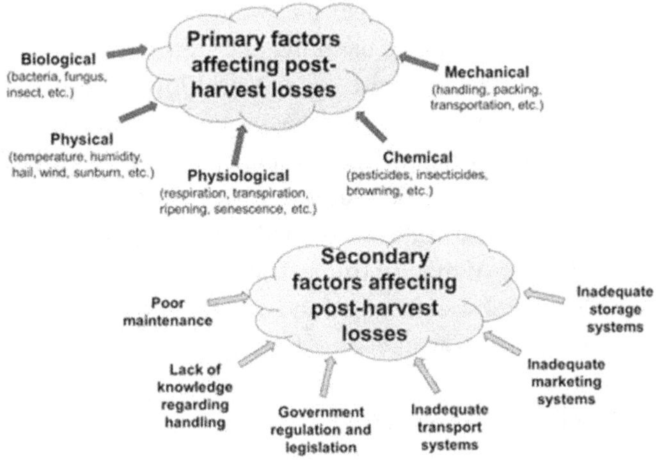

Microbial factors also cause significant post-harvest losses in fruits and vegetables. Post-harvest infections cause 10 to 30% of agricultural yields to be lost, and in some perishable crops, particularly in impoverished nations, they can cause more than 30 percent of crop yields to be lost (Ogedengbe et al., 2022). Fresh fruits and vegetables in particular are extremely perishable, and post-harvest handling, transportation, storage, and marketing can degrade their quality (Elik et al., 2019). Fruits and vegetables are highly perishable because of their high water content (around 93%) which is the main cause of rapid decay by fungi and bacteria (Singh et al., 2021). Pathogens and microbial growth encompass the proliferation of bacteria, molds, and fungi, which can lead to decay and spoilage (Dukare et al., 2022). The most significant and common disease is fungi, which infect a variety of host plants and cause the majority of fresh fruits and vegetables to be lost destructively and economically during storage and transportation (John et al., 2024). The main fungi that cause post-harvest losses are *Alternaria*, *Aspergillus*, *Colletotrichum*, *Botrytis*, *Monilinia*, *Diplodia*, *Penicillium*, *Rhizopus*, *Phomopsis*, *Mucor*, and *Sclerotinia*, and bacteria, such as *Pseudomonas* and *Erwinia* (Mudaliar et al., 2023). Fruits are particularly vulnerable to pathogenic fungi because of their low pH, high moisture content, and nutrient composition. These fungi can cause rots in fruit and produce mycotoxins, which render the fruit unfit for consumption (Alegbeleye et al., 2022). While vegetables are more vulnerable to bacteria due to their high pH which is suitable for the growth of bacteria. Ethylene gas, emitted by certain fruits and vegetables has the potential to expedite the ripening process and hasten decomposition in adjacent produce if appropriate segregation measures are not implemented (Chew et al., 2021).

Packaging is also responsible for post-harvest losses in fruits and vegetables. In India, fresh food was handled in packages that were too large, too abrasive, and too frail to safeguard the produce during handling and transportation (Singh & Singh, 2018). Package protection ratings were consistently low for all commodities and nations; 8% had no packaging, 31% were packaged in open baskets, and 46% of produce was packed in big sacks or fabric bundles (DLADLA, 2020). Mangoes and tomatoes were packed in plastic crates that were unclean and abrasion-prone in India, while Ghana's tomato crates were made of wood, had no vents, and could store between 60 and 70 kg of mature fruits. The majority of the fruit at the bottom of the massive wooden crates was smashed and usually thrown away before purchase (Ahmad & Siddiqui, 2015; Awasthi et al., 2022).

Impact of Post-Harvest Losses

Post-harvest losses can have a significant financial and environmental impact. Food loss and waste thereby affect nutrition, income production, economic growth, and hunger and poverty reduction (Manzoor et al., 2024). Food losses are a sign of ineffective and poorly functioning value chains and food systems. Thus, they signify a reduction in the financial worth of the participants in these chains (Cattaneo et al., 2021). Food availability and prices in one region of the world could be impacted by produce traded on global markets and wasted in other areas (Thyberg & Tonjes, 2016). Hence, they have a significant effect on the economy **(Figure4)**.

Similarly, post-harvest losses also harm the environment because food waste and losses contain resources like energy, biodiversity, greenhouse gasses, water, soil, and other elements that no one uses (Suryana et al., 2024). Producing food that will not be eaten results in wasteful CO_2 emissions as well as a reduction in the food's nutritional value (Conrad & Blackstone, 2021). Rotting fruits and vegetables releases methane and carbon dioxide, contributing to climate change and causing environmental pollution (Kumari et al., 2021). Hence, they negatively affect the environment and economic worth **(Figure4)**.

Figure 4. Major impacts of post-harvest losses

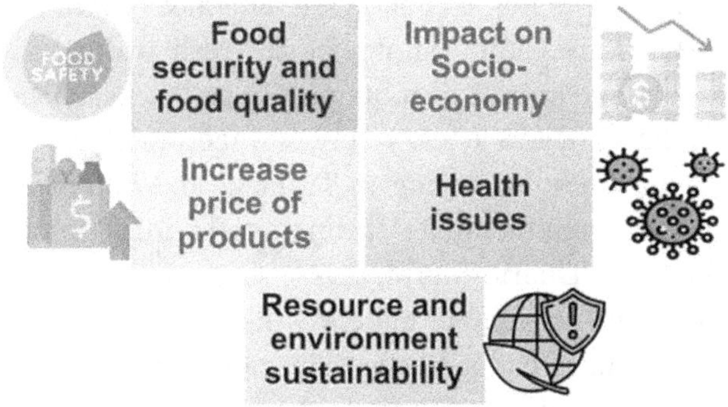

As farming is a labor-intensive and time-consuming process, it requires funds for every activity like buying fields, applying fertilizer, irrigating land, harvesting, and transporting to the market, particularly when it occurs on a scale that supports the market economy (Grgic, 2024). About 173 billion cubic meters of water are consumed annually as a result of food loss and waste, accounting for 24% of the total amount of water needed for agriculture (Camargo et al., 2022). To grow this lost and discarded food, 28 million tons of fertilizers are utilized annually. In addition to these quantifiable effects, they also negatively impacted natural landscapes, ecosystem services, and resources used to produce this lost and waste (Junaid & Gokce, 2024). This implies farmers must recoup their expenses and get a profit. Thus, post-harvest management is necessary to reduce food loss or waste along the supply chain from farm to consumer without compromising food quality (Wunderlich & Martinez, 2018). Hence, understanding these losses and finding ways to mitigate them are essential for ensuring food security and sustainability (Mayanja & Oluk, 2023).

METHODS FOR THE PRESERVATION OF FRUITS AND VEGETABLES AND THEIR PROS AND CONS

Preservation of produce is essential for several reasons. Nutrient preservation guarantees that even when produce is not freshly plucked, consumers can still benefit from it (Tadesse, 2024). Additionally, produce can be available throughout the year despite seasonal fluctuations in output, which is particularly beneficial for regions where some produce is only available during specific seasons or is not grown (Ruppenthal, 2008). Preservation also gives chance to farmers to sell their produce after harvesting and encourages employment in processing and packaging businesses and hence plays a significant role in the economy (Behera & France, 2016).

The post-harvest loss reduction strategies in fresh fruits and vegetables for small-scale farmers include improved containers or packages, change in field packing methods during harvest, low energy cool storage practices (zero energy cool chambers) (ZECC), use of solar drying in Tropical and Subtropical regions, hot water treatment, calcium treatments for firmness retention, use of ethylene action inhibitors, etc. (Gouda & Duarte-Sierra, 2024; Jarman et al., 2023). The use of low-cost, lightweight, and recyclable materials like corrugated fiber-board liners for cartons and plastic crates can save the values of perishables till consumption. Higher prices per kilogram were offered by the purchasers who visited the farm as a result of superior grading and sorting as well as the usage of better packaging (Teutsch & Kitinoja, 2019). ZECC, a low-cost on-farm cooling technique has several benefits as it runs entirely without electricity or power (Victory et al., 2023). Additionally, the necessary supplies, such as bricks, sand, bamboo, etc., are easy and reasonably priced to obtain. Cool chambers lower the temperature by 10 to 15 °C and keep the humidity level around 95% which prolongs shelf life and preserves the quality of horticulture produce (Kumar et al., 2018; Nayak et al., 2021). To preserve the nutritional content of fruits, it is best to keep them fresh. Nevertheless, the majority of storage techniques call for low temperatures, which are challenging to sustain along the chain, particularly in poor nations (Jain et al., 2023). One of these methods that can extend the shelf life of fruits is solar drying. Additionally, it lowers transportation costs, minimizes seasonal gluts, and raises dietary standards. Since fresh produce can contain up to 95% water, it is sufficiently wet to sustain the growth of microbes as well as enzyme activity. Drying aims to get the produce's water content down to a point where it isn't enough for enzyme activity or microbe development (Sánchez et al. 2020). The crucial moisture content ranges from 10 to 15% based on the produce. Thus, removing too much water could cause the product to become brittle and break easily. It's crucial to treat some produce with hot water to stop bacteria from spreading during storage and delivery. For example, fresh-cut mango slices made from unwashed fruit had substantially larger microbial populations than those made from washed fruit both just after preparation and seven days later at 5°C (Erkan and Yildirim 2017). Similarly, the treatment of $CaCl_2$ reduced the browning and softening of fruits and vegetables. The firmness of mango cubes treated with 1% $CaCl_2$ was significantly higher than those treated with 0.5% $CaCl_2$ or water (control). Ripening inhibitors, high CO_2, low O_2, and waxing are used together to extend the storage life of fresh produce (Moradinezhad, 2021). Various chemical formulations have been attempted to maintain the ethylene concentration below the threshold. When used in a regulated storage environment, ethylene absorbents such as calcium chloride ($CaCl_2$) and potassium permanganate ($KMnO_4$)

have a significant commercial potential that small-scale growers might not be able to achieve (Bisht & Singh, 2024).

The most common preservation methods are described below and depicted in **Figure5**.

Figure 5. The most common post-harvest preservation approaches for fruits and vegetables.

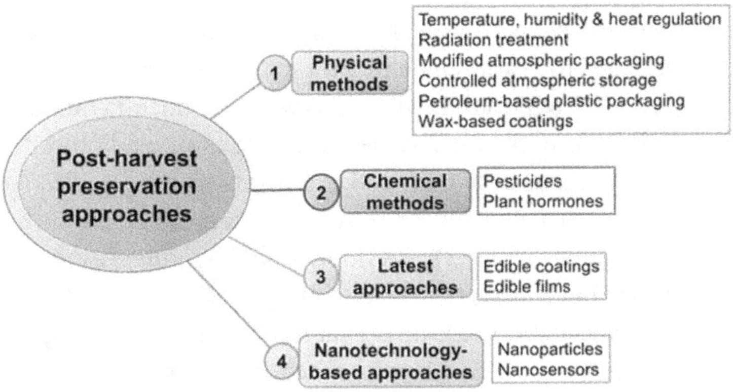

Physical Methods

The most commonly used physical methods nowadays are regulating temperature, regulating humidity, controlling atmospheric conditions, modified atmospheric packaging, radiation treatments, petroleum-based plastic packaging, and wax-based coatings for post-harvest preservation of fruits and vegetables (**Figure5**) (Khalid et al. 2022; Basumatary et al. 2022).

Regulation of Temperature

Low-temperature storage is the most popular physical preservation technique for fruits and vegetables. Fresh-cut fruits and vegetables must be stored in a chilled environment to slow down microbial development and prevent high enzyme activity. The physiological metabolism rate like respiratory and ethylene synthesis rates slows down at low temperatures, hence preserving the produce market value (Giannakourou & Tsironi, 2021). Fresh horticulture product sectors often use precooling as a standard treatment step because of its affordability, ease of use, and convenience (Vivek et al., 2019).

Although cooling has several advantages, the main limitation is that extremely low temperatures or prolonged refrigerated storage can cause chilling injuries, particularly in tropical and subtropical produce (Zainalabidin et al., 2019).

Regulation of Humidity and Heat

Short-term heat treatment is one of the earliest and most widely used methods of preservation which tries to lower the number of microorganisms and stop the action of certain enzymes. The main issue with perishable produce is that heat is linked to a considerable loss of nutritional value, flavor, texture, and color (Amit et al., 2017). The primary methods used on freshly cut fruits and vegetables include brief immersion in hot water that controls the growth of microbes on the surface of fruits and vegetables (Nicola et al., 2022). Additional heat treatments include moderate heat shock, hot water blanching, hot water rinsing and brushing (HWRB), etc. In the fresh-cut vegetable sector, short-time heat treatments are gaining popularity as an efficient way to counteract the adverse effects of enzymatic browning and as a substitute for chemical preservation (Sharma et al., 2021).

Water vapor moves between food that has been stored and the atmosphere around it until the water activity in the food and the atmosphere is balanced. While dry food will take moisture from the air, moist food will release moisture into the atmosphere (Syamaladevi et al., 2016). Except for onions and garlic, fresh horticulture items must be stored in an environment that minimizes relative moisture loss and wilting due to their high moisture content (Tapia et al., 2020). Items that are dehydrated or dried out must be kept in low relative humidity environments to prevent moisture from adhering to the location of mold growth. Thus, by regulating the humidity, the metabolism process can be controlled in fruits and vegetables (Wu et al., 2021).

Radiation Treatment

Irradiation is the most commonly used physical preservation technique for fruits and vegetables. It is effective in preserving the quality and lengthening the shelf life of produce and can replace chemical goods whose usage is prohibited (Bisht et al., 2021). Ultraviolet radiations are commonly used for the post-harvest preservation of fresh produce. They improve quality mainly by direct sterilization and induction of resistance. Nonionizing UV-C radiation is most commonly used for both whole and little processed fruits and vegetables by decontaminating and cleaning the produce (Bhatnagar et al., 2022). Gamma rays are the most significant ionizing radiation for preserving minimally processed fruits and vegetables. Gamma rays released by the irradiation source are capable of penetrating both microbes and structures. This deactivates the cell by damaging the microbial DNA (Indiarto et al., 2020). Foods contaminated with radiation do not turn radioactive and are regarded as safe. While the maximum dosage permitted for industrial application is 10 kGy, significantly smaller dosages can eradicate pathogens (1-5 kGy) and postpone the ripening and senescence of fruits and vegetables (0.2–0.5 kGy). For example, the application of 2% ascorbic acid (w/v) and 1 kGy of gamma-irradiation in eggplant prevented polyphenol oxidase (PPO) activity and surface browning by up to 6 days (Sharma et al., 2020). Similarly, the most effective dose for preserving microbiological quality in minimally processed carrots and lettuce was 2 kGy, which also decreased the number of harmful organisms such as *Bacillus cereus*, *Cronobacter sakazakii*, *Staphylococcus aureus*, and *Klebsiella* spp (Zhao et al., 2022).

This technique's primary drawbacks are the high implementation costs and the product-consumer interaction. Customers are frequently cautious and hesitant to accept goods produced by innovative technology, particularly when they lack sufficient knowledge about the procedure. Thus, a lack of information causes people to perceive dangers incorrectly, which leads to a rejection of foods contaminated with radiation (Castell-Perez & Moreira, 2021).

Modified Atmospheric Packaging (MAP)

MAP has been widely used for preserving the quality of fruits and vegetables. It enhances the shelf life by providing ideal storage and processing conditions. MAP provides a modified environment with low O_2 and high CO_2 levels to reduce metabolism, respiration, oxidative stress, tissue senescence, and ethylene synthesis (Gorris & Peppelenbos, 2020). MAP significantly increases moisture retention, which may be more important for maintaining quality than O_2 and CO_2 concentrations (Kumar et al., 2024). Additionally, packaging keeps the product away from the outside environment and contributes to the least amount of pathogen and pollutant exposure (Mangaraj & Goswami, 2009). Active and passive approaches can be used to modify the atmosphere. Passive MAP is more economical than active MAP. The gaseous equilibrium within packaging with a passively changed atmosphere is the result of a dynamic process that is influenced by the product's metabolism, the permeability of the polymer film utilized, and the storage environment (relative humidity and temperature) (Rennie & Sunjka, 2017). Packaging with active MAP depends on the displacement or replacement of gases within the package, or the application of absorbers or gas scavengers to provide the intended initial combination of N_2, oxygen O_2, and CO_2 (Oliveira et al. 2020). The effectiveness of this process depends on the produce's metabolism, gas compositions, and the permeability of the applied polymer layer. For e.g., the use of MAP has increased the shelf of cherries to 30-40 days in comparison to the 7-14 days in natural conditions. Similarly, the shelf life of tomatoes increased to 60 days compared to 30 days in the natural atmosphere. The use of gibberellic acid (GA_3) along with this technique enhanced the quality and shelf life of green peppers (Ozturk et al., 2022).

Controlled Atmospheric Storage

It is also the most commonly used method for preservation. The respiration and metabolism process mainly depends on the carbon dioxide-to-oxygen ratio, which determines the aging process of produce. Hence, by altering this atmospheric CO_2 to O_2 ratio, the respiration and metabolism processes can be controlled (Umeohia & Olapade, 2024). The aerobic respiration reduces as the CO_2 level increases but anaerobic respiration occurs when there is too low O_2 amount. A hermetically sealed storage space with O_2 and CO_2 analyzers, as well as gas removal and injection systems, is necessary for regulating the atmosphere (Oliveira et al., 2020). This method uses adsorption by pressure oscillation gas separation system that absorbs O_2 through a membrane with a filtration system to lower the O_2 pressure by infusing N_2 while increasing the CO_2 pressure by injecting this gas. Thus, the proper gas proportion must be to decrease respiration without causing low oxygen damage (Gorris & Peppelenbos, 2020). It also delays the amount of oxygen in the packing, which slows down the ripening process. However, maintaining this packaging oxygen content can frequently encourage the growth of anaerobic pathogens on the produce surface (Rothy 2023). This technique has been effective in prolonging the shelf life of certain vegetables like apples, broccoli, lettuce, and cauliflower maintaining their market quality for an extended period (Ebrahimi et al., 2022).

Although this technique has been used for more than 200 years, its cost is more than the modified atmospheric packaging system (Wei et al., 2021).

Petroleum-Based Plastic Packaging

Packaging materials are essential for safeguarding the produce during handling, transportation, and storage. Plastic is the most widely used material for packaging due to its low cost. However, because they are not entirely recyclable or biodegradable, environmental conservation laws have severely limited their use as packaging materials (Gupta et al., 2022). The most commonly used petroleum-derived polymeric materials are low-density polyethylene (LDPE), linear low-density polyethylene (LLDPE), high-density polyethylene (HDPE), polystyrene, polyvinyl chloride (PVC), polypropylene, and polyester, i.e. polyethylene terephthalate (PET), polyvinylidene chloride (PVDC), polyamide (Nylon) (Awad et al., 2020; Lambert, 2013). They are highly used for packaging due to their affordability, high barrier properties, stability, transparency, and lightweight qualities (Merino et al., 2022).

However, just 2% of plastic packaging is recycled worldwide, and recycling some packaging materials is difficult (Hahladakis et al., 2024). The manufacturing and careless handling of plastic packaging materials releases carbon dioxide into the atmosphere and also contaminates seas and land. Most likely, the earliest material utilized for packing was glass (Dziadowiec et al., 2023). Its broad application stems from its odorless nature and chemical inertness. These days, glass-based packaging is thin and resilient to sterilizing processes under extreme pressure and temperature (Hossain et al., 2024). Metals are the most adaptable material for packaging because they are completely recyclable and have a high degree of resistance (Pascall, 2020). The two most prevalent metals are found to be steel and aluminum. Steel is solely utilized as a container; aluminum can be molded to fit the goods. Metal gaskets typically provide protection from air, moisture, smells, and bacteria (Driscoll & Rahman, 2020).

The use of environmentally friendly substitute materials for packaging is becoming more popular as a result of concerns about petroleum-based products' non-biodegradable nature and the depletion of petroleum oil reserves. Hence, biodegradable materials are in high demand for packaging produce due to environmental concerns, and health perspectives (Jha et al., 2024; Ncube et al., 2020).

Wax-Based Coatings

The edible coatings originate from wax-based coatings. These waxes are made from petroleum materials, natural sources, or high molecular weight esters of monohydroxy alcohols and carboxylic acid. The coating gives the produce a glossy appearance and improves resistance to air, moisture, and microorganisms increasing consumer appeal (Jose et al., 2020). This has been mainly applied to fruits and vegetables like apples, oranges, grapes, avocados, watermelons, lime, lemons, eggplants, cucumbers, tomatoes, muskmelons, turnips, bell peppers, etc. Fruit may get a white, waxy coating on the surface due to heat exposure (Bai & Plotto, 2012). However, waxy coatings also have some disadvantages over edible coatings. These limitations are mostly influenced by the thickness, concentration, and preservatives that are incorporated into the wax to change the qualities of the coating (Pashova, 2023). Wax is emulsified with hazardous substances like morpholine to change the characteristics of the coatings. This morpholine reacts with the nitrate present in the diet and produces N-nitroso morpholine (NMOR) by nitrosation reaction. NMOR is formed in aqueous solutions of nitrite or by a reaction of gaseous nitrogen oxides, e.g., N_2O_3, N_2O_4, and NOx in aqueous solutions even under normal environmental conditions (Kumar, 2016). However regular consumption of morpholine may harm internal organs as it has carcinogenic properties. Morphine oleate is used in most fruit and vegetable coatings produced in the US due to its superior plasticizing and emulsifying qualities (Miller & McGinity, 2008). Nevertheless, the European Community has not shown any proof to support this addition. As a result, the application of these transporters to fruit and vegetable covering, results in adulteration. Whereas the coatings are regulated by regulatory bodies like the US FDA, European Commission, etc. and they are "Generally Recognized as Safe (GRAS)" for human ingestion on a global scale (Wang et al., 2019). Some regulatory bodies have allowed the minimal number of commercial coatings put on fruits and vegetables- roughly 0.2 ml (1 drop) of wax per piece but more thorough risk assessment research is required to see whether NMOR is accumulated over time following intake of coated produce (Maftoon Azad, 2006). The UK Food Standards Agency conducted a preliminary risk evaluation of morpholine-coated fruits and vegetables in October 2010 and discovered that 0.03-0.3 ppm of morpholine probably has little damage to consumer health (Kumar, 2016). Therefore, the use of morpholine is forbidden in the United Kingdom (UK) and the European Union (EU) countries that do allow its use are fully aware of these limitations and have tight protocol limits of 0.03-0.3 ppm (Kumar & Kapur, 2016). In Europe, morpholine is prohibited due to its precursor, NMOR, which is a carcinogen. Because the apple was found to contain roughly 2ppm of morpholine, the UK Food Standards Agency prohibited the import of Chilean apples and recommended against selling the impacted fruits in the country (Rastogi & Samyn, 2015). Hence, edible coatings/films containing natural additives could be a solution to wax-based coatings for preserving fruits and vegetables.

Chemical Methods

Instead of physical methods, several chemical preservatives are used for post-harvest preservation of fruits and vegetables. These chemical methods are mainly pesticides and plant hormones (**Figure5**) (Ogunnupebi et al., 2020).

Chemical Fungicides/Pesticides

Around 20-30% of losses occur every year in the agriculture sector because of pests, diseases, and weeds. Thus, pesticides are used to manage pests, weeds, insects, and diseases in agriculture that could otherwise ruin crops or decrease their yield (Begum et al., 2017). The most commonly used pesticides are insecticides (e.g., Chlorpyriphos, Malathion, Quinalphos, etc.), fungicides (e.g., Sulphur, Mancozeb, Carbendazim, Propineb, Imazalil (IMZ), Thiabendazole (TBZ), Sodium ortho-phenyl phenate (SOPP), Zineb, etc.) (Al-Alam, 2017), rodenticides (e.g., Zinc Phosphide, Aluminium Phosphide, Methyl Bromide, Warfarin, etc.), biopesticides (e.g., Neem based insecticides, Trichoderma viride, etc.), and weedicide (e.g., Glyphosate, 2,4-D Amine Salt, Pretilachlor, Atrazine, Butachlor, etc.) (Basu & Rao, 2020; Larramendy & Soloneski, 2022). Herbicides account for 40% of pesticide use worldwide, insecticides for 17%, and fungicides for 10%. The annual pesticide consumption in India is 40 million kg (Nayak & Solanki, 2021). Chemical fungicides have potent antifungal action that can prevent the growth of different fungi on the surface of fruits and vegetables. Dried fruit is often treated with methyl formate (Erinol), ethyl formate, and (in certain nations) ethylene oxide to eradicate mold and insect infestations (Sharma et al., 2019). To stop yeasts and molds from growing, sulfur dioxide, benzoic acid, propionic acid, ascorbic acid, and sorbic acid are added to processed fruit products, particularly juices. Rottenness can be immediately delayed by using chemical fungicides (Singh et al., 2019). Excessive usage of pesticides can have negative impacts on human health, although it increases agricultural productivity. The use of these chemicals degrades the environment, especially water pollution, and lowers human health, resulting in everything from cancer to nerve damage (Zhang, 2018). Pesticide residues can lead to 4,000-20,000 cancer patients per year. Pesticide residue levels should be below maximum residual limits (MRLs) for safe consumption (Zaller & Zaller, 2020).

Plant Hormones

The majority of the other chemical preservatives are plant hormones that directly control the ripening and aging processes of fruits and vegetables. Currently, the most widely used post-harvest preservative for fruits is 1-methyl cyclopropane (1-MCP), a plant ethylene inhibitor that functions as an ethylene receptor to prevent fruit from producing ethylene (Shahbaz et al., 2022). 1-MCP slows down the senescence process by decreasing ethylene formation, which is a key signaling molecule for fruit ripening and senescence. Other chemo insulators belonging to endogenous hormones are methyl jasmonate, salicylic acid, and melatonin (Pillai et al., 2024). Similarly, plant extracts like phenolics can effectively postpone the post-harvest senescence of fruits and vegetables (Shahbaz et al., 2022).

LATEST APPROACHES FOR THE PRESERVATION OF FRUITS AND VEGETABLES

Edible films/coatings are an alternative to synthetic or petroleum-based coatings, gaining attention as a sustainable technology for preserving fruits and vegetables (Pillai et al., 2024). Synthetic polymers including propylene, polystyrene, polyethylene, and polyolefin can build up as microplastic in the environment, which can lead to bioaccumulation and the photo-degradation of hazardous volatile chemicals like propanal and acrolein. Hence, it is unsuitable for humans and the environment (Zhang et al., 2023). The green solution could be the applications of edible coatings and edible films (**Figure5**).

Edible Coatings

Edible coatings are thin layers, usually less than 0.3 mm in size that are applied directly over the surface of fruits and vegetables. Hence, they are generally flavorless, colorless, and odorless. Edible coatings are a potential solution for preserving produce quality (Sapna et al., 2024). They are made of natural or food-grade ingredients such as polysaccharides (chitosan, alginate, and starch), lipids (beeswax, carnauba wax, and oils), proteins (gluten and gelatin), and composites. These substances act as physical barriers to the exchange of gases and water that suppress the growth of microbes (Khalid et al., 2022).

Polysaccharides are commonly employed in the fabrication of edible coatings to prolong the shelf life and preserve the quality of food. They are an excellent choice for surface applications because they are colorless, oil-free in appearance, and have low caloric content (Ferreira et al., 2016; Kocira et al., 2021). They possess a robust ability to block oxygen but they do not possess a sufficient ability to block moisture, hence their combination with other compounds to increase their resistance to moisture has been undertaken. They are biocompatible, and biodegradable, improved mechanical properties, and effective barrier properties against oxygen (Shiekh et al., 2021). The most commonly used polysaccharides for these coatings include chitosan, starch, alginate, pectin, gums, and cellulose. Chitosan, derived from chitin is most commonly used for increasing the shelf life of perishable fruits and vegetables due to its antibacterial properties (Feng et al., 2024; Guzdemir, 2024). Alginate, another hydrophilic polysaccharide derived from brown seaweed also stops produce from being dehydrated and losing its nutritious value (Khan et al., 2023).

The suitability of proteins for formulating coatings depends on the configuration of amino acids along the polypeptide chain and the subsequent interactions between chains. To a large extent, the coating's strength, gas, and liquid permeability are determined by these interactions (Mohamed et al., 2020). Protein-based edible coatings are advantageous because of their inherent properties, which allow them to effectively block oxygen even at low relative humidity levels (Chen et al., 2019). They have excellent film-forming abilities and nutritional properties. Soy, whey, gelatin, casein, gluten, and zein corn proteins are just some of the proteins that have been used for the production of edible coatings (Wu et al., 2023). The most commonly used lipids are paraffin, carnauba wax, and candelilla wax for coating fruits and vegetables. Lipids act as efficient moisture barriers due to their inherent capacity to hinder water migration (Sapna et al., 2024; Wu et al., 2023). They are hydrophobic and hence coating formed is notoriously thick and brittle. Lipids are often blended with polysaccharides or proteins to enhance the mechanical and barrier properties of composite coatings (Yousuf et al., 2022).

Composite coatings are preferred as they provide properties of two or more components. Polysaccharide and protein-based coatings have excellent gas barrier properties but they have weak water barrier properties due to their hydrophilic nature (Li et al., 2017). Thus, this limitation can be overcome by using lipid-based components that have better moisture barrier properties. However, the anaerobic conditions may cause off-flavor (Himanshu & Bhavani, 2023). Hence, the composite coatings are more suitable for improving the properties and thus preserving the fruits and vegetables by enhancing shelf life and nutritional properties (Vieira et a., 2021; Jahangiri et al., 2024.

The formulation of edible coatings/ films requires matrix and plasticizers as a necessary component, and crosslinkers and functionalized bioactive components as additional components that enhance the properties of necessary components as shown in **Figure6**.

Figure 6. Necessary (matrix & plasticizers/emulsifiers) and additional (functionalized & crosslinkers) components required for the formulation of edible coating/films

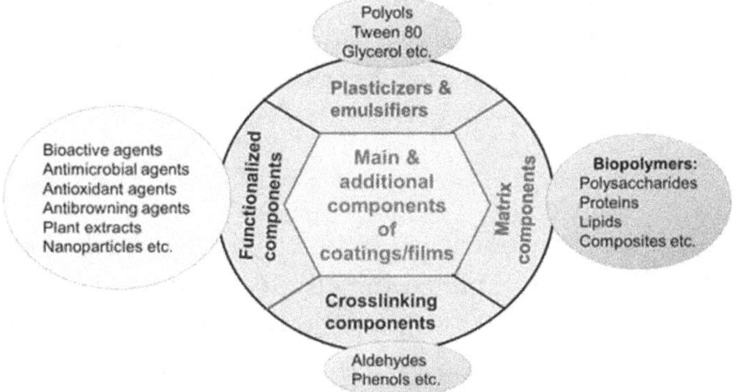

Several bioactive components like flavorings, texturizers, plasticizers, and nutrients are being incorporated into edible coating materials to lengthen the shelf life of the produce **Figure6**. (Aayush et al., 2022). The incorporation of natural antioxidants and antimicrobial agents may further improve the properties of coatings (Vieira et al., 2022). Incorporating natural compounds like plant extracts, herbal extracts, and essential oils into these coatings improves their antibacterial and antioxidant properties and satisfies customer desire for sustainable and organic food preservation techniques (Deshmukh & Gaikwad, 2022). The major categories of natural antibacterial agents encompass those derived from animals (like chitosan, protamine, and propolis), microbial sources (like nisin, lysozyme, and natamycin), and plant-origins, offering both antibacterial and antioxidant properties (such as tea polyphenols, plant essential oils, and Chinese herbal medicines) (Bhuvaneswari et al., 2023; Boukhatem & Setzer, 2020). The most commonly used extracts are ginger, garlic, aloe vera, tea, and propolis which prolong the longevity of fruits and vegetables (Deshmukh & Gaikwad, 2022). These substances are safe to eat due to the presence of several active ingredients (flavonoids, terpenoids, and polyphenols) that prevent oxidation and browning reactions in fruits and vegetables (Ajose et al., 2022).

The most common methods used for applying edible coatings are dipping, spraying, electro-spraying, panning, layer-by-layer, spreading, and vacuum infusion (Sapna et al., 2024). Dipping is the most widely used method for coating fruits and vegetables because of its simplicity, lack of equipment requirements, and uniformity. It is an inexpensive, manual method and hence is the most prevalent method for coating fruits and vegetables (Yadav et al., 2022). Edible coatings regulate physicochemical processes (respiration rate, weight loss, pH, and total dissolved matter), and preserve phytonutrients (antioxidants, phenolics, and pigments) properties of fruits and vegetables for a longer period (Pérez-Santaescolástica et al., 2022).

The main goals of applying edible coatings on fruits and vegetables are to increase their shelf life by reducing moisture loss, microbial growth inhibition, and respiration rate modulation (Kocira et al., 2021). As they are eco-friendly, biodegradable, and sustainable they can be eaten along with fruits and vegetables and hence are the best alternative to traditional approaches (Saxena et al., 2020; Tzia et al., 2016).

Edible Films

The edible films are prepared ahead of time and then attached to the produce, while the edible coatings are applied directly to the produce. Edible films are primarily manufactured as sheets or thin wrap films and later used to cover food intact (Jafarzadeh et al., 2021). The overall film strength, solubility, surface activity, appearance, flavor, and texture in the mouth may be determined by the nature of the edible films (Bai & Plotto, 2012).

Edible films are mainly formulated from biomacromolecules as they have a natural origin with less toxicity and GRAS by the US FDA. These biomacromolecules have good biocompatibility and biodegradability (Mohamed et al., 2020). Hence, they are highly used for formulating edible films as shown in **Figure6**. Biomacromolecules are majorly divided into three classes: hydrocolloids (polysaccharides and proteins); lipid colloids (fatty acids, acylglycerol, waxes); and composites (Tebar et al., 2023).

Although polysaccharides and protein-based materials have high gas barrier capabilities and mechanical and organoleptic properties they have limited water barrier properties due to their hydrophilic nature (Kocira et al., 2021). While lipid-based products offer superior moisture barriers, in anaerobic environments, they could cause off-flavors (Wu et al., 2023). The most commonly used methods are solvent casting (wet process) and melt extrusion (dry process) for formulating edible films (Matloob et al., 2023).

Edible films from agro-industrial by-products have been increasingly studied because they can combine the film-forming properties of biopolymers (mainly polysaccharides such as pectin and cellulose derivatives) with the color and flavor provided by the pigments and volatile compounds of the fruit (Saxena et al., 2020). Concerning fruit and vegetable by-products, it is a consensus that food processing operations may generate high amounts of residues. Most of these, though, are underutilized because of their low market value (Nirmal et al., 2023). Apple processing, for instance, generates over 9000 tons of apple peel each year. The seeds of avocados, grapes, lemon, jackfruit, and mango exhibit a phenolic content that is approximately 15% higher than that found in the fruit pulp (Basri et al. 2021).

The use of such wastes to produce edible films stands out as a feasible alternative to reduce the production cost of edible films and to add value to food by-products. Hence agro-wastes are employed as an alternative for packaging food, contributing to the advancement of sustainable agricultural objectives (Mondal et al., 2022; Pérez-Santaescolástica et al., 2022). Still, when the by-product is rich in lipids or low water-soluble hydrocolloids, it makes sense to extract the compounds of interest with any food-grade solvent-such as water add them to the waste, although lipids may be desired for providing the final material with an improved barrier to moisture (Montes and Muñoz 2021).

Properties of Edible Coatings and Films

The ability of edible coatings and films to prolong the shelf life of fruits depends on their functional properties like barrier, mechanical, and biological characteristics. These characteristics give new dimensions to sustainable food preservation techniques and preserve the post-harvest quality of fruits and vegetables (Bizymis & Tzia, 2022). The main properties of edible coatings/films are described below and shown in **Figure7 (A)**:

Figure 7. **(A).** *Main properties of edible coatings/films and* **(B).** *Benefits of edible coatings/films.*

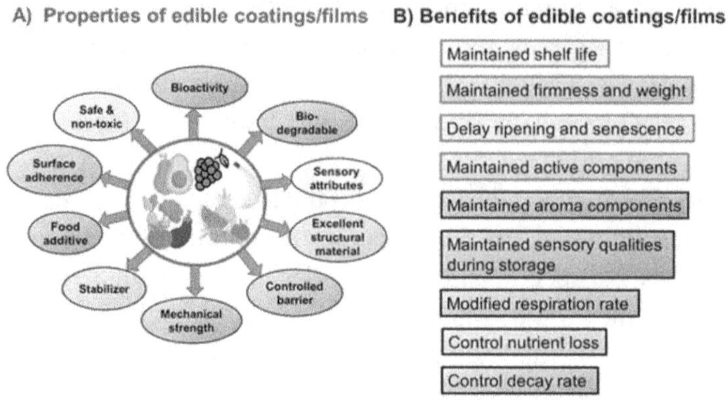

Mechanical Properties

The physical integrity of coated produce during handling and storage is affected by the mechanical strength, elasticity, and force required to break them, such as tensile strength (TS), elongation at break (E), and elasticity modulus (EM), and flexibility of edible coatings/films (Mouzakitis et al., 2022). They exhibit comparable mass transfer phenomena, including diffusion, adsorption, and penetration, which are connected to the solutes' movement between food and the atmosphere. These properties depend on the composition and structure of the used polymer and can be tailored to match the application needs (Tomoda et al., 2020). For example, proteins and polysaccharides are combined to formulate composite films that have increased the tensile strength and flexibility in fragile fruits like berries. The inclusion of plasticizers, like glycerol, can improve these materials' extensibility and flexibility of coatings/films that have improved adherence to the fruit and vegetable surface (Mihalca et al., 2021).

Barrier Properties

The primary function of edible coatings/films is to form a barrier to avert the loss of water vapor, volatiles, and exchange of CO_2, O_2, C_2H_4, and moisture during processing, handling, and storage, thus changing the internal atmosphere around fruits and vegetables (Martins et al., 2024). The presence of oxygen is the main enemy in reducing shelf life and nutritional components (Khalid et al. 2022). Oxygen is responsible for various harmful reactions like browning of tissue, quick oxidation of fats or lipids, and growth of fungi, yeast, mold, and bacteria (Zhao et al., 2022). Thus, edible coatings/films in fruits and vegetables play a key role in providing a gaseous exchange barrier by modifying the environment (Singh et al., 2023). They reduce the respiration process which is O_2 uptake and CO_2 production. Hence, they prevent microorganisms' growth, reduce their deterioration, and enhance the shelf life of fresh produce (Gardas et al., 2017). The adverse reactions like the generation of ethylene and respiration which have a degrading effect can be slowed down by providing a barrier. This reduces the rate of respiration and weight loss during storage and transportation, slowing down the aging process of fruits and vegetables (Moeini et al., 2022).

Shelf-Life Extension Property

The main goals of applying edible coatings/films on fruits and vegetables are to increase their shelf life by reducing moisture loss, microbial growth inhibition, and respiration rate modulation (Moreno et al., 2021). Chitosan-based coatings enhanced the shelf life, freshness, and marketability of strawberries and raspberries by delaying the ripening process and minimizing degradation (Lee et al., 2022). Similarly, alginate-based films have been used to maintain the color and stiffness of freshly cut apples, thus preserving the quality of minimally processed fruit (Parreidt et al., 2018).

Bioactive Properties

The edible coatings/films have an impact on the external appearance like glossiness, firmness, softness, physiological parameters like weight, ethylene production and respiration, and biochemical parameters like total phenols, antioxidants, lipid peroxidation, and enzyme activities of fruits and vegetables (Moeini et al., 2022). The edible coatings/films can transport gases (such as oxygen, carbon dioxide, nitrogen, and ethylene), water vapor, organic vapors (such as fragrances and solvents), and solutes (such as salts, additives, and pigments) between produce and the atmosphere (Martins et al., 2024; Perez-Vazquez et al., 2023). It is possible to broaden these traits by incorporating additional components such as essential oils, antimicrobials, antioxidants, antibrowning agents, ethylene scavengers (Sharma et al., 2021; Wei et al., 2021), moisture absorbers, nanoemulsions, nanoparticles, and bio-nanocomposites as shown in **Figure8** (Sapna et al., 2024).

Figure 8. Mechanisms of action of matrix and bioactive components of edible coatings/films (Sapna et al., 2024).

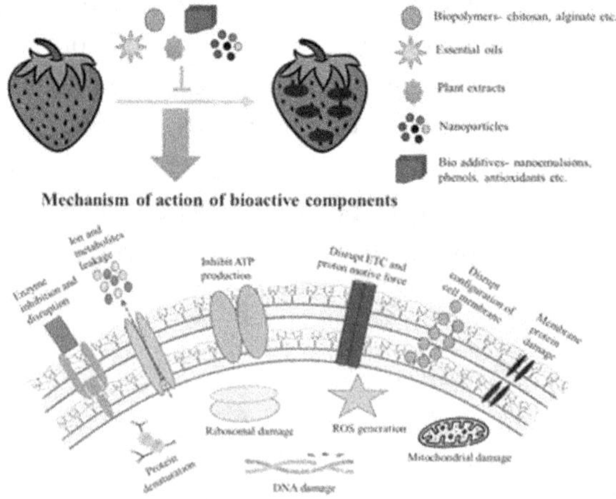

Hence, the incorporation of natural antimicrobial agents like essential oils in coating/film solutions may safeguard against microorganisms that cause spoilage as shown in **Figure8**. Essential oils are naturally occurring highly volatile aromatic compounds that can be extracted from various parts of plants like seeds, stems, leaves, flowers, and fruits (Basavegowda & Baek, 2021). The most commonly used plants are cinnamon, clove, coriander, eucalyptus, lavender, ginger, oregano, thyme, palmarosa, mint, fennel, lemongrass, and citrus fruits that have specific chemical compounds like eugenol, geraniol, menthol, limonene, carvacrol, and linalool (Teixeira et al., 2022). They are considered to be Generally Regarded as Safe (GRAS) antimicrobial additives due to their ability to inhibit the growth of microorganisms. The most common role of bioactive components in edible coatings/films involves microorganisms' enzyme inhibition, protein denaturation, disrupting cell membranes, inhibiting ATP production, etc., (Li et al., 2022; Sapna et al., 2024) (**Figure8**).

Sustainability

Edible coatings/films have become environmentally friendly substitutes for synthetic packaging materials that harm the environment. The materials used for formulating edible coatings/films complement the worldwide movement toward sustainability as they are derived from renewable resources (Kathpalia et al., 2022; Vittuari et al., 2019). The incorporation of natural substances into these coatings/films, including plant extracts and essential oils, not only increases their antioxidant and antibacterial properties but also satisfies consumer requirements for sustainable and organic food preservation techniques (Aqilah et al., 2023). These approaches promote environmental sustainability by assuring the quality and safety of preserved fruits and vegetables and reducing the reliance on nonbiodegradable packaging (Panchal et al., 2022).

Important Functions of Edible Coatings/Films

Edible coatings/films preserved and improved the shelf life of fruits and vegetables. They act as barriers and provide a regulated environment to produce, thus preventing the deterioration of fruits and vegetables (Aayush et al., 2022). They maintain the shelf-life, safety, and quality of fruits and vegetables during handling, transportation, and storage. The most common benefits of edible coatings/films are shown in **Figure7 (B)**. Edible coatings/films protect the produce from the damages that occur during transportation and handling by acting as a physical layer. They filter hazardous UV rays and protect the appearance, quality, and nutritional content of fruits and vegetables (Bizymis & Tzia, 2022; Sharma et al., 2019). They mitigate mechanical injuries and abrasions, thus preserving the produce and improving consumer appeal. They keep produce hydrated by limiting excessive moisture loss, which lowers the risk of dehydration and shrinking (Montes & Muñoz, 2021; Santaescolástica et al., 2022). They also maintain the penetration of oxygen and carbon dioxide amount through the layer and reduce respiration rate, thus enhancing the shelf life of produce as shown in **Figure8**. Thus, it prevents the deterioration of produce by maintaining gas exchange and enhancing freshness (Wu et al., 2023). They also provide barrier properties against volatile organic components (VOCs) and prevent organic vapor loss like additives, aromas, pigments, solvents, etc. This preserves the aroma, flavor, quality, and sensory properties of fruits and vegetables (Cloete et al., 2023; Yousuf et al., 2022).

However, the formulation and application of edible/films are also challenging. The main limitations are:

a) **Expensive production cost:** The usage of natural biopolymers and other bioactive components for formulating edible coatings/films leads to more expensive production costs than synthetic coatings (Gaspar & Braga, 2023).
b) **Restricted shelf life:** Although edible coatings/films enhanced the shelf life of fruits and vegetables. They could also have a limited shelf life, necessitating cautious storage and handling to keep produce fresh. prevent spoilage (Yousuf et al., 2018).
c) **Variability in efficacy:** The efficacy of edible coatings/films may differ based on the fruit type, variety, composition of the coating, and application technique (Avramescu et al., 2020).

d) **Potential for allergic responses:** Some people may experience allergic reactions to some of the biopolymers used for formulating edible coatings/films. E.g., chitosan may cause allergy in some people (Peerzada et al., 2023).
e) **Effect on sensory attributes:** The sensory attributes like texture, flavor, and aroma of coated produce may be affected by the components used for fabricating edible coatings/films. The change in sensory properties might influence the choices of consumers (Nunes et al., 2023).
f) Effect on the environment: Although edible coatings/films are biodegradable, their formulation and discarding could still have an effect on the environment, especially if used in large quantities (Gaspar & Braga, 2023).

Overall, even though edible coatings/films for preserving fruits and vegetables have numerous advantages over synthetic chemicals. But more research is required to resolve these restrictions and guarantee their security, and efficiency for customers (Zhang et al., 2021).

NANOTECHNOLOGY-BASED APPROACHES FOR PRESERVING FRUITS AND VEGETABLES

Nanotechnology offers the potential to enhance agricultural productivity in terms of quality and quantity. Every aspect of the food industry like food processing, food packaging, food monitoring, and the creation of functional foods may benefit from the use of nanotechnology (Saleh & Yousaf, 2018). This technology can change the color, flavor, or nutritional value of fruits and vegetables according to a person's dietary requirements in addition to the manufacturing of more potent colorants, flavorings, and nutritional food-related ingredients (Sagar et al., 2022). Additionally, this technology could extend shelf life and reduce the price of food additive components. Utilizing nanoagroparticles could also aid in reducing the negative environmental effects of agriculture. Commonly used nanoagroparticles include nano pesticides and nano fertilizers (Raj et al., 2022). The most common approaches include the use of nanoparticles and nanosensors as described below:

Role of Nanoparticles

Green nanotechnology, which synthesizes nanoparticles (NPs) through the utilization of plants and microorganisms such as fungi, algae, and bacteria, is more used due to its simplified manufacturing process and enhanced environmental sustainability (Salem & Fouda, 2021). NPs are composed of materials that are organic, inorganic, polymeric, or hybrid and their nanoscale sizes range from 1 to 100 nm (Saleh, 2020).

Metallic nanoparticles have attracted significant attention as a substitute or supplementary material due to their exceptional mechanical properties, barrier capabilities, compatibility with living organisms, and wide-ranging antimicrobial effects. Most metal-based nanoparticles, such as Ag, Au, ZnO, CuO, CuO_2, SiO_2, and TiO_2, possess potent antioxidant and antimicrobial properties (Ettadili et al., 2022). They have excellent properties like size, morphology, coating homogeneity, and more surface area to volume distribution that makes them more biologically active and stable. A key feature of nanoparticles is that the addition of a minimal quantity is required to establish enduring surface interaction with the base polymer used for fabricating coatings/films (Kadi, 2023). Because of their hydrophilic nature, they have no negative impact on the transparency and appearance of the fresh produce. Moreover, nanoparticles used for the fabrication of edible coatings/films could be derived from agro-industrial wastes. Therefore, these nanoparticle-derived coatings/films prepared using an optimized quantity of GRAS polymers and agro-waste materials, are safe for consumption (Litter & Ahmad, 2023; Sapna et al., 2024; Sarkar et al., 2022). The incorporation of nanoparticles has significantly extended the shelf life, water vapor barrier properties, sensory, and biological attributes and reduced the growth of microbes and weight loss of fruits and vegetables (Vieira et al., 2022).

For example, Silver nanoparticles (AgNPs) kill viruses, bacteria, yeast, and fungus, and are hence used for the preservation of produce. Hence, AgNPs are most commonly blended with biopolymers to improve their properties (Kraśniewska et al., 2020). Similarly, zinc oxide (ZnO) exhibits a diverse array of applications owing to its robust optical, electrical, piezoelectric, semiconducting, and chemical sensing properties (Nxumalo et al., 2022). Hence, ZnO NPs are utilized in the context of packaging fruits and vegetables due to their antimicrobial properties (Vinitha et al., 2021). Titanium dioxide particles smaller than 100 nm are commonly used as food additives and antimicrobials in food storage and packaging materials (Zhang & Rhim, 2022).

Nanosensors

Nanosensors are useful for identifying any changes in the color and gasses released as a result of deterioration. Typically, the sensors are sensitive to gases including ammonia, nitrogen oxides, sulfur dioxide, hydrogen, and hydrogen sulfide (Javaid et al., 2021). Nanosensors are novel devices that can satisfy customer demands for safety and higher-quality food due to their ability to quickly detect pathogens. A nanosensor is a device that is made up of a sensing layer and an electronic data processing component that is capable of interpreting signals from a variety of sources, including light, gas, chemicals, and moisture content in a storage space (Adam & Gopinath, 2022). Bio-analytical devices or nanosensors constitute integrated systems made up of various nanostructured and biological receptors. They have gained popularity in processing and packaging due to their integrity, speed, and affordability (Mamun et al., 2021). A nanosensor may identify pollutants at the molecular or even atomic level because of its sensitivity. The use of various nanomaterials like nanoparticles, nanorods, nanofibers, carbon nanotubes, and nanowires has improved the electrical and optical characteristics of nanosensors (Karim et al., 2020).

Nanosensors are commonly employed to detect ailments, chemical compounds, poisons, and pesticides. Most nanosensors use electronic and sensing devices to monitor variations in light, gas, heat, and chemicals transforming these changes into electrical signals (Javaid et al., 2021). Metal nanoparticles including zinc, gold, copper, silver, and platinum are often employed in sensor production. Different nanosensors are employed, such as nanocantilevers, electronic noses, nano-test strips, gas sensors, nanoparticle-based sensors, and array biosensors (Babu, 2022). The electronic nose is more efficient at detecting odors compared to human sense. It uses chemical sensors attached to data processing systems for chemical detection (John et al., 2021). Similarly, gas sensors made up of metals like palladium, platinum, and gold are used to check the characteristics of the odor. Etongue is a flavor-sensitive sensor made of a silicon chip with microbeads that reacts to several analytes. Traditionally gas chromatograph was used to measure the efflux of ethylene for measuring the metabolism of fruits and vegetables in a respiratory chamber (Zaidi et al., 2017). However, the latest iteration of the analyzer offers the ability to automatically calculate the rate of fruit production with each reading. Hence, it is a novel and promising technique for the detection of ethylene in apples, avocados, pears, and kiwis. Nanowires made of platinum and gold are employed as biosensors to enhance food analysis (Karim et al., 2020; Munawar et al., 2019).

FOOD SAFETY ORGANIZATIONS COPING WITH QUALITY AND QUANTITY OF FRUITS AND VEGETABLES

The implementation of food labeling has been proposed as a potential method to convert credibility qualities into easily identifiable features, hence facilitating the search for information and reducing uncertainty (Wünsche & Fernqvist, 2022). Numerous food entities, including the Bureau of Indian Standards (BIS) and assessment systems like Agmark, Indian Standards Institute mark (ISI), and Food Safety and Standards Authority of India (FSSAI), are responsible for upholding food quality and standards (Darbari & Nayar, 2020). Low-income people can receive financial assistance from government programs, such as the Supplemental Nutrition Assistance Program (SNAP) in the US to buy fruits and vegetables (Saxena et al., 2022).

A combination of reliable infrastructure, efficient supply chain management, and good agricultural practices is needed to reduce post-harvest losses (Fauziana et al., 2023). For these strategies to be implemented successfully, cooperation between government agencies, farmers, researchers, and the private sector is frequently necessary (Sonka et al., 2023). By putting these tactics into practice, waste can be decreased and the supply of safe, fresh fruits and vegetables can increased (Facchini et al., 2022).

Moreover, the employment of edible coatings/films governs strict safety regulations as they are directly applied over the surface of fruits and vegetables. Therefore, before their application, they need safety approval from the authorities (Sapna et al., 2024). Using materials that have already received FDA approval and are generally recognized as safe (GRAS) is a safety precaution. In addition to the FDA, the European Union (EU) and the International Organization for Standardization (ISO) are recognized regulatory bodies for the secure applications of edible coatings/films (Panchal et al., 2022). All the components must follow the four basic standards: (a) they must be as per goods manufacturing practices; (b) must not change the odor, color, texture, or aroma; (c) must not have any adverse impact on the composition of products; (d) not dangerous to human health (Priya et al., 2023; Sapna et al., 2024).

It is imperative to follow Good Manufacturing Practices (GMP) when developing novel edible coatings/films. Food producers are required to follow the regulations set forth according to Codex Alimentarius by the countries of origin and destination when creating packaging materials (Perveen et al., 2023). FDA has approved the use of paraffin wax on raw fruits and vegetables. Paraffin, carnauba, beeswax, candelilla, and polyethylene waxes are widely used in food-related applications alone or in combination (Pashova, 2023). European Directive has approved the use of shellac, pectins, carnauba wax, candelilla wax, polysorbate, lecithin, fatty acids, beeswax, and fatty acid salts for edible coatings/films (Khalid et al., 2022). The advancement of pectin-based edible coatings was facilitated by environmentally conscious packaging efforts, particularly those led by the US Environmental Protection Agency, to improve the durability and longevity of produce (Davis et al., 2023).

The coating for fresh fruit and vegetables is sold by the company flagship that has been fabricated from silk protein. Similarly, an Israeli company has also developed coatings for fruits and vegetables that enhance preservation. A 100% food-grade organic coating Shel-Life® is one of the tops that enhances the shelf life of fruits and vegetables. Since its implementation in January 2022, PolyNatural's Shel-Life® has prevented the waste of over 840 tons of fruit (Ungureanu et al., 2023). The other most commercially available fresh produce coatings in the USA are with trade names Nature seal, Fry shield, FreshSeal, Opta-Glaze, Z-Coat, Natural Shine, etc.

Similarly, the UK has coatings with the trade names Semperfresh, Nutri-save, Tal Prolong, Brilloshine, etc. (Sapna et al., 2024).

The government has taken measures to mitigate post-harvest losses via the following steps and missions- The Mission for Integrated Development of Horticulture (MIDH) of the Department of Agriculture & Farmers Welfare funds the construction, expansion, and modernization of cold storage facilities with a capacity of up to 5000 MT nationwide (Sapna et al., 2024).

In our nation, building a resilient biomanufacturing ecosystem to develop bio-based products that accelerate cutting-edge innovations is necessary. The recently approved 'BioE3' policy by the Union Cabinet will address issues like food security, human health, and farming issues by integrating advanced biotechnological processes. According to the Ministry of Science and Technology (MOST), the policy broadly focuses on sectors like biopolymers & enzymes, smart proteins & functional foods, high-value bio-based chemicals, precision biotherapeutics, carbon capture & its utilization; climate resilient agriculture; & marine & space research to address these issues. The policy will also accelerate technology development and commercialization by establishing Biomanufacturing and Bio-AI hubs and Biofoundry (dbtindia.gov.in).

The government also provides credit-linked back-ended subsidies to entrepreneurs in this component. The National Horticulture Board (NHB) runs a "Capital Investment Subsidy for Construction/Expansion/Modernization of Cold Storages and Storages for Horticulture Products". This initiative provides a credit-linked back-ended subsidy of 35% of the project's capital cost in general areas and 50% in Northeast, hilly, and scheduled areas. Cold storage and controlled atmosphere (CA) storage facilities with capacities over 5000 MT and up to 10000 MT can be built, expanded, and modernized with the subsidy. In the North East, units over 1000 MT can receive assistance (Sapna et al., 2024). The Ministry of Food Processing Industries (MOFPI) is implementing a Scheme for Integrated Cold Chain, Value addition, and Preservation Infrastructure under the Pradhan Mantri Kisan Sampada Yojana (Chopra & Mishra, n.d.). Reducing post-harvest losses of non-horticultural and horticultural produce and paying farmers fairly are the main goals. The government takes all these measures to mitigate the post-harvest losses in fruits and vegetables (Rutta, 2022).

To improve the food supply chains in developing nations, small farmers must be encouraged to band together, diversify, and increase the volume of food they produce and sell. Infrastructure, transportation, the food, and packaging industries all need investments. The public and private domains have a part to play in making this happen (Christiaensen et al., 2021).

CONCLUSIONS AND FUTURE PERSPECTIVES

Fruits and vegetables are the key components of our diet as they provide us with several benefits by providing minerals, vitamins, antioxidants, water content, dietary fibers, nutritional values, immunity, etc. Hence, they play a key role in the economy, environment, and human health. However, they face post-harvest losses that reduce the quality and quantity of fruits and vegetables. Poor harvesting, transportation, storage, and processing infrastructure in many developing nations is the main cause of fresh produce loss. Hence, reducing these post-harvest losses helps fight poverty, ensure food security, and preserve collected produce. The most common traditional preservation approaches were physical and chemical methods. Physical approaches include the regulation of temperature, humidity, irradiations, modified atmospheric packaging, controlled atmospheric storage, plastic-based packaging, and wax-based coatings. Although all these approaches preserve the fruits and vegetables and mitigate post-harvest losses, they have limitations like high cost, not easily available, toxic, non-biodegradable, and harmful to environment and humans. While, the chemical approach includes the use of chemical pesticides like fungicides, herbicides, weedicides, and plant hormones that also negatively impact the environment and health. These conventional techniques are not safe for humans and the environment.

Thus, an alternative could be using natural packaging systems like edible coatings and edible films. Formulation of edible coatings and films from biopolymers is a green solution to the above problems. These coatings/films are safe, eco-friendly, biodegradable, and non-toxic, hence safe for consumption. The major used biopolymers are polysaccharides, proteins, and lipids. They provide barrier properties against gases, water, and moisture, and hence reduce decay and enhanced the shelf life of produce. The properties of these biopolymers can be further improved by incorporating bioactive components like plant extracts, essential oils, antioxidants, and antimicrobial agents.

The nanotechnological approaches like nanosensors for monitoring and nanoparticles for preservation also play a key role in providing a holistic solution to address the multifaceted challenges associated with post-harvest management. This chapter will prove a step towards the achievement of the United Nations Sustainable Development Goals 2, 3, 7, and 12 which aim to end hunger, achieve food security, improve nutrition, and promote sustainable consumption, production, agriculture patterns, and good health and well-being. Concurrent nanoscience packaging research boosts post-harvest solutions. Nanomaterials reduce transportation, storage, and distribution losses by improving barrier qualities and adding unique functions. The combination of natural active agents and nanoscience is a proactive step toward solving present and future food preservation and transportation needs. The intersection of science, sustainability, and consumer preferences highlights the potential of these technologies to change the narrative of post-harvest losses and create a more robust and efficient global food supply chain.

Innovative uses of artificial intelligence (AI) in conjunction with post-harvest losses have the potential to pave up new possibilities for environmentally responsible farming. AI in the supply chain and output links reduces losses by predicting crop yield, detecting pests and diseases during storage and transportation, precision farming, harvest optimization, and overall risk management. Innovations for fabrication and applications of edible coatings/films for fruits and vegetables include the incorporation of smart, digital technologies, clustered regularly interspaced short palindromic repeats (CRISPR) technologies, the use of plant dyes and the use of novel solid natural residue in appropriate combination. More research should be done on the utilization of edible coatings/films and nanotechnology in the food processing sector that would serve as a novel model for the circular economy.

REFERENCES

Aayush, K., McClements, D. J., Sharma, S., Sharma, R., Singh, G. P., Sharma, K., & Oberoi, K. (2022). Innovations in the development and application of edible coatings for fresh and minimally processed Apple. *Food Control*, *141*, 109188. DOI: 10.1016/j.foodcont.2022.109188

Abdel-Karim, R., Reda, Y., & Abdel-Fattah, A. (2020). Nanostructured materials-based nanosensors. *Journal of the Electrochemical Society*, *167*(3), 37554. DOI: 10.1149/1945-7111/ab67aa

Adam, T., & Gopinath, S. C. B. (2022). Nanosensors: Recent perspectives on attainments and future promise of downstream applications. *Process Biochemistry*, *117*, 153–173. DOI: 10.1016/j.procbio.2022.03.024

Ahmad, M. S., & Siddiqui, M. W. (2015). *Postharvest quality assurance of fruits*. Springer. DOI: 10.1007/978-3-319-21197-8

Ajose, D. J., Abolarinwa, T. O., Oluwarinde, B. O., Montso, P. K., Fayemi, O. E., Aremu, A. O., & Ateba, C. N. (2022). Application of plant-derived nanoparticles (PDNP) in food-producing animals as a bio-control agent against antimicrobial-resistant pathogens. *Biomedicines*, *10*(10), 2426. DOI: 10.3390/biomedicines10102426 PMID: 36289688

Al-Alam, J. (2017). *Polluants organiques: analyse, application au {\guillemotleft} biomonitoring{\guillemotright} environnemental et introduction des biopesticides (algues marines) comme alternative*. Université de Strasbourg; Université Libanaise.

Al Mamun, M., Wahab, Y. A., Hossain, M. A. M., Hashem, A., & Johan, M. R. (2021). Electrochemical biosensors with aptamer recognition layer for the diagnosis of pathogenic bacteria: Barriers to commercialization and remediation. *Trends in Analytical Chemistry*, *145*, 116458. DOI: 10.1016/j.trac.2021.116458

Alegbeleye, O., Odeyemi, O. A., Strateva, M., & Stratev, D. (2022). Microbial spoilage of vegetables, fruits and cereals. *Applied Food Research*, *2*(1), 100122. DOI: 10.1016/j.afres.2022.100122

Amit, S. K., Uddin, M. M., Rahman, R., Islam, S. M. R., & Khan, M. S. (2017). A review on mechanisms and commercial aspects of food preservation and processing. *Agriculture \&. Food Security*, *6*, 1–22.

Aqilah, N. M. N., Rovina, K., Felicia, W. X. L., & Vonnie, J. M. (2023). A review on the potential bioactive components in fruits and vegetable wastes as value-added products in the food industry. *Molecules (Basel, Switzerland)*, *28*(6), 2631. DOI: 10.3390/molecules28062631 PMID: 36985603

Ariwaodo, C. A. (2022). Handling strategies and facilities for horticultural crops. *OAlib*, *9*(5), 1–29. DOI: 10.4236/oalib.1108577

Armghan Khalid, M., Niaz, B., Saeed, F., Afzaal, M., Islam, F., Hussain, M., Mahwish, , Muhammad Salman Khalid, H., Siddeeg, A., & Al-Farga, A. (2022). Edible coatings for enhancing safety and quality attributes of fresh produce: A comprehensive review. *International Journal of Food Properties*, *25*(1), 1817–1847. DOI: 10.1080/10942912.2022.2107005

Arora, N. K., & Mishra, I. (2022). Current scenario and future directions for sustainable development goal 2: A roadmap to zero hunger. *Environmental Sustainability*, *5*(2), 129–133. DOI: 10.1007/s42398-022-00235-8 PMID: 37521584

Arumugam, T., Sona, C. L., & Maheswari, M. U. (2021). Fruits and vegetables as Superfoods: Scope and demand. *Journal of Pharmaceutical Innovation*, *10*, 119–129.

Avramescu, S. M., Butean, C., Popa, C. V., Ortan, A., Moraru, I., & Temocico, G. (2020). Edible and functionalized films/coatings—Performances and perspectives. *Coatings*, *10*(7), 687. DOI: 10.3390/coatings10070687

Awad, S., Zhou, Y., Katsou, E., & Fan, M. (2020). Polymer matrix systems used for date palm composite reinforcement. *Date Palm Fiber Composites: Processing, Properties and Applications*, 119–159.

Awasthi, P., Joshi, D., Chataut, G., Bogati, S., & Bhandari, R. (2022). Biochemical Changes Due To Storable Packing Materials In Litchi (Litchi Chinensis Sonn) Fruit. [RFNA]. *Reviews in Food and Agriculture*, *3*(2), 80–86. DOI: 10.26480/rfna.02.2022.80.86

Babu, P. J. (2022). Nanotechnology mediated intelligent and improved food packaging. *International Nano Letters*, *12*(1), 1–14. DOI: 10.1007/s40089-021-00348-8

Bai, J., & Plotto, A. (2012). Coatings for fresh fruits and vegetables. *Edible Coatings and Films to Improved Food Quality. Second Edition. Taylor \& Francis Group, United States*, 185–242.

Basavegowda, N., & Baek, K.-H. (2021). Synergistic antioxidant and antibacterial advantages of essential oils for food packaging applications. *Biomolecules*, *11*(9), 1267. DOI: 10.3390/biom11091267 PMID: 34572479

Basu, S., & Rao, Y. V. (2020). Environmental effects and management strategies of the herbicides. *International Journal of Bio-Resource and Stress Management, 11*(6), 518–535. DOI: 10.23910/1.2020.2069d

Basumatary, I. B., Mukherjee, A., Katiyar, V., & Kumar, S. (2022). Biopolymer-based nanocomposite films and coatings: Recent advances in shelf-life improvement of fruits and vegetables. *Critical Reviews in Food Science and Nutrition, 62*(7), 1912–1935. DOI: 10.1080/10408398.2020.1848789 PMID: 33249872

Begum, A., Alam, S. N., & Jalal Uddin, M. (2017). Management of pesticides: Purposes, uses, and concerns. *Pesticide Residue in Foods: Sources, Management, and Control*, 53–86.

Behera, U. K., & France, J. (2016). Integrated farming systems and the livelihood security of small and marginal farmers in India and other developing countries. *Advances in Agronomy, 138*, 235–282. DOI: 10.1016/bs.agron.2016.04.001

Bhatnagar, P., Gururani, P., Bisht, B., Kumar, V., Kumar, N., Joshi, R., & Vlaskin, M. S. (2022). Impact of irradiation on physico-chemical and nutritional properties of fruits and vegetables: A mini review. *Heliyon, 8*(10), e10918. DOI: 10.1016/j.heliyon.2022.e10918 PMID: 36247116

Bhuvaneswari, P., Steffi, P. F., Thirumalaiyammal, B., & Mishel, P. F. (2023). 9 Potential antimicrobial uses agents of natural and their applications as bio-preservatives. *Antimicrobials in Food Science and Technology*, 154.

Bisht, A., & Singh, S. P. (2024). Postharvest Losses and Management of Horticultural Produce: A Review. *Journal of Scientific Research and Reports, 30*(3), 305–320. DOI: 10.9734/jsrr/2024/v30i31881

Bisht, B., Bhatnagar, P., Gururani, P., Kumar, V., Tomar, M. S., Sinhmar, R., Rathi, N., & Kumar, S. (2021). Food irradiation: Effect of ionizing and non-ionizing radiations on preservation of fruits and vegetables--a review. *Trends in Food Science \& Technology, 114*, 372–385.

Bizymis, A.-P., & Tzia, C. (2022). Edible films and coatings: Properties for the selection of the components, evolution through composites and nanomaterials, and safety issues. *Critical Reviews in Food Science and Nutrition, 62*(31), 8777–8792. DOI: 10.1080/10408398.2021.1934652 PMID: 34098828

Boukhatem, M. N., & Setzer, W. N. (2020). Aromatic herbs, medicinal plant-derived essential oils, and phytochemical extracts as potential therapies for coronaviruses: Future perspectives. *Plants, 9*(6), 800. DOI: 10.3390/plants9060800 PMID: 32604842

Bvenura, C., & Sivakumar, D. (2017). The role of wild fruits and vegetables in delivering a balanced and healthy diet. *Food Research International*, *99*, 15–30. DOI: 10.1016/j.foodres.2017.06.046 PMID: 28784471

Calín-Sánchez, Á., Lipan, L., Cano-Lamadrid, M., Kharaghani, A., Masztalerz, K., Carbonell-Barrachina, Á. A., & Figiel, A. (2020). Comparison of traditional and novel drying techniques and its effect on quality of fruits, vegetables and aromatic herbs. *Foods*, *9*(9), 1261.

Camargo, P. V. M., Gladness, E. K., Clech, M., Milutinovic, S., Bustamate, E. L., & Chau, T. T. H. (2022). Group H. *Youth Actions and Innovations towards Climate Change Adaptation and Mitigation to Promote Sustainable Agriculture in Their Communities.*, 109.

Castell-Perez, M. E., & Moreira, R. G. (2021). Irradiation and consumers acceptance. *Innovative Food Processing Technologies*, 122.

Cattaneo, A., Sánchez, M. V., Torero, M., & Vos, R. (2021). Reducing food loss and waste: Five challenges for policy and research. *Food Policy*, *98*, 101974. DOI: 10.1016/j.foodpol.2020.101974 PMID: 33012956

Chen, H., Wang, J., Cheng, Y., Wang, C., Liu, H., Bian, H., Pan, Y., Sun, J., & Han, W. (2019). Application of protein-based films and coatings for food packaging: A review. *Polymers*, *11*(12), 2039. DOI: 10.3390/polym11122039 PMID: 31835317

Chew, C. L., Tan, B. A., Low, J. Y. S., Mohd Hakimi, N. I. N., Kua, S. F., & Lim, C. M. (2021). Exogenous ethylene application on postharvest oil palm fruit bunches improves crude palm oil quality. *Food Science \& Nutrition, 9*(10), 5335–5343.

Chopra, M. A., & Mishra, R. (n.d.). *ROLE OF FOOD PROCESSING INDUSTRY IN FOOD AND NUTRITIONAL SECURITY IN INDIA.*

Christiaensen, L., Rutledge, Z., & Taylor, J. E. (2021). The future of work in agri-food. *Food Policy*, *99*, 101963. DOI: 10.1016/j.foodpol.2020.101963 PMID: 33071436

Cloete, L., Picot-Allain, C., Ramasawmy, B., Neetoo, H., Ramful-Baboolall, D., & Emmambux, M. N. (2023). Drivers and barriers for commercial uptake of edible coatings for fresh fruits and vegetables industry-A review. *Food Reviews International*, *39*(6), 3481–3514. DOI: 10.1080/87559129.2021.2012795

Conrad, Z., & Blackstone, N. T. (2021). Identifying the links between consumer food waste, nutrition, and environmental sustainability: A narrative review. *Nutrition Reviews*, *79*(3), 301–314. DOI: 10.1093/nutrit/nuaa035 PMID: 32585005

Darbari, A., & Nayar, H. S. (2020). *Indian spice exports and sanitary and phytosanitary measures related challenges: a study of the legal and policy framework.*

de Siqueira Oliveira, L., Eça, K. S., de Aquino, A. C., & da Silva, L. M. R. (2020). Modified and controlled atmosphere packaging. In *Fresh-cut fruits and vegetables* (pp. 151–164). Elsevier. DOI: 10.1016/B978-0-12-816184-5.00007-0

Debebe, S. (2022). Post-harvest losses of crops and its determinants in Ethiopia: Tobit model analysis. *Agriculture \&. Food Security*, *11*(1), 1–8. PMID: 35529169

Deshmukh, R. K., & Gaikwad, K. K. (2022). Natural antimicrobial and antioxidant compounds for active food packaging applications. *Biomass Conversion and Biorefinery*, ●●●, 1–22.

Díaz-Montes, E., & Castro-Muñoz, R. (2021). Edible films and coatings as food-quality preservers: An overview. *Foods*, *10*(2), 249.

Dladla, S. (2020). Effects of different packaging materials and storage conditions on the quality attributes of tomato fruits across the south african supply chain. (Doctoral dissertation, University of KwaZulu-Natal Pietermaritzburg).

Dong, J., Gruda, N., Li, X., Cai, Z., Zhang, L., & Duan, Z. (2022). Global vegetable supply towards sustainable food production and a healthy diet. *Journal of Cleaner Production*, *369*, 133212. DOI: 10.1016/j.jclepro.2022.133212

Driscoll, R. H., & Rahman, M. S. (2020). Types of Packaging Materials Used for Foods. In *Handbook of Food Preservation* (pp. 905–922). CRC Press. DOI: 10.1201/9780429091483-59

Dukare, A. S., Singh, R. K., Jangra, R. K., & Bhushan, B. (2022). Non-fungicides-based promising technologies for managing post-production penicillium induced spoilage in horticultural commodities: A comprehensive review. *Food Reviews International*, *38*(3), 227–267. DOI: 10.1080/87559129.2020.1727497

Dziadowiec, D., Matykiewicz, D., Szostak, M., & Andrzejewski, J. (2023). Overview of the cast polyolefin film extrusion technology for multi-layer packaging applications. *Materials (Basel)*, *16*(3), 1071. DOI: 10.3390/ma16031071 PMID: 36770077

Ebrahimi, A., Zabihzadeh Khajavi, M., Ahmadi, S., Mortazavian, A. M., Abdolshahi, A., Rafiee, S., & Farhoodi, M. (2022). Novel strategies to control ethylene in fruit and vegetables for extending their shelf life: A review. *International Journal of Environmental Science and Technology*, *19*(5), 4599–4610. DOI: 10.1007/s13762-021-03485-x

Elik, A., Yanik, D. K., Istanbullu, Y., Guzelsoy, N. A., Yavuz, A., & Gogus, F. (2019). Strategies to reduce post-harvest losses for fruits and vegetables. *Strategies (La Jolla, Calif.)*, *5*(3), 29–39.

Erkan, M., & Yıldırım, I. (2017). Postharvest quality and safety of fresh-cut vegetables. Minimally processed refrigerated fruits and vegetables, 271-326.

Ettadili, F. E., Aghris, S., Laghrib, F., Farahi, A., Saqrane, S., Bakasse, M., Lahrich, S., & El Mhammedi, M. A. (2022). Recent advances in the nanoparticles synthesis using plant extract: Applications and future recommendations. *Journal of Molecular Structure*, *1248*, 131538. DOI: 10.1016/j.molstruc.2021.131538

Facchini, F., Silvestri, B., Digiesi, S., & Lucchese, A. (2022). Agri-food loss and waste management: Win-win strategies for edible discarded fruits and vegetables sustainable reuse. *Innovative Food Science \& Emerging Technologies*, 103235.

Fauziana, D. R., Suwarsinah, H. K., & Prasetio, E. A. (2023). What factors impact the adoption of postharvest loss-reduction technologies in mangosteen supply chain? *Journal of Open Innovation*, *9*(3), 100102. DOI: 10.1016/j.joitmc.2023.100102

Feng, X., Li, S., Sun, Z., Yuan, H., Li, R., Yu, N., Zhang, Y., & Chen, X. (2024). The preservation effect of chitosan-hawthorn leaf extract coating on strawberries. *Journal of Food Protection*, *87*(4), 100244. DOI: 10.1016/j.jfp.2024.100244 PMID: 38378071

Ferdousi, J., Hossain, M. I., Saha, S. R., Rob, M., Pramanik, T. A. S., Islam, M. R., & Nath, D. D. (2024). POSTHARVEST PHYSIOLOGY OF FRUITS AND VEGETABLES AND THEIR MANAGEMENT TECHNOLOGY: A REVIEW. *The Journal of Animal and Plant Sciences*, *34*(2), 291–303. DOI: 10.36899/JAPS.2024.2.0717

Ferreira, A. R. V., Alves, V. D., & Coelhoso, I. M. (2016). Polysaccharide-based membranes in food packaging applications. *Membranes*, *6*(2), 22. DOI: 10.3390/membranes6020022 PMID: 27089372

Gardas, B. B., Raut, R. D., & Narkhede, B. (2017). Modeling causal factors of post-harvesting losses in vegetable and fruit supply chain: An Indian perspective. *Renewable & Sustainable Energy Reviews*, *80*, 1355–1371. DOI: 10.1016/j.rser.2017.05.259

Gardas, B. B., Raut, R. D., & Narkhede, B. (2018). Evaluating critical causal factors for post-harvest losses (PHL) in the fruit and vegetables supply chain in India using the DEMATEL approach. *Journal of Cleaner Production*, *199*, 47–61. DOI: 10.1016/j.jclepro.2018.07.153

Garza-Juárez, A., Pérez-Carrillo, E., Arredondo-Espinoza, E. U., Islas, J. F., Benítez-Chao, D. F., & Escamilla-García, E. (2023). Nutraceuticals and Their Contribution to Preventing Noncommunicable Diseases. *Foods*, *12*(17), 3262. DOI: 10.3390/foods12173262 PMID: 37685194

Gaspar, M. C., & Braga, M. E. M. (2023). Edible films and coatings based on agrifood residues: A new trend in the food packaging research. *Current Opinion in Food Science*, *50*, 101006. DOI: 10.1016/j.cofs.2023.101006

Giannakourou, M. C., & Tsironi, T. N. (2021). Application of processing and packaging hurdles for fresh-cut fruits and vegetables preservation. *Foods*, *10*(4), 830. DOI: 10.3390/foods10040830 PMID: 33920447

Gorris, L. G. M., & Peppelenbos, H. W. (2020). Modified-atmosphere packaging of produce. In *Handbook of food preservation* (pp. 349–362). CRC Press. DOI: 10.1201/9780429091483-26

Gouda, M. H. B., & Duarte-Sierra, A. (2024). An Overview of Low-Cost Approaches for the Postharvest Storage of Fruits and Vegetables for Smallholders, Retailers, and Consumers. *Horticulturae*, *10*(8), 803. DOI: 10.3390/horticulturae10080803

Grgic, I. (2024). *Unfair Trading Practices in Contract Farming-Ensuring Commercial Justice and Contract Enforcement*. University of Pittsburgh.

Gupta, P., Toksha, B., & Rahaman, M. (2022). A review on biodegradable packaging films from vegetative and food waste. *Chemical Record (New York, N.Y.)*, *22*(7), e202100326. DOI: 10.1002/tcr.202100326 PMID: 35253984

Guzdemir, O. (2024). Sustainable Food Packaging Solutions: Polysaccharide-Based Films and Coatings. In *Sustainable Engineering: Concepts and Practices* (pp. 73–95). Springer.

Hahladakis, J. N., Iacovidou, E., & Gerassimidou, S. (2024). Plastic waste in a circular economy. In *Environmental Materials and Waste* (pp. 99–134). Elsevier. DOI: 10.1016/B978-0-443-22069-2.00003-6

Himanshu, D. D. S., & Bhavani, M. (2023). *Polysaccharide, protein, lipid-based coatings and their impact on fruit crops*.

Hossain, R., Ghinangju, B. S., Biswal, S., Schandl, H., & Sahajwalla, V. (2024). Current technological options for recycling packaging waste: Challenges and opportunities in India. *Wiley Interdisciplinary Reviews. Energy and Environment*, *13*(1), e500. DOI: 10.1002/wene.500

Indiarto, R., Pratama, A. W., Sari, T. I., & Theodora, H. C. (2020). Food irradiation technology: A review of the uses and their capabilities. *International Journal of Engineering Trends and Technology, 68*(12), 91–98. DOI: 10.14445/22315381/IJETT-V68I12P216

Iñiguez-Moreno, M., Ragazzo-Sánchez, J. A., & Calderón-Santoyo, M. (2021). An extensive review of natural polymers used as coatings for postharvest shelf-life extension: Trends and challenges. *Polymers, 13*(19), 3271. DOI: 10.3390/polym13193271 PMID: 34641086

Ishangulyyev, R., Kim, S., & Lee, S. H. (2019). Understanding food loss and waste—Why are we losing and wasting food? *Foods, 8*(8), 297. DOI: 10.3390/foods8080297 PMID: 31362396

Jackson-Davis, A., White, S., Kassama, L. S., Coleman, S., Shaw, A., Mendonca, A., Cooper, B., Thomas-Popo, E., Gordon, K., & London, L. (2023). A review of regulatory standards and advances in essential oils as antimicrobials in foods. *Journal of Food Protection, 86*(2), 100025. DOI: 10.1016/j.jfp.2022.100025 PMID: 36916569

Jafarzadeh, S., Nafchi, A. M., Salehabadi, A., Oladzad-Abbasabadi, N., & Jafari, S. M. (2021). Application of bio-nanocomposite films and edible coatings for extending the shelf life of fresh fruits and vegetables. *Advances in Colloid and Interface Science, 291*, 102405. DOI: 10.1016/j.cis.2021.102405 PMID: 33819726

Jahangiri, F., Mohanty, A. K., & Misra, M. (2024). Sustainable biodegradable coatings for food packaging: Challenges and opportunities. *Green Chemistry, 26*(9), 4934–4974. DOI: 10.1039/D3GC02647G

Jain, S., Saxena, S., Minz, V., Behera, S. D., Harini, K., Mishra, S., & Nidhi, N. (2023). Post Harvest Handling of Fruit Crops. *International Journal of Environment and Climate Change, 13*(11), 1990–1999. DOI: 10.9734/ijecc/2023/v13i113357

Jarman, A., Thompson, J., McGuire, E., Reid, M., Rubsam, S., Becker, K., & Mitcham, E. (2023). Postharvest technologies for small-scale farmers in low-and middle-income countries: A call to action. *Postharvest Biology and Technology, 206*, 112491. DOI: 10.1016/j.postharvbio.2023.112491

Javaid, M., Haleem, A., Singh, R. P., Rab, S., & Suman, R. (2021). Exploring the potential of nanosensors: A brief overview. *Sensors International, 2*, 100130. DOI: 10.1016/j.sintl.2021.100130

Jha, S., Akula, B., Enyioma, H., Novak, M., Amin, V., & Liang, H. (2024). Biodegradable Biobased Polymers: A Review of the State of the Art, Challenges, and Future Directions. *Polymers*, *16*(16), 2262. DOI: 10.3390/polym16162262 PMID: 39204482

John, A., Isaac, W.-A., Daley, O., & Kahramano\uglu, \.Ibrahim. (2024). Biological and Environmental Factors Affecting Postharvest Quality of Fruits and Vegetables. In *Postharvest Physiology and Handling of Horticultural Crops* (pp. 36–64). CRC Press.

John, A. T., Murugappan, K., Nisbet, D. R., & Tricoli, A. (2021). An outlook of recent advances in chemiresistive sensor-based electronic nose systems for food quality and environmental monitoring. *Sensors (Basel)*, *21*(7), 2271. DOI: 10.3390/s21072271 PMID: 33804960

Jose, A., Pareek, S., & Radhakrishnan, E. K. (2020). Advances in edible fruit coating materials. *Advances in Agri-Food Biotechnology*, 391–408.

Joshi, A., Gupta, A. K., Semwal, S., Deoli, N., Rather, M. A., Naik, B., Kumar, V., Jha, A. K., Rustagi, S., & Preet, M. S. (2024). Sustainable solutions for food security: Evaluating pre-treatment technologies in the growing fruits and vegetables industry of India. *Sustainable Chemistry and Pharmacy*, *39*, 101580. DOI: 10.1016/j.scp.2024.101580

Joy, G.-F. F., Sunday, O. Z., Remilekun, A., & Olusegun, A. D. (2022). Waste-to-wealth; nutritional potential of five selected fruit peels and their health benefits: A review. *African Journal of Food Science*, *16*(7), 172–183. DOI: 10.5897/AJFS2021.2138

Junaid, M. D., & Gokce, A. F. (2024). Global agricultural losses and their causes. *Bulletin of Biological and Allied Sciences Research*, *2024*(1), 66. DOI: 10.54112/bbasr.v2024i1.66

Kadi, R. H. (2023). Development of zinc oxide nanoparticles as safe coating for the shelf life extension of grapes (Vitisvinifera L., Red Globe) fruits. *Materials Express*, *13*(1), 182–188. DOI: 10.1166/mex.2023.2334

Kathpalia, A., Arora, S., & Sharma, J. G. (2022). Post-Harvest Pest Management of Perishables. *RE:view*.

Khan, O. A., Zaidi, S., Islam, R. U., Naseem, S., & Junaid, P. M. (2023). Enhanced shelf-life of peach fruit in alginate based edible coating loaded with TiO2 nanoparticles. *Progress in Organic Coatings*, *182*, 107688. DOI: 10.1016/j.porgcoat.2023.107688

Kirci, M., Isaksson, O., & Seifert, R. (2022). Managing perishability in the fruit and vegetable supply chains. *Sustainability (Basel), 14*(9), 5378. DOI: 10.3390/su14095378

Kocira, A., Kozłowicz, K., Panasiewicz, K., Staniak, M., Szpunar-Krok, E., & Hortyńska, P. (2021). Polysaccharides as edible films and coatings: Characteristics and influence on fruit and vegetable quality—A review. *Agronomy (Basel), 11*(5), 813. DOI: 10.3390/agronomy11050813

Kraśniewska, K., Galus, S., & Gniewosz, M. (2020). Biopolymers-based materials containing silver nanoparticles as active packaging for food applications—A review. *International Journal of Molecular Sciences, 21*(3), 698. DOI: 10.3390/ijms21030698 PMID: 31973105

Kumar, N., Devgan, K., Kumar, A., Kaur, P., & Mahajan, P. (2024). Active and Passive Modified Atmosphere Packaging: Recent Advances. *Nonthermal Food Engineering Operations*, 225–259.

Kumar, R. (2016). Health effects of morpholine based coating for fruits and vegetables. *International Journal of Medical Research & Health Sciences, 5*(9), 32–38.

Kumar, R., Chandra, S., Singh, B., Kumar, R., & Kumar, A. A. (2018). Zero energy cool chamber for food commodities: Need of eco-friendly storage facility for farmers: A review. *Journal of Pharmacognosy and Phytochemistry, 7*(5), 2293–2301.

Kumar, R., & Kapur, S. (2016). Morpholine: A glazing agent for fruits and vegetables coating/waxing. *Int. J. Sci. Technol. Eng, 2*(11), 694–697.

Kumari, M., Kumar, M., Solankey, S. S., & Tomar, S. (2021). Effect of Green House Gases on Vegetable Production. *Advances in Research on Vegetable Production Under a Changing Climate, 1*, 211–219. DOI: 10.1007/978-3-030-63497-1_11

Lambert, S. (2013). *Environmental risk of polymer and their degradation products*. University of York.

Larramendy, M. L., & Soloneski, S. (2022). *Pesticides: Updates on Toxicity, Efficacy and Risk Assessment*. BoD--Books on Demand.

Lee, D., Shayan, M., Gwon, J., Picha, D. H., & Wu, Q. (2022). Effectiveness of cellulose and chitosan nanomaterial coatings with essential oil on postharvest strawberry quality. *Carbohydrate Polymers, 298*, 120101. DOI: 10.1016/j.carbpol.2022.120101 PMID: 36241324

Li, L., Sun, J., Gao, H., Shen, Y., Li, C., Yi, P., He, X., Ling, D., Sheng, J., Li, J., Liu, G., Zheng, F., Xin, M., Li, Z., & Tang, Y. (2017). Effects of polysaccharide-based edible coatings on quality and antioxidant enzyme system of strawberry during cold storage. *International Journal of Polymer Science, 2017*, 1–8. DOI: 10.1155/2017/9746174

Li, Y., Erhunmwunsee, F., Liu, M., Yang, K., Zheng, W., & Tian, J. (2022). Antimicrobial mechanisms of spice essential oils and application in food industry. *Food Chemistry, 382*, 132312. DOI: 10.1016/j.foodchem.2022.132312 PMID: 35158267

Liguori, G., Greco, G., Salsi, G., Garofalo, G., Gaglio, R., Barbera, M., Greco, C., Orlando, S., Fascella, G., & Mammano, M. M. (2024). Effect of the gellan-based edible coating enriched with oregano essential oil on the preservation of the 'Tardivo di Ciaculli' mandarin (Citrus reticulata Blanco cv. Tardivo di Ciaculli). *Frontiers in Sustainable Food Systems, 8*, 1334030. DOI: 10.3389/fsufs.2024.1334030

Litter, M. I., & Ahmad, A. (2023). The world of nanotechnology. In *Industrial applications of nanoparticles* (pp. 1–15). CRC Press. DOI: 10.1201/9781003183525-1

Maftoon Azad, N. (2006). *Evaluation of edible films and coatings for extending the postharvest shelf life of avocado.*

Mangaraj, S., & Goswami, T. K. (2009). Modified atmosphere packaging of fruits and vegetables for extending shelf-life-A review. *Fresh Produce, 3*(1), 1–31.

Manzoor, S., Fayaz, U., Dar, A. H., Dash, K. K., Shams, R., Bashir, I., Pandey, V. K., & Abdi, G. (2024). Sustainable Development Goals Through Reducing Food Loss and Food Waste: A Comprehensive Review. *Future Foods : a Dedicated Journal for Sustainability in Food Science, 9*, 100362. DOI: 10.1016/j.fufo.2024.100362

Martins, V. F. R., Pintado, M. E., Morais, R. M. S. C., & Morais, A. M. M. B. (2024). Recent Highlights in Sustainable Bio-Based Edible Films and Coatings for Fruit and Vegetable Applications. *Foods, 13*(2), 318. DOI: 10.3390/foods13020318 PMID: 38275685

Matloob, A., Ayub, H., Mohsin, M., Ambreen, S., Khan, F. A., Oranab, S., Rahim, M. A., Khalid, W., Nayik, G. A., Ramniwas, S., & Ercisli, S. (2023). A review on edible coatings and films: Advances, composition, production methods, and safety concerns. *ACS Omega, 8*(32), 28932–28944. DOI: 10.1021/acsomega.3c03459 PMID: 37599927

Mayanja, I. K., & Oluk, I. (2023). Intermediate Technologies: The Key to Eradicating Post-harvest Losses (PHLs) in Sub-Saharan Africa (SSA). *Journal of Advances in Food Science \& Technology, 10*(4), 64–74.

Merino, D., Quilez-Molina, A. I., Perotto, G., Bassani, A., Spigno, G., & Athanassiou, A. (2022). A second life for fruit and vegetable waste: A review on bioplastic films and coatings for potential food protection applications. *Green Chemistry*, *24*(12), 4703–4727. DOI: 10.1039/D1GC03904K

Mesterházy, Á., Oláh, J., & Popp, J. (2020). Losses in the grain supply chain: Causes and solutions. *Sustainability (Basel)*, *12*(6), 2342. DOI: 10.3390/su12062342

Mihalca, V., Kerezsi, A. D., Weber, A., Gruber-Traub, C., Schmucker, J., Vodnar, D. C., Dulf, F. V., Socaci, S. A., Fărca , A., Mure an, C. I., Suharoschi, R., & Pop, O. L. (2021). Protein-based films and coatings for food industry applications. *Polymers*, *13*(5), 769. DOI: 10.3390/polym13050769 PMID: 33801341

Miller, D. A., & McGinity, J. W. (2008). Aqueous polymeric film coating. In *Pharmaceutical Dosage Forms-Tablets* (pp. 415–454). CRC Press.

Moeini, A., Pedram, P., Fattahi, E., Cerruti, P., & Santagata, G. (2022). Edible polymers and secondary bioactive compounds for food packaging applications: Antimicrobial, mechanical, and gas barrier properties. *Polymers*, *14*(12), 2395. DOI: 10.3390/polym14122395 PMID: 35745971

Mohamed, S. A. A., El-Sakhawy, M., & El-Sakhawy, M. A.-M. (2020). Polysaccharides, protein and lipid-based natural edible films in food packaging: A review. *Carbohydrate Polymers*, *238*, 116178. DOI: 10.1016/j.carbpol.2020.116178 PMID: 32299560

Mohan, A., Krishnan, R., Arshinder, K., Vandore, J., & Ramanathan, U. (2023). Management of postharvest losses and wastages in the Indian tomato supply chain—A temperature-controlled storage perspective. *Sustainability (Basel)*, *15*(2), 1331. DOI: 10.3390/su15021331

Mohd Basri, M. S., Abdul Karim Shah, N. N., Sulaiman, A., Mohamed Amin Tawakkal, I. S., Mohd Nor, M. Z., Ariffin, S. H., Abdul Ghani, N. H., & Mohd Salleh, F. S. (2021). Progress in the valorization of fruit and vegetable wastes: Active packaging, biocomposites, by-products, and innovative technologies used for bioactive compound extraction. *Polymers*, *13*(20), 3503. DOI: 10.3390/polym13203503 PMID: 34685262

Mondal, K., Bhattacharjee, S. K., Mudenur, C., Ghosh, T., Goud, V. V., & Katiyar, V. (2022). Development of antioxidant-rich edible active films and coatings incorporated with de-oiled ethanolic green algae extract: A candidate for prolonging the shelf life of fresh produce. *RSC Advances*, *12*(21), 13295–13313. DOI: 10.1039/D2RA00949H PMID: 35520137

Moradinezhad, F. (2021). Quality improvement and shelf life extension of minimally fresh-cut mango fruit using chemical preservatives. *Journal of Horticulture and Postharvest Research, 4*(Special Issue-Fresh-cut Products), 13–24.

Mouzakitis, C.-K., Sereti, V., Matsakidou, A., Kotsiou, K., Biliaderis, C. G., & Lazaridou, A. (2022). Physicochemical properties of zein-based edible films and coatings for extending wheat bread shelf life. *Food Hydrocolloids*, *132*, 107856. DOI: 10.1016/j.foodhyd.2022.107856

Mudaliar, K., Sharma, V., Agnihotri, C., Agnihotri, S., Deora, A., & Singh, B. P. (2023). Microbiological impact and control strategies to monitor postharvest losses in fruits and vegetables. In *Postharvest Management of Fresh Produce* (pp. 113–147). Elsevier. DOI: 10.1016/B978-0-323-91132-0.00003-4

Munawar, A., Ong, Y., Schirhagl, R., Tahir, M. A., Khan, W. S., & Bajwa, S. Z. (2019). Nanosensors for diagnosis with optical, electric and mechanical transducers. *RSC Advances*, *9*(12), 6793–6803. DOI: 10.1039/C8RA10144B PMID: 35518460

Muñoz-Tebar, N., Pérez-Álvarez, J. A., Fernández-López, J., & Viuda-Martos, M. (2023). Chitosan edible films and coatings with added bioactive compounds: Antibacterial and antioxidant properties and their application to food products: A review. *Polymers*, *15*(2), 396. DOI: 10.3390/polym15020396 PMID: 36679276

Nayak, P., & Solanki, H. (2021). Pesticides and Indian agriculture—A review. *Int J Res Granthaalayah*, *9*(5), 250–263. DOI: 10.29121/granthaalayah.v9.i5.2021.3930

Nayak, S. L., Sharma, R. R., & Sethi, S. (2021). Traditional/Conventional Cooling Systems. In *Cold Chain Management for the Fresh Produce Industry in the Developing World* (pp. 33–44). CRC Press. DOI: 10.1201/9781003056607-5

Ncube, L. K., Ude, A. U., Ogunmuyiwa, E. N., Zulkifli, R., & Beas, I. N. (2020). Environmental impact of food packaging materials: A review of contemporary development from conventional plastics to polylactic acid based materials. *Materials (Basel)*, *13*(21), 4994. DOI: 10.3390/ma13214994 PMID: 33171895

Nicola, S., Cocetta, G., Ferrante, A., & Ertani, A. (2022). Fresh-cut produce quality: Implications for postharvest. In *Postharvest Handling* (pp. 187–250). Elsevier. DOI: 10.1016/B978-0-12-822845-6.00007-5

Nirmal, N. P., Khanashyam, A. C., Mundanat, A. S., Shah, K., Babu, K. S., Thorakkattu, P., Al-Asmari, F., & Pandiselvam, R. (2023). Valorization of Fruit Waste for Bioactive Compounds and Their Applications in the Food Industry. *Foods*, *12*(3), 556. DOI: 10.3390/foods12030556 PMID: 36766085

Nunes, C., Silva, M., Farinha, D., Sales, H., Pontes, R., & Nunes, J. (2023). Edible Coatings and Future Trends in Active Food Packaging—Fruits' and Traditional Sausages' Shelf Life Increasing. *Foods, 12*(17), 3308. DOI: 10.3390/foods12173308 PMID: 37685240

Nur Rothy, J. N. (2023). *Effectiveness of master bag system packaging with oxygen scavengers and CO2/N2 modified atmosphere to extend bison meat shelf life.*

Nxumalo, K. A., Fawole, O. A., & Oluwafemi, O. S. (2022). Evaluating the efficacy of gum arabic-zinc oxide nanoparticles composite coating on shelf-life extension of mandarins (cv. kinnow). *Frontiers in Plant Science, 13*, 953861. DOI: 10.3389/fpls.2022.953861 PMID: 35937367

Ogedengbe, T. C., Malomo, O. J., & Akanji, N. E. (2022). Post-Harvest Losses and Reduction Techniques in Crop Production: A Review. *International Journal of Agricultural Science, Research and Technology in Extension and Education Systems, 12*(4), 225–233.

Ogunnupebi, T. A., Oluyori, A. P., Dada, A. O., Oladeji, O. S., Inyinbor, A. A., & Egharevba, G. O. (2020). Promising natural products in crop protection and food preservation: Basis, advances, and future prospects. *International Journal of Agronomy, 2020*(1), 8840046. DOI: 10.1155/2020/8840046

Oluwole, O., Ibidapo, O., Arowosola, T., Raji, F., Zandonadi, R. P., Alasqah, I., Lho, L. H., Han, H., & Raposo, A. (2023). Sustainable transformation agenda for enhanced global food and nutrition security: A narrative review. *Frontiers in Nutrition, 10*, 10. DOI: 10.3389/fnut.2023.1226538 PMID: 37599683

Ozturk, B., Aglar, E., Saracoglu, O., Karakaya, O., & Gun, S. (2022). Effects of GA3, CaCl2 and modified atmosphere packaging (MAP) applications on fruit quality of sweet cherry at cold storage. *International Journal of Fruit Science, 22*(1), 696–710. DOI: 10.1080/15538362.2022.2113597

Palumbo, M., Attolico, G., Capozzi, V., Cozzolino, R., Corvino, A., de Chiara, M. L. V., Pace, B., Pelosi, S., Ricci, I., Romaniello, R., & Cefola, M. (2022). Emerging postharvest technologies to enhance the shelf-life of fruit and vegetables: An overview. *Foods, 11*(23), 3925. DOI: 10.3390/foods11233925 PMID: 36496732

Panchal, N., Das, K., Prabhakar, P. K., & Ghanghas, N. (2022). Edible films and coatings for fruits and vegetables: Composition, functions, and regulatory aspects. *Edible Food Packaging: Applications, Innovations and Sustainability*, 191–216.

Pascall, M. A. (2020). The role and importance of packaging and labeling in assuring food safety, quality \& compliance with regulations I: Packaging basics. In *Food safety and quality systems in developing countries* (pp. 261–283). Elsevier. DOI: 10.1016/B978-0-12-814272-1.00006-1

Pashova, S. (2023). Application of Plant Waxes in Edible Coatings. *Coatings, 13*(5), 911. DOI: 10.3390/coatings13050911

Peerzada Gh, J., Sinclair, B. J., Perinbarajan, G. K., Dutta, R., Shekhawat, R., Saikia, N., Chidambaram, R., & Mossa, A.-T. (2023). An overview on smart and active edible coatings: Safety and regulations. *European Food Research and Technology, 249*(8), 1–18. DOI: 10.1007/s00217-023-04273-2

Pérez-Santaescolástica, C., Munekata, P. E. S., Feng, X., Liu, Y., Bastianello Campagnol, P. C., & Lorenzo, J. M. (2022). Active edible coatings and films with Mediterranean herbs to improve food shelf-life. *Critical Reviews in Food Science and Nutrition, 62*(9), 2391–2403. DOI: 10.1080/10408398.2020.1853036 PMID: 33251820

Perez-Vazquez, A., Barciela, P., Carpena, M., & Prieto, M. A. (2023). Edible coatings as a natural packaging system to improve fruit and vegetable shelf life and quality. *Foods, 12*(19), 3570. DOI: 10.3390/foods12193570 PMID: 37835222

Perveen, S., Anwar, M. J., Ismail, T., Hameed, A., Naqvi, S. S., Mahomoodally, M. F., Saeed, F., Imran, A., Hussain, M., Imran, M., Ur Rehman, H., Khursheed, T., Tufail, T., Mehmood, T., Ali, S. W., & Al Jbawi, E. (2023). Utilization of biomaterials to develop the biodegradable food packaging. *International Journal of Food Properties, 26*(1), 1122–1139. DOI: 10.1080/10942912.2023.2200606

Piližota, V. (2023). Fruits and vegetables (including herbs). In *Food Safety Management* (pp. 235–268). Elsevier. DOI: 10.1016/B978-0-12-820013-1.00039-5

Pillai, A. R. S., Eapen, A. S., Zhang, W., & Roy, S. (2024). Polysaccharide-Based Edible Biopolymer-Based Coatings for Fruit Preservation: A Review. *Foods, 13*(10), 1529. DOI: 10.3390/foods13101529 PMID: 38790829

Priya, K., Thirunavookarasu, N., & Chidanand, D. V. (2023). Recent advances in edible coating of food products and its legislations: A review. *Journal of Agriculture and Food Research, 12*, 100623. DOI: 10.1016/j.jafr.2023.100623

Raj, R., Dhiman, S., Shakya, L., & others. (2022). *Nanotechnological interventions in sustainable agriculture.*

Ramjan, M. D., & Ansari, M. T. (2018). Factors affecting of fruits, vegetables and its quality. *Faslnamah-i Giyahan-i Daruyi, 6*(6), 16–18.

Rastogi, V. K., & Samyn, P. (2015). Bio-based coatings for paper applications. *Coatings*, *5*(4), 887–930. DOI: 10.3390/coatings5040887

Rennie, T. J., & Sunjka, P. S. (2017). Modified atmosphere for storage, transportation, and packaging. In *Novel postharvest treatments of fresh produce* (pp. 433–480). CRC Press. DOI: 10.1201/9781315370149-14

Ruppenthal, R. J. (2008). *Fresh Food from Small Spaces: The Square-inch Gardener's Guide to Year-round Growing, Fermenting, and Sprouting*. Chelsea Green Publishing.

Rutta, E. W. (2022). *Barriers Impeding the Deployment and Uptake of Solar-Powered Cold Storage Technologies for Postharvest Loss Reduction in Tomato Value Chain in Africa: Empirical Evidence from Tanzania*. Queen's University.

Sagar, N. A., Kumar, N., Choudhary, R., Bajpai, V. K., Cao, H., Shukla, S., & Pareek, S. (2022). Prospecting the role of nanotechnology in extending the shelf-life of fresh produce and in developing advanced packaging. *Food Packaging and Shelf Life*, *34*, 100955. DOI: 10.1016/j.fpsl.2022.100955

Saleh, N., & Yousaf, Z. (2018). Tools and techniques for the optimized synthesis, reproducibility and scale up of desired nanoparticles from plant derived material and their role in pharmaceutical properties. In *Nanoscale Fabrication, Optimization, Scale-Up and Biological Aspects of Pharmaceutical Nanotechnology* (pp. 85–131). Elsevier. DOI: 10.1016/B978-0-12-813629-4.00003-6

Saleh, T. A. (2020). Nanomaterials: Classification, properties, and environmental toxicities. *Environmental Technology \& Innovation*, *20*, 101067.

Salem, S. S., & Fouda, A. (2021). Green synthesis of metallic nanoparticles and their prospective biotechnological applications: An overview. *Biological Trace Element Research*, *199*(1), 344–370. DOI: 10.1007/s12011-020-02138-3 PMID: 32377944

Sarkar, M. R., Rashid, M. H., Rahman, A., Kafi, M. A., Hosen, M. I., Rahman, M. S., & Khan, M. N. (2022). Recent advances in nanomaterials based sustainable agriculture: An overview. *Environmental Nanotechnology, Monitoring \& Management*, *18*, 100687.

Saxena, A., Sharma, L., & Maity, T. (2020). Enrichment of edible coatings and films with plant extracts or essential oils for the preservation of fruits and vegetables. In *Biopolymer-based formulations* (pp. 859–880). Elsevier. DOI: 10.1016/B978-0-12-816897-4.00034-5

Saxena, P., Caldwell, J., Ramos, A., Flores, D., Shah, D., & Kuo, T. (2022). Food Security, Fruit and Vegetable Intake, and Chronic Conditions among Supplemental Nutrition Assistance Program Education Participants Attending Free Food and Produce Events. *Obesities*, *2*(3), 264–275. DOI: 10.3390/obesities2030021

Senturk Parreidt, T., Müller, K., & Schmid, M. (2018). Alginate-based edible films and coatings for food packaging applications. *Foods*, *7*(10), 170. DOI: 10.3390/foods7100170 PMID: 30336642

Shahbaz, M. U., Arshad, M., Mukhtar, K., Nabi, B. G., Goksen, G., Starowicz, M., Nawaz, A., Ahmad, I., Walayat, N., Manzoor, M. F., & Aadil, R. M. (2022). Natural plant extracts: An update about novel spraying as an alternative of chemical pesticides to extend the postharvest shelf life of fruits and vegetables. *Molecules (Basel, Switzerland)*, *27*(16), 5152. DOI: 10.3390/molecules27165152 PMID: 36014396

Sharma, A., Kumar, V., Shahzad, B., Tanveer, M., Sidhu, G. P. S., Handa, N., Kohli, S. K., Yadav, P., Bali, A. S., Parihar, R. D., Dar, O. I., Singh, K., Jasrotia, S., Bakshi, P., Ramakrishnan, M., Kumar, S., Bhardwaj, R., & Thukral, A. K. (2019). Worldwide pesticide usage and its impacts on ecosystem. *SN Applied Sciences*, *1*(11), 1–16. DOI: 10.1007/s42452-019-1485-1

Sharma, B., & Kumari, A. (2020). Role of Post-Harvest Physiology in Evolution of Transgenic Crops. In *Plant Breeding-Current and Future Views*. IntechOpen.

Sharma, C., Pathak, P., Yadav, S. P., & Gautam, S. (2024). Potential of emerging "all-natural" edible coatings to prevent post-harvest losses of vegetables and fruits for sustainable agriculture. *Progress in Organic Coatings*, *193*, 108537.

Sharma, P., Sharma, S. R., & Mittal, T. C. (2020). Effects and application of ionizing radiation on fruits and vegetables: A review. *Journal of Agricultural Engineering*, *57*(2), 97–126. DOI: 10.52151/jae2020572.1708

Sharma, P., Shehin, V. P., Kaur, N., & Vyas, P. (2019). Application of edible coatings on fresh and minimally processed vegetables: A review. *International Journal of Vegetable Science*, *25*(3), 295–314. DOI: 10.1080/19315260.2018.1510863

Sharma, R. R., Raghavendra, H. R., & Sethi, S. (2021). Postharvest Treatments for Horticultural Produce. In *Postharvest Handling and Diseases of Horticultural Produce* (pp. 43–56). CRC Press. DOI: 10.1201/9781003045502-4

Sharma, S., Nakano, K., Kumar, S., & Katiyar, V. (2024). Edible packaging to prolong postharvest shelf-life of fruits and vegetables: A review. *Food Chemistry Advances*, *4*, 100711. DOI: 10.1016/j.focha.2024.100711

Sharma, S., Sudhakara, P., Singh, J., Ilyas, R. A., Asyraf, M. R. M., & Razman, M. R. (2021). Critical review of biodegradable and bioactive polymer composites for bone tissue engineering and drug delivery applications. *Polymers, 13*(16), 2623. DOI: 10.3390/polym13162623 PMID: 34451161

Shiekh, K. A., Ngiwngam, K., & Tongdeesoontorn, W. (2021). Polysaccharide-based active coatings incorporated with bioactive compounds for reducing postharvest losses of fresh fruits. *Coatings, 12*(1), 8. DOI: 10.3390/coatings12010008

Singh, B., & Singh, S. (2018). *Advances in postharvest technologies of vegetable crops*. CRC Press. DOI: 10.1201/9781315161020

Singh, D., Sharma, R. R., & Kesharwani, A. K. (2021). Postharvest Losses of Horticultural Produce. In *Postharvest Handling and Diseases of Horticultural Produce* (pp. 1–24). CRC Press. DOI: 10.1201/9781003045502-1

Singh, M., Kaul, A., Pandey, V., & Bimbraw, A. S. (2019). Weed management in vegetable crops to reduce the yield losses. *International Journal of Current Microbiology and Applied Sciences, 8*(7), 1241–1258. DOI: 10.20546/ijcmas.2019.807.148

Singh, S., Dubey, A., Gangwar, V., Kumar, A., Kumar, A., Kumar, M., & Wamiq, M. (2023). Edible coatings for improving the storability of fresh fruits and vegetables: A review. *The Pharma Innovation, 12*(6), 3992–4002.

Singh, T. (2022). Economic growth and the state of poverty in India: Sectoral and provincial perspectives. *Economic Change and Restructuring, 55*(3), 1251–1302. DOI: 10.1007/s10644-021-09345-5

Sjah, T., & Zainuri, Z. (2020). Agricultural supply chain and food security. In *Zero Hunger* (pp. 79–88). Springer. DOI: 10.1007/978-3-319-95675-6_82

Sonka, S., Lee, H., & Shah, S. (2023). The YieldWise Approach to Post-Harvest Loss Reduction: Creating Market-Driven Supply Chains to Support Sustained Technology Adoption. *Agriculture, 13*(4), 910. DOI: 10.3390/agriculture13040910

Springmann, M., & Freund, F. (2022). Options for reforming agricultural subsidies from health, climate, and economic perspectives. *Nature Communications, 13*(1), 82. DOI: 10.1038/s41467-021-27645-2 PMID: 35013212

Suresh, B. V., Shireesha, Y., Kishore, T. S., Dwivedi, G., Haghighi, A. T., & Patro, E. R. (2023). Natural energy materials and storage systems for solar dryers: State of the art. *Solar Energy Materials and Solar Cells, 255*, 112276. DOI: 10.1016/j.solmat.2023.112276

Suryana, E. A., Mauludyani, A. V. R., Sumedi, S., Ariani, M., & Munarso, S. J. (2024). Reducing post-harvest loss and waste along the rice supply chain and consumption. *BIO Web of Conferences, 119*, 1013.

Syamaladevi, R. M., Tang, J., Villa-Rojas, R., Sablani, S., Carter, B., & Campbell, G. (2016). Influence of water activity on thermal resistance of microorganisms in low-moisture foods: A review. *Comprehensive Reviews in Food Science and Food Safety, 15*(2), 353–370. DOI: 10.1111/1541-4337.12190 PMID: 33371598

Tadesse, K. T. (2024). The Role of Post-Harvest Management in Ensuring Food Security in a Changing World: Review Article. *J Clin Res Case Stud, 2*(3), 1–14. DOI: 10.61440/JCRCS.2024.v2.41

Tapia, M. S., Alzamora, S. M., & Chirife, J. (2020). Effects of water activity (aw) on microbial stability as a hurdle in food preservation. *Water Activity in Foods: Fundamentals and Applications*, 323–355.

Teixeira, R. F., Balbinot Filho, C. A., & Borges, C. D. (2022). Essential oils as natural antimicrobials for application in edible coatings for minimally processed apple and melon: A review on antimicrobial activity and characteristics of food models. *Food Packaging and Shelf Life, 31*, 100781. DOI: 10.1016/j.fpsl.2021.100781

Teutsch, B., & Kitinoja, L. (2019). *100 under $100: Tools for reducing postharvest losses. The Postharvest Education Foundation*. PEF.

Thyberg, K. L., & Tonjes, D. J. (2016). Drivers of food waste and their implications for sustainable policy development. *Resources, Conservation and Recycling, 106*, 110–123. DOI: 10.1016/j.resconrec.2015.11.016

Tomoda, B. T., Yassue-Cordeiro, P. H., Ernesto, J. V., Lopes, P. S., Péres, L. O., da Silva, C. F., & de Moraes, M. A. (2020). Characterization of biopolymer membranes and films: Physicochemical, mechanical, barrier, and biological properties. In *Biopolymer membranes and films* (pp. 67–95). Elsevier. DOI: 10.1016/B978-0-12-818134-8.00003-1

Tzia, C., Tasios, L., Spiliotaki, T., Chranioti, C., & Giannou, V. (2016). 16 Edible Coatings and Films to Preserve Quality of Fresh Fruits and Vegetables. *Food Preservation, 531*.

Umeohia, U. E., & Olapade, A. A. (2024). Physiological Processes Affecting Postharvest Quality of Fresh Fruits and Vegetables. *Asian Food Science Journal, 23*(4), 1–14. DOI: 10.9734/afsj/2024/v23i4706

Ungureanu, C., Tihan, G., Zgârian, R., & Pandelea, G. (2023). Bio-Coatings for Preservation of Fresh Fruits and Vegetables. *Coatings*, *13*(8), 1420. DOI: 10.3390/coatings13081420

Vicente, A. R., Manganaris, G. A., Darre, M., Ortiz, C. M., Sozzi, G. O., & Crisosto, C. H. (2022). Compositional determinants of fruit and vegetable quality and nutritional value. In *Postharvest Handling* (pp. 565–619). Elsevier. DOI: 10.1016/B978-0-12-822845-6.00019-1

Victory, I., Adetunji, C. O., Olaniyan, O. T., Dauda, W. P., Adetuyi, B. O., Adetunji, J. B., Inobeme, A., Habib, A. I., Adeyomoye, O., & Adeniyi, M. J. (2023). Zero-energy evaporative cooling technology and the prospective for fresh produce postharvest handling most especially for the preservation of fruits and vegetable. In *Engineering principles, modeling and economics of evaporative coolers* (pp. 95–102). Elsevier. DOI: 10.1016/B978-0-323-90039-3.00018-9

Vieira, I. R. S., de Carvalho, A. P. A., & Conte-Junior, C. A. (2022). Recent advances in biobased and biodegradable polymer nanocomposites, nanoparticles, and natural antioxidants for antibacterial and antioxidant food packaging applications. *Comprehensive Reviews in Food Science and Food Safety*, *21*(4), 3673–3716. DOI: 10.1111/1541-4337.12990 PMID: 35713102

Vieira, T. M., Moldão-Martins, M., & Alves, V. D. (2021). Composite coatings of chitosan and alginate emulsions with olive oil to enhance postharvest quality and shelf life of fresh figs (Ficus carica L. cv.'Pingo De Mel'). *Foods*, *10*(4), 718. DOI: 10.3390/foods10040718 PMID: 33805309

Vinitha, V., Preeyanghaa, M., Vinesh, V., Dhanalakshmi, R., Neppolian, B., & Sivamurugan, V. (2021). Two is better than one: Catalytic, sensing and optical applications of doped zinc oxide nanostructures. *Emergent Materials*, *4*(5), 1093–1124. DOI: 10.1007/s42247-021-00262-x

Vittuari, M., De Menna, F., García-Herrero, L., Pagani, M., Brenes-Peralta, L., & Segrè, A. (2019). Food systems sustainability: The complex challenge of food loss and waste. In *Sustainable Food Supply Chains* (pp. 249–260). Academic Press.

Vivek, K., & Singh, S. S., & RC, P. (2019). A review on postharvest management and advances in the minimal processing of fresh-cut fruits and vegetables. *Journal of Microbiology, Biotechnology and Food Sciences*, *8*(5).

von Braun, J., Chichaibelu, B. B., Laborde Debucquet, D., & Torero, M. (2024). *Cost of ending hunger: Consequences of complacency, and financial needs for SDG2 achievement.*

Wallace, T. C., Bailey, R. L., Blumberg, J. B., Burton-Freeman, B., Chen, C. O., Crowe-White, K. M., Drewnowski, A., Hooshmand, S., Johnson, E., Lewis, R., Murray, R., Shapses, S. A., & Wang, D. D. (2020). Fruits, vegetables, and health: A comprehensive narrative, umbrella review of the science and recommendations for enhanced public policy to improve intake. *Critical Reviews in Food Science and Nutrition*, *60*(13), 2174–2211. DOI: 10.1080/10408398.2019.1632258 PMID: 31267783

Wang, W., Wang, Y., Liu, Z., Han, Y., & Wang, C. (2019). Study on application performance of oxidized polyethylene wax in powder coatings. *Progress in Organic Coatings*, *136*, 105294. DOI: 10.1016/j.porgcoat.2019.105294

Wani, N. R., Rather, R. A., Farooq, A., Padder, S. A., Baba, T. R., Sharma, S., Mubarak, N. M., Khan, A. H., Singh, P., & Ara, S. (2024). New insights in food security and environmental sustainability through waste food management. *Environmental Science and Pollution Research International*, *31*(12), 17835–17857. DOI: 10.1007/s11356-023-26462-y PMID: 36988800

Wei, H., Seidi, F., Zhang, T., Jin, Y., & Xiao, H. (2021). Ethylene scavengers for the preservation of fruits and vegetables: A review. *Food Chemistry*, *337*, 127750. DOI: 10.1016/j.foodchem.2020.127750 PMID: 32861166

Wijewardane, N. (2022). Sustainable Postharvest Management: The Key to Attaining Food Security. *78th 78th*, *11*, 15.

Workineh, M., & Enyew, M. (2021). *Review the Extent and Cause of Post Harvest Loss of Fruits and Vegetables in Ethiopia.*

Wu, W., Ni, X., Shao, P., & Gao, H. (2021). Novel packaging film for humidity-controlled manipulating of ethylene for shelf-life extension of Agaricus bisporus. *Lebensmittel-Wissenschaft + Technologie*, *145*, 111331. DOI: 10.1016/j.lwt.2021.111331

Wu, Y., Wu, H., & Hu, L. (2023). Recent Advances of Proteins, Polysaccharides and Lipids-Based Edible Films/Coatings for Food Packaging Applications: A Review. *Food Biophysics*, •••, 1–17.

Wunderlich, S. M., & Martinez, N. M. (2018). Conserving natural resources through food loss reduction: Production and consumption stages of the food supply chain. *International Soil and Water Conservation Research*, *6*(4), 331–339. DOI: 10.1016/j.iswcr.2018.06.002

Wünsche, J. F., & Fernqvist, F. (2022). The potential of blockchain technology in the transition towards sustainable food systems. *Sustainability (Basel), 14*(13), 7739. DOI: 10.3390/su14137739

Yadav, A., Kumar, N., Upadhyay, A., Sethi, S., & Singh, A. (2022). Edible coating as postharvest management strategy for shelf-life extension of fresh tomato (Solanum lycopersicum L.): An overview. *Journal of Food Science, 87*(6), 2256–2290. DOI: 10.1111/1750-3841.16145 PMID: 35502679

Yahia, E. M., Fonseca, J. M., & Kitinoja, L. (2019). Postharvest losses and waste. *Postharvest Technology of Perishable Horticultural Commodities*, 43–69.

Yousuf, B., Qadri, O. S., & Srivastava, A. K. (2018). Recent developments in shelf-life extension of fresh-cut fruits and vegetables by application of different edible coatings: A review. *Lebensmittel-Wissenschaft + Technologie, 89*, 198–209. DOI: 10.1016/j.lwt.2017.10.051

Yousuf, B., Sun, Y., & Wu, S. (2022). Lipid and lipid-containing composite edible coatings and films. *Food Reviews International, 38*(sup1), 574–597.

Zaidi, N. A., Tahir, M. W., Vellekoop, M. J., & Lang, W. (2017). A gas chromatographic system for the detection of ethylene gas using ambient air as a carrier gas. *Sensors (Basel), 17*(10), 2283. DOI: 10.3390/s17102283 PMID: 28991173

Zainalabidin, F. A., Sagrin, M. S., Azmi, W. N. W., & Ghazali, A. S. (2019). Optimum postharvest handling-effect of temperature on quality and shelf life of tropical fruits and vegetables. [JTRSS]. *Journal of Tropical Resources and Sustainable Science, 7*(1), 23–30. DOI: 10.47253/jtrss.v7i1.505

Zaller, J. G., & Zaller, J. G. (2020). What is the problem? Pesticides in our everyday life. *Daily Poison: Pesticides-an Underestimated Danger*, 1–125.

Zhang, W. (2018). Global pesticide use: Profile, trend, cost/benefit and more. *Proceedings of the International Academy of Ecology and Environmental Sciences, 8*(1), 1.

Zhang, W., & Rhim, J.-W. (2022). Titanium dioxide (TiO2) for the manufacture of multifunctional active food packaging films. *Food Packaging and Shelf Life, 31*, 100806. DOI: 10.1016/j.fpsl.2021.100806

Zhang, X., Ismail, B. B., Cheng, H., Jin, T. Z., Qian, M., Arabi, S. A., Liu, D., & Guo, M. (2021). Emerging chitosan-essential oil films and coatings for food preservation-A review of advances and applications. *Carbohydrate Polymers, 273*, 118616. DOI: 10.1016/j.carbpol.2021.118616 PMID: 34561014

Zhang, Z., Zou, S., & Li, P. (2023). Aging of plastics in aquatic environments: Pathways, environmental behavior, ecological impacts, analyses and quantifications. *Environmental Pollution*, •••, 122926. PMID: 37963513

Zhao, P., Ndayambaje, J. P., Liu, X., & Xia, X. (2022). Microbial spoilage of fruits: A review on causes and prevention methods. *Food Reviews International, 38*(sup1), 225–246.

KEY TERMS AND DEFINITIONS

Post-harvest losses: Post-harvest losses are the losses that degrade the quality and quantity of produce after harvesting to consumption stages. Quality losses affect the nutrient content and edibility of the produce, while quantity losses are reductions in the amount of the produce.

Food waste: Any of the food lost by waste/ deterioration. The food that is appropriate for human consumption but is being discarded. This waste occurs at the retailer and consumer level. Food waste includes both food waste and food loss.

Food security: Food security is defined as having physical and financial access to enough safe, nourishing food that satisfies dietary demands and food choices for an active and healthy life for all people at all times. The main pillars of food security are food availability, food access, stability, and food utilization.

Food safety: Food safety refers to the procedures and methods used to handle, prepare, and store food in a way that prevents foodborne infections and is often known as food hygiene. The main aim of food safety is to ensure that the food we are eating is safe and free of ingredients that could be harmful to our health.

Sustainability: Sustainability meets the demands of the present without compromising the future generations' ability to meet their own needs. The main pillars of sustainability are economy, environment, and society.

Sustainable Development Goal 2: SDG 2 aims to achieve zero hunger or end hunger and is one of the 17 SDGs of the United Nations.

Sustainable Development Goal 3: SDG 2 aims for good health and well-being and is one of the 17 SDGs of the United Nations. It ensures healthy lives and promotes the well-being of all ages.

Sustainable Development Goal 12: SDG 12 aims to ensure sustainable consumption and production patterns and is one of the 17 SDGs of the United Nations. It ensures good use of resources, improves energy efficiency and sustainable infrastructure, provides access to basic services, creates green and decent jobs, and ensures a better quality of life for all.

BioE3 policy: The recently approved 'BioE3' policy by the Union Cabinet will address issues like food security, human health, and farming issues by integrating advanced biotechnological processes. The policy broadly focuses on sectors like biopolymers & enzymes, smart proteins & functional foods, high-value bio-based chemicals, precision biotherapeutics, carbon capture & its utilization; climate resilient agriculture; & marine & space research to address these issues.

Preservation: Preservation is an approach that prevents food from microbial attack or going bad after harvesting. Preservation slows down the decomposition, oxidation, and enzymatic browning processes. Thus, by preserving food, food waste can be reduced and food nutrition and security can be improved contributing towards environmental sustainability.

Nanotechnology: Nanotechnology includes a material that ranges in size from 1 to 100 nanometers (nm) in at least one dimension or nanoscale. Nanotechnology modifies atoms and molecules and can give structures new chemical and physical characteristics.

Nanosensors: Nanosensors are chemical or mechanical sensors that can be used to monitor physical parameters like temperature, heat, gases, humidity, flavor, etc. at the nanoscale or to identify the presence of chemical species and nanoparticles.

Chapter 10
Energy Efficiency in the Fruits and Vegetables Sector

Jyoti Rani
Department of Botany, Chaudhary Devi Lal University, Sirsa, India

Mamta Rani
Deaprtment of Botany, Chaudhary Devi Lal University, Sirsa, India

Suresh Kumar
Department of Botany, Chaudhary Devi Lal University, Sirsa, India

Vibhuti Gulia
https://orcid.org/0009-0005-4718-186X
Department of Botany, Chaudhary Devi Lal University, Sirsa, India

Neelam Dhanda
Department of Botany, Chaudhary Devi Lal University, Sirsa, India

ABSTRACT

Improving energy efficiency in the fruit and vegetable sector helps reduce production costs, environmental impact, and supports food security, especially during periods of high energy price volatility. The sector is energy-intensive, particularly in processing activities like drying, freezing, and canning, which cost about $800 million annually in electricity and fuel. This chapter examines energy consumption patterns and technologies that improve efficiency at different stages of fruit and vegetable production to enhance food security and minimize environmental impacts.

DOI: 10.4018/979-8-3693-8342-1.ch010

1. INTRODUCTION

As emphasized by Smith et al. (2019) energy efficiency is pivotal within the fruit as well as the vegetable industry in order to minimize costs and environmental impact. This industry employs energy-consuming operations like planting, scaling, processing, packing, transporting and storing fruits and vegetables. To this end, systems of production have developed that enhance the efficiency of resources, minimize waste, and protect the environment (Jones & Mark, 2021). Farming practices such as the use of synthetic or organic fertilizers can change the energy invested in production and post-harvest operations, these differences occur for different types of crops (*i.e.* fruits and vegetables) and also from one territory to another. For example, to cultivate vegetables greenhouses need a great deal of energy for lighting, heating and cooling (Gonzalez et al., 2017). According to Thompson et al. (2020), temperature is the factor most important to postharvest shelf life of perishable products. Ice has long been used as a method of retarding bacterial production that causes food to spoil on a consistent basis but the first widespread means was refrigeration.

Advanced technologies and alternative methods must be utilized for the further realization of energy efficiency within this industry for which cutting edge technologies are employed. The wind is not the only energy resource which has applications in the agricultural sector for RE as it is less perilous than fossil fuels and the costs of solar panels, wind, and battery technology have gone down (Lal et al, 2018). Also, efficient energy generation requires accurate farming methods, energy-saving equipment and mechanism of energy distribution (Rana et al, 2022). It is possible to discern the primary EMS functions and appreciate their evolution in order to achieve more energy savings. Shifting these methods reduces costs and enhances the resource use efficiency of agricultural activities (Ahmed et al., 2019). Also, policies and incentives encouraging energy save and renewable energy in agriculture make farmers and producers more willing to adopt such environmentally friendly methods (Kumar and Sharma, 2020). Higher standards of consumer awareness related with the benefits of responsibly grown fruits and vegetables are propelling the industry's transformation to more energy saving processes. At the very basic, the food market seeks to supply the food that we consume but this aspect also contributes to sustainable development globally. In response to increasing consumer demands and legal requirements, it is becoming common for manufacturers (Williams et al., 2021) to produce less carbon and more energy.

Reduction of non-renewable energy dependency by carrying out agricultural activities using solar panels, wind turbines, and other renewable energy sources (Lal et al, 2018). The typical source variations brought about by contests in wind electricity generation that prompt a need to manage the generation and consumption of power at all times. Based on the research, moving machinery and agricultural implements consumed a lot of energy in planting, harvesting, and processing and therefore energy efficient homes ware prioritized especially with the current trends in energy usage (Rana et al., 2022). The environmental aspects that would come along with the energy generation activities, call for the energy efficiency to be integrated into all operations to enable Sustainable Production. Efficient and economical agriculture employing data techniques involves utilizing the maximum of water, fertilizer and pesticides and minimizing the energy used (Ahmed et al 2019). There's need to rethink farm production and farming movements, simple mechanization irrespective of irrigation or presence of good nurtured soil seems less important. The development of state-of-the-art energy-efficient refrigeration and storage techniques to maintain the quality of the affected product (Thompson et al., 2020). Further expanding the literature in regard to optimal ESS size for more efficient merging with such green energy resources.

Improving the management of the supply chain and logistics to reduce energy expenditure in moving goods from the farm land to the markets (Jones & Mark, 2021). Evaluating energy consumption over set durations and identifying wastages as well as developing countermeasure plans (Ahmed et al., 2019). Adoption of such practices of agriculture that is energy-efficient, elastic and conducive to the environment through programs and cash support from the government (Kumar & Sharma, 2020). By the application of relevant current technologies, optimization of processes, and utilization of alternate energy, the industry can significantly achieve cost reduction, ecological impact minimization, and energy reduction. Such activities enhance the resilience and sustainability of the food system in light of global objectives on climate change and sustainable development.

2. PATTERNS OF ENERGY CONSUMPTION

It is important to study and investigate the energy usage habits of the fruits and vegetable sector to identify potential efficiency improvements and reduction of environmental impact. This industry involves several stages such as cultivation, harvesting, processing, packaging transporting and storage which have different energy requirements.

Figure 1. Patterns of energy consumption

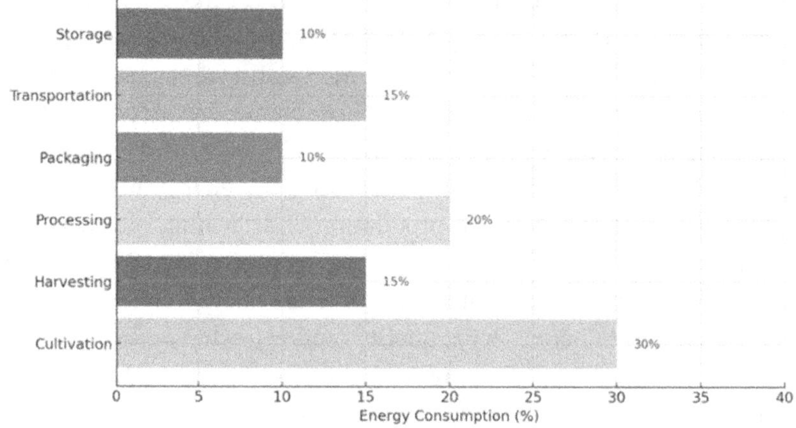

The industry can lessen its overall energy usage, operating expenses, and environmental effect by focusing on these areas for energy efficiency improvements.

1.Cultivation

The energy expenditure in tillage practices tends to differ drastically based on the crop type, methods used in production, and the region within which the farming is done. The traditional farming system is largely dependent on fuel powered appliances which function during land preparation, crop planting, crop irrigating, and pest control (Gonzalez et al., 2017). Nonetheless, in a bid to achieve optimal growing conditions, controlled environment agriculture, particularly the greenhouse facilities do however consume considerable amount of energy to provide artificial light, heat and cooling (Smith et al., 2019).

2. Harvesting

The harvesting activities may be mechanized or manual and depending on the methods applied the level of energy use will vary. Most harvesting tasks are however conducted in a mechanized fashion as this often involves cost effective power employing fuel burning devices, drastically increasing the energy footprint of the industry (Jones & Mark, 2021). On the other hand, it is true that manual harvesting would generally take more time but approximately less energy would be used during the process. Among these technologies, machine harvesting is seen as very promising in lowering harvesting cost and contributing to its efficiency.

3. Processing

Post-harvest operations also involve lots of energy when washing, sorting, grading, and packaging fruits and vegetables. Conveyance systems clean water supplies, and machines are among the tools that require energy for their daily activities (Rana et al., 2022). Severe energy requirements of some industrial processing methods include dehydration and freezing (Thompson et al., 2020). In fruits and vegetables various operations such as peeling and trimming, washing and blanching, heat processing, and filling and packaging are performed and both quantitative and qualitative losses stem from these activities.

4. Packaging

More energy is utilized during the packaging of the fruits and vegetables industry due to the use of package materials and their processes. Energy is applied in the production of packing materials such as cardboard, plastic, glass, and the actual packing process often requires the use of machinery such as automation (Williams et al., 2021). Measures have been taken to address this ecological impact by designing alternative solutions ideal for non-risk packaging.

5. Transportation

The aspects of how far away the market is, what means of transport will be used, and whether it is necessary to keep it cool and fresh will all determine the energy consumption in transit. Transportation over long distances in time adds a considerable burden on the energy and emissions footprint of the operation especially where it includes the use of trucks or airborne carriers (Jones & Mark, 2021). The strategies could be adopted in order to save on costs and time. Special attention is required when moving the harvested fruit from the farm through the grading and packing house to the market.

6. Storage

Most of the fruits and vegetables require refrigeration to prolong shelf life and retain quality. Storage can be considered one important activity which uses the most energy in the supply chain since the concerned organizations require thermal electricity which is at very low temperatures at all times (Thompson et al., 2020). Such efforts should be able to make it possible for less energy use in the future. Pests like insects, rodents and moulds cause deterioration to the food components both in the amount of food and food quality.

2.1 Energy Consumption in Production and Processing

Many energy-consuming food and vegetable industry because of the nature of processing as well as production activities (Adamu et al., 2023). This industry has three phases in terms of energy utilization which are consumed more and the less energy utilization.

2.1.1 Energy Use in Cultivation

In cultivation, energy use goes into pest management, fertilizer application, irrigation, and land preparation (Xinru et al., 2023). Chemical pesticides and fertilizers are the prime sources of chemical energy, electricity is consumed for irrigation pumps, while diesel fuel is consumed for tractors and other machinery. There are also agricultural-related factors that affect energy consumption, such as crop type, farming practices (for instance, conventional or organic farming), and regional climate (Kim et al., 2022. For instance, while Greenhouse farming allows more controlled environmental conditions making it possible to grow crops throughout the year and prevent damage to crops from adverse weather conditions it uses more energy in lighting, heating, and air conditioning than open field farming.

2.1.2 Energy Use in Harvesting and Post-Harvest Handling

Energy is needed during harvesting phase in terms of both manpower and the equipment. Harvesting before maturity, increases the chances of postharvest losses due to moulds and decay. There are energy absorbing tasks that are involved during the post-harvest of crops and these include washing, sorting, grading and packaging these crops (Majeed et al., 2023). As they increase efficiency, the new robotic harvesting machines consume more energy than the old ones (Nasso et al., 2024). Energy losses from post-harvest losses necessitate the need for thorough and proper handling and processing in terms of energy use. When this occurs, however, soaking wet grains pose with some risks. If the grains are over dried, then they become brittle and may shatter when-hullers or mills are employed, particularly due to threshing.

2.1.3 Energy Use in Processing

Energy intensive operations particularly for fruits and vegetables include a number of processes including their cleaning, peeling, chopping as well as cooking, drying and preservation (Corato & Cancellara et al, 2019). Energy input depends on the type of product and on the processing technology. Fruit dehydration e. g. through the use of hot air is also energy demanding particularly in most low-income countries where local methods are used. The grains do not only get to be rendered undesirable on the consumption surface but are also susceptible to insects such as the rice moth to mention one. The choice of the processing as well as other parameters such as the use of renewable energy or energy efficient machinery affects a large amount of energy consumption (Engler & Krarti et al., 2021).

2.2 Energy Consumption in Storage and Distribution

Another important aspect of energy use is the handling and transport as well as storage of fruits and vegetables since these goods have to be kept under controlled conditions and are also sensitive to spoilage (Liu et al., 2024).

2.2.1 Energy Use in Cold Storage

Cold storage is the most common method for bulk handling of perishables between production and marketing. Fruits and vegetables are refrigerated under sub-optimal temperatures for preservation purposes including overseas markets. The storage phase of refrigeration accounts for a calculation more significant than its fair share in the energy consuming processes associated with refrigeration. Fruits and vegetables normally have a very short shelf life after the harvest and within the range of normal harvesting temperatures. Different factors can influence energy consumption, including the refrigerant type used, the insulation quality, and the cooling systems' performance (Zhang et al., 2022). Also, the energy input is also borne by the period of storage plus the specific temperature regimes of particular items. For example, in the case of vegetables there are those that consume more energy than others; leafy vegetables such as lettuce or spinach require lower temperatures than root vegetables such as carrots (Xiaofeng et al., 2017).

2.2.2 Energy Use in Transportation and Distribution

If refrigeration is critical during transit, then distribution and transport are both processes that consume energy (Casson et al., 2024). Within a truck, the distribution of temperature (or other variable) depends on the inequality of air distribution from a fan system and the arrangement of the cargo inside the trailer. Since many perishable products are carried around using refrigerated trucks, also called reefer trucks, these tend to be very power-hungry trucks. Geographical position (GPS) of the truck is also an important piece of information. Even in such a case where some work is done remotely, the same location for a longer period of time might indicate technical failure, stopover with open doors, stopover for rest or one with the cooling installation turned off. The energy utilized at this point is shaped by the existing infrastructure or the absence of it, nearness to the market and transport means (air, sea, train or road). There has also been an increase in the energy needs as a result of the need for faster deliveries so as to prevent spoilage being that this is being done by air (Pang et al., 2024).

2.2.3 Energy Use in Retail and Consumption

The respective fruits and vegetables require energy for their refrigerated storage and display as well as for them to be stored at the shop level (Porat et al, 2018). With the rising levels of living, there is a growing need for interesting products namely, fresh, natural, health and nutritional fruit and vegetable products. Among retailing establishments, this refrigeration happens to be the most energy consuming activity as it accounts for the major share of energy expended in maintaining these products in toss such environments. The energy footprint of the industry can also be further aggravated by some behavioral aspects of consumers which cause increased energy consumption and increased food wastage such as opening refrigerated displays frequently or storing food inappropriately at the household level (Coudard et al, 2021).

3. TECHNOLOGIES USED FOR ENERGY EFFICIENCY

The energy use in the fruit and vegetable sector is substantial and therefore, it is mostly in large scale at that phase of the value chain that has to do with production includes processes of fresh cut matching, processing, and post-harvest vending storage (Joshi et al., 2024). To be able to mitigate these problems in order to minimize negative impact on the environment, lower expenses, improve energy efficiency many ways have been developed and enabled to practice (Bin Li et al., 2024). Here some technologies used for Energy efficiency:

1. Efficient Irrigation Systems

Drip irrigation and precision sprinklers are efficient systems with less energy consumption in fruit and vegetable farming (Bhavsar et al., 2023). Water resources discussion in agriculture is relevant in the context of sustainability of food systems, especially now when they are faced with the threat of water scarcity and climate variation. There is also a reduced amount of energy required to pump the water as these systems apply water at the root zone and prevent wastage of water as a result of evaporation or runoff. In order to allow for maximum efficiency and water conservation, advanced irrigation systems usually have automation controls and sensors, which for the most part increase energy efficiency further (Valero et al., 2020).

2. Renewable Energy Integration

Other energy forms such as solar and wind are steadily being adopted into farm operations (Sai et al., 2023). Solar panels are often positioned on farms to satisfy the energy needs of processing plants, greenhouses, and irrigation. Wind energy is also harnessed through turbines to generate electricity for agricultural activities. It's further possible to reap substantial financial benefits from the use of renewable energy as it reduces the dependence on fossil fuels and greenhouse gas emissions (Dumitran et al., 2024). This is because there is a huge volume of fruit and vegetables waste, they are heterogeneous and can be transformed into added value useful resources.

3. Energy-Efficient Cooling and Refrigeration

While refrigeration and cooling play an important role in prolonging the shelf life of freshly harvested fruits and vegetables, they tend to consume a lot of energy (Martinez et al. 2022). Also, some advanced technologies are a lot better than others, for instance, energy-efficient refrigerants, high-E insulation, variable-speed compressors, etc. In addition, phase change materials (PCMs) and evaporative cooling methods are also implemented to reduce energy demand while maintaining optimal storage conditions as described in literature (Xiang et al. 2023). Within the cold food supply chain system, refrigeration systems have been very helpful in the management of the cold chain due to the high degree of control required. By looking after temperature of the food, it safeguards this food for a population that is increasingly increasing.

4. Controlled Atmosphere Storage

Controlled atmosphere (CA) storage (Bodbodak et al. 2016), which moderates storage conditions by controlling the amount of carbon dioxide, oxygen, and humidity, allows to extend the shelf life of products, including fruits and vegetables. CA storage prevents the development of fruits and vegetables by reducing their activity metabolism, hence retarding putrefaction. While it is true that the use of CA storage means expenditure on energy to maintain the controlled environment, modern CA systems are built in such a way that they aim at energy management, rely on automation, and deploy advanced sensors to the core (Singh et al., 2024). Modified atmosphere packaging (MAP) inhibited litchi fruits browning under cold storage.

5. Energy-Efficient Transportation

Another domain in which the energy economy is significant is the movement of fruits and vegetables from the fields to markets and processing locations (Noor et al., 2024). Technologies like energy-efficient engines, route optimization software, and suitably insulated chilled transport units, etc., can help cut down fuel use and emissions (James et al., 2019). Additionally, hybrid cars and electric cars are growing in the acceptance rate to the farmers as an alternative transportation means which is safer for the environment. Fruits and vegetables can be characterized by perishable nature, temperature sensitivity and cover a long-distance during distribution that places higher technical requirements on their transportation and distribution.

6. Processing Technology Enhancements

Processes can be energy efficient by embracing modern technologies such as heat recovery technologies, high efficiency motors and advanced drying techniques (Niemira & Fan et al., 2014). For instance, microwave and infrared drying techniques use less energy than most of the conventional drying methods. In addition, the use of automation and process control systems ensures that processing operations are done in a manner that enhances energy efficiency (Koulouris et al., 2021). Microbial encumberment related to food deterioration would be the major reason for the food loss and the waste generated from agricultural produce. The deprivation of appropriate means of storage and preservation compounds the global food security problem to a great extent.

7. Smart Greenhouses

As a result, smart greenhouses are an advancement trend in energy conservation, as such buildings incorporate energy-efficient heating, ventilation and lighting systems (Kang et al. 2024). As for example, LED lighting requires less energy to operate when compared to conventional lighting systems but emulates the light spectrum that is ideal for plant growth. The automatic control system enhances the energy efficiency of the processes and the crops in terms of preserving optimal growth status by controlling moisture and heat depending on the data collected in real time (Ahmed et al. 2020). Automated greenhouse is a system which continuously and precisely monitors crop growth environment and automatically controls and regulates internal climate for the best crop performance by a highly sophisticated environmental control system.

8. Biomass and Biogas Utilization

Agricultural ventures can utilize energy generated from renewable forms including biomass and biogas (Chatterjee et al, 2024). Agricultural biomass such as fruit and vegetables peels and overstock can also be used to produce biogas through anaerobic processes (Bulak etal, 2023). Then biogas can be employed either as a fuel for heating buildings, generating electricity or as fuel for vehicles. Together with providing a renewable energy source, this also brings down waste generation Magama et al 2022 and Green House gas emissions. Anaerobic digestion presents a great opportunity in the recycling of food waste into biogas that is renewable and rich in methane.

9. Energy Management Systems

There is an increasing trend among companies operating in the fruit and vegetable industry to implement energy management systems (EMS) to monitor and control energy consumption (Lee et al, 2016). One of the most burning problems is restructuring systems of Agriculture so that they would face the challenges of climate changing policies. This system provides effective energy consumption tracking by X-ray exposing, in real-time to the energy consumptions of various operational units, thereby enabling corrective measures to be taken if any wastage is detected. In order to ensure that energy is fully utilized in the entire supply chain, EMS can also be used together with heads other than Venkatesh et al 2023 automated processing equipment and other domestic renewable energy solutions.

3.1 Advance Techniques in Cultivation Techniques

Changes in production techniques leading to increased yields, better quality and improved sustainability were key factors in the evolution of the fruit and vegetable sector (Brito et al., 2024). These countries have the capacity to apply scientific and technological progress and traditional tools of agriculture in the most effective ways in order to improve all conditions and the growth processes (Erekath et al., 2024). Several techniques used in cultivation in effective way given in table 1.

Table 1. Techniques used in cultivation in effective way

Practice	Description	Benefits	References
Precision Agriculture	Utilizes various sensors, GPS and data analysis for real-time soil moisture and crop health	Decreases application of herbicide, fertilizer and water and upsurges crop productivity.	Kumar et al., 2024
Vertical Farming	Produces crops in layers, in controlled environments such as hydroponics and aeroponics	Saves water and space. Reduces transportation cost and distances.	Rathor et al., 2024
Integrated Pest Management (IPM)	To minimize environmental pollution by using biological, cultural, physical, and chemical methods for pest control, and social risks.	Decreases usage of dangerous chemical and promotes sustainable farming practices.	Kariyanna and Sowjanya et al., 2024
GMOs and CRISPR Technology	Genetic engineering and CRISPR use for crop improvement like biotic and abiotic stress tolerance with better nutrition.	Crops that are resilient to environmental conditions and meet consumer needs.	Nayeri et al., 2023
Cover Cropping and Crop Rotation	Special crops are implanted to reduce land degradation and increase soil health. Crop rotation helps to reduce pest/disease.	Improves soil health, decreases chemical inputs, and prevents weeds.	Tsai and Lee et al., 2023
Agroforestry	Use trees and shrubs into crop fields for long-term productivity and sustainability.	Prevents soil erosion, increases productivity, and provides income from tree products.	Mankhin et al., 2023
Conservation Tillage	Minimizes soil disturbance during planting to protect soil structure and moisture.	Reduces erosion, preserves beneficial organisms.	Bhunia at al., 2021
Smart Irrigation Systems	Uses automation, sensors and weather forecasts to enhance water use.	Reduces energy and water wastage, improves productivity.	Kasera at al., 2024
Climate-Controlled Greenhouses	Manipulates temperature, light and moisture for optimal crop growth.	Enables year-round production of high-value crops in regions unsuitable for outdoor farming.	Rizwan et al., 2024

3.2 Energy-Efficient Harvesting and Post-Harvest Handling

Changing how crops are harvested and stored post-harvest, as well as how they are harvested energy efficiently, is important in the perspective of sustainable agriculture. This is possible at these stages because Farmers and processors may try to reduce their impact on the environment, save costs in running the business, and most of all maintain the safety and quality of their produce.

1. Mechanized Harvesting

The process of harvesting crops occurs with the aid of mechanical and energy conserving implements and machines. The devices in these categories work on the principle of using minimal motive power to perform majority of the tasks. Sophisticated harvesting equipment can be fitted with sensors and automation technologies to help reduce wastage and ensure that crops are harvested at the most optimal time, thereby preserving quality and reducing the need for later, more energy demanding practices such as sorting. Agricultural mechanization, as a recent innovation in agriculture, enhances productivity of agriculture through replacing human and animal power with machines.

2. Smart Sorting and Grading Systems

After the harvesting processes, uniformity is usually maintained through sorting and grading of fruits and vegetables. Systems fully utilizing machine vision and even artificial intelligence are designed to sort produce based on size and shape, color, firmness or other predefined attributes saving on energy. Apart from reducing food losses, such systems are designed to use less energy compared to conventional practices and require less manual labor.

3. Low-Energy Refrigeration and Cold Storage

Cooling of fruits and vegetables is crucial after the harvesting processes in extending their shelf life. Energy savings refrigeration systems include advanced insulating materials, variable-speed motors and high-efficiency compressors. With these systems, it is possible to achieve stable temperature control but at the functional minimum energy level. What's more, solar-operated cold building forklift units, which are useful for storage of goods after harvest, particularly in areas with erratic electricity sources are gaining wider use. Refrigerated display chillers as the last apparatus in the chain are instrumental in maintaining the quality of the perishable foods and are widely found in malls and supermarket chains.

4. Modified Atmosphere Packaging (MAP)

Employing MAP technology helps to extend the shelf life of a product by altering the composition of the packaging cash flow. Modified atmosphere packaging technology reduces oxygen and increases carbon dioxide levels to decrease the respiration rate of fruits and vegetables which thus reduces the amount of energy used for refrigeration. Energy Efficient MAP offers systems aimed at optimization of gas mixtures and packaging materials that assure food freshness with the least possible energy use.

5. Vacuum Cooling

Vacuum cooling is a fast method employed for evaporative cooling which is primarily done by causing evaporation of some moisture of the product under low pressure. Farmers would appreciate vacuum cooling, an energy- efficient post-harvest chilling design that instantaneously reasons the core temperature of harvested produce. When a produce is vacuum packed, the steam inside the vacuum chamber needs to evaporate which subsequently lowers temperature very fast. This approach employs very little energy in comparison with the standard cooling methods and is very effective on blossoms and green leafy vegetation's.

6. Energy-Efficient Transportation

High fuel expenditures are generally a common characteristic of transportation of harvested fruits and vegetables, particularly that which involves large distances. Energy efficient transportation approaches include employing vehicles with high fuel economy, efficient routing to reduce the lengths of journeys, as well as utilizing non fossil fuels like biodiesel or other less polluting duty trucks. At the same time, the energy efficiency of cold chain logistical processes, which consist of temperature-controlled transportation, is improved thanks to better insulations and more efficient cold storage refrigeration units.

7. Solar Drying

Sun drying of Vegetables and Fruits is the last stage of the postharvest operations. The use of solar dryers reduces the energy demand for drying farm products in particular fruits by solar thermal drying, when compared to convention electric and fuel ovens. Ineffective electric dryers are a thing of the past thanks to modern solar dryers which help the big and small businesses cut down drying time and construction cost because lean and efficient design and materials are employed in their construction.

8. Energy Recovery Systems

Wasted energy systems have been developed for the purposes of recapturing for other uses energy that would typically be expended in the processes of post-harvest handling. As an illustration of this, waste heat from domestic refrigeration systems can be efficiently reused for heating of air or water for other applications inside the building. Such applications reduce the total energy demand of the post-harvest processes which consequently yields lower costs as well as benefits to the environment in general. the influence of different substrates on hydrogen and biogas production and, hence, the amount of bioenergy produced.

9. Efficient Packaging Materials

Packing materials utilized for fresh produce packaging can significantly affect the amount of energy consumed at the postharvest stage. It uses low energy during manufacturing and transportation since the packaging products are compostable, recyclable, and lightweight. Additional energy savings can be achieved from improved packaging systems that protect the products but utilize less material.

10. Automation and Robotics

Post-harvest handling is becoming more and more crowded with the use of automation and robotics in order to conserve energy and make the processes more efficient. Robotic packaging arms, quality control systems based on AI technologies, and telehandler units with low precision can all perform their functions efficiently and with little energy consumption. These systems help to achieve lower postharvest energy consumption by creating less reliance on human labor and more efficient processes. There arises the concept of a robot that is ‘intelligent induced’ whereby the robot had the vision to see a defect on a product without a programmer having to control it.

3.3 Processing and Storage Technologies

The processes of processing and storing of fruits and vegetables are processes that are critical in terms of cost, energy efficiency and quality of the product. It can also be enhanced and waste, as well as greenhouse gas emissions, can be minimized by using energy saving measures in different components of the supply chain.

1. Advanced Refrigeration Systems

In the case of processing and collaboration of the fruits and vegetables, refrigeration has proven to be essential in prolonging their shelf life. Modern refrigeration includes the use of high efficiency compressors, variable speed motors auxiliary materials with updated development on conservation and refrigeration of energy. To cut down on the conventional energy consumption further, these systems can be used together with other renewable energy systems like the solar energy to support processes that require a constant temperature yet with low energy levels.

2. Cryogenic Freezing

Cryogenic freezing is techniques which are used to preserves fruits and vegetables rapidly by using extreme low temperature that usually uses either carbon dioxide or liquid nitrogen. This process conserves energy as compared to conventional freezing and freeze-drying processes while preserving the structure and quality of the fruit by minimizing the destruction of ice crystals formation. This technique is extremely important in protecting important agricultural products where quality is critical. Freezing is extensively employed as a method of conservation in the food industry due its ability to ensure a long preservation without so much deterioration of the product.

3. High-Pressure Processing (HPP)

Using extreme pressure to inactivate pathogens and enzymes in fruits and vegetables is called High-Pressure Processing (HPP), which is classified under non-thermal methods of food preservation. The method makes optimum use of energy since it extends shelf life without heat and uses energy lower than that used in thermal pasteurization or sterilization. HPP makes it possible to store food products safely preserving their taste, color, and don't compromising nutrition. Preservation technologies turn out to be quite effective in minimizing the accumulation of BAs in food since they improve both quality and safety of the product. From the available preservation technologies high pressure processing (HPP) is likely to lessen the number of BAs that would accumulate on food.

4. Modified Atmosphere Storage (MAS)

To slow down the rate of metabolic processes in fruits and vegetables, Modified Atmosphere Storage (MAS) changes the air inside storage Web. By controlling oxygen, carbon dioxide and nitrogen levels, MAS can increase the shelf life of products and reduce spoilage and the need for frequent refrigeration. To support gas-tight controlled atmosphere in the containers, energy efficient mass air systems (MAS) are designed to deliver the maximum gas mixes minimizing the energy consumption.

5. Energy-Efficient Drying Techniques

Fruits and vegetables are routinely dehydrated for storage, but methods used in external conventional drying are hight in energy consumption. Energy saving alternatives includes:

Solar Drying: This is a method in which moisture is removed from the crop using solar radiation. Modern solar dryers incorporate advanced features and designs that enhance the efficiency of the drying process and reduce the time taken for drying.

Heat Pump Drying: This type of drying is more efficient than the conventional electric or gas dryer as heat pumps are used to raise the temperature of the air and remove moisture from the air.

Microwave-Assisted Drying: This process uses both microwave energy and normal drying methods, thereby reducing both the time and energy/resources absorbed in the drying processes.

6. Vacuum Cooling

Once processed, fruits and vegetables can be rapidly cooled by vacuum cooling which is a method that does not require so much energy. This method relies on pressure in a vacuum chamber where water in the produce evaporates thereby cooling the produce. Because this technique reduces the time for cooling the products greatly reduces the amount of energy required and thus is effectively used on leafy vegetables and other products which perish easily.

7. Energy-Efficient Packaging

The packing materials should be cost effective in manufacturing, disposal, transport and use less energy in their processes. Some of the packaging advancements are:

Thin Out Sheet Structures: Cut down energy needed for ferries. Do not affect the environment negatively. This consists of the sensors that monitor the climatic conditions such as moisture, and temperature as well as freshness. It ensures the better utilization of storage thus minimizing wastage.

8. Cold Chain Optimization

Cold chain logistics are energy dependent as it involves temperature control in the supply chain. Other factors that contribute to the management of temperature during cold chain operations are:

Improved Insulation: Reduces the heat gained in transport and storage.

Synergy Appliance: Uses lower energy without compromising the temperature stability.

Route Optimization: The use of fewer fuel and time making deliveries due to well strategized chronological order of all the stops.

9. Energy Recovery Systems

Energy recovery systems recycle spent energy that is generated during the processing and use of such systems. For example, used warmth from refrigeration thermos can be applied inside the premises to heat air or even water for other purposes. This method is cost effective and reduces the amount of energy consumed.

10. Automated Processing Systems

The automation of processing lines appears to be a great opportunity for saving energy. Automated systems that work accurately and efficiently include conveyor belts, sorting devices, and packing machines resulting in reduction in physical work and energy use. These systems adjust their parameters in real-time to minimize energy consumption.

4. ECONOMIC AND ENVIRONMENTAL IMPACT

As a consequence, such measures bring substantial profit and allow protecting the environment within the fruit and vegetable production sector. Such impacts can be observed throughout the complete supply chain from production and processing to distribution and storage.

Economic Impact

1. Cost Reduction

The evident advantage of energy-efficient solutions is their ability to reduce the Operating Costs of the business. For example, as buildings replace older, less energy-efficient machinery, lighting, and refrigeration systems with more energy-efficient ones, the facilities use less gas, electric and other energy sources, making savings. Processes such as waste cutting, rationalization of work processes, and labor costs are made even more cost-effective via automation and advanced processing technology. Companies manage to increase their profit margins through reduction of energy and other operating expenses without cost increase to their customer bases.

2. Enhanced Competitiveness

Investing in energy-saving technologies enables firms to brand their products as environmentally benign, carving a market out of the emerging eco-friendly consumer base. This may be beneficial in terms of increased market growth as well as client retention. Numerous countries offer tax rebates, subsidies as well as low-interest loans for the purchase of energy-efficient technologies to businesses. These incentives may serve to enhance return on investment (ROI) while minimizing the burden of the first capital outlay. Energy efficient technologies will almost definitely imply higher capital outlay but in the long run incur lesser operational costs because of lower maintenance services as well energy bills. This may make them less responsive to volatility in energy prices or economic shifts.

3. Job Creation

New jobs resulting from the use of these technologies will include the fabrication, installation and servicing of these energy efficient equipment's. Also, the growing demand for energy efficiency services such as energy audits, efficiency assessments, and sustainability consulting could also contribute positively to employment in these areas. The increasing demand for energy-efficient technology workers will create demand for such skills and enhance workforce renewal.

Environmental Impact

1. Reduction in Greenhouse Gas Emissions

Panels made up of energy-efficient technologies reduce the amount of energy required for operation thus decreasing carbon dioxide (CO2) emissions. CO2 emissions from carbon storage facilities are lowered significantly by switching to renewable energy sources or energy efficient refrigeration technologies. The fruit and vegetable sector can also be a contributing force to help countries make progress in regional and global climate goals, such as those in the Paris Agreement, by reducing energy consumption and emissions. This reduces the emissions of harmful gases into the atmosphere caused by the industry.

2. Conservation of Natural Resources

Measured energy efficiency means less fuel and electricity are required and therefore reliance on the earth's resources such as coal, oil and gas are enlightened. Finding such guidelines of sustainable management may help carry on the wider picture for future generations. Most energy saving technology also help in reducing the use of water, which is generally a scarce resource in most parts. For example, cooling modules with energy-saving technologies quite often require less water as compared to conventional methods.

3. Waste Minimization

Food waste and waste reduction are possible to some extent with methods that increase the storage life of fruits and vegetables and use modern refrigeration and packaging methods. This decreases the amount of waste that has to be disposed of which amounts to reduction in pollution and the use of mass resources. Assisted by advanced technologies e.g. has high pressure processing, which is enhanced by the efficient use of energy intake drying of materials, the level of rejection of raw material protection can be reduced since a higher percentage of the harvested raw produce would be processed into saleable finished product.

4. Promotion of Sustainable Agriculture

Thus, by minimizing such impacts, energy-efficient technologies can improve environmental impact on the region by fostering more renewable agricultural practices. For example, with the use of geospatial precision", less fertilizer will be applied resulting in less chemical drainage into surrounding water bodies. Energy-saving technologies help to promotion of Conservation agriculture through resource use minimization and efficient management of impact to sensitive ecological zones.

5. Reduction in Pollution

Energy efficient technical methods reduce the number of pollutants in the normal environment served by employing appropriate methods. Such energies as solar and wind appeal less to pollution when producing electric energy than conventional methods. Such methods of energy efficiency in agricultural production and handling should enhance the soil health and reduce the rate of deterioration of the soils by reducing the need for chemical inputs.

5. FUTURE TRENDS AND CHALLENGES

The energy performance of the fruit and vegetable sector is likely to change significantly with the alteration in the industry trends, progress in technology and global environmental concerns. This way, these openings do also offer opportunities however there are some challenges that need to be addressed in order to be applied effectively.

Future Trends

1. Integration of Smart Technologies

With the increasing use of the internet of things (IoT) in agriculture, it has become easier to monitor and control the energy used in the entire supply chain in real-time. Automated systems can help reduce energy wastage and increase efficiency in agriculture by optimizing energy utilization on processing, refrigeration, irrigation and other applications. There is a great potential that energy management will undergo extensive changes as AI and machine learning will be highly embedded in it. Due to their capacity for modification based on energy predictions determination, predictive analytics will reduce the use of energy and improve the efficiency.

2. Renewable Energy Adoption

It is expected that systems of wind turbines and solar panels will be in wide use, particularly in resource-rich regions and countries. These technologies can offset the consumption of fossil fuels and reduce carbon emissions by providing power to buildings for manufacturing, processing or storage. Fruits and vegetables' processing waste can also be recovered in the bioenergy, which enhances the energy potential of the industry. This waste minimization approach improves cost effectiveness since energy is being generated through waste energy in addition, to using energy.

3. Advanced Cooling and Refrigeration Technologies

Instead of toxic refrigerants, the new technology uses magnetic fields which offers an environmentally and energy-saving form of cooling. Phase Change Materials (PCMs) are being developed to enhance refrigeration systems that absorb and release energy when you go from one state of the compound to another state (e.g. solid to liquid). This technique is optimal for storage units which only require minimal energy usage to maintain the cooled storage conditions.

4. Precision Agriculture

Precision agriculture is mainly concerned with the efficient utilization of water, fertilizers, and energy by digitalizing the information from several sources from sensors and drones to satellite photos. Resources are better managed hence energy used is less and less harm on environment. Indoor crops are grown with the novel method of vertical farming by making use of stacked layers of crops rather than ordinary farming practices which have been highly ineffective and wasteful in time and space. Such systems usually contain controlled environment and usually use LED lighting, which is extremely energy efficient.

5. Sustainable Packaging and Transportation

The use of eco-friendly packaging materials which are easily Disposable and biodegradable is fast gaining storm. This reduces the overall negative effects of this industry to the environment. It is expected that with the use of electric vehicles (EVs) and alternative fuels, the energy used for and emissions from the distribution of fruits and vegetables will be lower.

6. Regenerative Agriculture

More agricultural practices that are called regenerative are being adopted. They focus on the restoration of soil health and the capturing of 'greenhouse gasses' such as carbon dioxide. Such methods can enhance energy efficiency by reducing the need for fertilizers and chemicals and improving the ability of the crops to adapt to excessive heat.

Challenges

1. High Initial Costs

Energy saving technologies is beneficial in that they yield cost saving benefits in the long run however an initial burden of such investment may be too high for many small medium enterprises (SME). With regard to widespread uptake, finance options and business capital remain key challenges. This is because companies could be dissuaded from committing resources to adopting energy efficient practices because of the Great Weighted Average Cost of Capital structure (GWACC) payback period especially when operating in low margined industries.

2. Technological Adoption and Integration

There can be challenges in adopting the new technology with the existing institutions and without necessity of interaction and cooperation much modification to the staff, operations and infrastructure may take place. Resistance to change and lack of technical know-how can also delay adoption. One may find it challenging to assure that new technologies can be integrated harmoniously into systems where other such technologies are introduced as more and more such technologies are incorporated. This can lead to inefficiencies and extra costs due to the absence of uniformity in the functioning of IoT devices and the effective presentation of data.

3. Energy Supply and Infrastructure

Clean energy is one of the trending issues it remains unrealizable at times or everywhere. Consequently, enormous investments are required to improve grid reliability as well as to invest in energy storage. In some regions, the mature energy system cannot cope with the increasing spinning reserve of renewable energies and new technologies. Difficulty in redesigning the grid to include these technologies is one of the major challenges.

4. Regulatory and Policy Challenges

Even if some parts of the world are beginning to evolve energy laws and regulations in one way or another, still there is a market failure for the corporations intending to embark on any energy-efficient endeavours. Worthy innovative development may be hampered by the absence of a strong consistent policy. Firms may have to bear excessive costs so as to adhere to the new demands concerning energy, especially where those demands require drastic changes in operations or advanced equipment where necessary.

5. Consumer Awareness and Demand

Still more presentation is required over the definition of energy efficiency within the lifecycle of food supply chains due to their sustainability tendency. Making consumers aware and creating a market for energy savers is the principal means of reforming change. The two determinants that normally predict the modification of energy efficient items are consumer preferences and the premium consumers are willing to pay for such environmentally friendly solutions. A shift in consumer patterns is required for the sustained interest in energy efficiency initiatives.

6. Environmental and Climate Factors

However, it is imperative to say that in this period of global warming, it has become more critical to adopt energy saving techniques but it has also become more difficult to do so as well. Perfect storms can destruction of available energy sources for people; when storms subside, energy will be too expensive to produce. There might be resurfacing water and energy market for water and energy since certain resources will become scarce which will be an uphill task to retain and improve on the aspects of energy efficiency in the sector.

6. CONCLUSION

Energy efficiency in the fruit and vegetable sector is a prerequisite for making the sector more environmentally friendly, more secure from market volatility and more efficient overall. The sector can apply modern methods to reduce its energy use significantly during production processing, storage, and distribution by using, among other technologies, IoT, AI, renewable energy, and precision agriculture.

Although there are many advantages to going green, there's a downside as well. To support large-scale implementation, large upfront investment, technical integration issues, and legal barriers need to remain under control. In fact, such efforts, in the long term, bear positive effects in terms of energy safety, decreased energy prices, reduced greenhouse gas emissions, and better competitive stance in the green-oriented markets.

Future energy efficiency improvements will call for enduring creativity, bolstering policies, and increasing consumer education as the industry progresses. There is potential for the vegetable and fruit sector to be actively involved in the reformation of fresh supplies by adopting new technologies to assist the global process in becoming more resilient and sustainable.

REFERENCES

Adamu, H., Bello, U., Yuguda, A. U., Tafida, U. I., Jalam, A. M., Sabo, A., & Qamar, M. (2023). Production processes, techno-economic and policy challenges of bioenergy production from fruit and vegetable wastes. *Renewable & Sustainable Energy Reviews*, *186*, 113686. DOI: 10.1016/j.rser.2023.113686

Ahmed, C. B., & Rouina, B. B. (2019) Tunisia: Water Resources' Management and Sustainability of Agriculture. Encyclopedia of Environmental Health page: 170-184. https://doi.org/DOI: 10.1016/B978-0-12-409548-9.11854-8

Ahmed, H. A., Yu-Xin, T., & Chang, Y. (2020). Optimal control of environmental conditions affecting lettuce plant growth in a controlled environment with artificial lighting: A review. *South African Journal of Botany*, *130*, 75–89. DOI: 10.1016/j.sajb.2019.12.018

Ahmed, I. O., Abdelkader, A., Farrell, C., Rooney, D., & Morgan, K. (2019). Reusing, recycling and up-cycling of biomass: A review of practical and kinetic modelling approaches. *Fuel Processing Technology*, *192*, 179–202. DOI: 10.1016/j.fuproc.2019.04.026

Bhavsar, D., Limbasia, B., Mori, Y., Aglodiya, M. I., & Shah, M. (2023). A comprehensive and systematic study in smart drip and sprinkler irrigation systems. *Smart Agricultural Technology*, *5*, 100303. DOI: 10.1016/j.atech.2023.100303

Bhunia, S., Karmakar, S., Bhattacharjee, S., Roy, K., Kanthal, S., Pramanick, M., Baishya, A., & Mandal, B. (2021). Optimization of energy consumption using data envelopment analysis (DEA) in rice-wheat-green gram cropping system under conservation tillage practices. *Energy*, *236*, 121499. DOI: 10.1016/j.energy.2021.121499

Bodbodak, S., & Moshfeghifar, M. (2016). Advances in modified atmosphere packaging of fruits and vegetables. In *Eco-friendly technology for postharvest produce quality* (pp. 127–183). Academic Press.

Brito, I. P. C., & Silva, E. K. (2024). Pulsed electric field technology in vegetable and fruit juice processing: A review. *Food Research International*, *184*, 114207. DOI: 10.1016/j.foodres.2024.114207 PMID: 38609209

Bulak, P., Pietrycha, K., Kaczor, M., Pawłowska, M., & Bieganowski, A. (2023). Post-breeding waste from Zophobas morio for biogas and energy generation via anaerobic methane fermentation. *Fuel*, *351*, 128844. DOI: 10.1016/j.fuel.2023.128844

Casson, A., Ferrazzi, G., Guidetti, R., Bellettini, C., Narote, A. D., Rollini, M., Piccardo, A., Volturo, E., Cosentino, M., & Limbo, S. (2024). Wholesale fruit and vegetable market in Milan: Turning food surpluses into environmental gains. *Journal of Cleaner Production*, *462*, 142625. DOI: 10.1016/j.jclepro.2024.142625

Chatterjee, B., & Mazumder, D. (2024). Valorization of fruit and vegetable waste in a novel three-stage hybrid anaerobic digester for enhanced biogas production: Performance study and microbial community analysis. *Biochemical Engineering Journal*, *209*, 109403. DOI: 10.1016/j.bej.2024.109403

Corato, U. D., & Cancellara, F. A. (2019). Measures, technologies, and incentives for cleaning the minimally processed fruits and vegetables supply chain in the Italian food industry. *Journal of Cleaner Production*, *237*, 117735. DOI: 10.1016/j.jclepro.2019.117735

Coudard, A., Corbin, E., de Koning, J., Tukker, A., & Mogollon, J. M. (2021). Global water and energy losses from consumer avoidable food waste. *Journal of Cleaner Production*, *326*, 129342. DOI: 10.1016/j.jclepro.2021.129342

Dumitran, G. E., Vuta, L. I., Negrusa, E., & Birdici, A. (2024). Reducing greenhouse gas emissions in Romanian agriculture using renewable energy sources. *Journal of Cleaner Production*, *467*, 142918. DOI: 10.1016/j.jclepro.2024.142918

Engler, N., & Krarti, M. (2021). Review of energy efficiency in controlled environment agriculture. *Renewable & Sustainable Energy Reviews*, *141*, 110786. DOI: 10.1016/j.rser.2021.110786

Erekath, S., Seidlitz, H., Schreiner, M., & Dreyer, C. (2024). Food for future: Exploring cutting-edge technology and practices in vertical farm. *Sustainable Cities and Society*, *106*, 105357. DOI: 10.1016/j.scs.2024.105357

Gonzalez, M. A., Hernandez, L., Toro, C. R., Ruiz, H. A., Valdes, J. A., Herrera, R., & Aguilar, C. N. (2017). Rhizopus oryzae–Ancient microbial resource with importance in modern food industry. *International Journal of Food Microbiology*, *257*, 110–127. DOI: 10.1016/j.ijfoodmicro.2017.06.012 PMID: 28651077

Ho An, C., Jong Ri, H., Han, T., Kim, S., & Song Ju, U. (2022). Feasibility of winter cultivation of fruit vegetables in a solar greenhouse in temperate zone; experimental and numerical study. *Solar Energy*, *233*, 18–30. DOI: 10.1016/j.solener.2022.01.024

James, C. (2019). Food transportation and refrigeration technologies—Design and optimization. In *Sustainable Food Supply Chains* (pp. 185–199). Academic Press.

Jones, S., & Mark Gillott, T. (2021). A multi-energy system optimisation software for advanced process control using hypernetworks and a micro-service architecture. *Energy Reports*, *7*(4), 167–175. DOI: 10.1016/j.egyr.2021.08.159

Joshi, A., Gupta, A., Mansi, S., Semwal, S., Deoli, N., Rather, M. A., Naik, B., Kumar, V., Jha, A. K., Rustagi, S., & Preet, M. S. (2024). Sustainable solutions for food security: Evaluating pre-treatment technologies in the growing fruits and vegetables industry of India. *Sustainable Chemistry and Pharmacy*, *39*, 101580. DOI: 10.1016/j.scp.2024.101580

Kang, M., Lee, J. E., Sohn, S., Kim, J., Cho, S., Park, J., Kang, N., & Park, K. S. (2024). Estimation of cucumber net primary production using environmental and control information in a smart multi-span plastic greenhouse. *Computers and Electronics in Agriculture*, *219*, 108819. DOI: 10.1016/j.compag.2024.108819

Kariyanna, B., & Sowjanya, M. (2024). Unravelling the use of artificial intelligence in management of insect pests. *Smart Agricultural Technology*, *8*, 100517. DOI: 10.1016/j.atech.2024.100517

Kasera, R. K., & Acharjee, T. (2024). A Comprehensive IoT edge based smart irrigation system for tomato cultivation. *Internet of Things : Engineering Cyber Physical Human Systems*, *28*, 101356. DOI: 10.1016/j.iot.2024.101356

Koulouris, A., Misailidis, N., & Petrides, D. (2021). Applications of process and digital twin models for production simulation and scheduling in the manufacturing of food ingredients and products. *Food and Bioproducts Processing*, *126*, 317–333. DOI: 10.1016/j.fbp.2021.01.016

Kumar, V., Sharma, K. V., Kedam, N., Patel, A., Kate, T. R., & Rathnayake, U. (2024). A comprehensive review on smart and sustainable agriculture using IoT technologies. *Smart Agricultural Technology*, *8*, 100487. DOI: 10.1016/j.atech.2024.100487

Lal, R., & Yadav, S. S. (2018). Vulnerability of women to climate change in arid and semi-arid regions: The case of India and South Asia. *Journal of Arid Environments*, *149*, 4–17. DOI: 10.1016/j.jaridenv.2017.08.001

Lee, D., & Cheng, C. (2016). Energy savings by energy management systems: A review. *Renewable & Sustainable Energy Reviews*, *56*, 760–777. DOI: 10.1016/j.rser.2015.11.067

Li, B. (2024). Leading role of natural resources, eco-efficiency assessment, and energy transition in environmental sustainability: A depth of digital transformation. *Resources Policy*, *94*, 105145. DOI: 10.1016/j.resourpol.2024.105145

Liu, X., Liu, J., Chi, Y., & Yang, Y. (2024). A coordinated planning and management framework for transmission and distribution systems with novel bilateral sharing energy storage model and time-phased consumption subsidy strategy. *Journal of Energy Storage*, *95*, 112377. DOI: 10.1016/j.est.2024.112377

Magama, P., Chiyanzu, I., & Mulopo, J. (2022). A systematic review of sustainable fruit and vegetable waste recycling alternatives and possibilities for anaerobic biorefinery. *Bioresource Technology Reports*, *18*, 101031. DOI: 10.1016/j.biteb.2022.101031

Majeed, Y., Khan, M., Waseem, M., Zahid, U., Mahmood, F., Majeed, F., Sultan, M., & Raza, A. (2023). Renewable energy as an alternative source for energy management in agriculture. *Energy Reports*, *10*, 344–359. DOI: 10.1016/j.egyr.2023.06.032

Mankhin, B., & Md, A. (2023). Market attractiveness of pineapple and banana agroforestry systems of Madhupur Sal (Shorea robusta) forest: A sustainable way of generating income and conserving forests. *Journal of Agriculture and Food Research*, *11*, 100475. DOI: 10.1016/j.jafr.2022.100475

Martinez, C. I. P., & Poveda, A. C. (2022). Characterization of cooling equipment in the food industry: Case study of the Colombian meat, dairy, and fruit and vegetable sectors. *Environmental Development*, *41*, 100693. DOI: 10.1016/j.envdev.2021.100693

Nasso, M., Blasi, E., Pezzoli, F., & Cicatiello, C. (2024). Investigating Co-Innovation strategies to prevent food loss in the fruits and vegetables sector. *Journal of Cleaner Production*, *467*, 142984. DOI: 10.1016/j.jclepro.2024.142984

Nayeri, S., Dehghanian, Z., Lajayer, B. A., Thomson, A., Astatkie, T., & Price, G. W. (2023). CRISPR/Cas9-Mediated genetically edited ornamental and aromatic plants: A promising technology in phytoremediation of heavy metals. *Journal of Cleaner Production*, *428*, 139512. DOI: 10.1016/j.jclepro.2023.139512

Niemira, B. A., & Fan, X. (2014) FRUITS AND VEGETABLES | Advances in Processing Technologies to Preserve and Enhance the Safety of Fresh and Fresh-Cut Fruits and Vegetables. Encyclopedia of Food Microbiology. (Pages) 983-991. https://doi.org/DOI: 10.1016/B978-0-12-384730-0.00428-6

Noor, A., Moyle, P. M., Malik, A., Ziora, Z. M., & Pant, K. K. (2024). Transformative upcycling of fruit-vegetable waste for nutraceutical and pharmaceutical breakthroughs and circular economy evolution. *Process Safety and Environmental Protection*, *187*, 1022–1036. DOI: 10.1016/j.psep.2024.05.002

Pang, Q., Qiu, M., Zhang, L., & Chiu, Y. (2024). Congestion effects of energy and its influencing factors: China's transportation sector. *Socio-Economic Planning Sciences*, *92*, 101850. DOI: 10.1016/j.seps.2024.101850

Porat, R., Lichter, A., Terry, L. A., Harker, R., & Buzby, J. (2018). Postharvest losses of fruit and vegetables during retail and in consumers' homes: Quantifications, causes, and means of prevention. *Postharvest Biology and Technology*, *139*, 135–149. DOI: 10.1016/j.postharvbio.2017.11.019

T. Qi, J. Ji, X. Zhang, L. Liu, X. Xu, K. Ma, Y. Gao (2022) Research progress of cold chain transport technology for storage fruits and vegetables. Journal of Energy Storage 56(B): 105958. https://doi.org/DOI: 10.1016/j.est.2022.105958

Rana, A. (2022). Potential of renewable energy in water-energy-food nexus in Jordan. *Energy Nexus*, *7*, 100140. DOI: 10.1016/j.nexus.2022.100140

Rana, M., Arshad, R. N., Malek, Z., Roobab, U., Ranjha, M., Jambrak, A., Muhammad, I. Q., Khan, N., & Lorenzo, J. M. (2022). Nonthermal food processing: A step towards a circular economy to meet the sustainable development goals. *Food Chemistry: X*, *16*, 100516. DOI: 10.1016/j.fochx.2022.100516 PMID: 36439936

Rathor, A., Choudhury, S., Sharma, A., Nautiyal, P., & Shah, G. (2024). Empowering vertical farming through IoT and AI-Driven technologies: A comprehensive review. *Heliyon*, *10*(15), e34998. DOI: 10.1016/j.heliyon.2024.e34998 PMID: 39157372

Rizwan, A., Khan, A. N., Ibrahim, M., Ahmad, R., Iqbal, N., & Kim, D. H. (2024). Optimal environment control and fruits delivery tracking system using blockchain for greenhouse. *Computers and Electronics in Agriculture*, *220*, 108889. DOI: 10.1016/j.compag.2024.108889

Sai, M., Singh, S., Gupta, M., Nimdeo, Y. M., Raushan, R., Deorankar, A. V., Ananda, T. M., Rout, P., Chanotiya, C. S., Pakhale, V. D., & Nannaware, A. (2023). Solar energy: A promising renewable source for meeting energy demand in Indian agriculture applications. *Sustainable Energy Technologies and Assessments*, *55*, 102905. DOI: 10.1016/j.seta.2022.102905

Sharma, H. B., Kumar, R. V., Cheela, V. R. S., Ranjan, V. P., Jaglan, A. K., Dubey, B., Goel, S., & Bhattacharya, J. (2020). Challenges, opportunities, and innovations for effective solid waste management during and post COVID-19 pandemic. *Resources, Conservation and Recycling*, *162*, 105052. DOI: 10.1016/j.resconrec.2020.105052 PMID: 32834486

Singh, D., Tembhare, M., Dikshit, A., Dangi, M. B., & Kumar, S. (2024). Technical and operational challenges in setting up a decentralized biogas plant: Opportunities and future perspective toward sustainable nation. *Process Safety and Environmental Protection*, *185*, 392–407. DOI: 10.1016/j.psep.2024.02.079

Smith, P., & Fitton, N. (2019). The role of renewable energy in the global energy transformation. *Energy Strategy Reviews*, *24*, 38–50. DOI: 10.1016/j.esr.2019.01.006

Thompson, R. B., Gallardo, M., & Elia, A. (2020). Decision support systems and models for aiding irrigation and nutrient management of vegetable crops. *Agricultural Water Management*, *240*, 106209. DOI: 10.1016/j.agwat.2020.106209

Thompson, T. M., Young, B. R., & Baroutian, S. (2020). Pelagic Sargassum for energy and fertiliser production in the Caribbean: A case study on Barbados. *Renewable & Sustainable Energy Reviews*, *118*, 109564. DOI: 10.1016/j.rser.2019.109564

Tsai, H.-W., & Lee, Y.-C. (2023). Effects of land use change and crop rotation practices on farmland ecosystem service valuation. *Ecological Indicators*, *155*, 110998. DOI: 10.1016/j.ecolind.2023.110998

Venkatesh, J., Lee, S. Y., Back, S., Kim, T. G., Kim, G. W., Kim, J. M., Kwon, J. K., & Kang, B. C. (2023). Update on the genetic and molecular regulation of the biosynthetic pathways underlying pepper fruit color and pungency. *Current Plant Biology*, *100303*(35-36), 100303. DOI: 10.1016/j.cpb.2023.100303

Walmsley, T. G., Varbanov, P. S., Su, R., Ong, B., & Lal, N. (2018). Frontiers in process development, integration and intensification for circular life cycles and reduced emissions. *Journal of Cleaner Production*, *201*, 178–191. DOI: 10.1016/j.jclepro.2018.08.041

Williams, P. M., & Arezoo Dadrasnia, I. (2021). Sustainable nutrient recovery from animal manure: A review of current best practice technology and the potential for freeze concentration. *Journal of Cleaner Production*, *315*, 128106. DOI: 10.1016/j.jclepro.2021.128106

B. Xiang, X. Zhang (2023) Advancements in the development of field precooling of fruits and vegetables with/without phase change materials. Journal of Energy Storage 73(B): 109007. https://doi.org/DOI: 10.1016/j.est.2023.109007

Xiaofeng, X., Xuelai, Z., & Munyalo, J. M. (2017). Simulation Study on Temperature Field and Cold Plate Melting of Cold Storage Refrigerator Car. *Energy Procedia*, *142*, 3394–3400. DOI: 10.1016/j.egypro.2017.12.476

Yu, X., Pu, H., & Sun, D. W. (2023). Nano light conversion agents: Recent advances, future prospects and challenges in enhancing plant green cultivation. *Journal of Cleaner Production*, *426*, 138919. DOI: 10.1016/j.jclepro.2023.138919

Zinkernagel, J., Valero, J. F., Seresti, S. Y., & Intrigliolo, D. S. (2020). New technologies and practical approaches to improve irrigation management of open field vegetable crops. *Agricultural Water Management*, *242*, 106404. DOI: 10.1016/j.agwat.2020.106404

Chapter 11
Energy-Efficient Technologies for Enhancing Food Safety, Quality, and Security: Innovations and Impacts

Vishal Jain
https://orcid.org/0000-0003-1126-7424
Sharda University, India

Archan Mitra
https://orcid.org/0000-0002-1419-3558
NITTE University, India

ABSTRACT

The integration of energy-efficient technologies in the food sector presents a transformative approach to addressing global challenges related to food safety, quality, and security. These technologies, which include solar-powered cold storage, LED lighting, and precision agriculture, reduce energy consumption and enhance food preservation, production efficiency, and resource management. This research explores the impact of these innovations on food safety and environmental sustainability, with particular emphasis on their ability to support Sustainable Development Goals (SDGs) like Zero Hunger and Climate Action. By analyzing case studies, the study demonstrates how energy-efficient solutions can mitigate environmental degradation while improving the resilience of food systems.

DOI: 10.4018/979-8-3693-8342-1.ch011

1. INTRODUCTION

Food safety, quality, and security remain pivotal challenges that significantly influence global health outcomes, economic stability, and environmental sustainability. The United Nations Food and Agriculture Organization (FAO) has consistently emphasized the critical need for addressing these challenges as they affect billions worldwide, particularly in low and middle-income countries (FAO, 2021). The interconnections between food production systems and environmental sustainability are increasingly recognized, necessitating innovations that can both enhance food security and reduce environmental impacts. This paper explores the potential of energy-efficient technologies in revolutionizing the food industry by improving food safety, quality, and security while simultaneously addressing the urgent need for sustainable environmental practices.

The burgeoning global population, projected to reach 9.7 billion by 2050, underscores the urgency for enhanced food production systems that are not only efficient but also sustainable (United Nations, Department of Economic and Social Affairs, Population Division, 2019). The traditional food supply chain is fraught with inefficiencies—significant energy consumption and substantial waste generation—that exacerbate environmental degradation and contribute to climate change (Smith & Gregory, 2013). Thus, the development and integration of energy-efficient technologies within the food industry present a promising pathway to mitigating these issues. Such technologies not only aim to reduce the energy footprint of food production but also enhance the safety and quality of food, thereby supporting the United Nations' Sustainable Development Goals (SDGs), particularly SDG 2 (Zero Hunger) and SDG 12 (Responsible Consumption and Production).

This paper will delineate the scope of energy-efficient technologies in the food sector, ranging from advanced cooling and heating systems to innovative food preservation methods and optimized supply chain logistics that minimize energy use while maximizing food safety and quality. The objectives of this research include:

1. Identifying and describing energy-efficient technologies that are currently employed or have potential applicability in the food industry.
2. Evaluating the impacts of these technologies on food safety, quality, and security.
3. Analyzing the broader environmental, economic, and social implications of adopting such technologies in the food sector.

The significance of this research lies in its potential to contribute to a more sustainable food system that can meet current and future demands without compromising the ability of future generations to meet their own needs. By integrating the principles of energy efficiency, this research aligns with broader environmental objectives, including reducing greenhouse gas emissions, minimizing energy expenditures, and conserving natural resources. Additionally, by improving food safety and quality, energy-efficient technologies also play a critical role in public health and food security, directly impacting human well-being and productivity.

The methodology of this research will involve a comprehensive literature review and analysis of existing case studies to draw insights into the current landscape and future potential of energy-efficient technologies in the food industry. This approach will provide a solid foundation for understanding the complex interplay between technology, food safety, and environmental sustainability, setting the stage for more detailed investigations in subsequent sections of the paper.

In conclusion, the introduction sets the stage for a nuanced exploration of how energy-efficient technologies can address some of the most pressing challenges in the food sector. The subsequent sections will build on this foundation, delving deeper into specific technologies, their applications, and their impacts, thereby providing a thorough understanding of this critical area of research.

2. LITERATURE REVIEW

This literature review synthesizes existing research on the nexus of energy efficiency, food safety, and security. By examining a broad array of sources, this section outlines the current landscape of traditional and modern food preservation technologies, assesses the impact of energy efficiency on the carbon footprint of food systems, and reviews practical applications of energy-efficient technologies in enhancing food safety and quality. Additionally, this review identifies gaps in the current research, justifying the need for further study and refinement of the research questions.

2.1 Traditional and Modern Food Preservation Technologies

Food preservation technologies have evolved from traditional methods like drying, smoking, and fermenting to modern techniques such as refrigeration, pasteurization, and irradiation. Traditional methods have been noted for their low energy requirements and minimal environmental impact but often fall short in terms of efficiency and scalability (Kumar & Kalita, 2017). Modern food preservation technologies, while effective at extending the shelf life and safety of food products, typically require significant energy inputs, raising concerns about their sustainability (James & James, 2010). The development of innovative energy-efficient preservation technologies is therefore crucial. Recent studies have focused on advancements such as vacuum cooling, ultrasonic waves, and pulsed electric fields, which offer potential for reducing energy use while maintaining or improving the safety and quality of food products (Bhat, et al., 2018).

2.2 Energy Efficiency in Food Production and Distribution

The role of energy efficiency in mitigating the environmental impact of food production and distribution is critical. Food systems are responsible for a considerable portion of global energy use and greenhouse gas emissions, primarily through conventional farming practices, food processing, and logistics (Garnett, 2011). Enhancing energy efficiency in these areas can significantly reduce the carbon footprint of food systems. For example, optimized refrigeration systems and the integration of renewable energy sources in food processing and transportation can decrease energy consumption and associated emissions (Foster et al., 2014). Moreover, energy-efficient technologies such as precision agriculture, which utilizes GPS and IoT technologies, have shown promise in reducing the inputs needed for crop production, thus saving energy and reducing environmental impact (Tzounis et al., 2017).

2.3 Case Studies on Energy-Efficient Technologies

Several case studies highlight how energy-efficient technologies have been implemented to improve food safety and quality. One such example is the adoption of solar-powered cold storage units in India, which have helped reduce spoilage and improve the quality and safety of perishable food items, particularly in rural areas where electricity is unreliable (Mukherjee, 2016). Another case study from the Netherlands illustrates the use of geothermal energy in greenhouse horticulture, which has not only reduced energy consumption but also increased the yield and quality of produce (Van Os, 2019).

These case studies not only demonstrate the practical benefits and feasibility of such technologies but also illustrate the broader implications for global food security and sustainability.

2.4 Identification of Research Gaps

While existing literature provides substantial insights into the benefits and applications of energy-efficient technologies in food preservation, production, and distribution, several gaps remain. First, there is a lack of comprehensive data quantifying the specific energy savings and cost-effectiveness of various energy-efficient food technologies across different scales of operation. Additionally, more research is needed to explore the long-term impacts of these technologies on food safety and quality, particularly under varying operational and environmental conditions. Lastly, there is a need for further exploration of the socio-economic barriers to adopting these technologies, especially in low-resource settings.

The reviewed literature underscores the importance of integrating energy-efficient technologies in food systems to enhance food safety, quality, and security while reducing environmental impacts. The gaps identified through this review suggest avenues for future research, which could provide deeper insights into optimizing food systems for sustainability and health.

3. RESEARCH QUESTIONS

The research will focus on the following questions:

- How do energy-efficient technologies impact food safety and quality across different stages of the food supply chain?
- What are the innovative energy-efficient technologies currently being developed or implemented to enhance food security?
- What are the environmental, economic, and social impacts of adopting these technologies?

4. METHODOLOGY

4.1 Research Design

The methodology of this study employs a mixed-methods approach, integrating both qualitative and quantitative research to provide a comprehensive analysis of the impacts of energy-efficient technologies on food safety, quality, and security. This dual-methodology approach is particularly beneficial in addressing complex research questions where multiple variables influence the outcomes. The quantitative aspect aims to provide objective measurements and generalize results from a larger sample, while the qualitative component seeks to offer contextual insights and deeper understanding (Creswell & Creswell, 2018).

4.2 Data Collection

4.2.1 Quantitative Data Collection

Quantitative data will be collected through structured surveys and analysis of existing databases. The surveys will be designed to capture data on the prevalence, types, and perceived effectiveness of energy-efficient technologies across different sectors of the food industry. This will include variables such as cost savings, energy consumption, and changes in food waste levels. Respondents will be selected through stratified sampling to ensure representation across various food industry sectors, including production, processing, distribution, and retail.

The use of existing databases, such as those maintained by government agencies, industry associations, and international organizations, will allow for the analysis of trends and the identification of patterns in the adoption of energy-efficient technologies over time. These databases often contain comprehensive data on energy usage, food safety incidents, and compliance with food safety standards, providing a robust secondary source of information to support survey findings (Smith et al., 2020).

4.2.2 Qualitative Data Collection

Qualitative data will be collected through semi-structured interviews with a diverse range of stakeholders, including technology developers, food industry experts, and representatives from regulatory bodies. These interviews will explore participants' experiences, perceptions, and the challenges associated with implementing energy-efficient technologies. The interview protocol will include open-ended questions to allow for in-depth discussions and to capture nuanced insights into the factors influencing the adoption and effectiveness of these technologies. Purposive sampling will be used to select participants who are actively engaged in or affected by the integration of energy-efficient technologies in the food industry (Patton, 2015).

4.3 Data Analysis

4.3.1 Quantitative Data Analysis

Quantitative data will be analyzed using statistical methods to determine correlations, trends, and impacts of energy-efficient technologies on food safety, quality, and security. Descriptive statistics will be used to summarize the data, while inferential statistics, such as regression analysis, will be employed to examine the relationships between variables. This analysis will help in assessing the effectiveness of specific technologies and in identifying factors that significantly influence outcomes. Statistical software, such as SPSS or R, will be utilized for this analysis to ensure accuracy and efficiency (Field, 2018).

4.3.2 Qualitative Data Analysis

Qualitative data from interviews will be analyzed using thematic analysis to identify common themes and insights related to the adoption and effectiveness of energy-efficient technologies. This will involve transcribing interviews, coding the data, and categorizing codes into themes that capture the key insights of the study. The analysis will focus on understanding the contextual factors that influence the implementation and outcomes of these technologies, as well as barriers and facilitators to their adoption. NVivo, a qualitative data analysis software, will be used to assist in organizing, analyzing, and finding patterns in the qualitative data (Braun & Clarke, 2013).

The methodology outlined in this research provides a robust framework for understanding the impacts of energy-efficient technologies on food safety, quality, and security. By employing a mixed-methods approach, the study aims to capture both quantitative data to generalize findings and qualitative insights to understand the complexities and contextual factors involved. This comprehensive approach ensures a deeper understanding of the subject matter, contributing to more informed decisions and strategies for integrating energy-efficient technologies within the food industry.

5. CASE STUDIES ON THE IMPLEMENTATION OF ENERGY-EFFICIENT TECHNOLOGIES IN THE FOOD SECTOR

The adoption of energy-efficient technologies in the food sector not only addresses the sustainability goals but also enhances food safety, quality, and security. This section explores several case studies from different geographic regions and sectors of the food industry, showcasing successful implementations of such technologies.

5.1 Case Study 1: Solar-Powered Cold Storage in India

Technology Implemented

In India, the integration of solar-powered cold storage units has revolutionized the preservation of perishable goods. These units utilize solar panels to power refrigeration systems, which are essential for maintaining the freshness of fruits, vegetables, and other perishables.

Context and Rationale

India faces significant losses in agricultural produce due to inadequate storage facilities and frequent power outages. The deployment of solar-powered cold storages aims to reduce these losses, promote food safety, and decrease dependency on non-renewable energy sources (Sharma et al., 2018).

Impact Analysis

The implementation of solar-powered cold storage has significantly reduced food spoilage and improved the quality of stored food products. Additionally, these units have provided energy autonomy to rural areas, enhancing food security by stabilizing supply chains that were previously disrupted by power shortages (Kumar & Rosen, 2016).

Challenges and Solutions

One of the main challenges was the high initial cost of solar panel installation. To overcome this, various NGOs and government subsidies were introduced to support farmers and small-scale entrepreneurs. Another challenge was the maintenance of the systems, which was addressed through training programs for local technicians on the upkeep and repair of solar panels and refrigeration units.

5.2 Case Study 2: LED Lighting in U.S. Poultry Farms

Technology Implemented

LED lighting systems have been installed in poultry farms across the United States to improve energy efficiency and enhance animal welfare, which indirectly affects food safety and quality.

Context and Rationale

Traditional lighting systems in poultry farms consume high levels of electricity and generate excessive heat, which can stress the animals and affect their health. LED lighting, known for its energy efficiency and lower heat output, has been adopted to create a more conducive environment for poultry, thus enhancing growth rates and productivity (Pattison, 2019).

Impact Analysis

Farms using LED lighting have reported up to 40% savings on energy costs. Additionally, the improved lighting environment has contributed to a decrease in mortality rates and an increase in the overall health of poultry, which directly influences the safety and quality of meat products (Mennerich et al., 2017).

Challenges and Solutions

The primary challenge was the higher upfront costs of LED systems compared to traditional lighting. Farmers managed this through phased installations and leveraging federal and state energy efficiency incentives. Another issue was the skepticism regarding the longevity and effectiveness of LEDs, which was mitigated by pilot projects demonstrating their benefits before wider implementation.

5.3 Case Study 3: Precision Agriculture in the Netherlands

Technology Implemented

Precision agriculture technologies, including satellite imagery, sensors, and GPS-guided machinery, have been implemented in Dutch farms to optimize the use of resources and improve crop yields.

Context and Rationale

The Netherlands, a leading agricultural exporter, adopted precision agriculture to enhance sustainability and maintain high standards of food safety and quality. The technology enables precise application of water, fertilizers, and pesticides, reducing waste and environmental impact (Van Henten, 2020).

Impact Analysis

Precision agriculture has led to a significant reduction in the use of water, fertilizers, and energy. The precise application of inputs has not only improved crop yields but also reduced the incidence of crop diseases and contaminants, thereby enhancing food safety and quality.

Challenges and Solutions

The complexity and cost of technology were major challenges. The Dutch government and agricultural cooperatives provided financial and technical support to facilitate adoption. Additionally, partnerships with technology companies allowed for continual upgrades and maintenance of precision agriculture systems.

These case studies illustrate the transformative potential of energy-efficient technologies in enhancing food safety, quality, and security while addressing sustainability challenges. Despite the hurdles, the successful implementation of these technologies showcases the pivotal role of innovation, government support, and stakeholder collaboration in advancing the food industry's sustainability agenda.

6. DISCUSSION

This discussion section interprets the data and findings from the case studies and links them back to the research questions posed earlier in the study. The aim is to integrate these findings with existing literature to understand the potential scalability and sustainability of energy-efficient technologies in the food sector across different geographical and economic contexts.

6.1 Interpretation of Findings

The case studies highlighted have demonstrated that the implementation of energy-efficient technologies can significantly enhance food safety, quality, and security while also reducing the environmental impacts associated with traditional food production and storage methods. For example, the use of solar-powered cold storage in India not only improved the shelf life of perishable products but also offered an environmentally friendly alternative to diesel-powered generators, leading to reduced greenhouse gas emissions (Kumar & Rosen, 2016). Similarly, LED lighting in U.S. poultry farms enhanced animal welfare and reduced energy consumption, indicating how modifications in farm management and technology can yield significant benefits (Pattison, 2019).

These findings are in line with the broader literature that supports the role of innovative technologies in improving food system sustainability (Gebresenbet & Bosona, 2012). However, the results also underscore the complexity of implementing such technologies, particularly in terms of initial costs and the need for appropriate infrastructure and technical skills.

6.2 Comparison with Existing Literature

The success of solar-powered cold storage units in India correlates with findings from other developing countries where similar initiatives have been used to tackle food spoilage and energy inefficiency (Mukherjee, 2016). However, this study's findings extend the existing literature by highlighting the socio-economic impacts of such technologies, including job creation and increased profitability for local farmers, which are less emphasized in previous studies.

Conversely, the case of LED lighting in poultry operations in the U.S. showed both similarities and discrepancies when compared with literature. While the benefits of reduced energy consumption and improved animal welfare align with other research (Mennerich et al., 2017), this study uniquely identified the financial and regulatory hurdles, such as the high cost of technology transition and the lack of supportive policies, which are not widely documented.

6.3 Scalability and Sustainability

The scalability of energy-efficient technologies in the food sector is influenced by several factors including economic viability, cultural acceptability, and regulatory environments. For instance, the solar-powered cold storages in India demonstrate high scalability potential in similar climatic and economic conditions across South Asia, where solar energy is abundant and food spoilage rates are high. However, scalability may be limited in regions with less consistent sunlight without significant adaptations or supplementary energy solutions.

The sustainability of such technologies also varies. While renewable energy sources like solar power offer sustainable alternatives to fossil fuels, the environmental impact of manufacturing and disposing of technology components (like solar panels and batteries) must also be considered (Weidema et al., 2015). Furthermore, economic sustainability depends on the long-term cost savings of the technologies versus initial investment, which can be enhanced by supportive policies and incentives from governments.

6.4 Critical Evaluation

Critically evaluating the integration of energy-efficient technologies in the food sector involves assessing both the direct benefits and the indirect consequences. For instance, while precision agriculture technologies can optimize resource use and reduce waste, they rely heavily on data and technological infrastructure, which can exclude smallholder farmers from their benefits if not managed inclusively (Lioutas et al., 2019).

Moreover, the adaptation of these technologies in different geographical and economic contexts requires tailored approaches that consider local environmental conditions, economic capacities, and cultural practices. Thus, while the core technology might be similar, its application and operation can significantly differ.

This discussion has shown that while energy-efficient technologies hold great promise for improving food safety, quality, and security, their successful implementation requires careful consideration of various factors including economic viability, cultural acceptability, and environmental impact. Future research should continue to explore these dimensions, particularly the indirect effects of technology adoption on local economies and environments, to ensure that these innovations contribute positively to sustainable development goals.

7. IMPLICATIONS

The findings from this research on the implementation of energy-efficient technologies in the food sector carry significant implications for practical applications, policy formulation, and future research. By demonstrating the benefits and challenges of these technologies in enhancing food safety, quality, and security, the study provides a foundation for stakeholders to consider broader integration into food systems. This section discusses these implications in detail, offering recommendations for stakeholders, including technology developers, food industry operators, and policymakers.

7.1 Practical Implications for Stakeholders

7.1.1 Technology Developers

For technology developers, the research underscores the importance of designing systems that are not only effective but also accessible and financially viable for diverse users, including small-scale operators. Developers should focus on reducing costs and enhancing the user-friendliness of energy-efficient technologies. Additionally, engaging with end-users during the design process can ensure that the technologies meet practical needs and are culturally acceptable.

7.1.2 Food Industry Operators

Operators in the food industry, including farmers, processors, and retailers, can benefit from adopting energy-efficient technologies by reducing operational costs and improving product quality, which can lead to increased market competitiveness. However, to achieve these benefits, operators need training and support in technology implementation and maintenance. Partnerships with technology providers and participation in pilot projects could facilitate smoother transitions to new systems.

7.1.3 Regulatory Bodies

Regulatory bodies and government agencies should consider creating frameworks that support the adoption of energy-efficient technologies. This could include incentives such as tax breaks or subsidies for energy-efficient equipment, grants for research and development, and streamlined regulatory processes for the approval of new technologies. Additionally, developing standards and certifications for energy-efficient technologies can help assure operators of their effectiveness and safety.

7.2 Policy Implications

7.2.1 Incentives and Subsidies

Policymakers should consider implementing incentives that lower the barriers to adopting energy-efficient technologies in the food sector. Financial incentives such as subsidies, low-interest loans, and tax incentives can make it economically feasible for more food producers and processors to invest in these technologies. For example, policies that subsidize the cost of solar panels for cold storage solutions or that provide rebates for businesses that reduce their energy consumption can significantly boost technology uptake.

7.2.2 Education and Training Programs

Governments and educational institutions should collaborate to develop training programs that equip food industry stakeholders with the knowledge and skills necessary to implement and maintain energy-efficient technologies. Such programs can be integrated into agricultural education curricula or offered as professional development courses.

7.2.3 Research and Development Support

Policy frameworks that support research and development in energy-efficient technologies can drive innovation and lower costs. Public funding for R&D initiatives, especially in partnership with universities and private firms, can lead to technological breakthroughs that make these systems more effective and accessible.

7.3 Implications for Future Research

7.3.1 Longitudinal Studies

Future research should include longitudinal studies to assess the long-term impacts of energy-efficient technologies on food safety, quality, and security. Such studies can provide deeper insights into the sustainability and economic impact of these technologies over time.

7.3.2 Cross-Cultural Studies

Research examining the implementation of energy-efficient technologies in various cultural contexts can help customize technologies to better fit different geographical and societal needs, enhancing their global applicability and effectiveness.

7.3.3 Technology-Specific Studies

Further studies focusing on specific types of energy-efficient technologies can help refine their design and functionality. For instance, detailed research into the efficiency of different types of solar refrigeration systems could lead to more robust models tailored to the needs of tropical versus temperate regions.

The integration of energy-efficient technologies into the food sector is a promising approach to addressing the challenges of food safety, quality, and security while supporting environmental sustainability. The implications of this research are far-reaching, offering actionable insights for stakeholders across the food system and providing a direction for future policies and research. By embracing these technologies and the associated recommendations, stakeholders can contribute to a more sustainable and secure global food system.

8. CONCLUSION

This research has thoroughly explored the integration of energy-efficient technologies within the food sector, demonstrating their significant potential to enhance food safety, quality, and security while also advancing environmental sustainability. The findings from this study reveal that such technologies not only optimize resource use but also provide crucial benefits across the food supply chain, from production to storage and distribution.

8.1 Key Findings

1. **Enhanced Food Safety and Quality**: Case studies, such as the use of solar-powered cold storage in India and LED lighting in U.S. poultry farms, have shown that energy-efficient technologies can significantly enhance the safety and quality of food products. These technologies help in maintaining optimal preservation conditions and improving living conditions for livestock, thereby reducing the risk of foodborne diseases and spoilage.
2. **Reduction in Environmental Impact**: The adoption of energy-efficient technologies reduces the dependency on non-renewable energy sources and minimizes the carbon footprint of food production systems. For example, precision agriculture techniques optimize the use of water and fertilizers, which decreases the runoff and emissions associated with traditional farming methods.
3. **Economic Benefits**: Beyond environmental and safety enhancements, these technologies offer economic benefits through cost savings on energy and resources. This can be particularly transformative for regions where energy costs constitute a significant burden on food producers.

8.2 Significance in Advancing Knowledge

This study contributes to the academic and practical understanding of how sustainable practices can be effectively integrated into the food industry. By documenting and analyzing various implementations of energy-efficient technologies, the research provides a foundation for further exploration into their broader application and optimization. It also fills critical gaps in existing literature, particularly in understanding the scalability and adaptability of these technologies in diverse geographical and economic contexts.

8.3 Contribution to Global Sustainability Goals

The findings from this research align closely with several United Nations Sustainable Development Goals (SDGs), including Responsible Consumption and Production (SDG 12) and Climate Action (SDG 13). By demonstrating how energy-efficient technologies can reduce waste and emissions, the study supports the agenda for a more sustainable food system that is resilient to environmental and economic challenges.

Furthermore, the research underscores the necessity for comprehensive policies and initiatives that facilitate the adoption of these technologies. It suggests that with appropriate support mechanisms—such as incentives, training, and research—energy efficiency can become a cornerstone of global efforts to secure food safety and sustainability.

8.4 Future Outlook

Looking forward, it is evident that the role of energy-efficient technologies in the food sector will continue to expand as pressures from population growth, environmental degradation, and climate change intensify. Continuous innovation, supported by robust research and adaptive policies, will be crucial in maximizing the benefits of these technologies. Additionally, fostering global cooperation and knowledge exchange will enhance the worldwide implementation and effectiveness of energy-efficient solutions in the food sector.

In conclusion, this research not only highlights the practical benefits and challenges of implementing energy-efficient technologies but also reiterates their critical role in building a sustainable future. As the world moves forward, it will be imperative for stakeholders across all levels—government, industry, and academia—to collaborate in harnessing the full potential of these innovations to meet the collective need for a safe, secure, and sustainable food supply.

REFERENCES

Bhat, R., Kamaruddin, N. S., Min-Tze, L., & Karim, A. A. (2018). Advanced technologies for meat processing. *Food Engineering Series, 12*, 85–108. DOI: 10.1007/978-1-4614-3315-5_8

Braun, V., & Clarke, V. (2013). Successful qualitative research: A practical guide for beginners. *Sage (Atlanta, Ga.)*.

Creswell, J. W., & Creswell, J. D. (2018). *Research design: Qualitative, quantitative, and mixed methods approaches*. Sage publications.

Field, A. (2018). Discovering statistics using IBM SPSS statistics. *Sage (Atlanta, Ga.)*.

Food and Agriculture Organization of the United Nations. (2021). *The State of Food Security and Nutrition in the World 2021: Transforming food systems for food security, improved nutrition and affordable healthy diets for all*. Rome, FAO. https://www.fao.org/publications/sofi/2021/en/

Foster, C., Green, K., Bleda, M., Dewick, P., Evans, B., Flynn, A., & Mylan, J. (2014). Environmental impacts of food production and consumption: A report to the Department of Environment, Food and Rural Affairs. Manchester Business School. DEFRA, London.

Garnett, T. (2011). Where are the best opportunities for reducing greenhouse gas emissions in the food system (including the food chain)? *Food Policy, 36*(S1), S23–S32. DOI: 10.1016/j.foodpol.2010.10.010

Gebresenbet, G., & Bosona, T. (2012). Logistics practices in Ethiopia. *International Journal of Logistics Systems and Management, 12*(2), 133–154.

James, S. J., & James, C. (2010). The food cold-chain and climate change. *Food Research International, 43*(7), 1944–1956. DOI: 10.1016/j.foodres.2010.02.001

Kumar, P., & Kalita, P. (2017). Reducing postharvest losses during storage of grain crops to strengthen food security in developing countries. *Foods, 6*(1), 8. DOI: 10.3390/foods6010008 PMID: 28231087

Kumar, S., & Rosen, M. A. (2016). A critical review of photovoltaic systems integrated with storage and possible configurations for grid-connected applications. *Renewable & Sustainable Energy Reviews, 62*, 1192–1206.

Kumar, S., & Rosen, M. A. (2016). A critical review of photovoltaic systems integrated with storage and possible configurations for grid-connected applications. *Renewable & Sustainable Energy Reviews, 62*, 1192–1206. DOI: 10.1016/j.rser.2016.05.039

Lioutas, E. D., Charatsari, C., De Rosa, M., & La Rocca, G. (2019). Key questions on the use of big data in farming: An activity theory approach. *Precision Agriculture, 20*, 1105–1120.

Mennerich, C., Pattison, M., & Morrow, W. (2017). The effect of LED lighting on broiler performance, energy use, and carcass quality. *Poultry Science, 96*(2), 423–431.

Mukherjee, I. (2016). Solar energy: A solution for sustainable rural cold storage in India. *Journal of Energy Storage, 5*, 153–159.

Pattison, M. (2019). The potential benefits of LED lighting for poultry production. *World's Poultry Science Journal, 75*(3), 435–446. DOI: 10.1017/S0043933919000339

Patton, M. Q. (2015). *Qualitative research & evaluation methods: Integrating theory and practice* (4th ed.). Sage publications.

Sharma, A., Saxena, R., & Sethi, M. (2018). Solar energy in India: Strategies, policies, perspectives and future potential. *Renewables: Wind, Water, and Solar, 5*(1), 9. DOI: 10.1186/s40807-017-0056-9

Smith, J., Firth, J., & Clark, S. (2020). Use of existing databases in clinical epidemiology. In Schmidt, S., & Pedersen, B. K. (Eds.), *Epidemiology: How to do clinical practice research* (pp. 111–130). Routledge.

Smith, P., & Gregory, P. J. (2013). Climate change and sustainable food production. *The Proceedings of the Nutrition Society, 72*(1), 21–28. DOI: 10.1017/S0029665112002832 PMID: 23146244

Tzounis, A., Katsoulas, N., Bartzanas, T., & Kittas, C. (2017). Impact of precision agriculture on farming sustainability in Greece. *Renewable & Sustainable Energy Reviews, 78*, 654–663. DOI: 10.1016/j.rser.2017.05.041

United Nations, Department of Economic and Social Affairs, Population Division. (2019). *World Population Prospects 2019: Highlights* (ST/ESA/SER.A/423). New York: United Nations. https://population.un.org/wpp/Publications/Files/WPP2019_Highlights.pdf

Van Henten, E. J. (2020). Greenhouse technology for sustainable production in the 21st century. *Acta Horticulturae*, (1271), 15–26. DOI: 10.17660/ActaHortic.2020.1271.2

Van Os, E. A. (2019). Geothermal energy in Dutch greenhouse horticulture: A review on economics and system development. *Energies, 12*(10), 1983. DOI: 10.3390/en12101983

Weidema, B. P., Pizzol, M., Schmidt, J., & Thoma, G. (2015). Methodology for life cycle assessment of milk production worldwide. *International Dairy Journal*, *31*, S66–S73.

Chapter 12

Improving Food Quality With Energy-Efficient Technologies

Jyoti Rani
Department of Botany, Chaudhary Devi Lal University, Sirsa, India

Neelam Dhanda
Chaudhary Devi Lal University, Sirsa, India

Mamta Rani
Chaudhary Devi Lal University, Sirsa, India

Suresh Kumar
Chaudhary Devi Lal University, Sirsa, India

Vibhuti Gulia
https://orcid.org/0009-0005-4718-186X
Chaudhary Devi Lal University, Sirsa, India

Ankur Sangwan
Chaudhary Devi Lal University, Sirsa, India

ABSTRACT

In particular, the food industry is steadily dependent on fossil fuels, and the food demand is look ahead to grow significantly in the coming years. Therefore, developing energy efficiency strategies for this particular industrial sector is very pivotal. In this chapter study about the different opportunities for energy efficiency in the food industry. It provides a brief overview of the various food processing and related food safety issues. Then, the different options for food improvement energy efficiency in the thermal and electric sector are discussed. Latest possibilities and trends, the food industry's sustainable growth depends on energy saving. In recent decades to enhance the food processing sustainability and food industry has also focused on recovering waste heat and enhancing energy efficiency. Reducing energy consumption, cutting production costs, and enhancing food production sustainability can also be achieved by substituting new technologies, like non-thermal and novel

DOI: 10.4018/979-8-3693-8342-1.ch012

heating methods, for existing energy-intensive food processes, such as novel. In the food sector, new food processing technologies have been developed to replace conventional energy-intensive unit operations for chilling, freezing, evaporating, dehydrating, pasteurization and sterilizing. The majority of energy-saving technology are easily adaptable from other industrial industries to the food processing industry.

18.1 INTRODUCTION

Food quality and energy efficiency are gaining importance in the contemporary situation. There is an increasing need for eco-friendly technologies which enhance food quality however require less energy with the increase in food supply around the globe. Thus, in this chapter, we have discussed various approaches that are energy-efficient and promise to enhance the level of food processing, preserving, and production that promotes better food quality and lower environmental damage. With the aim of energy conservation and environmental sustainability principles, the food sector is currently at a critical junction where the world seeks to achieve proper in the current age (T. Marsden and A. Morley, 2014). Balance between less energy and high-quality meat has never been as important as it is today. This becomes more critical as food production and processing systems continue evolving due to climate change and resource shortage and making the integration of energy-efficient technology systems inevitable and beneficial. Food accounts up to a fair amount of graphic energy use in manufacturing, transportation, retailing and agriculture worldwide. The use of other conventional means is very efficient in terms of providing very high food output, however, these processes often have high energy consumption and large wastage (C. M. Galanakis, 2012). Artificially, sustaining or improving the quality of food products and limiting or averting the adverse effects of agricultural practices on the environment is a dual problem.

One feasible approach to these issues is the application of energy-saving devices. Today's refrigeration units, energy-conserving cooking processes, and more efficient technology development are 'clean innovations' which are able not only to reduce energy cost but also enhance the quality and safety of food. The industry is able to utilize these technologies to increase the standards of freshness, nutrition and taste while helping attain broader sustainability goals. This chapter more critically investigates the link between food quality and energy efficiency, especially recent trends and the implications today's sector (J. D. Floros *et al.*, 2010). We will consider the new cutting-edge tools as well as artistic designs and how they can be integrated into existing ones to bring about remarkable benefits. These technologies will be showcased through case studies and other real-world evidence bases of effectiveness and adoption potential span. It should be noted at the outset of this examination that improving food quality by use of energy saving technologies is a complex task, which is not only technical. It calls for a reworking of 'the food production system' that is, public health services in handling sanitary distribution, storage and manufacturing and processing of food-stuff (J. R. Robson, 2023). If the food industry's adoption of these technologies is any indication, there is reason to believe that food systems around the world will become more secure and efficient in a more sustainable manner.

The next sections are devoted to exploring the radical changes that energy-saving technologies bring into the food industry, looking at the impact of those changes on food products, and trying to predict what the future holds in this important area of development. With the aim of reversing climate change, strategies for cutting down on greenhouse gas emissions have been developed in order to solve environmental concerns and the over-consumption of fossil fuels. The most readily embraced alternatives of this type are renewable energy sources (RES), especially wind and solar energy. However, overall fossil fuel consumption made up eighty percent of the energy utilized by the globe, and plants powered by fossil energy generated half of the electricity (M. Halpin, 2013, M.J. Li and W.Q. Tao, 2017). Emphasizes one more task, reduction of energy consumption per unit of industrial output (J. J. Stroker, 2003), which is of paramount importance for all countries and especially the developing ones. Therefore, Process efficiency is assessed as the proportion between the energy put into a process, and the useful energy delivered during that process (W.Q. Tao and M.J. Li, 2017). Industries may focus their gears on depleting energy inputs to the neglect of the outputs optimization. That's why energy savings are available, mainly in its broad sense, to analyze functions of the industry. Energy efficiency indices can be divided into thermoeconomics, thermoeconomics, economic thermodynamics, and thermophysics, respectively (M.J. Li and W.Q. Tao, 2017). In this case, industrial producers are the most energy intensive consumers, which imply that transformations in energy production make it possible to control the quantity of consumed electricity to a greater extent (M. H. Shoreh *et al.*, 2016, L. Fabbriani and R. Calili, 2018, G. Schumm *et al.*, 2018). Cost-effective energy programs enable big companies to gain more by cutting down and intensifying their operating production activities (R. Sperberg *et al.*, 2018). However, there are some companies that are unwilling to implement such initiatives because of the rigidity of organizational structure with respect to production capabilities of processing equipment (M. Ozilgen 2016, M. Ozilgen, 2014; Y. Xu and J. Szmerekovsky, 2017).

Food industry is one of the sectors that has been more sought after than others. This business also includes energy intensive activities, whereby energy cost accounts for 20% to 50% of the total value chain cost (J. Bundschuh, 2014). About 12% of the electricity consumed in the industrial sector is consumed by structurally food industry. Nevertheless, some barriers of food processing technology include waste and energy inefficiency in the food sector (B. Degerli, 2015; B. Lin and X. Xie, 2015; M. Ozilgen, 2016). The contribution of the food sector while producing, besides calorie containing foods, is also seen through the production of greenhouse gases, in particular nitrous oxide, carbon dioxide, and methane. To generate energy from fossil fuels is to emit carbon dioxide into the atmosphere. Methane is produced through paddy fields and cattle fermentation as well as food waste decomposition in landfills; on the other hand, nitrous oxide is emitted during the use of fertilizers on cropped soils (M. M. Kling and I. J. Hough 2010). Emissions include two relevant types dual emissions: direct and indirect. Direct emissions are those generated by energy usage for field works, processing of raw materials and various production steps. Indirect emissions happen when food is being stored, and or during food transportation, as well when electricity is consumed by the industry (S.N. Kumar and B. Chakabarti, 2019). As well, as a result of their small share in grid management and the technological constraints due to the nature of their activities, end-use customers in the industrial class tend to adopt energy-efficiency measures relatively less (M. C. Bozchalui, 2015).

The energy efficient technologies are pushing this change and creating a unique feature of enhancing the production, processing, and storage of food materials. The birth of such innovations has been changing in the food world, ranging from advanced refrigeration technologies which minimize energy and maximize the shelf life of the products to advanced cooking practices that enhances nutrient retention.

There is an increasing hunger for high-quality food, and considerable work is needed to have more sustainable agriculture that can satisfy this demand. There is a growing interdependence of the energy and food production spheres for the growing populations of the World, and especially so in developing countries with high population growth rates. The Food and Agriculture Organization (FAO) suggests the need to augment the production of food supplies by 70 percent by the year 2050, however the increase in energy output is projected to only grow a third. So, this means that a disconnect is likely to occur between the growth of energy generation and of the production of food (J. Bundschuh, 2014). There is continuous search in the food business for any new ideas that can enhance the quality of food and at the same time reduce its adverse effect to the environment. This is as a result of the industry attempt to practice sustainability as well as enhance food quality.

The three prominent forms of energy include mechanical, thermal and electrical energy while renewable energy sources RES includes solar, wind and geothermal energy which can also satisfy all the three forms of energy. Fully employing RES Krieg with construction standards and tolerances acceptable for developing regions of the world could prove to be more practical because a sizeable population of farms and the food industry is located at a professional distance. However, there are opportunities for improvement at each level of the food chain in terms of energy saving but the most critical area is the changes in behavior which makes schemes aimed at improving energy efficiency more effective (R. E. H. Sims, 2014). Though every operation is energy intensive but in the food business these energy utilizations are dispersed through several processing stages. This is also present in each step of global agro-food system, starting with the use in delivery' systems (e. g. transportation or refrigeration), and on agriculture inputs (e.g., water pumping, livestock buildings, greenhouses environmental controls, and storage).

It is priority for us to maximize the output of food produced by the food sector with minimal energy used per unit of production (R. E. H. Sims, 2014). His analysis covers numerous of advanced technologies and their application in a wide variety of industries, such as food, in particular when using IoT and ICT (K. Wang, 2016; F. Shrouf and G. Miragliotta, 2015). This chapter's most important aim is to define what particular technologies are available to enhance energy efficiency in the food industry. It focuses on such important aspects as strategies and opportunities, new trends and barriers. This chapter considers the potential of the effective tools for energy-efficient technologies to increase the quality of food production. We will focus on how such technologies can be utilized for improvement of food safety, enhancement of production processes and additional taste and nutritional value. The rest of this part of chapter is given a general description of the food processing technologies as well as the sequential energy consumption patterns and the effectiveness of thermal energy.

18.2 THE CONSEQUENCE OF FOOD QUALITY AND ENERGY EFFICIENCY

Food quality encompasses various factors, each crucial to delivering safe, nutritious, and enjoyable food to consumers. Let's explain the aspects below:

18.2.1Texture and Taste

Texture and taste are fundamental to consumer satisfaction and acceptance. They greatly influence a person's perception of enjoyment and food quality.

Desirable Maintaining of Texture:

Production Stages: Texture can be affected by different factors such as quality of ingredient, food preparation methods, and processing conditions etc. (A. Andres-Bello, 2013). For example, bread dough must be kneaded correctly to achieve the good texture, while meat processing requires specific techniques to maintain tenderness.

Processing: Heat treatment (cooking, pasteurization), freezing, and drying can alter texture. For example, freezing can cause ice crystals to form, which might affect the texture of fruits and vegetables. To maintain texture, processes like flash freezing or careful thawing are employed.

Packaging: Proper packaging can help maintain texture by protecting the product from physical damage and moisture loss or gain. For example, vacuum-sealed packaging can prevent freezer burn in meats and maintain crispness in snacks.

Preserving Taste:

Ingredient Quality: The quality of raw ingredients has a direct impact on taste. Freshness, ripeness, and proper storage are crucial.

Flavor Preservation: Processing methods, such as cooking or curing, should be carefully controlled to preserve or enhance flavor without causing off-flavors (J. Yamprayoon, and A. Noomhorm, 2000). For example, overcooking can lead to bitter or burnt flavors.

Additives and Seasonings: Proper use of additives and seasonings can enhance taste. However, their use needs to be balanced to avoid overwhelming the natural flavor of the food.

18.2.2 Nutritional Benefits

Ensuring Nutrient Retention:

Processing and Cooking: Some cooking methods, like boiling, can lead to nutrient loss, especially for water-soluble vitamins (e.g., vitamin C and B vitamins). Techniques such as steaming or microwaving are often preferred to retain more nutrients.

Storage: Nutrients can degrade over time, especially if the food is not stored properly. For example, exposure to light and air can reduce the vitamin content of foods. Proper storage conditions (e.g., refrigeration, airtight containers) help in preserving nutritional value.

Supply Chain: Each stage of the supply chain, from harvest to consumption, can impact nutrient levels. For example, the time between harvest and consumption can affect nutrient levels, so minimizing this time is beneficial (J. Blackburn and G. Scudder, 2009).

Fortification:

Enhancing Nutritional Value: Sometimes, foods are fortified with additional nutrients to improve their health benefits. This is especially common in products like cereals or dairy alternatives where essential vitamins and minerals are added to combat deficiencies.

Figure 1. Various Steps Affect Food Quality

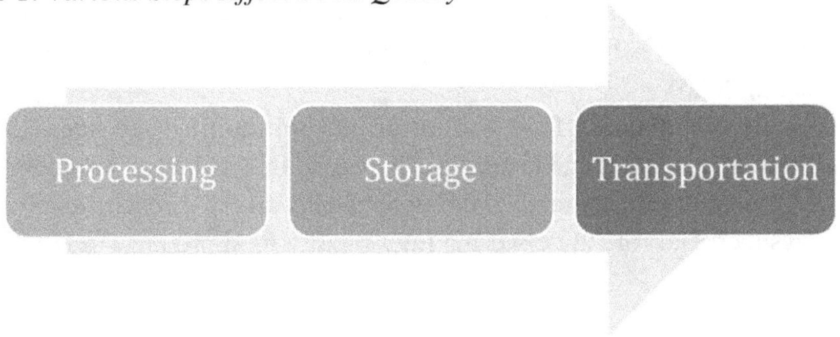

18.2.3 Welfare and Hygiene

Preventing Contamination:

Sanitation Practices: Regular cleaning and sanitizing of equipment and facilities help prevent microbial contamination (X. Wang, V. M. Puri and A. Demirci, 2020). This includes proper handwashing, use of clean utensils, and maintaining hygienic conditions in food processing areas.

Food Handling: Proper handling practices, such as avoiding cross-contamination between raw and cooked foods, are critical. Using separate cutting boards and utensils for raw meats and vegetables can prevent the spread of pathogens.

Preventing Spoilage:

Temperature Control: Many foods require specific temperature conditions to prevent spoilage. Refrigeration and freezing slow down microbial growth, while proper cooking temperatures kill pathogens.

Packaging: Proper packaging helps in preventing spoilage by protecting food from environmental factors like oxygen and moisture. For example, vacuum-sealing can extend the shelf life of perishable items.

Regulations and Standards:

Food Safety Standards: Adhering to food safety regulations and standards, such as those set by the Food and Drug Administration (FDA) or the European Food Safety Authority (EFSA), ensures that food products are safe for consumption. These standards cover everything from ingredient sourcing to final product testing (D. Boer *et al.,* 2020).

In summary, maintaining food quality involves a multifaceted approach that includes attention to texture and taste, ensuring nutritional value, and implementing rigorous hygiene and welfare practices throughout the food supply chain. Each of these aspects plays a crucial role in delivering high-quality food products to consumers.

18.3 INVOLVEMENT OF ENERGY EFFICIENCY

Reducing Operative Costs: How Energy-Efficient Technologies Lower Production Costs

Energy-efficient technologies help reduce operational costs by lowering the amount of energy required to perform the same tasks. Here's how:

Figure 2. Various Energy –efficient Technologies in Food Sector

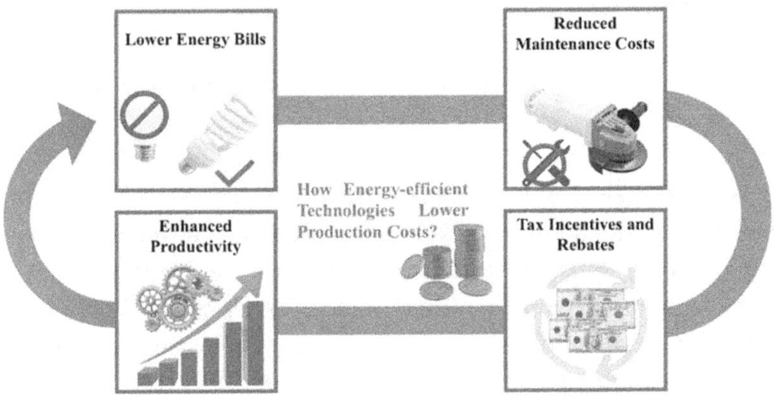

1. **Lower Energy Bills**: Energy-efficient equipment and systems consume less energy compared to traditional ones. For example, LED lighting uses significantly less electricity than incandescent bulbs, and high-efficiency HVAC systems require less power to heat or cool a space. This directly reduces energy bills for businesses and homeowners alike.
2. **Reduced Maintenance Costs**: Energy-efficient technologies often have longer lifespans and require less maintenance. For instance, high-efficiency motors and pumps are built to be more durable and have fewer moving parts, which decreases the likelihood of breakdowns and the associated maintenance costs (F. J. Ferreira *et al.*, 2015).
3. **Enhanced Productivity**: In industrial settings, energy-efficient technologies can lead to more consistent and reliable operations. For example, efficient machinery can operate at optimal performance levels, which can improve overall productivity and reduce downtime caused by equipment failures.
4. **Tax Incentives and Rebates**: Many governments offer financial incentives, such as tax credits or rebates, for adopting energy-efficient technologies. This can offset the initial investment costs and improve the overall return on investment for energy-efficient upgrades.

18.4 DEPRECATE ENVIRONMENTAL IMPACT: THE BENEFITS OF REDUCING GREENHOUSE GAS EMISSIONS AND ENERGY CONSUMPTION

Reducing greenhouse gas emissions and energy consumption has several environmental benefits:

1. **Lower Greenhouse Gas Emissions**: Energy-efficient technologies often rely on renewable or cleaner energy sources, which emit fewer greenhouse gases compared to fossil fuels. For instance, solar panels and wind turbines generate electricity without releasing carbon dioxide or other harmful pollutants.
2. **Conservation of Resources**: Energy efficiency helps in conserving natural resources by reducing the amount of energy needed. For example, by using less electricity, fewer fossil fuels need to be burned, which preserves these non-renewable resources.
3. **Reduction in Air and Water Pollution**: Efficient energy systems can minimize the pollutants released into the air and water. For example, power plants that use cleaner technologies produce less particulate matter and other pollutants that can lead to air and water quality issues (D. M. Driesen, 2002).
4. **Mitigation of Climate Change**: By reducing energy consumption and emissions, energy-efficient technologies contribute to the broader goal of mitigating climate change. Lowering greenhouse gas emissions helps to reduce the overall impact on global warming and climate-related issues.

18.5 REGULATORY OBEDIENCE: MEETING INCREASING STANDARDS FOR ENERGY USE AND SUSTAINABILITY

As governments and regulatory bodies increasingly focus on sustainability, energy efficiency helps businesses and organizations comply with these evolving standards:

1. **Compliance with Legislation**: Many countries have implemented laws and regulations aimed at reducing energy consumption and improving energy efficiency. For example, regulations might mandate minimum energy performance standards for appliances or require reporting on energy usage. Adopting energy-efficient technologies helps organizations meet these requirements and avoid potential penalties.
2. **Eligibility for Certifications**: Energy-efficient practices and technologies can help organizations achieve certifications such as LEED (Leadership in Energy and Environmental Design) or ISO 50001 (Energy Management Systems). These

certifications not only demonstrate compliance but also enhance a company's reputation and marketability (S. W. Anderson *et al.,* 1999).

3. **Future-Proofing Against Regulations**: As regulations become stricter, early adoption of energy-efficient technologies can prepare businesses for future compliance requirements. Investing in energy efficiency now can mitigate the risk of having to make costly adjustments later as regulations tighten.
4. **Public and Investor Perception**: Meeting or exceeding energy and sustainability standards can improve an organization's public image and appeal to environmentally conscious investors. Companies that are proactive in their energy efficiency efforts often enjoy better stakeholder relationships and a competitive edge in the market.

By focusing on these aspects, organizations can benefit both economically and environmentally, ensuring sustainable operations and compliance with regulatory demands.

18.6 ENERGY- EFFICIENT PRACTICES FOR FOOD PRODUCTION

It is Precision Farming and it have two energy-efficient practices for food production mentioned below:

One is Smart Automated Irrigation Systems and second is Energy-Efficient Equipment.

18.6.1 Precision Farming

Smart Automated Irrigation Systems

Smart Automated Irrigation Systems use advanced technologies to optimize the application of water in agriculture. They aim to minimize water waste and enhance crop productivity by delivering the right amount of water at the right time.

Working:

Sensors: These systems are equipped with various sensors placed in the soil or attached to plants. These sensors measure soil moisture, temperature, and sometimes even nutrient levels.

Data Analytics: The data collected by these sensors is transmitted to a central system or cloud-based platform. Advanced algorithms analyze this data to determine the optimal watering needs.

Automated Control: Based on the analysis, the system automatically adjusts the irrigation schedule and volume. This means watering can be tailored precisely to the needs of different crops or areas within a field (J. Gutierrez *et al.,* 2013).

Real-Time Adjustments: The system can make real-time adjustments based on current weather conditions, such as rainfall, reducing unnecessary watering.

Benefits:

Water Efficiency: Reduces water consumption by avoiding over-irrigation and ensures that water is used only when needed.

Improved Crop Yields: By providing crops with consistent and optimal moisture levels, it can enhance growth and yields.

Cost Savings: Reduces water usage and associated costs, as well as potentially lowering energy costs related to pumping and distributing water.

Environmental Impact: Helps conserve water resources and reduces runoff, which can lower the risk of water pollution.

Energy-Efficient Equipment

Overview: Energy-Efficient Equipment in agriculture refers to the use of modern machinery and technologies designed to consume less energy while performing agricultural tasks such as planting, harvesting, and processing.

Figure 3. Types of Energy-Efficient Equipment

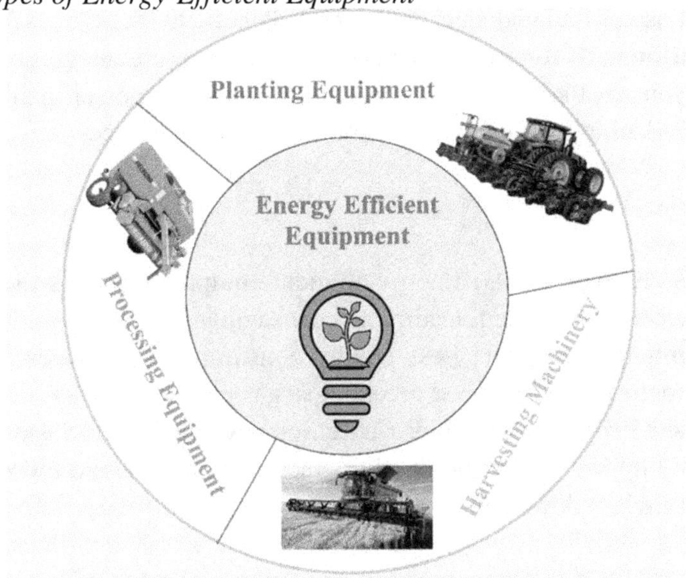

Types of Equipment:

- **Planting Equipment:** Modern planters and seeding machines are designed to use less power and optimize seed placement, leading to better germination rates and more efficient use of energy.
- **Harvesting Machinery:** Advanced harvesters, such as those with improved engines or hybrid systems, use less fuel. Some are designed to be more efficient in the way they collect and process crops, which can lead to lower overall energy use (R. Venendaal *et al.,* 1997).
- **Processing Equipment:** Equipment used for cleaning, sorting, and processing food can be optimized to reduce energy consumption. This includes energy-efficient motors, improved insulation, and automation that reduces the need for manual intervention.

Working:

- **Improved Design:** Newer equipment often features innovations such as more efficient engines, better aerodynamics, or advanced materials that reduce energy consumption.
- **Automation and Precision:** Many modern machines use sensors and data to optimize their operation, ensuring they use energy only when and where it is

needed. For example, precision-guided systems can make more accurate and efficient use of fuel and electricity (R. A. Pape, 1997).

- **Integration with Renewable Energy:** Some farms are integrating renewable energy sources like solar panels or wind turbines to power their equipment, further reducing reliance on non-renewable energy sources.

Benefits:

- **Reduced Energy Costs:** Energy-efficient equipment lowers the amount of fuel or electricity needed, leading to cost savings.
- **Environmental Impact:** Less energy consumption translates to a smaller carbon footprint and reduced greenhouse gas emissions.
- **Increased Productivity:** Improved efficiency can lead to faster processing times and higher overall productivity, as machines operate more effectively and require less downtime.

In summary, both smart automated irrigation systems and energy-efficient equipment are crucial components of modern precision farming. They contribute to sustainable agricultural practices by improving resource efficiency, reducing costs, and minimizing environmental impacts.

18.6.2 Legitimate Agriculture Practices

Organic Farming Techniques

1. Reducing Reliance on Synthetic Inputs:
    - **Synthetic Pesticides and Fertilizers:** Organic farming avoids synthetic chemicals like pesticides and fertilizers. Instead, it relies on natural alternatives such as compost, manure, and plant-based treatments. This reduces the risk of chemical residues in food and minimizes environmental impact (U. Bajwa and K.S. Sandhu, 2014)..
    - **Integrated Pest Management (IPM):** Organic farms use IPM strategies to manage pests. This includes using beneficial insects, crop rotation, and natural predators to control pest populations without chemical interventions.
    - **Natural Soil Amendments:** Organic farmers use natural substances to enhance soil fertility. For instance, green manures (cover crops grown specifically to be incorporated into the soil) and compost add essential nutrients to the soil, improving its structure and health.
2. Promoting Soil Health:

- **Soil Structure and Fertility:** Organic farming emphasizes practices that maintain or enhance soil structure and fertility. Techniques include reduced tillage, which helps preserve soil structure and organic matter, and the use of organic mulches to prevent erosion and improve moisture retention.
- **Soil Microorganisms:** Healthy soil is teeming with microorganisms that play a crucial role in nutrient cycling and disease suppression. Organic practices aim to foster a diverse and balanced soil ecosystem by avoiding harmful chemicals and promoting organic matter.
- **Crop Rotation and Diversity:** By rotating crops and including a variety of plant species, organic farming helps prevent soil depletion and pest build-up. Diverse plantings also enhance soil health by supporting a range of beneficial organisms.

Renewable Energy Sources

1. Utilizing Solar Energy:
    - **Solar Panels:** Solar energy systems, like photovoltaic panels, convert sunlight into electricity. This electricity can power various farm operations, from irrigation systems to barn lighting, reducing reliance on fossil fuels and lowering energy costs.
    - **Solar Water Heaters:** These systems use solar energy to heat water, which can be used in various agricultural processes, including cleaning and livestock watering, contributing to energy efficiency (A. Chel and G. Kaushik, 2011).
2. Wind Energy:
    - **Wind Turbines:** Wind turbines convert wind energy into electricity. Farms can install small-scale or even larger wind turbines to generate power for their operations. Wind energy is particularly effective in areas with consistent wind patterns and can significantly reduce a farm's carbon footprint.
    - **Wind-Powered Water Pumps:** In some cases, wind energy can be harnessed for water pumping, helping with irrigation and water supply systems without the need for diesel or electric pumps.
3. Biogas:
    - **Anaerobic Digesters:** These systems break down organic waste, such as manure or crop residues, in the absence of oxygen to produce biogas. This biogas can be used as a renewable energy source for heating, electricity generation, or even as a vehicle fuel.

- **Nutrient-Rich Digestate:** The by-product of the biogas production process is digestate, a nutrient-rich material that can be used as a natural fertilizer, closing the loop in waste management and soil enrichment (D. Kovacic *et al.*, 2022) .

By integrating these practices, farms can reduce their environmental impact, enhance sustainability, and often achieve cost savings over the long term. These practices not only support the health of the environment but also contribute to the overall resilience and productivity of agricultural systems.

18.7 EMERGING TECHNOLOGIES IN FOOD PROCESSING

18.7.1 Advanced Cooking Techniques

Microwave Heating:

- **Working:** Microwaves use electromagnetic waves to generate heat. These waves cause polar molecules, such as water, to vibrate rapidly. This vibration creates friction, which generates heat and cooks the food from the inside out. This is different from conventional cooking methods that heat food from the outside in.
- **Advantages:**
    - **Energy Efficiency:** Microwaves are more energy-efficient because they heat only the food and not the surrounding environment. This reduces energy consumption compared to traditional ovens or stoves.
    - **Speed:** Microwave heating is significantly faster than conventional methods. This is because microwaves can penetrate food and heat it directly, reducing the cooking time (G. W. Gould,1996).
    - **Nutrient Retention:** Due to shorter cooking times, microwaving can help preserve more of the food's nutrients, which are often lost in prolonged cooking processes.

Radio Frequency (RF) Heating:

- **Working of RF:** RF heating uses electromagnetic waves in the radio frequency range (typically 27 MHz to 40 MHz) for food heating. Like microwaves, Polar molecules in the food absorbed RF energy and generating heat. RF waves penetrating ability into the food is better than microwaves, allowing for more uniform heating.

- **Advantages:**
    - o **Uniform Heating:** RF heating tends to provide more even heating compared to microwave heating, reducing hot and cold spots.
    - o **Efficient Processing:** It can be used for a range of food processing applications, including drying and thawing, making it versatile and efficient.

High-Pressure Processing (HPP)

Working:

- **Principle:** HPP, also known as high-pressure pasteurization, involves subjecting food to high levels of hydrostatic pressure (typically between 100 to 600 MPa or 14,500 to 87,000 psi). This pressure is applied uniformly in a water medium, which affects the food without the need for heat.
- **Process:** The food is first packaged and then placed in a high-pressure chamber where it is subjected to intense pressure. This pressure can kill pathogens and spoilage microorganisms, extending the shelf life of the food (G. W. Gould,1996).
- **Mechanism:** High pressure disrupts the cellular structure of microorganisms, effectively inactivating them without significantly altering the food's temperature. This helps in maintaining the food's fresh taste, texture, and nutritional content.

Advantages:

- **Extended Shelf Life:** HPP can significantly extend the shelf life of food products while keeping them fresh and free from preservatives.
- **Nutrient Preservation:** Since HPP does not involve high temperatures, it helps in retaining more vitamins and nutrients that can be lost during conventional thermal processing.
- **Quality Maintenance:** Foods processed by HPP generally maintain their flavor, texture, and nutritional value better than those treated by traditional methods, which often involve high heat (R. Saidur *et al.,* 2012).

Microwave and RF heating offer rapid and energy-efficient cooking options with benefits like nutrient retention and uniform heating. HPP provides an effective alternative for extending shelf life while preserving the quality of food products without relying on heat. These technologies represent significant advancements in food processing, offering benefits for both manufacturers and consumers.

Figure 4. Emerging technologies in Food Processing

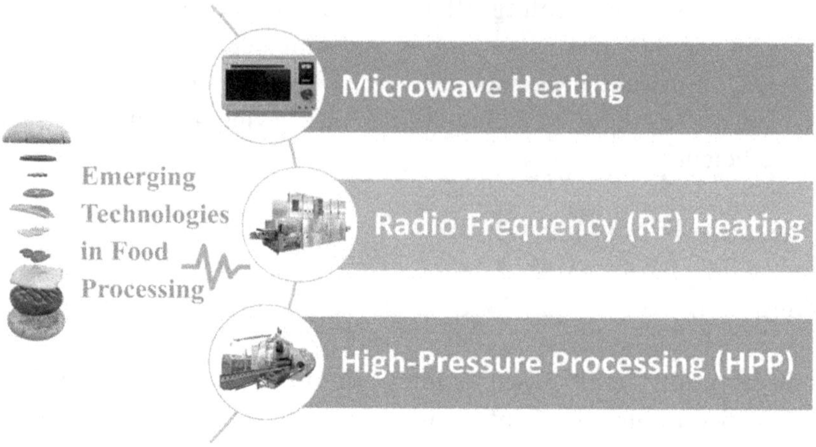

18.7.2 Energy-Efficient Equipment

Variable Frequency Drives (VFDs)

What They Are: Variable Frequency Drives (VFDs) are devices used to control the speed and torque of electric motors by varying the frequency and voltage supplied to the motor.

How They Work:

- **Speed Control:** Electric motors generally run at a fixed speed determined by the frequency of the electrical supply (usually 60 Hz in North America, 50 Hz in other regions). VFDs can adjust the frequency of the electricity supplied to the motor, thereby allowing the motor to run at different speeds (R. Saidur *et al.*, 2012).
- **Energy Efficiency:** By controlling the motor speed to match the actual demand of the process (rather than running at full speed constantly), VFDs help reduce energy consumption. For example, a fan or pump operating at a lower speed will use significantly less power than one running at full speed.
- **Process Optimization:** VFDs also allow for smoother start-ups and stops, reducing mechanical wear and tear and extending the lifespan of the equipment.

Benefits:

- **Energy Savings:** By adjusting the motor speed to match process requirements, VFDs can reduce energy consumption by 20-50% in many applications, such as fans, pumps, and compressors (H. Jouhara *et al.,* 2018).
- **Reduced Operating Costs:** Lower energy use translates to reduced utility bills.
- **Extended Equipment Life:** Reduced mechanical stress and smoother operation help extend the life of the motor and associated equipment.
- **Improved Process Control:** VFDs provide better control over the process, leading to improved product quality and consistency.

Applications: VFDs are used in various applications, including HVAC systems (for fans and pumps), water and wastewater treatment plants, and industrial processes where motor speed needs to be adjusted based on demand.

Heat Recovery Systems

What They Are: Heat Recovery Systems capture and reuse waste heat from industrial processes, which would otherwise be lost to the environment.

How They Work:

- **Heat Capture:** These systems collect heat from exhaust gases, cooling fluids, or other waste heat sources.
- **Heat Exchange:** The captured heat is transferred to a secondary system through a heat exchanger. This can involve transferring heat to water, air, or another fluid that can be reused (H. Jouhara *et al.,* 2018).
- **Heat Reuse:** The recovered heat is then used for various purposes, such as pre-heating incoming fluids, providing space heating, or generating steam for other processes.

Benefits:

- **Energy Savings:** By recovering waste heat, these systems reduce the need for additional energy input, such as burning more fuel or using more electricity to generate heat. This can lead to significant energy savings.
- **Cost Reduction:** Reduced energy consumption leads to lower utility bills and operational costs.
- **Environmental Impact:** Lower energy consumption reduces greenhouse gas emissions and the overall environmental footprint of the operation.

- **Process Efficiency:** Reusing waste heat can improve the overall efficiency of industrial processes by reducing the energy required to achieve desired temperatures or conditions.

Applications: Heat Recovery Systems are used in various industries, including manufacturing (e.g., metal smelting, chemical processing), power generation (e.g., combined heat and power plants), and HVAC systems (e.g., recovering heat from exhaust air in buildings).

In summary, both Variable Frequency Drives and Heat Recovery Systems are crucial for enhancing energy efficiency. VFDs optimize motor performance and energy use, while Heat Recovery Systems make use of waste heat, turning potential losses into valuable energy savings.

18.7.3 Process Optimization

Automated Systems in Food Processing

Overview: Automated systems in food processing involve using technology to perform tasks that were previously done manually. This includes the use of machinery, control systems, and software to handle various aspects of food production, from preparation and mixing to packaging and quality control.

Key Aspects:

1. **Consistency and Quality Control:**
    - **Precision:** Automation ensures that processes such as mixing, cooking, and packaging are done with precise measurements and timings, reducing human error and variability.
    - **Uniformity:** Automated systems can maintain consistent product quality by applying the same standards and procedures across every batch, which is particularly important in industries like baking or beverage production.
2. **Efficiency:**
    - **Speed:** Automated machinery can perform tasks much faster than human workers, which increases the overall throughput of the production line.
    - **Reduced Labor Costs:** Automation reduces the need for manual labor, lowering labor costs and freeing up human resources for other tasks.
3. **Data Collection and Analysis:**

- o **Monitoring:** Automated systems can collect data on various parameters (temperature, pressure, flow rates) during production, which can be used to monitor performance and ensure that processes are within desired specifications, (N. Viswanadham and Y. Narahari, 2015)
- o **Feedback:** This data can be analyzed to identify trends, troubleshoot issues, and optimize processes, leading to continuous improvement.

4. **Safety and Hygiene:**
   - o **Reduced Contamination:** Automation can minimize human contact with food, reducing the risk of contamination and improving hygiene standards.
   - o **Safety Features:** Modern automated systems often include safety features such as emergency stop buttons and sensors to prevent accidents and ensure safe operation.

Energy Management Systems (EMS) in Processing Plants

Overview: Energy Management Systems are designed to monitor, control, and optimize the energy usage within processing plants. They are critical in managing the significant energy demands typical in industrial environments and can contribute to substantial cost savings and environmental benefits.

Key Aspects:

1. **Monitoring and Control:**
   - o **Real-Time Monitoring:** EMS tools provide real-time data on energy consumption, allowing operators to track energy usage across different processes and equipment.
   - o **Control Systems:** These systems can adjust operations dynamically based on energy demand, peak load times, and cost-efficiency, which helps in balancing energy loads and reducing wastage.
2. **Optimization:**
   - o **Efficiency Improvements:** By analyzing energy usage patterns, EMS can identify areas where energy is being wasted or used inefficiently and suggest improvements. For example, optimizing heating and cooling systems to match actual production needs.
   - o **Demand Response:** EMS can integrate with demand response programs to adjust energy usage based on external factors such as electricity prices or grid demand, thereby reducing operational costs.
3. **Cost Savings:**

- o **Energy Bills:** Effective energy management can lead to significant reductions in energy bills by lowering consumption and avoiding peak rate charges.
- o **Maintenance Costs:** By optimizing energy use and ensuring equipment operates efficiently, EMS can also reduce maintenance costs and extend the lifespan of machinery (H. A. Trinh *et al.*, 2022).

4. **Environmental Impact:**
   - o **Reduced Emissions:** Lower energy consumption leads to reduced greenhouse gas emissions and supports sustainability goals.
   - o **Compliance:** EMS can help plants meet regulatory requirements and environmental standards by providing accurate reporting and ensuring efficient.

Implementing automated systems in food processing enhances consistency, efficiency, and safety, while energy management systems optimize energy use, reduce costs, and support sustainability. Both approaches contribute significantly to improving operational performance in processing plants.

18.8 ENERGY-EFFICIENT TECHNOLOGIES IN FOOD STORAGE AND DISTRIBUTION

18.8.1 Refrigeration and Freezing

1. **Low-Energy Refrigeration Units:** Modern low-energy refrigeration units are designed to minimize energy consumption while ensuring that food is kept at the appropriate temperatures. These systems use a variety of advanced technologies to achieve higher efficiency:
   - **Enhanced Compressor Technologies:** Many new refrigeration units feature more efficient compressors that use less energy to maintain the desired temperature. Variable-speed compressors, for instance, adjust their speed based on cooling needs, reducing energy use compared to traditional fixed-speed models.
   - **Advanced Refrigerants:** Traditional refrigerants can be harmful to the environment and less efficient. Newer systems use advanced refrigerants with lower global warming potential (GWP) and improved thermodynamic properties, which help achieve better cooling efficiency with less energy.
   - **Electronic Expansion Valves:** These valves precisely control the flow of refrigerant, improving the efficiency of the cooling process. By

regulating the amount of refrigerant entering the evaporator, they optimize the cooling cycle and reduce energy consumption.

- **High-Efficiency Fans and Motors:** Modern units often incorporate energy-efficient fans and motors that consume less electricity while maintaining effective airflow and heat exchange (K. J. Chua *et al.*, 2013).
- **Smart Controls and Sensors:** These technologies monitor and adjust the temperature and humidity in real-time. Sensors detect changes in conditions and make precise adjustments to maintain optimal performance, reducing unnecessary energy use.

2. **Improved Insulation:** Enhancing insulation materials in food storage facilities plays a crucial role in reducing energy consumption. Good insulation helps keep the cold air inside and the warm air out, which can significantly lower the amount of energy needed to maintain a stable temperature.
    - **High-Performance Insulation Materials:** Modern insulation materials, such as polyurethane foam or vacuum-insulated panels (VIPs), provide better thermal resistance compared to traditional materials. These materials have high R-values, meaning they are more effective at preventing heat transfer (A. R. Gentle *et al.*, 2011).
    - **Sealed Insulation:** Improved insulation techniques include the use of sealed and continuous insulation systems that minimize gaps and seams. This reduces heat loss and improves the overall energy efficiency of the refrigeration system.
    - **Reflective and Radiant Barriers:** In some cases, reflective barriers or radiant barriers are used to reduce the amount of heat entering the storage space. These materials reflect infrared radiation away from the building, helping to keep the internal temperature stable.
    - **Temperature and Humidity Control:** Advanced insulation also works in tandem with climate control systems that manage both temperature and humidity. By reducing the amount of energy required to control these factors, overall energy consumption is minimized.

Energy-efficient refrigeration and freezing technologies focus on optimizing the cooling process and improving insulation to reduce energy consumption. These advancements not only help lower operational costs but also contribute to environmental sustainability by reducing the carbon footprint associated with food storage and distribution.

18.8.2 Transportation Innovations

1.Energy-Efficient Vehicles

Electric Vehicles (EVs):

- **Power Source:** Electric vehicles are powered by electricity stored in batteries rather than by internal combustion engines that burn fossil fuels. This shift significantly reduces greenhouse gas emissions since electricity can be generated from renewable sources like wind, solar, and hydro (B. Tarroja *et al.*, 2014).
- **Benefits:** EVs have lower operating costs due to reduced fuel costs and fewer moving parts, which decreases maintenance needs. They also offer better energy efficiency; electric motors convert over 90% of the electrical energy from the battery to power the wheels, compared to only 20% to 30% efficiency in conventional gasoline engines.
- **Challenges:** The primary hurdles include battery range, charging infrastructure, and the time required to charge the batteries compared to refueling with gasoline. However, advancements are continually improving these aspects, with longer-range batteries and faster charging technologies.

Hybrid Vehicles:

- **Power Source:** Hybrid vehicles use a combination of an internal combustion engine and an electric motor. They can switch between or combine these power sources to optimize efficiency (V. Krithika and C. Subramani 2018).
- **Types:** There are various types of hybrids, including full hybrids (which can run solely on electric power at low speeds), mild hybrids (which assist the engine with electric power but cannot run on electricity alone), and plug-in hybrids (which have larger batteries that can be charged externally for extended electric-only driving).
- **Benefits:** Hybrids improve fuel economy by using electric power at low speeds and when idling, reducing the reliance on gasoline. They also offer extended driving ranges compared to pure EVs since they can rely on gasoline when the battery is depleted.
- **Challenges:** While hybrids are more efficient than conventional vehicles, they still rely on fossil fuels and are not as clean as pure EVs. They also require complex systems to manage the interaction between the electric and combustion engines.

2. Optimized Logistics

Route Planning:

- **Purpose:** Optimized route planning aims to determine the most efficient paths for transportation, minimizing travel distance and time. This reduces fuel consumption and emissions, leading to cost savings and environmental benefits.
- **Technologies Used:** Advanced algorithms and software are used to calculate optimal routes based on real-time traffic data, weather conditions, road closures, and delivery schedules. Technologies such as GPS and Geographic Information Systems (GIS) play a critical role in this process (J. K. Thakur *et al.*, 2017).
- **Benefits:** Effective route planning can significantly reduce the amount of fuel consumed by vehicles, lower transportation costs, and decrease emissions. For example, fewer miles traveled means less fuel burned and fewer pollutants released into the atmosphere.

3. Load Optimization:

- **Purpose:** Load optimization focuses on maximizing the efficiency of cargo space in vehicles. By properly arranging and consolidating goods, transportation companies can ensure that vehicles are carrying the maximum amount of cargo without exceeding weight limits or space constraints.
- **Techniques:** Techniques include using software to plan the best way to load items into containers, minimizing empty space, and combining shipments to reduce the number of trips needed. For instance, using algorithms to optimize the arrangement of packages within a truck can led to more efficient use of space (S. Onut *et al.,* 2008).
- **Benefits:** Load optimization helps reduce the number of trips required to transport goods, which lowers fuel consumption and emissions. It also leads to cost savings by increasing the efficiency of transportation operations and reducing the overall number of vehicles needed.

4. Integration of Technologies:

- **Real-Time Data:** Integrating real-time data for both route and load optimization can further enhance efficiency. For example, live traffic updates can adjust routes on the fly, and real-time cargo tracking can help manage inventory and optimize delivery schedules.

- **Automation and AI:** The use of artificial intelligence and machine learning can continuously improve route planning and load optimization by learning from historical data and adjusting strategies based on new patterns and trends (R. H. Jhaveri *et al.*, 2022).

Energy-efficient vehicles and optimized logistics represent key innovations in transportation aimed at reducing environmental impact and improving efficiency. By transitioning to electric and hybrid vehicles and employing advanced technologies for route and load optimization, we can significantly cut down on fuel consumption and emissions while also enhancing the overall effectiveness of transportation systems.

18.8.3 Smart Packaging

Active and Intelligent Packaging

1. Active Packaging:

Active packaging refers to packaging systems that actively interact with the contents inside to extend shelf life, improve quality, or enhance safety. There are several key technologies and functions involved:

- **Oxygen Scavengers:** These are substances incorporated into the packaging material that absorb oxygen from the package's interior. By removing oxygen, these scavengers can inhibit the growth of aerobic microorganisms and prevent oxidation, which can spoil food and degrade quality. This is especially useful for products like snacks, meat, and dairy.
- **Moisture Absorbers:** These materials absorb excess moisture within the package. Excess moisture can lead to microbial growth and product degradation. Moisture absorbers are commonly used in packaging for baked goods and dried products (P. V. Mahajan *et al.*, 2008).
- **Ethylene Absorbers:** Ethylene is a natural gas emitted by some fruits and vegetables that accelerates ripening. Ethylene absorbers help to remove this gas from the package, thereby slowing down the ripening process and extending the freshness of produce.
- **Antimicrobial Agents:** These substances can be added to packaging materials to inhibit the growth of bacteria, molds, and yeast. They can be particularly effective in meat, dairy products, and ready-to-eat meals.

2. Intelligent Packaging:

Intelligent packaging involves systems that monitor and provide information about the condition of the packaged product. This technology helps to ensure that the product remains in optimal condition throughout its lifecycle.

- **Sensors and Indicators:** Intelligent packaging often includes embedded sensors that can monitor various parameters like temperature, humidity, and gas composition. For example, temperature indicators can change color if the product has been exposed to temperatures outside of the safe range, signaling potential spoilage (S. Wang *et al.,* 2021).
- **Time-Temperature Indicators (TTIs):** TTIs are used to track the cumulative exposure of a product to temperature fluctuations over time. This is particularly important for perishable goods like vaccines and pharmaceuticals, which require strict temperature control.
- **RFID (Radio-Frequency Identification):** RFID tags can be used to track products throughout the supply chain. These tags can store data about the product's history and condition, which can be accessed by scanning devices to ensure quality and safety.

Eco-Friendly Materials

Eco-friendly packaging materials are designed to reduce the environmental impact of packaging by minimizing waste and using sustainable resources. Here are some of the key approaches:

1. Biodegradable Materials:
   - **Compostable Plastics:** Made from natural materials like corn starch or sugarcane, these plastics break down into natural substances in composting environments. They provide a viable alternative to traditional petroleum-based plastics.
   - **Biopolymers:** These are derived from renewable resources and are designed to break down more easily than conventional plastics. Examples include polylactic acid (PLA) and polyhydroxyalkanoates (PHA) (A. Z. Naser *et al.,* 2021).
2. Recycled Materials:
   - **Recycled Paper and Cardboard:** Packaging made from post-consumer or post-industrial recycled paper helps to reduce deforestation and energy use associated with virgin paper production.
   - **Recycled Plastics:** Using plastics that have been recycled from previous products helps to reduce the demand for new raw materials and lowers the overall environmental footprint.

3. Minimalist Packaging:
    - **Reducing Packaging Volume:** Using less material for packaging can help to decrease waste. This approach often involves designing packaging to be as thin and lightweight as possible without compromising functionality or protection (W. Jedlicka, 2008).
    - **Design for Reuse:** Some packaging designs are intended to be reused, such as glass jars or containers that can be refilled or repurposed. This reduces the need for single-use packaging.
4. Renewable Materials:
    - **Plant-Based Materials:** Packaging made from renewable resources such as bamboo, hemp, or algae can reduce reliance on fossil fuels and contribute to more sustainable practices.
    - **Edible Packaging:** Innovations in edible packaging involve creating packaging from food-grade materials that can be safely consumed or will naturally degrade after use, thereby minimizing waste.

In summary, active and intelligent packaging technologies enhance the functionality and safety of food products, while eco-friendly materials aim to reduce the environmental impact of packaging. Both approaches represent significant advancements in the quest for more sustainable and effective packaging solutions.

18.9 FUTURE TRENDS AND INNOVATIONS

18.9.1 Emerging Technologies

Artificial Intelligence (AI) and Machine Learning (ML) are revolutionizing various industries, including agriculture and food production. Here's how these technologies contribute to improving food quality and energy efficiency:

Optimizing Energy Use

1. **Precision Agriculture:**
    - o **Data-Driven Insights:** AI and ML algorithms analyze data from various sources, such as soil sensors, weather forecasts, and satellite images. This analysis helps farmers make informed decisions on irrigation, fertilization, and pest control, leading to more efficient energy use. For example, smart irrigation systems can use ML to determine the precise amount of water needed for crops, reducing water and energy waste (P. K. Kashyap *et al.*, 2021).

- **Automation:** AI-powered machinery, like drones and autonomous tractors, can optimize field operations. These machines use AI to perform tasks such as planting, harvesting, and applying fertilizers more efficiently, minimizing energy consumption and reducing the overall environmental footprint.

2. **Energy Management in Food Processing:**
   - **Predictive Maintenance:** AI can predict when equipment is likely to fail or require maintenance, preventing unexpected breakdowns and optimizing the energy use of machinery. This ensures that food processing plants run smoothly and efficiently, avoiding energy wastage (F. Chemat *et al.*, 2017).
   - **Energy Optimization:** AI algorithms can analyze energy usage patterns in food processing facilities and suggest improvements. For instance, AI can optimize heating, cooling, and lighting systems to minimize energy consumption while maintaining the quality of the food products.

Enhancing Food Quality

1. **Quality Control:**
   - **Visual Inspection:** AI-powered computer vision systems can inspect food products for defects or inconsistencies. Machine learning models trained on large datasets can identify subtle quality issues that might be missed by human inspectors. This ensures that only high-quality products reach consumers.
   - **Predictive Quality Analytics:** AI can predict how different variables, such as temperature and humidity, affect food quality during storage and transportation. By analyzing historical data and real-time conditions, AI helps maintain optimal conditions to preserve the quality of food products.
2. **Personalized Nutrition:**
   - **Customized Food Products:** AI can analyze individual dietary preferences and nutritional needs to recommend or create personalized food products. Machine learning algorithms can process data from various sources, including genetic information and lifestyle factors, to tailor food formulations that meet specific health goals and improve overall quality (R. Roy *et al.*, 2022).
   - **Recipe Optimization:** AI can assist in developing new recipes or improving existing ones by analyzing ingredient combinations, flavor

profiles, and nutritional content. This helps in creating healthier and more appealing food products.

3. **Supply Chain Optimization:**
   - **Demand Forecasting:** AI can predict consumer demand for different food products, helping manufacturers and retailers adjust production and inventory levels. This reduces food wastage and ensures that high-quality products are available when and where they are needed (N. Raak *et al.,* 2017).
   - **Traceability:** AI enhances food traceability through block chain and other technologies, ensuring that food products can be tracked throughout the supply chain. This improves food safety and quality by quickly identifying and addressing any issues that arise.

AI and ML contribute to both energy efficiency and food quality by optimizing resource use, improving quality control, enabling personalized nutrition, and enhancing supply chain management. These advancements lead to more sustainable and high-quality food production systems.

The deployment of energy-efficient technologies in the food industry is a key driver in advancing food quality while fostering environmental stewardship. The ability to maintain optimal processing conditions, enhance product consistency, and achieve cost savings underscores the value of these technologies. As the industry continues to evolve, embracing energy efficiency will be crucial not only for meeting current demands for high-quality food but also for securing a sustainable future for food production and consumption (C. M. Galanakis, 2024).

18.10 CONCLUSION

Energy-efficient technologies are reorganizing the food industry by improving food quality while reducing energy consumption and environmental impact on food. Espousing these technologies not only enhances the sustainability of food production but also encounter the growing demands for high-quality food. As we look to the future, continued innovation and adoption of these technologies will be crucial in achieving a more sustainable and efficient food system. The integration of energy-efficient technologies in the food industry represents a transformative opportunity to enhance food quality while simultaneously promoting sustainability. By adopting advanced energy-efficient systems, food production processes can be optimized to deliver superior quality products that meet both consumer expectations and environmental standards.

Energy-efficient technologies contribute significantly to maintaining optimal conditions throughout the food production and storage processes. For instance, innovations such as advanced refrigeration systems and precision-controlled help in preserving the nutritional value, freshness and safety of food products. These technologies minimize temperature fluctuations and decrease the risk of spoilage, ensuring that food quality remains consistently high from production through to the starting point of consumption. Additionally, energy-efficient technologies often facilitate more exact control over processing conditions, which can improve the consistency and quality of food products. Automated systems equipped with energy-efficient sensors and controls enable manufacturers to fine-tune processes with greater accuracy, leading to products that not only taste better but also retain their desired textures and nutritional profiles.

From an environmental point of view, the acquiring of energy-efficient technologies in the food industry also aligns with broader sustainable desired results. Reduced energy consumption translates into lower greenhouse gas emissions and less environmental impact, contributing to a more sustainable food production system. This shift not only supports ecological balance but also appeals to increasingly environmentally-conscious consumers. Moreover, energy-efficient technologies can result in important cost savings over time. By decreasing energy consumption, food producers can lower operational costs, which can be reinvested into further exploring innovative product development or improving food quality. This economic benefit reinforces the viability of acquiring energy-efficient practices as a sound business decision.

REFERENCES

Anderson, S. W., Daly, J. D., & Johnson, M. F. (1999). Why firms seek ISO 9000 certification: Regulatory compliance or competitive advantage? *Production and Operations Management*, *8*(1), 28–43. DOI: 10.1111/j.1937-5956.1999.tb00059.x

Andrés-Bello, A., Barreto-Palacios, V. I. V. I. A. N., García-Segovia, P., Mir-Bel, J., & Martinez-Monzo, J. (2013). Effect of pH on color and texture of food products. *Food Engineering Reviews*, *5*(3), 158–170. DOI: 10.1007/s12393-013-9067-2

Bajwa, U., & Sandhu, K. S. (2014). Effect of handling and processing on pesticide residues in food-a review. *Journal of Food Science and Technology*, *51*(2), 201–220. DOI: 10.1007/s13197-011-0499-5 PMID: 24493878

Blackburn, J., & Scudder, G. (2009). Supply chain strategies for perishable products: The case of fresh produce. *Production and Operations Management*, *18*(2), 129–137. DOI: 10.1111/j.1937-5956.2009.01016.x

Bozchalui, M. C., Canizares, C. A., & Bhattacharya, K. (2015). Optimal energy management of greenhouses in smart grids. *IEEE Transactions on Smart Grid*, *6*(2), 827–835. DOI: 10.1109/TSG.2014.2372812

Bundschuh, J., Chen, G., & Mushtaq, S. (2014). Towards a sustainable energy technologies based agriculture, in Sustain. *Energy Solution Agricultural ROUT LEDGEinassociation with GSEResearch*, *3*(15), 3–15.

Chel, A., & Kaushik, G. (2011). Renewable energy for sustainable agriculture. *Agronomy for Sustainable Development*, *31*(1), 91–118. DOI: 10.1051/agro/2010029

Chemat, F., Rombaut, N., Meullemiestre, A., Turk, M., Perino, S., Fabiano-Tixier, A. S., & Abert-Vian, M. (2017). Review of green food processing techniques. Preservation, transformation, and extraction. *Innovative Food Science & Emerging Technologies*, *41*, 357–377. DOI: 10.1016/j.ifset.2017.04.016

Chua, K. J., Chou, S. K., Yang, W. M., & Yan, J. (2013). Achieving better energy-efficient air conditioning–a review of technologies and strategies. *Applied Energy*, *104*, 87–104. DOI: 10.1016/j.apenergy.2012.10.037

De Boer, A., Krul, L., Fehr, M., Geurts, L., Kramer, N., Urbieta, M. T., & Hepburn, P. A. (2020). Animal-free strategies in food safety and nutrition: What are we waiting for? Part I: Food safety. *Trends in Food Science & Technology*, *106*, 469–484. DOI: 10.1016/j.tifs.2020.10.034

Degerli, B., Nazir, S., Sorgüven, E., Hitzmann, B., & Ozilgen, M. (2015). As sessment of the energy and exergy efficiencies of farm to fork grain cultivation and bread making processes in Turkey and Germany. *Energy*, *93*, 421–434. DOI: 10.1016/j.energy.2015.08.095

Driesen, D. M. (2002). Sustainable Development and Air Quality: The Need to Replace Basic Technologies with Cleaner Alternatives. *Envtl. L. Rep. News & Analysis*, *32*, 10277.

European Commission. data– Eurostat, Energy. balancesheets [Online]. (2017) Available: http://ec.europa.eu/eurostat/documents/3217494/8113778/KS-EN 17-001- EN-N.pdf/99cc20f1-cb11-4886-80f9-43ce0ab7823c

Fabbriani, L., & Calili, R. (2018). Proposal of energy efficiency policies for food and beverage industry in Brazil. *Journal of Renewable and Sustainable Energy*, *10*(6), 065903. DOI: 10.1063/1.5002088

Ferreira, F. J., Baoming, G., & de Almeida, A. T. (2015). Reliability and operation of high-efficiency induction motors. In *2015 IEEE/IAS 51st Industrial and Commercial Power Systems Technical Conference (I andCPS)* (pp. 1-13). IEEE. DOI: 10.1109/ICPS.2015.7266412

Floros, J. D., Newsome, R., Fisher, W., Barbosa-Cánovas, G. V., Chen, H., Dunne, C. P., & Ziegler, G. R. (2010). Feeding the world today and tomorrow: the importance of food science and technology: an IFT scientific review. *Comprehensive Reviews in Food Science and Food Safety*, *9*(5), 572–599. DOI: 10.1111/j.1541-4337.2010.00127.x PMID: 33467827

Galanakis, C. M. (2012). Recovery of high added-value components from food wastes: Conventional, emerging technologies and commercialized applications. *Trends in Food Science & Technology*, *26*(2), 68–87. DOI: 10.1016/j.tifs.2012.03.003

Galanakis, C. M. (2024). The future of food. *Foods*, *13*(4), 506. DOI: 10.3390/foods13040506 PMID: 38397483

Gentle, A. R., Aguilar, J. L. C., & Smith, G. B. (2011). Optimized cool roofs: Integrating albedo and thermal emittance with R-value. *Solar Energy Materials and Solar Cells*, *95*(12), 3207–3215. DOI: 10.1016/j.solmat.2011.07.018

Gould, G. W. (1996). Methods for preservation and extension of shelf life. *International Journal of Food Microbiology*, *33*(1), 51–64. DOI: 10.1016/0168-1605(96)01133-6 PMID: 8913809

Gutiérrez, J., Villa-Medina, J. F., Nieto-Garibay, A., & Porta-Gándara, M. A. (2013). Automated irrigation system using a wireless sensor network and GPRS module. *IEEE Transactions on Instrumentation and Measurement, 63*(1), 166–176. DOI: 10.1109/TIM.2013.2276487

Halpin, M. (2013). Advancing standards in energy efficiency. *IEEE Industry Applications Magazine, 19*(4), 78–79. DOI: 10.1109/MIAS.2013.2255426

Jedlicka, W. (2008). *Packaging sustainability: tools, systems and strategies for innovative package design*. John Wiley and Sons.

Jhaveri, R. H., Revathi, A., Ramana, K., Raut, R., & Dhanaraj, R. K. (2022). A Review on Machine Learning Strategies for Real-World Engineering Applications. *Mobile Information Systems, 2022*(1), 1833507. DOI: 10.1155/2022/1833507

Jouhara, H., Khordehgah, N., Almahmoud, S., Delpech, B., Chauhan, A., & Tassou, S. A. (2018). Waste heat recovery technologies and applications. *Thermal Science and Engineering Progress, 6*, 268–289. DOI: 10.1016/j.tsep.2018.04.017

Kashyap, P. K., Kumar, S., Jaiswal, A., Prasad, M., & Gandomi, A. H. (2021). Towards precision agriculture: IoT-enabled intelligent irrigation systems using deep learning neural network. *IEEE Sensors Journal, 21*(16), 17479–17491. DOI: 10.1109/JSEN.2021.3069266

Kling, M. M., & Hough, I. J. (2010). *The American Carbon Footprint: Un derstanding your food's impact on climate change, Bright*. Planet, Inc.

Kovacic, D., Loncaric, Z., Jovic, J., Samac, D., Popovic, B., & Tisma, M. (2022). Digestate management and processing practices: A review. *Applied Sciences (Basel, Switzerland), 12*(18), 9216. DOI: 10.3390/app12189216

Krithika, V., & Subramani, C. (2018). A comprehensive review on choice of hybrid vehicles and power converters, control strategies for hybrid electric vehicles. *International Journal of Energy Research, 42*(5), 1789–1812. DOI: 10.1002/er.3952

Kumar, S. N., & Chakabarti, B. (2019). nergy and carbon footprint of food industry, in Energy footprints food Text. *Section. Springer*, pp. 19 44.

Li, M. J., & Tao, W. Q. (2017). Review of methodologies and polices for evaluation of energy efficiency in high energy-consuming industry. *Applied Energy, 187*, 203–215. DOI: 10.1016/j.apenergy.2016.11.039

Lin, B., & Xie, X. (2015). Factor substitution and rebound effect in China's food industry. *Energy Conversion and Management, 105*, 20–29. DOI: 10.1016/j.enconman.2015.07.039

Mahajan, P. V., Rodrigues, F. A., Motel, A., & Leonhard, A. (2008). Development of a moisture absorber for packaging of fresh mushrooms (Agaricus bisporous). *Postharvest Biology and Technology*, *48*(3), 408–414. DOI: 10.1016/j.postharvbio.2007.11.007

Marsden, T., & Morley, A. (2014). Current food questions and their scholarly challenges: Creating and framing a sustainable food paradigm. In *Sustainable food systems* (pp. 1–29). Routledge. DOI: 10.4324/9780203083499-9

Naser, A. Z., Deiab, I., & Darras, B. M. (2021). Poly (lactic acid)(PLA) and polyhydroxyalkanoates (PHAs), green alternatives to petroleum-based plastics: A review. *RSC Advances*, *11*(28), 17151–17196. DOI: 10.1039/D1RA02390J PMID: 35479695

Onut, S., Tuzkaya, U. R., & Dogaç, B. (2008). A particle swarm optimization algorithm for the multiple-level warehouse layout design problem. *Computers & Industrial Engineering*, *54*(4), 783–799. DOI: 10.1016/j.cie.2007.10.012

Ozilgen, M. (2016). Energy utilization and carbon dioxide emission during production of snacks. *Journal of Cleaner Production*, *112*, 2601–2612. DOI: 10.1016/j.jclepro.2015.10.031

Ozilgen, M. (2018). Nutrition and production related energies and exergies of foods. *Renewable & Sustainable Energy Reviews*, *96*(August), 275–295. DOI: 10.1016/j.rser.2018.07.055

Pape, R. A. (1997). The limits of precision-guided air power. *Security Studies*, *7*(2), 93–114. DOI: 10.1080/09636419708429343

Raak, N., Symmank, C., Zahn, S., Aschemann-Witzel, J., & Rohm, H. (2017). Processing-and product-related causes for food waste and implications for the food supply chain. *Waste Management (New York, N.Y.)*, *61*, 461–472. DOI: 10.1016/j.wasman.2016.12.027 PMID: 28038904

Robson, J. R. (Ed.). (2023). *Famine: Its Causes, Effects and Management*. Taylor and Francis. DOI: 10.4324/9781003412366

Roy, R., Marakkar, S., Vayalil, M. P., Shahanaz, A., Anil, A. P., Kunnathpeedikayil, S., & Yadav, K. K. (2022). Drug-food interactions in the era of molecular big data, machine intelligence, and personalized health. *Recent Advances in Food, Nutrition & Agriculture*, *13*(1), 27–50. DOI: 10.2174/2212798412666220620104809 PMID: 36173075

Saidur, R., Mekhilef, S., Ali, M. B., Safari, A., & Mohammed, H. A. (2012). Applications of variable speed drive (VSD) in electrical motors energy savings. *Renewable & Sustainable Energy Reviews*, *16*(1), 543–550. DOI: 10.1016/j.rser.2011.08.020

Schumm, G., Philipp, M., Schlosser, F., Hesselbach, J., Walmsley, T. G., & Atkins, M. J. (2018). Hybrid heating system for increased energy efficiency and f lexible control of low temperature heat. *Energy Efficiency*, *11*(5), 1117–1133. DOI: 10.1007/s12053-017-9584-6

Shoreh, M. H., Siano, P., Shafie-khah, M., Loia, V., & Catalao, J. P. (2016). A survey of industrial applications of Demand Response. *Electric Power Systems Research*, *141*, 31–49. DOI: 10.1016/j.epsr.2016.07.008

Shrouf, F., & Miragliotta, G. (2015). Energy management based on Internet of Things: Practices and framework for adoption in production management. *Journal of Cleaner Production*, *100*, 235–246. DOI: 10.1016/j.jclepro.2015.03.055

Sims, R. E., Bundschuh, J., & Chen, G. (2014). Global energy resources, supply and demand, energy security and on-farm energy efficiency. *Sustainable Energy Solutions in Agriculture*, *1*, 19–52.

Sperberg, R., Kniss, R., & Ruegg, T. (2018). Energy Efficiency in California Cement Plants: Energy Savings Potential Realized by Accessing Utility Energy-Efficiency Program Resources. *IEEE Industry Applications Magazine*, *24*(4), 12–16. DOI: 10.1109/MIAS.2017.2740453

Stroker, J. J. (2003). What's the real cost of higher efficiency? *IEEE Industry Applications Magazine*, *9*(3), 32–37. DOI: 10.1109/MIA.2003.1195680

Tarroja, B., AghaKouchak, A., & Samuelsen, S. (2016). Quantifying climate change impacts on hydropower generation and implications on electric grid greenhouse gas emissions and operation. *Energy*, *111*, 295–305. DOI: 10.1016/j.energy.2016.05.131

Thakur, J. K., Singh, S. K., & Ekanthalu, V. K. (2017). Integrating remote sensing, geographic information systems and global positioning system techniques with hydrological modeling. *Applied Water Science*, *7*(4), 1595–1608. DOI: 10.1007/s13201-016-0384-5

Trinh, H. A., Truong, H. V. A., Do, T. C., Nguyen, M. H., & Ahn, K. K. (2022). Optimization-based energy management strategies for hybrid construction machinery: A review. *Energy Reports*, *8*, 6035–6057. DOI: 10.1016/j.egyr.2022.04.050

Venendaal, R., Jørgensen, U., & Foster, C. A. (1997). European energy crops: A synthesis. *Biomass and Bioenergy*, *13*(3), 147–185. DOI: 10.1016/S0961-9534(97)00029-9

Viswanadham, N., & Narahari, Y. (2015). *Performance modeling of automated systems*. PHI Learning Pvt. Ltd.

Wang, K., Wang, Y., Sun, Y., Guo, S., & Wu, J. (2016). Green Industrial Internet of Things Architecture: An Energy-Efficient Perspective. *IEEE Communications Magazine*, *54*(11), 48–54. DOI: 10.1109/MCOM.2016.1600399CM

Wang, S., Liu, X., Yang, M., Zhang, Y., Xiang, K., & Tang, R. (2015). Review of time temperature indicators as quality monitors in food packaging. *Packaging Technology & Science*, *28*(10), 839–867. DOI: 10.1002/pts.2148

Wang, X., Puri, V. M., & Demirci, A. (2020). Equipment cleaning, sanitation, and maintenance. Food safety engineering, 333-353.

Xu, Y., & Szmerekovsky, J. (2017). System dynamic modeling of energy savings in the US food industry. *Journal of Cleaner Production*, *165*, 13–26. DOI: 10.1016/j.jclepro.2017.07.093

Yamprayoon, J., & Noomhorm, A. (2000). Effects of preservation methods on geosmin content and off-flavor in Nile tilapia (Oreochromis niloticus). *Journal of Aquatic Food Product Technology*, *9*(4), 95–107. DOI: 10.1300/J030v09n04_09

Chapter 13
Sustainable Packaging Solutions:
Food Packaging

Muhammad Sibt-e-Abbas
https://orcid.org/0009-0003-5981-755X
MNS-University of Agriculture, Multan, Pakistan

Rimsha Umar
MNS-University of Agriculture, Multan, Pakistan

Salman Khalil
https://orcid.org/0009-0006-3058-8265
MNS-University of Agriculture, Multan, Pakistan

ABSTRACT

This chapter explores sustainable packaging solutions, highlighting the importance of reducing environmental impact through the adoption of renewable materials, energy-efficient processes, and cutting-edge technologies. It emphasizes the principles of reducing, reusing, and recycling, with a focus on biodegradable and compostable materials that mitigate the adverse effects of conventional packaging. The role of smart packaging in enhancing food safety and quality is also examined, presenting new opportunities in the industry. The stakeholders—manufacturers, retailers, policymakers, and consumers can collaborate to drive the adoption of eco-friendly packaging, overcome challenges, and ensure scalability and cost-effectiveness. Looking to the future, it envisions a path of continuous innovation, where advancements in biopolymers, intelligent systems, and digital technologies pave the way for packaging solutions that are both environmentally and economically sustainable.

DOI: 10.4018/979-8-3693-8342-1.ch013

1. INTRODUCTION

1.1. Overview of the Environmental and Economic Importance of Packaging in the Food Industry

Several official institutions denote a sustainable economy as a "green economy" which covers the three-dimensionality of sustainability including economy, environment, and human well-being. A major economic challenge is improving resource efficiency – a partial solution to enhance human well-being in fair burden-sharing, social equity, and ecosystem resilience. The principal role of food packaging is to provide foodstuffs with adequate protection from external damage and influences (Otto *et al.*, 2021).

However, it should also preserve the food quality and safety during the processing steps of transport, distribution as well as storage. This reduces food loss and waste thereby contributing to the economy. Moreover, food packaging represents the product before purchase. This makes it important to select the design and material of packaging thoughtfully (Truong, 2019).

An unlimited access to information for consumers has resulted in radical change of packaging lately (Sandu, 2014). Consumers opt more for food packaging that causes less waste, uses recycled materials and can be recycled after use. Furthermore, consumers prefer to buy foods in traditional packaging such as cardboard/paper, glass, recycled packaging, those with minimum packaging material, or simply unpackaged foods. Specifically, consumers avoid plastic packaging (Thangapandi, 2023).

1.2. Definition and Scope of Sustainable Packaging

Sustainable packaging focuses on minimizing environmental impact through the use of renewable resources, energy-efficient processes, and materials that can be reused, recycled, or composted. A key principle is reducing material usage by designing lighter packages and eliminating unnecessary components, which decreases waste and resource consumption. Reusable packaging, like refillable containers, extends product lifecycles and supports a circular economy by reducing waste (Coelho *et al.,* 2020).

Recycling is another critical element, emphasizing the use of single-material solutions to simplify the process. The use of renewable materials, such as bioplastics and sustainably sourced paper, reduces reliance on finite resources and supports sustainability. Biodegradable and compostable packaging offers eco-friendly alternatives, breaking down naturally or turning into nutrient-rich compost, reducing waste and pollution (Tajeddin, 2017).

Biodegradable options, such as mushroom-based packaging and cornstarch foam, decompose without harmful residues, while compostable packaging enriches soil in the right conditions. Reusable packaging, like refillable containers, reduces waste by encouraging multiple uses. Recyclable materials, designed with clear labeling, can easily be processed into new products, and minimalist packaging further reduces waste by using fewer materials (Varžinskas and Markevičiūtė, 2020).

2. ENVIRONMENTAL IMPACT OF CONVENTIONAL PACKAGING

Food packaging reduces food wastage across the supply chain, which has undeniable positive environmental consequences. However, it is impossible to overlook the fact that food packaging accounts for roughly two-thirds of all packaging in developed nations. While there are packaging design options that can reduce food waste and the packaging's environmental impact, the majority of packaging systems have apparent environmental impacts, particularly if they are for single time use. Because packaging systems require energy, water, and raw materials to produce, as well as the discharge of greenhouse gases (GHG), toxic pollutants, heavy metals, wastewater, and sludge, their detrimental effects on the environment begin at the point of creation. As a result of their disposal, further additional effects are produced at the end of their packaging life. The ratio of packaging to food's environmental impact varies greatly depending on the product (Shin and Selke 2014).

2.1. Analysis of Commonly Used Packaging Materials (Plastics, Glass, Metals, etc.)

Nowadays, a variety of materials are frequently employed in food packaging applications, and new materials are continually being created. Paper and paperboards, which are frequently used in corrugated boxes, milk cartons, folding cartons, paper plates and cups, bags and sacks, and wrapping paper; metals like steel and aluminium, which are used in cans, tubes, containers, films, caps, and closures; plastics like Polyolefins (PE and PP), Polyester (like PET), Polyvinyl Chloride, PolyVinylidene Chloride, Polystyrene, PA, and Ethylene Vinyl Alcohol are among the traditional materials used in packaging of food (Geueke *et al.*, 2018). Nowadays, bottles, bags, trays, foils, cups, pouches, pots, and bowls are all made of these plastics, which have emerged as the most widely used materials for food packaging (Marsh and Bugusu, 2007).

A mixture of silica (the glass former), sodium carbonate (the melting agent), and stabilizers such as alumina and limestone/calcium carbonate are heated to high temperatures in order to melt the components into a thick liquid mass that is put into moulds to create glass containers. Glass is basically a best option for food packing because it is odourless and chemically inert with almost all food ingredients. It keeps products fresh for a long time without compromising flavour or taste since it is invincible to gases and vapours. Glass can be used to heat sterilize both high-acid and low-acid foods because of its resistance to high processing temperatures. Manufacturers can utilize thinner glass due to improved break resistance, which lowers weight and makes it easier to dispose of and transport (Marsh and Bugusu, 2007).

Of all the packing materials, metal is the most adaptable. Excellent barrier and physical protection, formability and ornamental potential, recyclability, and customer acceptance are all included in this product. Steel and aluminium are the two metals most frequently used in packaging. Despite aluminum's ease of recycling, recycled aluminium cannot be used to make foils without the thin sheets developing pinholes. Tinplate, which is made of low-carbon steel, is far less expensive than aluminium and can be recycled again without compromising its quality (Fellows and Axtell, 2002).

Both condensation polymerization (poly-condensation) and addition polymerization (poly-addition) of monomer units are used to create plastics. The use of plastics in food packaging has a number of benefits. Plastics are fluid and mouldable, allowing for significant design flexibility in the creation of sheets, shapes, and structures. Plastics have a variety of physical and visual qualities and are affordable and lightweight due to their chemical resistance. They can be disposed of in three broad ways: landfilling, energy reclamation, and recycling (Lawrence, 2017).

2.2. Environmental Issues Associated With Conventional Packaging

A food packaging material's environmental effects begin with its design and conclude with its disposal. In addition to producing wastewater and/or sludge that may contain pollutants, the packaging production process releases gases (GHG, sulphur oxides, and nitrogen oxides), fine particulates, volatile organic compounds,PAH and heavy metals like lead and arsenic into the atmosphere. Packaging materials are thrown away after use, and improperly managed garbage can pollute the water, air, and land, which has detrimental consequences on people, animals, and plants (Šerešová and Kočí, 2020). Half of the polymers made from fossil fuels in the packaging industry are used in food packaging. The packaging that once held the food is discarded along with it. These plastics made from fossil fuels take several years to break down and are tenacious in the environment. They decompose into micro-plastics as they go, which can readily make their way into the food chain and cause bio accumulation when eaten by fish, for instance (Payne *et al.*, 2019).

- **Resource depletion:** The biggest and fastest-growing market for synthetic polymers made from fossil fuels is the packaging sector. The majority of plastic garbage that pollutes the environment comes from food packaging. Furthermore, as petroleum sources are limited and facing depletion, it is necessary to produce substitute materials that can perform the same function as traditional plastics (Ncube *et al.*, 2020).
- **Greenhouse gas emissions:** Irrespective of whether the packaging is biodegradable or not, food waste and its associated packaging account for more than 50% of greenhouse gas emissions in the life-cycle study of food packaging. This suggests that food waste production and disposal have a far greater influence than packaging (Barlow and Morgan, 2013).
- **Waste generation and pollution:** Although there are many different packaging materials available today, including bottles, cartons, cellophane, cans, plastics, etc., the issue of managing their waste has also grown as new packaging materials have been developed. The use of packing materials is rapidly growing in every country practically. The manufacturing, utilization, transportation, and disposal of packaging materials all have an direct influence on the environment. The packaging industry contributes significantly to soil, water, air, and land contamination (Varun *et al.*, 2016). After being thrown out, these materials frequently wind up in landfills or, worse, in rivers and oceans, where they linger for hundreds of years. Micro-plastics, which are created when these plastics degrade, have the potential to permeate ecosystems and the food chain, endangering both human health and wildlife (Porta *et al.*, 2020).

2.3. Case studies Highlighting the Impact of Conventional Packaging on Ecosystems

Lebreton *et al.* (2018) highlighted the rapid accumulation of waste of plastic, including packaging materials, in the Great Pacific Garbage Patch, leading to significant ecological damage. Over thousands of km, this enormous buildup of plastic waste (mostly from traditional packing materials) stretches across the North Pacific. Plastic bottles, bags, and micro-plastics from degraded packaging materials make up the majority of the debris. Furthermore, when plastics break down into micro-plastics, toxic compounds are released into the food chain, which eventually affects both human health and marine life. River Yamuna which is considered one of the most polluted rivers in India, is highly contaminated with waste of plastic from conventional packaging. Singh and Ramanathan (2019) highlighted the adverse impacts of plastic packaging waste on the pollution levels of the River Yamuna, negatively effecting aquatic biodiversity and local populations as well. The biodiversity of river also has declined significantly due to water pollution. Moreover, various communities living near the river also suffer from a lack of pure and clean drinking water.

3. SUSTAINABLE PACKAGING MATERIALS AND TECHNOLOGIES

3.1. Overview of Sustainable Packaging Concepts and Principles

Environmental factors are now becoming more significant in the worldwide packaging development trends. Tragically, once the packaging is utilized, it becomes waste and can have a big effect on the environment. This is particularly true for single-use packaging and packaging composed of different materials (Ma and Moultrie, 2017). International literature is utilizing the term "sustainable packaging" more and more. Theoretically, these packaging materials have better performance and quality characteristics, meet higher environmental, economic, and social criteria than traditional packaging, and open up new opportunities for waste management and recovery. All phases of the packaging life cycle, including manufacturing, packaging, distribution, transportation, utilization, and disposal, should be covered by these standards or guidelines. A procedural method that adds characteristics to a typical product that raise its economic, social, and environmental value throughout the course of its whole life cycle produces sustainable packaging. Packaging is a crucial component of sustainable development, which is based on the three pillars of economic development, social development, and environmental protection (Kozik, 2020).

- **Reduce, Reuse, Recycle (3Rs):** Collection and sorting are the first and most crucial phases in the recycling of plastics. Sorting means choosing waste that can be recycled through both automated and human methods. This can be accomplished at the garbage's origin by employing distinct collecting methods or at facilities by mechanically processing and classifying mixed waste. Packaging sorting facilities are therefore essential to material recovery and waste management systems (Cimpan *et al.*, 2015). The main goal of the reduction principle is to utilize less material in processes and products (Sæter *et al.*, 2020). In the context of food packaging, this means using less material to create a product while yet ensuring that it serves its intended purpose. Packaging products with significantly less of an impact on the environment can be produced using material reduction techniques (Siracusa *et al.*, 2014). Reusing means using waste from discarded plastic packaging multiple of times. Because it requires fewer effort and money to complete, this is the most popular way (Al-Salem *et al.*, 2009). Instead of being thrown of, waste materials can find an alternative way to gain value through reuse. Reusable plastic packaging is useful for primary, secondary, and tertiary packing and can help lower overall packaging cost (Gardas *et al.*, 2019). The packaging can be used as many times as it still fulfils its intended purpose. However, repurposed plastic packaging may be polluted due to its use; therefore, it may be necessary to wash the packaging before reusing it to prevent pathogens from migrating to food (Grigore, 2017).

- **Circular economy approach:** Exploring packaging's function within the circular economy, which is basically restorative and regenerative by design and seeks to extract as much value as possible from products, components, and materials over time, is possible given its impact on several ecosystems. In this context, more environmentally friendly methods of minimizing packaging use and controlling packaging waste through the separation, collection, recovery, and recycling of used packaging are replacing the linear take-make-use-dispose approach to packaging (Batista *et al.*, 2018). Therefore, it is imperative to comprehend the function of packaging as a sustainable practice in order to establish a circular economy, given the growing pressure on supply chains to be sustainable in all respects (Meherishi *et al.*, 2019).

3.2. Biodegradable and Compostable Materials

Nowadays, the food packaging sector is looking for lightweight, biodegradable packaging to cut down on waste, materials, and transportation expenses. Biopolymer-based plastics show promise in meeting this need. Heat, moisture, and/or microbial enzymes can start the defragmentation process known as biodegradation, which breaks down longer molecules into smaller ones. It can also be simply described as the process by which living things break down substances (Ashok *et al.*, 2016). Biopolymers, often known as biomaterials, are polymers made from renewable resources. Originally, biopolymers were made from plant-based raw materials, but more recently, animal-based ones have also been used. Their biodegradability is their primary characteristic (Davidovic and Savic, 2010).

- **Types (e.g., bioplastics, paper-based materials):**

Material made up of biopolymers can be divided into three main classes according to the origin and production methods:

i. Polymers extracted or isolated from biomass
ii. Polymers created by classical chemical synthesis and bio-monomers

iii. Polymers produced directly through natural or genetically modified organisms (Chiellini, 2008).

i. Polymers extracted/isolated directly from biomass

The majority of biopolymers on the market fall into this group. This class of polymers is found in domestic animals, marine life, and plants. Examples include polysaccharides, which can be utilized alone or in combination with synthetic polyesters like polylactic acid (PLA). These include cellulose, chitin and starch, whey protein, casein, collagen, soy protein, and myofibrillar proteins of animal muscle. Cellulose-based paper is the most widely used type in food packaging. The study of hemicellulose, the second most prevalent plant polymer in the world, as a biomaterial for food packaging is still in its early stages (Grondahl *et al.*, 2006).

ii. Polymers produced by classical chemical synthesis and bio-monomers

By chemical synthesis a wide variety of biopolyesters can be produced. Among these biopolymers, polylactic acid (PLA-polylactide) is the most well-known. Because PLA has characteristics with polystyrene, it is a biodegradable thermoplastic linear polyester. The fermentation of glucose or starch from another source provides the raw material for making lactic acid (Wackett, 2008). PLA is primarily used to make thermoformed pads, food packaging and service containers, films, transparencies, bottles, and other blown packaging (Seglia *et al.*, 2009).

iii. Polymers obtained directly from natural or genetically modified organisms

These polymers are accumulated by many bacteria as a carbon reserve and energy source. Polyhydroxyalkanoates (PHAs) and bacterial cellulose are members of this group. PHAs are polyesters that are hydrophobic and insoluble in water, and they are a component of the structure of living organisms. Their traits are primarily linked to the features of the monomer building blocks, from which microbial fermentation can produce a broad array of distinct biopolymers. The entire biochemical synthesis process is carried out in the cell by a sequence of enzyme reactions. Depending on the carbon source and the bacteria, the polyhydroxyalkanoate can be produced as a rubber like polymer, plastic, or a stiff, brittle polymer possess comparable qualities to those of polyethylene and propylene, as well as being elastic and thermoplastic (which is preserved when cooled) (Zivkovic, 2009).

- o **Environmental benefits and challenges:** Their qualities, processing, and cost were the primary issues with the most popular food packaging. Specifically, their application is limited by their brittleness, low processing resistance (except from PHA polymers), low temperatures that cause distortion, and their barrier qualities, especially with regard to water vapour. The price has decreased recently and will undoubtedly continue to do so, and plants that produce this material should become more efficient and optimize their processes over time. One of the most urgent issues pressing the development of such materials is still the restricted supply of raw ingredients (Ivonkovic *et al.*, 2017).

3.3. Edible Packaging

Researchers and manufacturers are being forced to embrace bio-based sustainable packaging materials and circular economy practices in the current period due to consumers' shifting attitudes towards food packaging, quality, disposal patterns, and environmental concerns. An innovative and environmentally responsible substitute for traditional food packaging may be edible packaging. Since edible packaging materials are made from natural food-grade polymers like polysaccharides, proteins, or lipids, or are sourced from plants, animals, or marine life, consumers can eat them (Mohamed *et al.*, 2020).Food quality is preserved, shelf life is increased, and waste is somewhat decreased using edible packaging. Because edible packaging is an essential component of the product, customers do not need to unpack it, which is a distinct advantage (Saklani *et al.*, 2019). Commercial packaging of fruits, vegetables, dairy goods, and meat items can be done with edible packaging materials (Jeya Jeevahan *et al.*, 2020).

3.4. Reusable and Refillable Packaging Systems

The literature uses a variety of words to describe different kinds of reusable packaging. For instance, reusable primary and tertiary packaging are referred to as "returnable packaging materials" and "returnable transport items" respectively. Examples of returnable packaging materials include steel coils, gas cylinders, chemical containers, refillable glass bottles for drinks and special packaging made for shipping medical equipment, wind turbine parts and other items. It is well known that reuse techniques lower the demand for raw resources and the amount of garbage that ends up in landfills (Mahmoudi and Parviziomran, 2020).

4. INNOVATIONS IN PACKAGING DESIGN AND MANUFACTURING

Food packaging is made to keep food safe and secure, to give the necessary information about the food, and to make handling food easier from delivery to the table of the customer. The preservation and safe transportation of food items until they are consumed are the main goals of food packaging. Apart from providing traceability, indicators of tampering, and portion control, packaging also serves marketing as a crucial secondary role (Marsh and Bugusu, 2007). The fundamental requirements and basic traits of good packaging include mass production, cost- effective and efficient packing materials, appropriate structure and form, ease of use, and disposal considerations. Based on these features, packaging design and development calls for the use of engineering and environmental science in addition to industrial design, creativity, and marketing strategies. Although packaging serves several fundamental purposes, such as ease and preservation, its disposal should also be taken into consideration when developing new packaging. One package development issue that might come up soon is this one. Reducing the weight and/or volume of materials has been addressed by packaging technology in an attempt to save money and resources. A number of trends have been noted in the development of food packaging, such as the decrease of sources, the enhancement of design for handling and convenience, and environmental concerns about the materials and procedures used in packaging. Research and development efforts have been directed towards developing various functions for food packaging systems beyond this straightforward barrier function, such as active packaging, modified atmosphere packaging (MAP), edible films and coatings, and packaging that tackles environmental concerns (Han, 2014).

4.1. Innovative Functions of Packaging Material:

Active packaging technologies have added numerous new features beyond a protective barrier, such as intelligence and oxygen-scavenging capabilities, antibacterial activity, atmosphere control, edibility, and biodegradability. Shelf-life extension, product quality maintenance (and in certain situations, quality enhancement), and safety assurance are the key motivations for the study, development, and commercial use of active packaging technologies (Lim, 2011).

4.2. Advance Packaging Technologies:

By shielding foods from infections and bioterrorism and by improving the convenience of food manufacturing, distribution, retailing, and consumption, active packaging solutions increase the shelf life of food goods. The food sector currently uses a number of commercially available active packaging technology applications. Commercially available oxygen-scavenging devices come in the form of oxygen-removing sachets. Food oxidation and rancidity, as well as the development of aerobic bacteria and mould, can be avoided in an oxygen-free environment. Packaging systems that absorb carbon dioxide stop packages from inflating because carbon dioxide is formed after the packaging process. For instance, non-enzymatic browning reactions in packaged coffee beans can cause carbon dioxide to be produced during storage. Products like coffee or natural cheeses that need to be aged or fermented, or those experience Maillard reactions and release gas after packaging, are typically good candidates for carbon dioxide-absorbing devices. Pharmaceuticals, electronic gadgets, dried meals, and foods that are sensitive to moisture have all long been packaged using moisture-scavenging techniques. In these systems, a sachet containing desiccant ingredients is part of the package. Food microbiological safety, bioterrorism, and shelf-life extension are all directly impacted by antimicrobial packaging solutions, which stop the growth of harmful and spoiling bacteria. In conclusion, creating functional packaging materials with chemically or physically active ingredients that release in a precise, regulated way is the main goal of active packaging technology research (Coma, 2008).

Intelligent packaging has been classified as a distinct entity as well as a component of active packaging. Intelligent features in this type of packaging facilitate food production and delivery while also enhancing food security and safety assurance (Rodrigues and Han, 2003). Fresh produce, meats, and fish are typically kept fresher longer by using modified atmosphere packaging (MAP), which regulates their metabolic metabolism, such as respiration. Recent years have seen the introduction of novel modified environment technologies, including high oxygen flushing for red meats, carbon monoxide injection for red meats and inert gas (such as argon) flushing for fruits and vegetables. The best packing materials with the right gas permeability characteristics must be chosen for a MAP system to function properly. The food industry finds MAP systems appealing due to the rapidly expanding demand for semi-processed bulk goods, non-frozen chilled meats, minimally processed vegetables and fruits, and ready-to-eat meals (Han, 2014).

5. LIFECYCLE ASSESSMENT AND SUSTAINABILITY METRICS

5.1. Lifecycle Assessment (LCA) of packaging materials

A Lifecycle Assessment (LCA) of packaging materials is a systematic method for evaluating the environmental impacts associated with all stages of a product's life, from raw material extraction through production, use, and disposal. The LCA methodology typically follows four key stages: goal and scope definition, inventory analysis, impact assessment, and interpretation. The goal and scope definition establishes the purpose of the assessment and the boundaries of the study, while the inventory analysis involves compiling data on energy and material inputs and outputs throughout the lifecycle. The impact assessment phase evaluates the potential environmental impacts based on the inventory data, often using indicators such as greenhouse gas emissions, energy consumption, water usage, and waste generation. Finally, the interpretation stage synthesizes the findings and provides recommendations based on the results (Civancik-Uslu *et al.*, 2019).

- **Methodology and key indicators:** Key indicators used in LCA include carbon footprint, which measures greenhouse gas emissions, and resource depletion, which assesses the use of non-renewable resources. Other important indicators may include ecotoxicity, which evaluates the potential harm to ecosystems, and human health impacts, which consider the effects of pollutants on human populations (Pešta *et al.*, 2020; Bock, 2023). The choice of indicators can be influenced by the specific materials being assessed and the environmental concerns associated with them, as highlighted in studies that emphasize the importance of considering biodiversity and the risks of littering, particularly for plastic packaging (Silvenius *et al.*, 2013).
- **Comparative analysis of traditional vs. sustainable packaging:** When comparing traditional packaging materials to sustainable alternatives, significant differences in environmental impact emerge across their lifecycles. Traditional materials, such as plastics, often have a high carbon footprint due to their petroleum-based origins and energy-intensive production processes. For instance, studies have shown that single-use plastics contribute significantly to waste and environmental degradation, with a notable increase in consumer dissatisfaction linked to excessive packaging (Escursell *et al.*, 2021; Park & Waqar, 2022). In contrast, sustainable packaging options, such as biodegradable materials or those made from renewable resources, tend to have lower environmental impacts. These materials can reduce waste and resource consumption, as they are designed to decompose more readily in natural environments (Bulantekin & Alp, 2023; Saboia & Oliveira, 2021).

Research indicates that sustainable packaging can lead to substantial reductions in environmental impacts. For example, a study on recyclable cosmetic packaging found that fully recyclable materials could reduce environmental impacts by up to 74% when reused (Gatt & Refalo, 2021). Similarly, LCA studies have demonstrated that alternatives like paper and biodegradable films can outperform traditional plastic in terms of carbon emissions and waste generation (Pasqualino *et al.*, 2011;

An *et al.*, 2014). The transition to sustainable packaging not only addresses environmental concerns but also aligns with consumer preferences for eco-friendly products, which can enhance brand loyalty and market competitiveness (Ren *et al.*, 2022; Rathore *et al.*, 2023).

5.2. Metrics for Evaluating Sustainability

To evaluate the sustainability of packaging materials, several key metrics are essential, including carbon footprint, water usage, and waste generation and management. Each of these metrics provides insights into the environmental impacts associated with packaging throughout its lifecycle.

- **Carbon Footprint:** The carbon footprint of packaging materials is a critical metric that quantifies the total greenhouse gas emissions (GHG) produced during the lifecycle of packaging, from raw material extraction to disposal. Measuring the carbon footprint involves a comprehensive assessment of emissions at each stage, including production, transportation, usage, and end-of-life disposal. Various methodologies, such as the ISO 14067 standard, provide frameworks for calculating carbon emissions associated with specific products or processes (Šerešová & Kočí, 2020)Patel *et al.*, 2023). To reduce the carbon footprint, companies can adopt several strategies. For instance, transitioning from conventional materials like polyethylene terephthalate (PET) to more sustainable alternatives, such as polylactic acid (PLA), can significantly lower emissions (Choi *et al.*, 2018). Additionally, optimizing production processes to enhance energy efficiency and utilizing renewable energy sources can further mitigate carbon emissions (McPherson *et al.*, 2019; Rizan *et al.*, 2021). Implementing closed-loop systems, where materials are recycled back into the production cycle, also contributes to reducing the overall carbon footprint by minimizing the need for virgin materials (Rizan *et al.*, 2021).
- **Water Usage:** Water usage is another critical metric in evaluating the sustainability of packaging. The production of packaging materials often requires significant amounts of water, impacting local water resources and ecosystems. Analyzing water usage involves assessing the total water consumed throughout the lifecycle of packaging materials, including extraction, processing, and disposal stages (Jursová *et al.*, 2019). Strategies for water conservation in packaging production include adopting water-efficient technologies and practices, such as recycling water within manufacturing processes and utilizing rainwater harvesting systems. Additionally, companies can engage in sustainable sourcing practices that prioritize materials requir-

ing less water for production, such as certain biodegradable options (Bock, 2023). By implementing these strategies, businesses can reduce their water footprint and contribute to the conservation of vital water resources.

- **Waste Generation and Management:** Waste generation and management practices are crucial for assessing the sustainability of packaging. The lifecycle of packaging materials often results in significant waste, particularly with single-use plastics that contribute to environmental pollution. Evaluating waste generation involves quantifying the amount of waste produced at each stage of the packaging lifecycle and identifying opportunities for reduction (Chowdhury & Kabir, 2023). Enhancing recycling and composting efforts is vital for effective waste management. Companies can implement take-back programs that encourage consumers to return packaging for recycling or reuse, thereby reducing landfill contributions (Bock, 2023; Chowdhury & Kabir, 2023). Additionally, designing packaging for recyclability or compostability can facilitate waste management and promote a circular economy (Šerešová & Kočí, 2020). Education and awareness campaigns targeting consumers can also play a significant role in improving recycling rates and encouraging responsible disposal practices (Fernqvist *et al.*, 2015).

6. ECONOMIC AND REGULATORY CONSIDERATIONS

6.1. Economic Analysis of Sustainable Packaging

Sustainable packaging has emerged as a pivotal aspect of modern business practices, driven by economic considerations and consumer demand.

- **Cost-Benefit Analysis of Sustainable Packaging:** The transition from traditional to sustainable packaging entails an intricate cost-benefit analysis. Initially, sustainable packaging often incurs higher production costs due to the use of eco-friendly materials and innovative designs. For instance, biodegradable polymers and recycled materials can be more expensive than conventional plastics, impacting the overall production expenses (Ma, 2024; Sharma & Ghoshal, 2018). However, these costs can be offset by long-term savings associated with reduced waste management expenses and potential regulatory compliance costs. Companies adopting sustainable practices may benefit from lower disposal costs and reduced environmental taxes, which are increasingly being implemented to discourage wasteful practices (Civancik-Uslu *et al.*, 2019; Gustavo *et al.*, 2018). Moreover, sustainable packaging can lead to enhanced brand loyalty and consumer trust, translating into increased sales. Research indicates that consumers are willing to pay a premium for products with sustainable packaging, as it aligns with their values and environmental concerns (Gunawan, 2024; Rebollar *et al.*, 2017). This willingness to pay can result in higher profit margins for companies that invest in sustainable packaging solutions. Additionally, the adoption of intelligent packaging technologies can minimize product spoilage and waste, further enhancing economic efficiency (Gigauri, 2023; Rahman, 2023). The environmental impact of sustainable packaging is another critical factor in the cost-benefit equation. Sustainable packaging typically results in lower carbon emissions and reduced landfill waste, contributing positively to a company's corporate social responsibility (CSR) goals (Koeijer *et al.*, 2019; Kozik, 2020). By mitigating environmental harm, companies can enhance their brand image and appeal to a growing segment of environmentally conscious consumers, thereby driving sales and market share (Mattia *et al.*, 2021; Granato *et al.*, 2022).

- **Market Trends and Consumer Demand:** Current market trends reflect a significant shift towards sustainable packaging, driven by evolving consumer preferences. A growing body of research indicates that consumers are increasingly prioritizing sustainability in their purchasing decisions. For example, studies show that consumers are more likely to choose products with sustainable packaging, perceiving them as more ethical and responsible (Herbes *et al.*, 2020; Lee & Park, 2020). This trend is particularly pronounced among younger consumers, who are more inclined to support brands that demonstrate a commitment to environmental sustainability (Gunawan, 2024; Kumar, 2024). The rise of e-commerce has also influenced packaging trends, as online retailers seek to minimize their environmental footprint while ensuring product safety during transit. This has led to innovations in packaging design that prioritize minimalism and recyclability (Rahman, 2023; Zhao, 2023). Companies are responding to these trends by integrating sustainable practices into their marketing strategies, recognizing that sustainability can serve as a competitive advantage in a crowded marketplace (Mattia *et al.*, 2021; Gustavo *et al.*, 2018). Furthermore, the COVID-19 pandemic has heightened awareness of the importance of packaging in protecting products, leading to increased consumer scrutiny of packaging materials (Barone *et al.*, 2021). As a result, brands are compelled to adopt sustainable packaging solutions not only to meet consumer demand but also to comply with emerging regulations aimed at reducing plastic waste and promoting circular economy principles (Guillard *et al.*, 2018; Han *et al.*, 2018).

6.2. Regulatory Frameworks and Compliance

Navigating the complex landscape of regulatory frameworks and compliance requirements in sustainable packaging involves understanding international and national regulations, key standards and certifications, as well as the incentives and penalties that drive industry practices.

- **International and National Regulations:** International and national regulations play a crucial role in shaping sustainable packaging practices. Various countries have implemented stringent regulations aimed at reducing plastic waste and promoting eco-friendly packaging solutions. For instance, the European Union's Single-Use Plastics Directive mandates the reduction of certain plastic products, thereby encouraging the adoption of sustainable alternatives (Afif *et al.*, 2021). Similarly, countries like Canada and several U.S. states have enacted laws to limit the use of single-use plastics, pushing manufacturers towards sustainable packaging options (Oloyede & Lignou, 2021). These regulatory frameworks not only impose restrictions but also spur innovation within the packaging industry. Companies are incentivized to develop new materials and technologies that comply with these regulations, leading to advancements in biodegradable and recyclable packaging solutions (Singh *et al.*, 2018). Furthermore, international agreements, such as the Paris Agreement, indirectly influence packaging practices by setting broader environmental goals that nations strive to achieve, thereby impacting local regulations and corporate strategies (Kumar *et al.*, 2021).
- **Standards and Certifications for Sustainable Packaging:** Standards and certifications are essential for ensuring compliance and promoting best practices in sustainable packaging. Key certifications include:
    - *FSC Certification*: The Forest Stewardship Council (FSC) certification ensures that products come from responsibly managed forests, promoting sustainable forestry practices (Norton *et al.*, 2022).
    - *Cradle to Cradle Certification:* This certification assesses the safety, circularity, and responsibility of products and materials, encouraging companies to design for sustainability from the outset (Afif *et al.*, 2021).

These certifications not only help companies demonstrate their commitment to sustainability but also enhance consumer trust and marketability. By adhering to recognized standards, businesses can differentiate themselves in a competitive market and align with consumer preferences for environmentally responsible products (Norton *et al.*, 2022).

- **Incentives and Penalties for Sustainable Practices:** Incentives for adopting sustainable packaging practices are increasingly prevalent, often provided by governments to encourage businesses to transition towards eco-friendly solutions. Common incentives include tax breaks, grants, and subsidies for companies that invest in sustainable packaging technologies (Afif *et al.*, 2021). For example, some jurisdictions offer financial incentives for businesses that implement recycling programs or use renewable materials in their packaging (Afif *et al.*, 2021). Conversely, penalties for non-compliance with sustainability regulations can be significant. Companies that fail to adhere to packaging regulations may face fines, legal action, and damage to their reputation (Afif *et al.*, 2021). For instance, the imposition of Extended Producer Responsibility (EPR) schemes requires manufacturers to bear the costs of recycling and disposal of their products, incentivizing them to design more sustainable packaging (Laubinger *et al.*, 2021). Such regulatory measures are designed to drive positive change in the industry by holding companies accountable for their environmental impact.

7. CONSUMER PERSPECTIVES AND BEHAVIORAL INSIGHTS

7.1. Consumer Awareness and Perception of Sustainable Packaging

The consumer side of sustainable packaging is a multifaceted area of research that encompasses consumer attitudes, preferences, and the various factors influencing their choices. This synthesis draws on a range of studies and surveys to explore these dimensions comprehensively.

- **Surveys and Studies on Consumer Attitudes:** Recent studies indicate that consumers are increasingly aware of and concerned about the environmental impacts of packaging. For instance, a study by Chirilli *et al.* highlights that consumers appreciate clear symbols and information regarding the environmental sustainability of packaging, suggesting that simple and impactful messaging can significantly influence consumer behavior towards eco-friendly options (Chirilli *et al.*, 2022). Similarly, research by Norton *et al.* demonstrates that the inclusion of recycling information on packaging can effectively shift consumer behavior towards more sustainable choices, indicating that well-designed labeling serves as a powerful tool for promoting sustainable consumption (Norton *et al.*, 2022). Moreover, a qualitative study by Farias emphasizes the interplay between pro-environmental attitudes and packaging preferences, revealing that consumers are more likely to favor sustainable packaging when they perceive it as beneficial for the environment (Farias, 2023). This aligns with findings from Oloyede and Lignou, who found that a significant portion of consumers prioritize environmentally labeled packaging when making purchasing decisions (Oloyede & Lignou, 2021). Collectively, these studies underscore a growing consumer inclination towards sustainable packaging, driven by increased awareness and the desire for clear, actionable information.

- **Factors Influencing Consumer Choices:** Several factors influence consumer decisions regarding sustainable packaging, including environmental concerns, cost, convenience, and brand reputation. Environmental concerns are a primary driver, as evidenced by research indicating that consumers are motivated by the desire to reduce their ecological footprint. For instance, studies show that consumers are more likely to choose products with sustainable packaging when they are informed about the environmental benefits (Allison *et al.*, 2021). Cost and convenience also play crucial roles in consumer choices. Bertoluci *et al.* argue that while consumers express a preference for sustainable packaging, the design must also meet their practical needs, such as ease of use and cost-effectiveness (Bertoluci *et al.*, 2014). This sentiment is echoed by Lee and Park, who note that while sustainability is valued, factors like functionality and aesthetic appeal significantly influence consumer preferences (Lee & Park, 2020).

Brand reputation further complicates the landscape; consumers are more inclined to trust and choose brands that demonstrate a commitment to sustainability. Magnier and Schoormans highlight that the visual appearance of sustainable packaging, combined with credible verbal claims, enhances purchase intentions and reinforces brand ethicality (Magnier & Schoormans, 2015). This suggests that consumers are not only influenced by the sustainability of packaging but also by the overall brand narrative surrounding environmental responsibility.

7.2. Strategies to Enhance Consumer Engagement

Increasing consumer engagement with sustainable packaging is critical for fostering eco-conscious behaviors and promoting environmental sustainability. Two effective strategies in this regard are education and awareness campaigns, and the provision of incentives for adopting sustainable practices.

- **Education and Awareness Campaigns:** Education and awareness campaigns play a pivotal role in informing consumers about the benefits of sustainable packaging. Research indicates that consumers often lack sufficient knowledge about sustainable packaging options and their environmental impacts, which can hinder their purchasing decisions (Chen *et al.*, 2021). For instance, the complexity and variety of eco-labels associated with recycled packaging can confuse consumers, leading to a negative impact on their sustainable consumption intentions (Chen *et al.*, 2021). Therefore, targeted educational initiatives that clarify these labels and the benefits of sustainable packaging are essential. Such campaigns can enhance consumer understanding and foster a sense of responsibility towards sustainable practices (Norton *et al.*, 2022; Chirilli *et al.*, 2022). Moreover, studies have shown that consumers who are more aware of environmental sustainability tend to exhibit more eco-friendly behaviors (Chirilli *et al.*, 2022). This suggests that increasing awareness through educational campaigns can lead to a more informed consumer base that is likely to make environmentally conscious choices. For example, campaigns that highlight the end-of-life attributes of packaging—such as biodegradability and recyclability—can significantly influence consumer perceptions and purchasing decisions (Norton *et al.*, 2022). Additionally, effective communication strategies that emphasize the eco-friendliness of packaging can bridge the gap between consumer attitudes and behaviors, thereby enhancing engagement with sustainable packaging (Anquez *et al.*, 2022; Boz *et al.*, 2020).

o **Incentives for Adopting Sustainable Practices:** Incentives are another crucial strategy for encouraging consumers to adopt sustainable packaging practices. Various forms of incentives, such as discounts, rewards, and loyalty programs, can motivate consumers to choose sustainable options over conventional ones. Research indicates that perceived future benefits of sustainable practices can significantly increase the likelihood of their adoption (Piñeiro *et al.*, 2020). For instance, loyalty programs that reward consumers for purchasing products with sustainable packaging can create a positive feedback loop, reinforcing eco-conscious behaviors (Dena *et al.*, 2023). Furthermore, the effectiveness of government incentives, when combined with stakeholder pressure, has been shown to enhance the adoption of sustainable practices (Touil, 2023). This suggests that a multi-faceted approach, incorporating both consumer incentives and regulatory support, can lead to greater engagement with sustainable packaging. Additionally, studies have demonstrated that consumers are often willing to pay a premium for products packaged sustainably, indicating that financial incentives can effectively drive positive behavioral changes (Cammarelle *et al.*, 2021).

8. CHALLENGES AND FUTURE DIRECTIONS

The transition to sustainable packaging is fraught with challenges that span technical, logistical, market, and consumer dimensions. Understanding these challenges is crucial for developing effective strategies and identifying future directions for sustainable packaging innovations.

8.1. Technical and Logistical Challenges

One of the primary technical challenges in sustainable packaging is scalability. Many sustainable materials, such as bioplastics and biodegradable polymers, are not yet produced at the scale necessary to meet global demand. This limitation often results in higher costs compared to conventional packaging materials, which can deter manufacturers from adopting these alternatives (Escursell *et al.*, 2021;Nilsen-Nygaard *et al.*, 2021). The performance of sustainable packaging solutions is another concern; they must meet the same safety and durability standards as traditional materials to be viable in the market (Salgado *et al.*, 2021; Perera *et al.*, 2023). For instance, while biobased and biodegradable materials show promise, they often struggle with moisture barriers and shelf-life extension, which are critical for food packaging (Nilsen-Nygaard *et al.*, 2021; Perera *et al.*, 2023).

Logistical issues also arise in the supply chain, particularly regarding the sourcing of sustainable materials and the infrastructure needed for their recycling or composting (Morashti *et al.*, 2022; Hamid & Samy, 2022). The integration of intelligent packaging technologies, which can enhance sustainability by reducing waste and spoilage, presents an innovative approach to overcoming these logistical hurdles (Gigauri, 2023; Drago *et al.*, 2020). However, the development and implementation of such technologies require significant investment and collaboration across the supply chain, which can be challenging to coordinate (Kaushani *et al.*, 2022).

8.2. Market and Consumer Barriers

Market acceptance of sustainable packaging is hindered by consumer behavior and perceptions. Many consumers exhibit resistance to new packaging types, often due to a lack of understanding or familiarity with sustainable materials (Lignou & Oloyede, 2021; Ren *et al.*, 2022). Research indicates that consumers with low environmental concerns may evaluate sustainable packaging negatively, particularly if it does not align with their expectations of conventional packaging (Lignou & Oloyede, 2021). This highlights the need for effective communication strategies that educate consumers about the benefits and functionalities of sustainable packaging (Ren *et al.*, 2022; Sheth & Parvatiyar, 2020).

Behavior change is another significant barrier; consumers must be motivated to alter their purchasing habits and preferences in favor of sustainable options (Lignou & Oloyede, 2021; Morashti *et al.*, 2022). Strategies such as providing clear labeling, demonstrating the environmental benefits of sustainable packaging, and offering incentives for choosing eco-friendly products can help facilitate this transition (Ren *et al.*, 2022; Sheth & Parvatiyar, 2020). Furthermore, addressing misconceptions about the quality and performance of sustainable packaging is essential for increasing consumer acceptance (Lignou & Oloyede, 2021; Morashti *et al.*, 2022).

8.3. Emerging Trends and Innovations

Looking ahead, several key areas for research and development in sustainable packaging are emerging. Innovations in materials science, particularly the development of biopolymers and edible packaging, hold significant potential for reducing environmental impact (Nair *et al.*, 2023; Perera *et al.*, 2023). Future research should focus on optimizing these materials for performance and cost-effectiveness, ensuring they meet industry standards while remaining accessible to manufacturers (Nilsen-Nygaard *et al.*, 2021; Perera *et al.*, 2023).

Disruptive technologies, such as smart and active packaging, are also poised to revolutionize the packaging industry. These technologies can enhance product safety, extend shelf life, and reduce waste by providing real-time information about the condition of packaged goods (Gigauri, 2023; Drago *et al.*, 2020). The integration of digital technologies into packaging can facilitate better supply chain management and consumer engagement, ultimately driving the adoption of sustainable practices (Gigauri, 2023; Kaushani *et al.*, 2022).

9. CONCLUSION

This chapter has provided an in-depth analysis of sustainable packaging solutions, emphasizing the importance of reducing environmental impact through the use of renewable materials, energy-efficient processes, and innovative packaging technologies. The principles of reducing, reusing, and recycling have been discussed alongside the benefits of biodegradable and compostable materials. We explored how these approaches can address the negative effects of conventional packaging, such as resource depletion and pollution, while promoting a circular economy. Additionally, technical advancements like smart packaging offer promising opportunities for improving food safety and quality.

To realize the full potential of sustainable packaging, stakeholders across the food industry—manufacturers, retailers, policymakers, and consumers—must work together. Businesses are encouraged to adopt eco-friendly packaging solutions to meet growing consumer demands and comply with regulatory frameworks. Investments in research, technology, and education are essential to overcome the existing challenges related to scalability, cost, and performance of sustainable packaging materials. Consumer engagement through awareness campaigns and incentives is also critical for driving behavior change and supporting a transition toward more sustainable practices.

Looking ahead, the future of sustainable packaging lies in continuous innovation and collaboration. Advances in biopolymer technology, intelligent packaging systems, and material science will enable the development of solutions that are not only environmentally friendly but also economically viable. The integration of renewable energy and digital technologies will further enhance packaging sustainability. With concerted efforts from all stakeholders, sustainable packaging will become a standard practice, contributing significantly to the global effort to reduce waste, conserve resources, and protect the environment.

REFERENCES

Abeliotis, K., Zachos, F., & Lasaridi, K. (2014). A database tool for raising awareness on the life cycle environmental impacts of household packaging waste. *Journal of Computational Environmental Sciences, 2014*, 1–6. DOI: 10.1155/2014/809102

A.E. (Ed.), *Comprehensive Biotechnology* (2nd ed., Vol. 4, pp. 629–644). Pergamon Press.

Afif, K., Rebolledo, C., & Roy, J. (2021). Drivers, barriers and performance outcomes of sustainable packaging: A systematic literature review. *British Food Journal, 124*(3), 915–935. DOI: 10.1108/BFJ-02-2021-0150

Al-Salem, S. M., Lettieri, P., & Baeyens, J. (2009). Recycling and recovery routes of plastic solid waste (PSW): A review. *Waste Management (New York, N.Y.), 29*(10), 2625–2643. DOI: 10.1016/j.wasman.2009.06.004 PMID: 19577459

Allison, A., Lorencatto, F., Michie, S., & Miodownik, M. (2021). Barriers and enablers to buying biodegradable and compostable plastic packaging. *Sustainability (Basel), 13*(3), 1463. DOI: 10.3390/su13031463

An, X., Liu, P., Meng, Q., Su, C., & Zhao, S. (2014). Research on life cycle environmental property of biomass biodegradable packaging material. *Applied Mechanics and Materials, 670-671*, 1620–1623. . DOI: 10.4028/www.scientific.net/AMM.670-671.1620

Anquez, E., Raab, K., Cechella, F., & Wagner, R. (2022). Consumers' perception of sustainable packaging in the food industry. *Revista Direitos Culturais, 17*(41), 251–265. DOI: 10.20912/rdc.v17i41.728

Ashok, A., Rejeesh, C., & Renjith, R. (2016). Biodegradable polymers for sustainable packaging applications: A review. *IJBB, 1*, 11.

Barlow, C.Y., Morgan, D.C., (2013). Polymer film packaging for food: an environmental assessment. Resour. Conserv. Recycl. 78, 74e80.

Barone, , Matheus, J. R. V., de Souza, T. S. P., Moreira, R. F. A., & Fai, A. E. C. (2021). Green-based active packaging: Opportunities beyond COVID-19, food applications, and perspectives in circular economy—A brief review. *Comprehensive Reviews in Food Science and Food Safety, 20*(5), 4881–4905. Advance online publication. DOI: 10.1111/1541-4337.12812 PMID: 34355490

Batista, L., Bourlakis, M., Smart, P., Maull, R., (2018). In search of a circular supply chain archetypeea content-analysis-based literature review. Prod. Plan. Control 29 (6), 438e451.

Bertoluci, , Leroy, Y., & Olsson, A. (2014). Exploring the environmental impacts of olive packaging solutions for the European food market. *Journal of Cleaner Production, 64*, 234–243. Advance online publication. DOI: 10.1016/j.jclepro.2013.09.029

Bertoluci, G., Leroy, Y., & Olsson, A. (2014). Exploring the environmental impacts of olive packaging solutions for the european food market. *Journal of Cleaner Production*, *64*, 234–243. DOI: 10.1016/j.jclepro.2013.09.029

Bock, M., & Meyerding, S. G. H. (2023). Consumer perception of food product packaging materials sustainability versus life cycle assessment results: The case of processed tomatoes—a quantitative study in Germany. *Sustainability (Basel)*, *15*(23), 16370. DOI: 10.3390/su152316370

Boz, Z., Korhonen, V., & Sand, C. (2020). Consumer considerations for the implementation of sustainable packaging: A review. *Sustainability (Basel)*, *12*(6), 2192. DOI: 10.3390/su12062192

Bulantekin, Ö. and Alp, D. (2023). Perspective chapter: development of food packaging films from microorganism-generated polyhydroxyalkanoates.. DOI: 10.5772/intechopen.108802

Cammarelle, A., Viscecchia, R., & Bimbo, F. (2021). Intention to purchase milk packaged in biodegradable packaging: Evidence from italian consumers. *Foods*, *10*(9), 2068. DOI: 10.3390/foods10092068 PMID: 34574179

Chen, C., Sujanto, R., Tseng, M., Chiu, A., & Lim, M. (2021). How is the sustainable consumption intention model in food industry under preference uncertainties? the consumer willingness to pay on recycled packaging material. *Sustainability (Basel)*, *13*(21), 11578. DOI: 10.3390/su132111578

Chiellini, E. (2008): Enviromentally compatible food packaging. Woodhead publishing limited, Cambridge England, 8–10

Chirilli, C., Molino, M., & Torri, L. (2022). Consumers' awareness, behavior and expectations for food packaging environmental sustainability: Influence of sociodemographic characteristics. *Foods*, *11*(16), 2388. DOI: 10.3390/foods11162388 PMID: 36010389

Choi, B., Yoo, S., & Park, S. (2018). Carbon footprint of packaging films made from ldpe, pla, and pla/pbat blends in south korea. *Sustainability (Basel)*, *10*(7), 2369. DOI: 10.3390/su10072369

Chowdhury, T., & Kabir, G. (2023). Life cycle assessment (lca) of package deliveries: Sustainable decision-making for the academic institutions. *Green and Low-Carbon Economy*, *2*(1), 1–13. DOI: 10.47852/bonviewGLCE3202824

Cimpan, C., Maul, A., Jansen, M., Pretz, T., & Wenzel, H. (2015). Central sorting and recovery of MSW recyclable materials: A review of technological state-of-the-art, cases, practice and implications for materials recycling. *Journal of Environmental Management*, *156*, 181–199. DOI: 10.1016/j.jenvman.2015.03.025 PMID: 25845999

Civancik-Uslu, D., Puig, R., Voigt, S., Walter, D., & Fullana–i–Palmer, P. (2019). Improving the production chain with lca and eco-design: Application to cosmetic packaging. *Resources, Conservation and Recycling*, *151*, 104475. DOI: 10.1016/j.resconrec.2019.104475

Coelho, P. M., Corona, B., ten Klooster, R., & Worrell, E. (2020). Sustainability of reusable packaging–Current situation and trends. In *Resources, Conservation and Recycling: X* (Vol. 6). Elsevier B.V. DOI: 10.1016/j.rcrx.2020.100037

Coma, V., (2008). Bioactive packaging technologies for extended shelf life of meat-based products. Meat Science. 78 (1 2), 90 103

Davidovic´, A., & Savic´, A. (2010). Microbial production of biodegradable polymer. [in Croatian]. *Tehnologica Acta*, *3*, 3–13.

Dena, A., Honrado, L., Lin, P., & Dotong, E. (2023). Promoting the usage of eco-friendly tertiary packaging: A market research on the perceived behavior of filipino consumers based on sustainability factors. *American Journal of Social Development and Entrepreneurship*, *2*(1), 26–31. DOI: 10.54536/ajsde.v2i1.1223

Drago, E., Campardelli, R., Pettinato, M., & Perego, P. (2020). Innovations in smart packaging concepts for food: An extensive review. *Foods*, *9*(11), 1628. DOI: 10.3390/foods9111628 PMID: 33171881

Escursell, S., Llorach-Massana, P., & Roncero, M. (2021). Sustainability in e-commerce packaging: A review. *Journal of Cleaner Production*, *280*, 124314. DOI: 10.1016/j.jclepro.2020.124314 PMID: 32989345

Farias, A., Lane, H., Killingsworth, J., Warden, J. M., & Wais, S. (2023). Exploring the interplay of pro-environmental attitudes, dietary choices, and packaging preferences: A virtual reality restaurant scenario study. *Challenges*, *15*(1), 1. DOI: 10.3390/challe15010001

Fernqvist, F., Olsson, A., & Spendrup, S. (2015). What's in it for me? food packaging and consumer responses, a focus group study. *British Food Journal*, *117*(3), 1122–1135. DOI: 10.1108/BFJ-08-2013-0224

Gardas, B. B., Raut, R. D., & Narkhede, B. (2019). Identifying critical success factors to facilitate reusable plastic packaging towards sustainable supply chain management. *Journal of Environmental Management*, *236*, 81–92. DOI: 10.1016/j.jenvman.2019.01.113 PMID: 30716694

Gatt, I., & Refalo, P. (2021). Life cycle assessment of recyclable, reusable and dematerialised plastic cosmetic packages. *IOP Conference Series. Materials Science and Engineering*, *1196*(1), 012022. DOI: 10.1088/1757-899X/1196/1/012022

Geueke, B., Groh, K., & Muncke, J. (2018). Food packaging in the circular economy: Overview of chemical safety aspects for commonly used materials. *Journal of Cleaner Production*, *193*, 491–505. DOI: 10.1016/j.jclepro.2018.05.005

Granato, G., Fischer, A., & Trijp, J. (2022). The price of sustainability: How consumers trade-off conventional packaging benefits against sustainability. *Journal of Cleaner Production*, *365*, 132739. DOI: 10.1016/j.jclepro.2022.132739

Grigore, M. E. (2017). Methods of recycling, properties and applications of recycled thermoplastic polymers. *Recycling*, *2*(4), 24. DOI: 10.3390/recycling2040024

Grondahl, M., Gustafsson, A., & Gatenholm, P. (2006). Gas-phase surface flourination of arabinoxlyan films. *Macromolecules*, *39*(7), 2718–2712. DOI: 10.1021/ma052066q

Guillard, V., Gaucel, S., Fornaciari, C., Angellier-Coussy, H., Buche, P., & Gontard, N. (2018). The next generation of sustainable food packaging to preserve our environment in a circular economy context. *Frontiers in Nutrition*, *5*, 121. Advance online publication. DOI: 10.3389/fnut.2018.00121 PMID: 30564581

Gunawan, A., Amalia, F. A., Ramadhan, M., & Bansah, P. F. (2024). Exploring the reasons of indonesian young adult consumers toward sustainably packaged food & beverages product. *Journal of Marketing Innovation (Jmi)*, *4*(1). Advance online publication. DOI: 10.35313/jmi.v4i1.106

Gustavo, J.Jr, Pereira, G., Bond, A., Viegas, C., & Borchardt, M. (2018). Drivers, opportunities and barriers for a retailer in the pursuit of more sustainable packaging redesign. *Journal of Cleaner Production*, *187*, 18–28. DOI: 10.1016/j.jclepro.2018.03.197

Hamid, L. and Samy, I. (2022). Fabricating natural biocomposites for food packaging.. DOI: 10.5772/intechopen.100907

Han, J., Ruiz-García, L., Qian, J., & Yang, X. (2018). Food packaging: A comprehensive review and future trends. *Comprehensive Reviews in Food Science and Food Safety*, *17*(4), 860–877. DOI: 10.1111/1541-4337.12343 PMID: 33350114

Han, J. H. (2014). A review of food packaging technologies and innovations. Innovations in food packaging, 3-12.

Herbes, C., Beuthner, C., & Ramme, I. (2020). How green is your packaging—A comparative international study of cues consumers use to recognize environmentally friendly packaging. *International Journal of Consumer Studies*, *44*(3), 258–271. DOI: 10.1111/ijcs.12560

Ivonkovic, A., Zeljko, K., & Talic, S., & Lasic, M. (2017). Biodegradable packaging in the food industry. *J. Food Saf. Food Qual*, *68*(2), 26–38.

Jeya Jeevahan, J., Chandrasekaran, M., Venkatesan, S. P., Sriram, V., Britto Joseph, G., Mageshwaran, G., & Durairaj, R. B. (2020). Scaling up difficulties and commercial aspects of edible films for food packaging: A review. *Trends in Food Science & Technology*, *100*, 210–222. DOI: 10.1016/j.tifs.2020.04.014

Jursová, S., Burchart-Korol, D., & Pustějovská, P. (2019). Carbon footprint and water footprint of electric vehicles and batteries charging in view of various sources of power supply in the czech republic. *Environments (Basel, Switzerland)*, *6*(3), 38. DOI: 10.3390/environments6030038

Kaushani, K., Rathnasinghe, N., Katuwawila, N., Jayasinghe, R., Nilmini, A., & Priyadarshana, G. (2022). Trends in smart packaging technologies for sustainable monitoring of food quality and safety. International Journal of Research and Innovation in Applied Science, 07(07), 07-30. DOI: 10.51584/IJRIAS.2022.7702

Koeijer, B., Gelhard, C., & Klooster, R. (2019). Sustainability priorities across the strategic and operational level in packaging development. *Packaging Technology & Science*, *32*(12), 618–629. DOI: 10.1002/pts.2480

Kozik, N. (2020). Sustainable packaging as a tool for global sustainable development. In *SHS web of conferences* (Vol. 74, p. 04012). EDP Sciences. DOI: 10.1051/shsconf/20207404012

Kumar, R., & Sharma, A. (2024). Review of pet food packaging in the us market: Future direction towards innovation and sustainability. *Annual Research & Review in Biology*, *39*(6), 16–30. DOI: 10.9734/arrb/2024/v39i62085

Kumar, R., Verma, A., Shome, A., Sinha, R., Sinha, S., Jha, P., & Prasad, P. (2021). Impacts of plastic pollution on ecosystem services, sustainable development goals, and need to focus on circular economy and policy interventions. *Sustainability (Basel)*, *13*(17), 9963. DOI: 10.3390/su13179963

Laubinger, F., Brown, A., Dubois, M., & Börkey, P. (2021). Modulated fees for extended producer responsibility schemes (epr).. DOI: 10.1787/19970900

Lawrence, M. J. (2017). Disposal of plastics. In *A guide to the manufacture, performance, and potential of plastics in agriculture* (pp. 187–195). Elsevier. DOI: 10.1016/B978-0-08-102170-5.00010-5

Lebreton, L., Slat, B., Ferrari, F., Sainte-Rose, B., Aitken, J., Marthouse, R., & Reisser, J. (2018). Evidence that the Great Pacific Garbage Patch is rapidly accumulating plastic. *Scientific Reports*, *8*(1), 1–15. DOI: 10.1038/s41598-018-22939-w PMID: 29568057

Lee, J. and Park, K. (2020). Because it is green or beautiful? consumer's perceived value of a unique type of sustainable packaging, store evaluation, and store patronage intentions.. DOI: 10.31274/itaa.12190

Lignou, S., & Oloyede, O. (2021). Consumer acceptability and sensory profile of sustainable paper-based packaging. *Foods*, *10*(5), 990. DOI: 10.3390/foods10050990 PMID: 34062787

Ma, W., & Li, L. (2024). Trends and prospects in sustainable food packaging materials. *Foods*, *13*(11), 1744. DOI: 10.3390/foods13111744 PMID: 38890972

Magnier, L., & Schoormans, J. (2015). Consumer reactions to sustainable packaging: The interplay of visual appearance, verbal claim and environmental concern. *Journal of Environmental Psychology*, *44*, 53–62. DOI: 10.1016/j.jenvp.2015.09.005

Mahmoudi, M., & Parviziomran, I. (2020). Reusable packaging in supply chains: A review of environmental and economic impacts, logistics system designs, and operations management. *International Journal of Production Economics*, *228*, 107730. DOI: 10.1016/j.ijpe.2020.107730

Marsh, K., & Bugusu, B. (2007). Food packaging—Roles, materials, and environmental issues. *Journal of Food Science*, *72*(3), R39–R55. DOI: 10.1111/j.1750-3841.2007.00301.x PMID: 17995809

Mattia, G., Leo, A., & Pratesi, C. (2021). Recognizing the key drivers and industry implications of sustainable packaging design: A mixed-method approach. *Sustainability (Basel)*, *13*(9), 5299. DOI: 10.3390/su13095299

McPherson, B., Sharip, M., & Grimmond, T. (2019). The impact on life cycle carbon footprint of converting from disposable to reusable sharps containers in a large us hospital geographically distant from manufacturing and processing facilities. *PeerJ*, *7*, e6204. DOI: 10.7717/peerj.6204 PMID: 30809428

Meherishi, L., Narayana, S. A., & Ranjani, K. S. (2019). Sustainable packaging for supply chain management in the circular economy: A review. *Journal of Cleaner Production*, *237*, 117–582. DOI: 10.1016/j.jclepro.2019.07.057

Mohamed, S. A. A., El-Sakhawy, M., & El-Sakhawy, M. A. M. (2020). Polysaccharides, protein and lipid -based natural edible flms in food packaging: A review. *Carbohydrate Polymers, 38*, 116178. DOI: 10.1016/j.carbpol.2020.116178 PMID: 32299560

Morashti, J., An, Y., & Hyunmi, J. (2022). A systematic literature review of sustainable packaging in supply chain management. *Sustainability (Basel), 14*(9), 4921. DOI: 10.3390/su14094921

Nair, S., Trafiałek, J., & Kolanowski, W. (2023). Edible packaging: A technological update for the sustainable future of the food industry. *Applied Sciences (Basel, Switzerland), 13*(14), 8234. DOI: 10.3390/app13148234

Ncube, L. K., Ude, A. U., Ogunmuyiwa, E. N., Zulkifli, R., & Beas, I. N. (2020). Environmental impact of food packaging materials: A review of contemporary development from conventional plastics to polylactic acid based materials. *Materials (Basel), 13*(21), 4994. DOI: 10.3390/ma13214994 PMID: 33171895

Nilsen-Nygaard, J., Fernández, E., Radusin, T., Rotabakk, B., Sarfraz, J., Sharmin, N., Sivertsvik, M., Sone, I., & Pettersen, M. (2021). Current status of biobased and biodegradable food packaging materials: Impact on food quality and effect of innovative processing technologies. *Comprehensive Reviews in Food Science and Food Safety, 20*(2), 1333–1380. DOI: 10.1111/1541-4337.12715 PMID: 33547765

Norton, V., Waters, C., Oloyede, O., & Lignou, S. (2022). Exploring consumers' understanding and perception of sustainable food packaging in the uk. *Foods, 11*(21), 3424. DOI: 10.3390/foods11213424 PMID: 36360036

Oloyede, O., & Lignou, S. (2021). Sustainable paper-based packaging: A consumer's perspective. *Foods, 10*(5), 1035. DOI: 10.3390/foods10051035 PMID: 34068639

Otto, S., Strenger, M., Maier-Nöth, A., & Schmid, M. (2021). Food packaging and sustainability–Consumer perception vs. correlated scientific facts: A review. *Journal of Cleaner Production, 298*, 126733. DOI: 10.1016/j.jclepro.2021.126733

Park, J., & Waqar, Z. (2022). Life cycle assessment of returnable mailers used for apparel electronic commerce: A case study in canada. *Packaging Technology & Science, 35*(9), 651–662. DOI: 10.1002/pts.2653

Pasqualino, J., Meneses, M., & Castells, F. (2011). The carbon footprint and energy consumption of beverage packaging selection and disposal. *Journal of Food Engineering, 103*(4), 357–365. DOI: 10.1016/j.jfoodeng.2010.11.005

Patel, N., Feofilovs, M., & Romagnoli, F. (2023). Carbon footprint evaluation tool for packaging marketplace. *Environmental and Climate Technologies*, *27*(1), 368–378. DOI: 10.2478/rtuect-2023-0027

Payne, J., McKeown, P., & Jones, M. D. (2019). A circular economy approach to plastic waste. *Polymer Degradation & Stability*, *165*, 170–181. DOI: 10.1016/j.polymdegradstab.2019.05.014

Perera, K., Jaiswal, A., & Jaiswal, S. (2023). Biopolymer-based sustainable food packaging materials: Challenges, solutions, and applications. *Foods*, *12*(12), 2422. DOI: 10.3390/foods12122422 PMID: 37372632

Pešta, J., Šerešová, M., & Kočí, V. (2020). Carbon footprint assessment of construction waste packaging using the package-to-product indicator. *Sustainability (Basel)*, *12*(23), 10094. DOI: 10.3390/su122310094

Piñeiro, V., Arias, J., Dürr, J., Elverdin, P., Ibáñez, A., Kinengyere, A., Opazo, C. M., Owoo, N., Page, J. R., Prager, S. D., & Torero, M. (2020). A scoping review on incentives for adoption of sustainable agricultural practices and their outcomes. *Nature Sustainability*, *3*(10), 809–820. DOI: 10.1038/s41893-020-00617-y

Porta, R., Sabbah, M., & Di Pierro, P. (2020). Biopolymers as food packaging materials. *International Journal of Molecular Sciences*, *21*(14), 4942. DOI: 10.3390/ijms21144942 PMID: 32668678

Rahman, Z., Johnson, J., Tripathi, S., & Krithika, D. J. (2023). The future of packaging: Minimalism in modern era. *Interantional Journal of Scientific Research in Engineering and Management*, *07*(10), 1–11. DOI: 10.55041/IJSREM26073

Rathore, S., Schuler, B., & Park, J. (2023). Life cycle assessment of multiple dispensing systems used for cosmetic product packaging. *Packaging Technology & Science*, *36*(7), 533–547. DOI: 10.1002/pts.2729

Razza, F., Innocenti, F.D., (2012). Bioplastics from renewable resources: the benefits of biodegradability. Asia Pac. J. Chem. Eng. 7, S301eS309

Rebollar, R., Gil, I., Lidón, I., Martín, J., Fernández, M., & Rivera, S. (2017). How material, visual and verbal cues on packaging influence consumer expectations and willingness to buy: The case of crisps (potato chips) in spain. *Food Research International*, *99*, 239–246. DOI: 10.1016/j.foodres.2017.05.024 PMID: 28784480

Ren, Z., Zhang, D., & Zhu, G. (2022). Sustainable design strategy of cosmetic packaging in china based on life cycle assessment. *Sustainability (Basel)*, *14*(13), 8155. DOI: 10.3390/su14138155

Rizan, C., Lillywhite, R., Reed, M., & Bhutta, M. (2021). Minimising carbon and financial costs of steam sterilisation and packaging of reusable surgical instruments. *British Journal of Surgery*, *109*(2), 200–210. DOI: 10.1093/bjs/znab406 PMID: 34849606

Rodrigues, E. T., & Han, J. H. (2003). Intelligent packaging. In Heldman, D. R. (Ed.), *Encyclopedia of Agricultural and Food Engineering* (pp. 528–535). Marcel Dekker.

Saboia, L. and Oliveira, R. (2021). Replacement of conventional plastic shampoo packaging for biodegradable ones – a survey.. Revista Eletrônica Em Gestão Educação E Tecnologia Ambiental, 25, e15. DOI: 10.5902/2236117065105

Sæter, F., Alvarado, I. O., & Pettersen, I. N. (2020). Reuse principle for primary packaging circularity in the food system. DS 101: *Proceedings of NordDesign 2020*, Lyngby, Denmark, 12th-14th August 2020, 1-12.

Saklani, P. S. (2019). A review of edible packaging for foods. *International Journal of Current Microbiology and Applied Sciences*, *8*(7), 2885–2895. DOI: 10.20546/ijcmas.2019.807.359

Salgado, P., Giorgio, L., Musso, Y., & Mauri, A. (2021). Recent developments in smart food packaging focused on biobased and biodegradable polymers. *Frontiers in Sustainable Food Systems*, *5*, 630393. Advance online publication. DOI: 10.3389/fsufs.2021.630393

Sandu, M. (2015). Food Safety-Component of National Security. *Pub. Sec. Stud.*, *4*, 15.

Seglia, D., Krasanova, I., Heidemane, G., Kampuse, S., Dukalska, L., & Kampus, K. (2009). Packaging tehnology influence on the shelf life extention of fresh raspberries. *Acta Horticulturae*, (877), 433–440.

Šerešová, M., & Kočí, V. (2020). Proposal of package-to-product indicator for carbon footprint assessment with focus on the Czech Republic. *Sustainability (Basel)*, *12*(7), 3034. DOI: 10.3390/su12073034

Sharma, R., & Ghoshal, G. (2018). Emerging trends in food packaging. *Nutrition & Food Science*, *48*(5), 764–779. DOI: 10.1108/NFS-02-2018-0051

Sheth, J., & Parvatiyar, A. (2020). Sustainable marketing: Market-driving, not market-driven. *Journal of Macromarketing*, *41*(1), 150–165. DOI: 10.1177/0276146720961836

Shin, J., & Selke, S. E. M. (2014). Food packaging. In Clark, S., Jung, S., & Lamsal, B. (Eds.), *Food processing: principles and applications* (pp. 249–273). Wiley., DOI: 10.1002/9781118846315.ch11

Silvenius, F., Grönman, K., Katajajuuri, J., Soukka, R., Koivupuro, H., & Virtanen, Y. (2013). The role of household food waste in comparing environmental impacts of packaging alternatives. *Packaging Technology & Science*, *27*(4), 277–292. DOI: 10.1002/pts.2032

Singh, R. P., & Ramanathan, A. L. (2019). Urbanization, River Yamuna Pollution, and the Impact of Plastic Waste on Water Quality. *Environmental Pollution*, *252*, 799–809.

Singh, S., Kumar, J., & Rao, P. (2018). Environmental impact assessment framework for product packaging. *Management of Environmental Quality*, *29*(3), 499–515. DOI: 10.1108/MEQ-10-2017-0105

Siracusa, V., Ingrao, C., Lo Giudice, A., Mbohwa, C., & Dalla Rosa, M. (2014). Environmental assessment of a multilayer polymer bag for food packaging and preservation: An LCA approach. *Food Research International*, *62*, 151–161. DOI: 10.1016/j.foodres.2014.02.010

Tajeddin, B. (2017). Packaging composite materials from renewable resources. In *Handbook of Composites from Renewable Materials* (Vols. 1–8, pp. 525–561). wiley. DOI: 10.1002/9781119441632.ch58

Thangapandi, P., Perception Of Sustainable Chai-Flasks & Its Influence On Gen Z's Buying Behavior Dr. N. Bargavi S. Bharhane Raghaventheran.

Touil, A., Ayad, K., El Hamidi, N., & Babounia, A. (2023). Toward greener supply chains: Analysis of the determining factors. *Environment and Ecology*, *14*(2), 114–126. DOI: 10.21511/ee.14(2).2023.09

Truong, Q., (2019). Food packaging process material's effect on environment and their control measures.

Varun, S. A., & Nautiyal, H. (2015). Environmental impacts of packaging materials. In Environmental Footprints of Packaging (pp. 115-137). Singapore: Springer Singapore.

Varžinskas, V., & Markevičiūtė, Z. (2020). Sustainable food packaging: Materials and waste management solutions. *Environmental Research, Engineering and Management*, *76*(3), 154–164. DOI: 10.5755/j01.erem.76.3.27511

Wackett, L. P. (2008). Polylactic acid (PLA) An annotated selection of World Wide Web sites relevant to the topics in Environmental Microbiology. *Microbial Biotechnology, 1*(5), 432–433. DOI: 10.1111/j.1751-7915.2008.00056.x

Williams, H., & Wikström, F. (2011). Environmental impact of packaging and food losses in a life cycle perspective: A comparative analysis of five food items. *Journal of Cleaner Production, 19*(1), 43–48. DOI: 10.1016/j.jclepro.2010.08.008

Zhao, J. (2023). Exploring the development trend of packaging design in the new era., 1272-1278. DOI: 10.2991/978-2-38476-092-3_162

Živkovic´, N. (2009). *Polyhydroxy alkanoates, Green Chemistry (seminar)*. Food Technology Zagreb. (in Croatian)

Chapter 14
Quality Enhancement in the Food Processing Sector With Sustainable Innovations

Ainee Tanwir
https://orcid.org/0009-0007-4286-5159
Pikky Tech Private Limited, India

Tatheer Fatima
https://orcid.org/0009-0003-6942-5197
Era Medical University, India

Fatima Noor
https://orcid.org/0009-0003-8966-3264
Aligarh Muslim University, India

ABSTRACT

This article delves into the recent advancements in food processing, exploring their impact on food safety, quality, and sustainability. Alongside safety and quality, the food industry also strongly emphasises sustainability and environmental impact. The adoption of renewable energy sources, the reduction of waste, and the implementation of circular economy principles in the food supply chain are becoming increasingly important. Furthermore, the integration of technologies with other preservation methods, such as antimicrobial agents or modified atmospheres, has shown promising results in enhancing the microbial stability and quality of food products. These innovations possess the capacity to transform the food sector.

DOI: 10.4018/979-8-3693-8342-1.ch014

1. INTRODUCTION

India is one of the important food producers in the world, and its contribution to world food trade is quite noteworthy. However, the food segment in India is still to be exploited, as it has immense potential for improvement. The proposition established that to be competitive, food processing industries must pay attention to food quality enhancement, along with expanded production and reduced costs. Food quality supplementation can be achieved through the right choice of raw materials, improved processing techniques, and producing value-added food products. Once quality specifications are kept in mind while designing food products and processes, the other classic practices for improving quality, like reductions in waste, rework, and variability content of the product, follow automatically. The thrust of the refurbishment campaign ought to be both resource and product quality enhancement. Keeping the resources and products at peak quality levels involves managing the knowledge for safe food processing. As companies get more serious about implementing food safety management systems, advanced technologies are being developed to produce long-lasting processed foods. Recall has prompted canneries and baby food manufacturers, especially, to look for and use improved processing and packaging methods to extend commercial sterility. Control technologies and software related to thermal processing have risen with the advancement of these technologies. Such advances bring them closer to the establishment of higher levels of preventive management during the process development.

2. IMPORTANCE OF QUALITY ENHANCEMENT IN THE FOOD PROCESSING SECTOR

The food processing industry has grown rapidly in the past few years and is ranked 5th in world exports. Considering the importance of the food processing sector, countries across the world have created regulatory bodies to ensure food quality and safety, thus increasing the actual costs of food processing. The sector has great potential to offer increasing employment opportunities, raise farm productivity, provide an incentive for weak economic regions, enhance resource usage, reduce retail food prices, and serve as a source of foreign exchange.

After meeting the mandatory requirements of quality and safety, the processor generally spends time and resources to protect the product for intended use during packaging, distribution, and storage, not only in order to meet the standards but also to satisfy the customer. This process of ensuring the satisfaction of the customer is what is addressed as quality enhancement. The level of satisfaction of the processed food by the customer depends on a host of product attributes such as flavor, aroma, crispness, texture, size, and appearance, amongst others, which are mostly sensory attributes and confer aesthetic value to the product. The product attributes will degrade over time, and it is the degree of degradation and the loss in sensory quality of food and variations in different attributes of fresh fruits and vegetables that denote their varying degrees of perishability. Extending the life of processed food by preventing or retarding these processes is what is addressed as shelf life extension.

In summary, the greater the shelf life, the greater its perceived value. In addition to maintaining and improving perceived value, process enhancement can generate extra revenue by creating value for raw products, by developing and manufacturing new, creative products. The overall objective of the food processor, then, is to create products that satisfy the customer's expectations of quality while meeting their safety requirements. To achieve this, the processor has the challenge of optimizing three characteristics that are related to the organized system for manufacturing: quality, productivity, and cost. Providing customers with foods that have a longer shelf life and better sensory attributes than competing products gives processors a unique marketing advantage. These foods command substantial price premiums and most often result in loyalty from customers. Clearly, value enhancement involves the management of product quality and attributes, particularly shelf life, the development of added-value gourmet and niche market products, and the identification and capture of the export markets.

3. FOOD PROCESSING TECHNOLOGY

The advancement of food processing technology has been driven by several variables, including consumer demands, which have unavoidably influenced the direction of new developments in food production, preservation, and control. Historical, demographic, economic, political, and environmental variables have influenced consumer lifestyles and eating habits. These important variables are reflected in the emerging developments in food processing, which include improvements in methods and procedures for food production, storage, and monitoring. New foods that are easy to maintain and handy are in high demand from consumers. Fresh or little processed food needs non-synthetic chemical preservatives since consumers are increasingly aware of the need for safer products (Señorans F J,2003).To increase food products' safety, stability, competitiveness, and efficiency while also promoting health, the past ten years have seen a focus on quality (Boekel V, 2010).

Ready-to-eat foods are currently experiencing a remarkable global displacement trend, taking the place of traditional cuisines that prioritise freshly prepared meals. Furthermore, ready-to-eat foods are a kind of food items that don't require any previous cooking or preparation; they're frequently packaged and prepared for consumption

(Huang L and Hwang, 2012). These foods are also processed so they don't require extra attention when consumed, suchas heating (Paudyal R and Karatzas K A 2016).

4. CONVENTIONAL PRESERVATION TECHNIQUES IN FOOD PROCESSING SECTOR

Conventional preservation techniques employed for less processed foods encompass:

Refrigeration: Maintaining food at low temperatures inhibits microbial proliferation and enzyme activity, hence prolonging shelf life. Nonetheless, refrigeration alone may be inadequate for all minimally processed items.

Freezing: Freezing protects food by deactivating enzymes and suppressing microbial proliferation. Nevertheless, improper execution may result in alterations to texture and a depletion of nutrients.

Pasteurization: The application of heat to eliminate pathogenic bacteria serves as an excellent preservation technique for certain less processed goods, such as juices and dairy products. Excessive heat can diminish nutrition and modify sensory characteristics.

5. INNOVATIVE PRESERVATION TECHNOLOGIES

To overcome the constraints of conventional techniques, innovative preservation technologies are being investigated:

- High-Pressure Processing (HPP) employs elevated pressure to deactivate bacteria and enzymes while preserving the nutritional and sensory attributes of food to a minimal extent. It has been effectively utilized in many minimally processed goods including as drinks, guacamole, and ready-to-eat meals.
- Pulsed Electric Fields (PEF) employ brief bursts of high-voltage electricity to deactivate microorganisms. It is especially efficacious for liquid foods and can maintain heat-sensitive nutrients.
- Ultraviolet (UV) light can effectively disinfect the surfaces of minimally processed goods, such as fresh produce. It is a non-thermal technique that little affects the nutritional or sensory characteristics of food.
- Biopreservation involves employing naturally existing antimicrobial agents, such as bacteriocins from lactic acid bacteria, to suppress the proliferation of harmful and spoilage microbes in minimally processed foods.

6. THERMAL PROCESSING

The thermal process is used in thermal food processing to inactivate harmful microorganisms and enzymes, with minimal damaging effect on product quality. Although this is the most common method, it may not be environmentally sound due to the lack of non-renewable resources and the risk of toxic component formation. However, because of critical factors, such as (i) the challenging future food demands; (ii) the expectations of consumers to lead healthier, safer, and more convenient lives; and (iii) the decreasing safety margins due to economic pressures, the productive and food processing sectors have engaged in active research and technological development aimed at advanced thermal processing technologies.

The developments in the state-of-the-art and recent applications of thermal, non-classical thermal, and combined techniques such as conventional heating, ohmic heating, microwave, microwave-assisted preservation processes, radiofrequency, infrared, high pressure, and pulsed electric field processing, ultraviolet, extrusion, and others are reported. These technologies are promising to overcome the critical factors that might hinder the traditional thermal method. The scientific and industrial challenge is related to the effective combination of the beneficial attributes that have already been recognized in these techniques.

Definition and Importance: The definition of food processing is that it involves the processes used to impart preservation and stability to food. Preservation is accomplished by chemical, physical, or enzymatic changes that destroy bacteria, yeasts, and moulds, or by creating physical barriers to such organisms. Preservation also involves the inhibition of chemical, enzymatic, or oxidative processes so that food will not deteriorate too rapidly. Food processing is needed to extend the food supply longer than what is available from freshly harvested agricultural products. New technologies have been applied together with traditional food preservation methods, such as thermal processing, freezing, drying, evaporation, canning, salting, sugaring, and pickling, to extend the shelf life of food products while maintaining quality and ensuring safety.

Thermal processing uses heat to preserve food, enhance its palatability, and alter its physical, chemical, and microbiological properties. The application of heat in various forms is a fundamental and widely-used technique in food processing, as it can effectively inactivate pathogenic microorganisms, modify food texture and flavor, and extend the shelf life of products. Thermal processing plays a crucial role in ensuring the safety, quality, and stability of a wide range of food products. Examples of thermal processing methods include:

- Sterilization: A more intense heat treatment that destroys all microorganisms, resulting in a longer shelf life for products like canned foods and bottled sauces. This method ensures the complete elimination of potential pathogens and spoilage organisms, allowing for extended storage and distribution of these shelf-stable food items.(Ricci et al., 2021)

- Blanching: A brief exposure to hot water or steam, often used for vegetables before freezing or canning, to inactivate enzymes and improve quality. Blanching helps to preserve the color, texture, and nutrient content of the food, while also preparing the vegetables for further processing.
- Baking: Cooking food using dry heat in an oven, a common method for breads, pastries, and some meat and vegetable dishes. Baking not only cooks the food but also imparts unique flavors and textures through the Maillard reaction and other heat-induced chemical changes.
- Roasting: Similar to baking but often performed at higher temperatures, roasting is frequently used for meats and certain vegetables. This method can enhance the caramelization and flavor development of the food, creating a more robust and appealing taste profile.
- Frying: Cooking food by immersing it in hot oil or fat, a technique that can add flavor and crispness to items like french fries and fried chicken. The high-temperature oil facilitates rapid heat transfer and the development of a crispy exterior, while the interior remains tender and moist.

There are many food preservation methods, but thermal processing, which includes canning and retort pouch processing, is the principal method today for the commercial sterilization of food products. Food is processed to create a product that is safe to eat for a specified shelf life and may include the addition of microbiological, chemical, and physical stability attributes within the safety and quality limitations built in by the selected process with the right container and packaging materials. Heat and chemical preservatives are widely used in foods associated with new nutritional attributes to help reduce the risk of food borne illness. Food borne pathogens are very resilient, and typical heating processes are not always effective at killing them. That is the reason for special food safety regulations have been implemented to ensure that the level of process lethality will kill pathogenic spore formers that may be present in baby food in numbers capable of producing toxins before spoilage safety issues alert parents. In brief, the preservation of low-acid food is the most important processing step to prevent food borne sickness and product recalls. The heating process is always designed to prevent spoilage and ensure the safety of the product and the presence of controlling pathogen spores.

While many thermal processes have been used for centuries, modern food processing has seen innovations and improvements focusing on efficiency, control, and quality. Here are some modern technologies used in thermal processing:

1 Pasteurisation- A mild heat treatment called pasteurization is applied to many different kinds of food products. Pasteurization has two main purposes: first, it eliminates harmful bacteria from food to prevent illness; second, it eliminates spoilage (souring) germs to enhance food storage. The main source of it is Louis Pasteur's discovery in 1857 that souring in milk could be postponed by heating it to 122–142°F (50.0–61°C). However, it wasn't until later that century that it was conclusively determined that bacteria were the cause of spoiling and disease. But even before this, foods were being delivered for public consumption for a substantial amount of time without knowledge of how they were being kept in sealed containers using a sterilizing process.

Pasteurization is defined as follows by the International Dairy Federation (IDF, 1986): ''Pasteurization is a process applied to a product with the objective of minimizing possible health hazards arising from pathogenic microorganisms associated with the product (milk) which is consistent with minimal chemical, physical, and organoleptic changes in the product. 'Fruit juices, creams, ice cream mixes, eggs, fermented goods, soups, and other beverages are among the goods that come under the above definition in addition to milk. In the past, pasteurization was done in a batch process called the Holder process at 63°C for 30 minutes. This was replaced with the introduction and acceptance of continuous HTST processes. However, not all microorganisms are inactivated by pasteurization; those that survive are known as thermodurics and those that survive a harsher treatment at 80–100°C for 30min are known as spore formers.

2 Retort- Using heat to increase the temperature of the containers in a commercial closed vessel known as a retort or autoclave is the commercial sterilization method. Foods that have been packaged and commercially sterilized can be kept for up to two years at room temperature in hermetically sealed containers (Featherstone, 2015). There are three steps in the sterilizing process: The first step, known as the "come-up time," is the amount of time needed for a high-flow heating medium to attain the necessary pressure of 15–20 psi (about 1–1.4 bar) above atmospheric pressure and the retort temperature of 240–250°F (about 115–121°C). Depending on the target microorganism or expected microbial community, the retort adjusts temperature and pressure during the second stage, Pt (holding or cooking stage), to ensure lethality. Cooling water is added during the third phase, known as CDT (come-down time), to maintain the temperature drop. By cooling food, food processing can be avoided, and thermophilic bacteria development is also inhibited. Nevertheless, retort pouches may rupture during the cooling process. Applying overpressure air throughout the cooling process will help prevent this by preserving package integrity and preventing container deformation (Mosna & Vignali, 2015).

3 Microwave- Over several decades, microwave heating has been used extensively in food processing. Microwave heating is used in food processing for various purposes, such as baking, pasteurisation, thawing, tempering, and drying of food ingredients (Gupta & Wong, 2007; Metaxas & Meredith, 1983).The ability to achieve fast heating rates, a significant reduction in cooking time, more uniform heating, safe handling, ease of operation, and inexpensive maintenance has made microwave heating popular in the food processing industry (Salazar-Gonzalez, San Martin-Gonzalez, Lopez-Malo, & Sosa-Morales, 2012; Zhang, Tang, Mujumdar, & Wang, 2006). Furthermore, compared to traditional heating during the cooking or reheating process, microwave heating may alter the flavour and nutritional value of food to a lesser amount (Vadivambal & Jayas, 2010).

4 Radio Frequency Heating: Similar to microwave heating, radio frequency heating uses electromagnetic waves to generate heat within the food. However, RF heating can penetrate deeper into the product, making it suitable for larger items or those with varying densities. It's used in baking, drying, and thawing processes. (Chacha et al., 2021)

5 Ohmic Heating: Passes an electrical current through the food to generate heat internally. This method provides rapid and uniform heating, minimizing quality degradation. It's particularly suitable for liquid and particulate foods like soups and sauces. (Chacha et al., 2021)

6 High-temperature short time (HTST)– It is thermal processing is generally recognized as any thermal process for food that involves the combination of high temperature (110-130°C) and short conventional time (i.e., less than 20 minutes) or with extended shelf-life products (5-30 minutes), all during which the thermal process is carried out. The application of such processes is mainly concentrated in the dairy industry for the production of cheese, evaporated or sweetened condensed milk, and milk powder. The use of high-temperature short-time processes for chilled food is also of interest, and in particular, has been assessed for fish and egg mixtures. These thermal processes are particularly interesting because of the minimal detrimental effect on the organoleptic and nutritional characteristics of these products and also because of the short time needed for the food to be exposed to high temperatures.

Given the fact that they could act as a useful alternative to ordinary thermal treatments, it is of interest to search for the basic microbiological and chemical factors that affect the killing rate of spores necessary to obtain the required food sterilization level. Since the lower heating medium of HTST is usually unsaturated steam, the process can sometimes be modified and used under pressure for selected food products. For this assessment, the major requirements for the application of pressure are the development and comprehension of the scientific principles involved, in combination with an understanding of the raw materials, the importance of the biological hazardous level of microorganisms, and clear detection of the chemical and physicochemical reactions taking place inside the food. It should be clear that further determination of the lethal effect of pressure at moderate and higher temperatures must be noted and defined before wide-scale production of these food products can be carried out.

7 Microwave-assisted pasteurisation (MAP)- MAP is a hybrid technique that combines microwave and pasteurization. It uses microwave electromagnetic energy and hot water from pasteurization. Using this technique, food products are heated uniformly such that the heat is absorbed by all of the food's constituent parts. Preheating, microwave heating, holding, and cooling are the four steps of MAP, all of which improve food quality and shorten processing times (Jain D,2018, Inanoglu S, 2021).One instance of a multiple-mode microwave heating cavity model in traditional microwave technology is 2,450 MHz.

According to reports, the MAP operates at a longer 915 MHz microwave wavelength, allowing single-mode cavities to process a wide variety of food packaging (Tang J,2018, Stanley R A, 2017).

The purpose of using MAP is to minimize quality loss and increase shelf lifeby lessening the number of bacteria or viruses in the food when compared to traditional microwave

8 Microwave-assisted ultraviolet sterilization (MWUV)- The combination of microwave and UV light, or MWUV, is more effective at lowering the number of microorganisms while preserving milk quality. Furthermore, MWUV showed a decrease in coliform bacteria and sterilised the milk without compromising its inherent qualities. It provides effective employment of specific absorption characteristics of metal nanoparticles in the optical or UV ranges, as well as selective and extremely efficient powder-based microwave heating over metals. In this way, it improved the performance of UV devices, primarily by enhancing the absorbance, which extends the practical range of microbiological sterilization. The working hypothesis is that the MWUV approach can simulate the action of a non-resonant MW substance or MW absorbent, a well-accepted system used for standard MW heating techniques in material and food science.

The theory and working principles of Non-resonant Microwave Heating assume the use of selective, high-resistance MW absorbents, responsible for generating a quick and uniform temperature rise in thermoplastic systems. In contrast, the properties and possible applications of gold and silver make them desirable components for MW-assisted UV disinfection. The results confirmed the main working hypotheses on the properties of gold and silver MWUV nanoparticles with a nano aggregated assembled characteristic, as well as the improvement of device performance with higher absorbance, increased release of energy in the UV range, and, most importantly, the preservation of the evolved shock regime with uniform energy release. This maintained uniform UV energy emission limited the effects of overheating-induced melting upon the internal or external MW absorption not seen in standard MW absorbing UV lamps.

9 Pressure-assisted thermal sterilisation (PATS)

A technique called pressure-assisted thermal sterilisation (PATS) makes better use of HPP by utilising the heat produced by compression itself. Practically speaking, cost and technological limitations on equipment prevent HPP from sterilizing food on its own. Compared to conventional thermal treatments, the clever fusion of HPP and heat energy yields extremely healthful foods of higher quality (Barbosa-Cánovas G V,2014).PATS is a low-temperature, short-operation method that necessitates the use of initial chamber temperatures between 60°C and 90°C. Internal compression heating at 600 MPa or higher allows the in-process temperatures to exceed any range between 90°C and 130°C (Matser A M,2004). Furthermore, because both pressure and compression heat add to the process's lethality, the approach offers a high-temperature, quick processing option (Barbosa-Cánovas G V,2014). In HPP, pathogenic and spoilage bacteria are inactivated by applying an isostatic pressure of 100–1000 MPa at room temperature to maintain the quality of food products (Raso J and Barbosa-Cánovas G V 2003).

These modern thermal processing technologies offer improved control, efficiency, and product quality compared to traditional methods. They contribute to producing safe, shelf-stable, and convenient food products.

7. **Non-Thermal processing**- Non-thermal food processing refers to an array of technologies or applications where heat is not used to prepare, form, or preserve foods. It involves the basic steps of preparing, cleaning, and categorizing raw foods. Novel methods for food processing meet new dietary standards with more nutritious, flavorful, safe, convenient, and cost-effective food products while extending shelf life through non-thermal forms of product handling and heat transfer. However, problems in texture, flavour, and nutrient value that are

generally associated with traditional thermal processing technologies are lacking in the application of these recent non-thermal food processing techniques. Compared to conventional thermal methods, non-thermal food processing is a developing field, and there are several choice criteria and styles to choose from, all designed to reduce the time and energy wasted.

Diverse technologies make use of the area of energy delivered at various frequencies; however, none intrinsically destroys pathogens or spoilage. It is more important to maintain a particular product's quality after processing.

The industry increasingly backs food research for beneficial brands providing almost instant consumer acceptance. Numerous classes of non-thermal food processing methods have evolved, for example, pulsed electric field, supercritical carbon dioxide, high pressure, ultraviolet, ultrasound, or irradiation.

Each technique preserves food by reported benefits that facilitate food enhancement or improved product quality, maintain the capacity to handle nutrient levels, extend the life of food on the shelf, impair microbiological implications, or destroy pathogens. Innovative processing techniques of this type tend to have the same characteristics in such terms but involve fewer constraints. This means that we should expect fruitful innovation that is commensurate with this study and that, as food researchers focused on the creation of new products, we should attempt to nourish and guide more technological success and creativity.

Non-thermal processing methods aim to minimize heat exposure to preserve the nutritional and sensory qualities of food. Some examples include:

1. **High-pressure processing (HPP)-** It is a non-thermal processing technology that has been commercialized with excellent success (Daryaei H, 2016). This minimal thermal technology, which is mostly used at room temperature, is applied to food goods at pressures between 100 and 1000 MPa (Guerrero-Beltrán J A,2005). The primary goal of this technique is to displace thermal processing technologies. With the help of this technology, food safety against microorganisms is ensured, and processed foods can maintain their inherent flavours and nutritional value. As a result, HPP technology is regarded as a minimal processing method that ensures food safety and flavour(Daryaei H and Balasubramaniam V M 2012). Nevertheless, there are several drawbacks to this technique, such as its high cost, the ability of food enzymes to survive in large quantities, and its restricted packing options (Marc R A 2020).
2. **Pulsed Electric Field:** This non-thermal method applies short, high-voltage electric pulses to food to inactivate microorganisms. The electric field causes pores to form in the cell membranes, leading to cell death and preservation of the food's natural flavors, textures, and nutrient content.(Evelyn et al., 2021)

3. **Pulsed Light:** Intense, short-duration flashes of broad-spectrum, high-intensity light are used to eliminate surface microorganisms on foods, such as fruits and vegetables. This method effectively kills pathogens while leaving the internal structure and nutrients of the food largely intact.(Evelyn et al., 2021)
4. **Ultrasound**- During food items' shelf life, non-thermal processes like ultrasound, ultraviolet (UV), cold plasma therapy, and high-pressure processing (HPP) can regulate the growth of spoilage and pathogens. Because ultrasound is helpful and can be used in conjunction with thermal procedures, it can maximize energy usage by speeding up the rate at which sterilisation occurs (Alarcon-Rojo A D, 2015). It is used in food processing for analysis and quality control, and based on its specified frequency range, it can be categorized as low or high energy. Ultrasound is used to extract bioactive compounds from plants or food, in addition to being helpful in processing and preservation procedures(FDA 2014, Soria A C 2010). To achieve greater efficacy, it is advised to be used in conjunction with other decontaminating treatments. It's also possible to unintentionally change the food's texture or content. Negative modifications are made to certain food attributes, such as taste, colour, or nutritional value(Marc R A 2020).
5. **Ultraviolet**- It has been widely accepted that UV radiation with wavelengths between 220 and 280 nm, especially at 254 nm, can stop bacteria from multiplying by causing DNA thymine dimer formation (Unluturk S2008). Moreover, solid food—especially packaged goods—can be disinfected with UV radiation. Because of its low penetration level, this method has been utilized to sterilise and reduce the number of microorganisms in food processing equipment, food surfaces, and packaging materials. Additionally, as a backup plan, fruit juice pasteurization is accomplished using this UV light technology (Marc R A 2020). However, the most important aspect that limits pulsed light is sample heating. The processing parameters used, such as the food's distance from the pulsed UV light source and exposure length, may promote lipid oxidation (Heinrich V, 2015, Wambura P, 2011). Enhancing the food pathogens' dosage response behavior in thick liquid food is also crucial. Further information on the kinetics of inactivation of pathogens and spoilage bacteria is required to forecast UV disinfection rates on food surfaces (Marc R A 2020).
6. **Cold Plasma Treatment-** This innovative technology utilizes ionized gas, or plasma, to inactivate microorganisms in food. The plasma generates highly reactive species that disrupt the cell walls and membranes of pathogens, effectively eliminating them while avoiding the detrimental effects of high temperatures. The choice of processing method, whether thermal or non-thermal depends on the specific food product, its characteristics, and the desired outcomes in terms of safety, quality, and shelf life. Because cold plasma therapy utilizes less energy and temperature during food processing for microbial inactivation and product

decontamination, it is regarded as a contemporary non-thermal method (Dong X, 2020, Chizoba Ekezie F G, 2018). After processing, cold plasma therapy is a quick method that produces no hazardous byproducts or exhaust fumes. It is necessary to take into account the food's overall consistency, colour, texture, and chemical changes in addition to its nutritional value. There isn't a commercial device on the market for cleaning and sanitizing food items. Moreover, in addition to the industry, it is embraced by several institutions and research organizations (Marc R A 2020).

7. **Ozone Treatment:** Ozone gas is employed for the disinfection and preservation of food, offering a non-thermal alternative to traditional methods. Ozone's strong oxidizing properties can effectively kill a wide range of microorganisms, making it a valuable tool for food safety and shelf-life extension.(Enríquez-Castro et al., 2021)
8. **Irradiation:** Exposing food to ionizing radiation, such as gamma rays or electron beams, can effectively kill pathogens and extend the shelf life of products without compromising their nutritional value. This non-thermal method disrupts the DNA of microorganisms, preventing their growth and proliferation.(Food Irradiation: What You Need to Know, 2022)

8. MECHANICAL PROCESSING

Food products can be physically manipulated through various mechanical processes to achieve desired changes in size, shape, texture, or other characteristics. Common mechanical processing methods used in the food industry include:

1. **Sorting and Grading:** Separating food items based on criteria such as size, shape, quality, or other attributes to ensure consistency and meet specific requirements. This process helps to standardize the appearance and characteristics of the final product. (Types Of Food Processing Equipment, 2021)
2. **Cleaning:** Removing dirt, debris, and other unwanted contaminants from the food to improve its cleanliness and safety. This step is crucial for maintaining food hygiene and meeting regulatory standards.
3. **Peeling and Cutting:** Removing the outer layers or skin of the food and then reducing it into smaller, more manageable pieces through techniques like slicing, dicing, or chopping. This can enhance the food's appearance, texture, and ease of preparation.(Jideani et al., 2017)
4. **Mixing and Blending:** Combining different food ingredients to create uniform, homogeneous mixtures or dispersions. This process ensures consistent flavor, texture, and composition throughout the product.

5. **Grinding and Milling:** Reducing the size of food particles, such as grains or spices, into finer, more consistent textures. This can improve the food's solubility, digestibility, and overall quality. (Margasahayam & Balraj, 2018)
6. **Extrusion:** Shaping food by forcing it through a die, a process commonly used to create products like pasta, snacks, and breakfast cereals. Extrusion allows for the production of a wide range of food items with specific shapes, textures, and characteristics.(Cotacallapa-Sucapuca et al., 2021)

9. CHEMICAL AND BIOLOGICAL PROCESSING

Chemical Processing refers to methods that utilize chemical reactions or changes to preserve, modify, or transform food. These methods often involve the addition of chemical substances or the control of chemical parameters like pH. Chemical processing techniques are used extensively in the food industry to achieve a wide range of objectives, such as enhancing food safety, extending shelf life, improving sensory properties, and modifying the physical and nutritional characteristics of food products.(Augusto, 2020) Some common examples of chemical processing methods include curing, smoking, the addition of preservatives, pH adjustment, and chemical modification of food components. Common Chemical processing methods used in the food industry include:

1. **Curing:** Uses salt, sugar, nitrates/nitrites, or other substances to preserve and flavor food, primarily meats, fish, and cheese. The chemical reactions that occur during curing inhibit microbial growth, enhance flavor and texture, and contribute to the development of desirable sensory properties. Curing involves the controlled application of these chemical substances, which interact with the food's components to produce specific effects. For instance, salt can draw out moisture, inhibit the growth of harmful bacteria, and contribute to the development of a firm texture. Nitrites and nitrates help preserve the food's color, prevent the growth of Clostridium botulinum, and impart a distinctive cured flavor. The combination of these curing agents and the chemical changes they induce can also lead to the production of unique aroma compounds, further enhancing the food's sensory appeal. (Food processing, 2023)
2. **Smoking:** Exposing food to smoke introduces a complex array of chemical compounds that contribute to the food's flavor, color, and preservation. The smoke contains a rich mixture of substances, including phenols, carbonyls, and organic acids, which can impart unique and desirable sensory properties to the food. (Drewnowski et al., 2022) These compounds not only enhance the flavor and appearance of the food but also exert antimicrobial and antioxidant effects,

helping to preserve the food and extend its shelf life. The chemical reactions that occur during the smoking process can lead to the formation of compounds that inhibit the growth of harmful microorganisms, further improving the food's safety and quality. This versatile processing method allows food producers to create distinctive, flavorful, and well-preserved products that appeal to consumers.

3. **Addition of Preservatives:** Involves adding chemical substances like antimicrobials, antioxidants, or chelating agents to inhibit microbial growth, prevent oxidation, and maintain the quality and shelf life of food products. These preservatives can include organic acids such as benzoic acid, sorbic acid, or acetic acid, as well as sulfites, nitrites, and other compounds that target specific spoilage and pathogenic microorganisms. The use of these preservatives helps to prevent the growth of harmful bacteria, mold, and yeast, which can lead to food spoilage and foodborne illnesses. Additionally, antioxidants like BHA, BHT, or ascorbic acid help to protect the food from oxidative deterioration, preserving its color, flavor, and nutritional value.(Fadhel et al., 2019)
4. **pH Adjustment:** Controlling the acidity or alkalinity of food can inhibit microbial growth and influence texture, flavor, and other physicochemical properties. Adjusting the pH can be achieved through the addition of acids like citric acid or bases like sodium hydroxide, which can help preserve the food, improve its safety, and modify its sensory characteristics. By lowering the pH, certain pathogenic bacteria like Clostridium botulinum can be inhibited, while adjusting the pH can also affect the solubility, gelation, and other functional properties of food components, leading to desired changes in texture and mouthfeel.(Coldea, 2019)
5. **Chemical Modification:** Includes processes like hydrogenation of oils or the use of chemical leavening agents to alter the physical and chemical characteristics of food, such as texture, viscosity, or reactivity. These modifications can improve the functional properties of the food, enabling the development of new and innovative food products with enhanced sensory attributes, improved nutritional profiles, and extended shelf life. For instance, hydrogenation of oils can transform liquid oils into solid or semi-solid fats, which are then used in the production of baked goods, margarine, and other processed foods. Chemical leavening agents, such as baking soda or baking powder, can create light and airy textures in baked goods by generating carbon dioxide through a chemical reaction. These types of chemical modifications allow food manufacturers to tailor the properties of food to meet consumer demands and industry standards. (Gioia et al., 2017)

Biological Processing employs biological agents, primarily microorganisms or enzymes derived from them, to transform food through fermentation, enzymatic reactions, and other biological processes. These methods can enhance the nutritional value, sensory properties, and shelf life of food products. Fermentation utilizes microorganisms to metabolically convert food components, resulting in improved digestibility, enhanced flavors, beneficial byproducts, and extended shelf life. Enzymatic modification harnesses the catalytic power of enzymes to alter the texture, appearance, and functionality of foods, such as tenderizing meat or clarifying juices. Additionally, processes like malting, a controlled germination of grains, can increase enzyme activity and improve the flavor profiles of products like beer. By leveraging the transformative capabilities of biological agents, food manufacturers can create a wide range of products with improved nutritional, sensory, and shelf-life attributes. (Sindhu et al., 2018).

Common biological processing methods used in the food industry include:

1. **Fermentation:** Fermentation is a crucial biological processing technique that utilizes the metabolic activities of microorganisms such as bacteria, yeasts, and molds to transform food. This ancient process has been employed for centuries to create a diverse array of food products, including yogurt, sauerkraut, kimchi, cheese, bread, beer, and wine. Through fermentation, the microorganisms break down and convert the food's components, leading to a range of beneficial outcomes. Fermentation can improve the digestibility of the food by breaking down complex molecules into more readily absorbable forms. It can also enhance the food's flavor profile by generating unique aroma compounds and acids that contribute to the desired taste. Additionally, fermentation can produce beneficial byproducts, like probiotics, which promote a healthy gut microbiome. Furthermore, the metabolic activities of the microorganisms can extend the shelf life of the final product by inhibiting the growth of spoilage and pathogenic microbes, thereby improving the food's safety and preserving its quality over time.(Dinkçi et al., 2019)
2. **Enzymatic Modification:** This technique utilizes enzymes, which are biological catalysts derived from microorganisms or plant sources, to catalyze specific reactions in food. It can be employed to achieve a variety of beneficial effects, such as tenderizing meat, clarifying juices, or modifying the properties of starches. By harnessing the catalytic power of enzymes, food manufacturers can improve the texture, appearance, and overall functionality of their products. For instance, enzymes can be used to break down complex proteins in meat, leading to a more tender and palatable texture. Similarly, enzymes can be utilized to clarify fruit juices by breaking down pectins and other compounds that can cause cloudiness. Additionally, enzymes can be employed to modify

the characteristics of starches, such as altering their viscosity or gelatinization properties, which can enhance the quality and performance of baked goods and other starch-based foods.(Souto et al., 2016)

3. **Malting:** Malting is a controlled germination process used in the production of grains, particularly for beer production. This process involves soaking the grains in water, allowing them to partially germinate, and then drying them. The germination process activates and enhances the enzyme activity within the grains, which can then be used to convert the starch present into fermentable sugars. This enzymatic conversion of starch to fermentable sugars is a crucial step in the brewing process, as it provides the necessary sugars for the yeast to convert into alcohol, ultimately improving the flavor profile and characteristics of the final beer product.(Bokulich & Bamforth, 2013)

In the food industry, it is common to combine chemical and biological processing techniques with other processing methods, such as thermal or mechanical processing, in order to achieve the desired outcomes in food production. This integrated approach allows food manufacturers to leverage the unique benefits and capabilities of different processing methods to enhance the safety, quality, and sensory characteristics of their products. By integrating these various processing techniques, food producers can optimize the functional properties, nutritional profiles, and shelf-life of their food items, thereby meeting consumer demands and industry standards more effectively.

10. ACTIVE AND INTELLIGENT PACKAGING

Beyond processing, packaging plays a crucial role in food safety. Active and intelligent packaging systems are revolutionizing food preservation.

Active packaging is a fascinating innovation in food preservation that goes beyond simply containing the product. It's designed to actively interact with the food or the environment within the packaging to extend shelf life, enhance safety, and maintain or even improve sensory characteristics. Here's how it works:

Active packaging incorporates specific compounds or components into the packaging material itself. These components are designed to perform a particular function, such as:

1. **Antimicrobial activity:** Active packaging can release antimicrobial agents like ethanol or essential oils, or incorporate antimicrobial substances directly into the packaging material. This inhibits the growth of bacteria, molds, and yeasts that can spoil food, thus extending its shelf life.(Perera et al., 2023)

2. **Oxygen scavenging:** Some active packaging systems utilize oxygen absorbers. These sachets or films absorb oxygen from the package, creating a modified atmosphere that inhibits the growth of aerobic microorganisms and prevents oxidation reactions that can lead to spoilage. This is particularly useful for extending the shelf life of oxygen-sensitive products like coffee, nuts, and baked goods.(Gupta, 2023)
3. **Moisture control:** Active packaging can also regulate moisture levels within the package. Some materials absorb excess moisture, preventing the growth of mold and mildew, while others release controlled amounts of moisture to prevent food from drying out.(Perera et al., 2023)
4. **Ethylene absorption:** Ethylene is a natural plant hormone that accelerates ripening and spoilage in certain fruits and vegetables. Active packaging can include ethylene-absorbing materials to slow down the ripening process and extend shelf life.(Pham et al., 2023)

Active packaging is a rapidly evolving field with ongoing research and development into new applications and materials. It represents a significant step towards more sustainable and efficient food preservation, reducing food waste and ensuring that consumers enjoy fresher, safer, and higher-quality products.

Intelligent packaging: Intelligent packaging, sometimes referred to as "smart" packaging, takes food packaging a step further than its traditional role. While standard packaging primarily protects and contains the product, intelligent packaging adds a layer of interaction and information. It utilizes various technologies to monitor the condition of the food or its surrounding environment, providing valuable data about the product's state throughout its journey from production to consumption.(Dodero et al., 2021)(Drago et al., 2020)(Alizadeh et al., 2020). The key feature of intelligent packaging is that in Intelligent packaging incorporates sensors and indicators within the packaging material or as separate components within the sealed package. These elements can detect, sense, and record changes in various parameters related to the food or its environment.

Types of Sensors and Indicators

Time-Temperature Indicators: These indicators provide a visual representation of the temperature a product has been exposed to over time. This is particularly useful for perishable goods, helping to ensure proper storage and handling.(Gao et al., 2020)

1. **Gas Sensors:** These sensors detect changes in the gas composition within the packaging, such as oxygen, carbon dioxide, or ethylene levels. This information can indicate spoilage or the freshness of the product.(Lee, 2021)

2. **Biosensors:** These sophisticated sensors can detect the presence of specific microorganisms or toxins produced by spoilage bacteria, providing a more direct indication of food safety.(Mustafa & Andreescu, 2018)
3. **RFID Tags:** Radio Frequency Identification tags can store and transmit information wirelessly. In intelligent packaging, RFID tags can be used for tracking products throughout the supply chain, monitoring storage conditions, and even providing consumers with detailed product information.(Zuo et al., 2022)

Benefits of Intelligent Packaging

1. **Enhanced Food Safety:** By providing real-time information about the product's condition, intelligent packaging helps to ensure food safety and reduce the risk of foodborne illnesses.
2. **Reduced Food Waste:** Consumers and retailers can make more informed decisions about food consumption and disposal based on the information provided by intelligent packaging, minimizing unnecessary waste.
3. **Improved Quality Control:** Throughout the supply chain, intelligent packaging allows for better monitoring of storage and transportation conditions, ensuring that products reach consumers in optimal condition.
4. **Enhanced Traceability:** RFID tags and other tracking technologies embedded in intelligent packaging enhance traceability, enabling better response in case of product recalls or safety concerns.
5. **Increased Consumer Trust:** By providing transparency and real-time information about the product's journey and condition, intelligent packaging fosters consumer trust and confidence.(Alizadeh et al., 2020)(Katti et al., 2019)

Examples of Intelligent Packaging

1. **Fresh Produce:** TTIs on fruit packaging can indicate if the cold chain has been broken, while ethylene-absorbing sachets can help to slow down ripening.(Muller & Schmid, 2019)
2. **Meat and Poultry:** Gas sensors can detect spoilage gases, and biosensors can identify the presence of harmful bacteria.(Wang et al., 2019)(Katti et al., 2019)
3. **Dairy Products:** TTIs can monitor temperature fluctuations, and pH sensors can indicate spoilage.(Muller & Schmid, 2019)

Intelligent packaging represents a significant advancement in food preservation and safety. As technology continues to evolve, we can expect even more sophisticated and integrated intelligent packaging solutions, further enhancing the quality, safety, and sustainability of our food supply.

11. ENHANCING FOOD QUALITY BY MINIMAL PROCESSING TECHNOLOGIES

11.1 Introduction

The investigation of minimal processing methods in improving food quality is becoming increasingly pertinent within the sustainability concerns confronting the global food chain. In their 2019 research, Picart-Palmade et al. examine the sustainable deficiencies present in agricultural production, emphasizing its substantial contributions to greenhouse gas emissions and water usage. The authors highlight the ecological impact of food processing, notably the production of organic water pollutants and the concerning proportion of food waste, particularly in fruits, vegetables, and cereals.

Their analysis indicates an urgent requirement for enhancement programs focused on energy usage, solid waste reduction, and efficient water management. The authors observe that increasing energy costs have compelled food industries to use improved energy management strategies, as traditional thermal procedures continue to be the most energy-consuming techniques in food processing. The article highlights the rise of green technologies that aim to transform raw resources into high-quality food products with improved efficiency while emphasizing that thermal approaches prioritize food safety.

In this context, non-thermal procedures are proposed as feasible sustainable alternatives, providing a means to diminish both energy and water usage during processing and storage. The research by Picart-Palmade et al. (2019) paves the way for further investigation into minimum processing methods, illustrating their capacity to improve food quality and tackle significant environmental issues in the agricultural sector. This literature analysis seeks to expand on these ideas by analyzing the extensive research related to minimal processing technologies and their effects on food quality improvement.

11.2 The Importance of Minimally Processed Foods in Modern Consumer Preferences

Rise of Minimally Processed Foods: The increase in minimally processed foods indicates a heightened understanding of the health consequences linked to highly processed and ultra-processed diets. As customers increasingly prioritize health, there is a significant trend towards foods that preserve their natural attributes while providing convenience. This initiative is essential in addressing the rising incidence of diet-related illnesses and enhancing overall health. (*Minimally Processed Food | TABLE Debates*, n.d.)

Understanding Minimally Processed Foods: The NOVA classification system defines minimally processed foods as those that have undergone minimal alterations from their original state. This category encompasses fresh fruits and vegetables, whole grains, and specific dairy products that have been processed to improve their shelf life or facilitate preparation without substantially modifying their nutritional value. Examples consist of:

- Fresh fruits and vegetables are frequently rinsed or chopped for the sake of convenience.
- Fruits and vegetables that have been frozen rapidly to maintain their nutritional value.
- Brown rice and whole wheat flour are examples of whole cereals.
- Pasteurized milk: A process that preserves the nutritional value of milk while eliminating detrimental bacteria.

In comparison to their ultra-processed counterparts, these foods are generally more nutrient-dense. The latter are frequently contaminated with artificial ingredients, unhealthy fats, and added carbohydrates, which can diminish their health benefits. (*Minimally Processed Food | TABLE Debates*, n.d.)

11.3The Rise of Minimally Processed Foods

1. Conversely, minimally processed meals offer a more nutritious option. They preserve a greater proportion of their inherent nutrients and typically contain less additives. The increasing popularity of these cuisines can be ascribed to multiple factors:
2. Heightened Health Consciousness: As public knowledge on nutrition expands, customers are increasingly predisposed to select foods that beneficially impact their health. Diets abundant in minimally processed foods correlate with reduced risks of chronic diseases and enhanced overall health outcomes.
3. Culinary Trends: The farm-to-table movement and the desire for organic vegetables have increased the need for minimally processed solutions. Consumers are progressively demanding transparency in food source and processing techniques.
4. Minimally processed foods frequently achieve a balance between convenience and nutritional value. Pre-cut veggies and frozen fruits preserve their nutritious worthwhile expediting meal preparation.

5. Environmental Concerns: There is an increasing recognition of the ecological consequences of food production. Minimally processed foods typically originate from sustainable sources and necessitate less energy for processing than ultra-processed alternatives.

11.4 Consumer Demand of minimally processed foods: The demand for minimally processed foods has been consistently rising, influenced by heightened health consciousness, evolving dietary tastes, and increased sustainability concerns. This transition signifies a wider trend in the food sector as customers want healthier and more organic alternatives. Comprehending the determinants affecting this demand might yield insights into prevailing market trends and customer behaviour. (Inguglia et al., 2023)

Understanding Consumer Demand

Definition of Demand: Demand is an economic concept that denotes consumers' inclination to acquire products and services at different price levels. It is affected by various factors, including pricing, consumer preferences, income levels, and the availability of substitute products. The law of demand posits that a fall in price generally leads to an increase in the quantity desired, and conversely. (Fardet, 2018)

11.5 Key Factors Influencing Demand for Minimally Processed Foods

Consumer interest in minimally processed foods is rising due to increasing awareness of the link between diet and health, and concerns about the potential negative impacts of highly processed foods. Here's a breakdown of the key factors driving this trend:

1. Health Awareness: Consumers are increasingly cognizant of the health consequences associated with their dietary selections. Studies demonstrate that diets rich in minimally processed foods correlate with reduced risks of chronic diseases, including obesity, diabetes, and cardiovascular conditions. This understanding compels consumers to pursue items regarded as healthier substitutes for ultra-processed alternatives. (Perera & Smith, 2013)
2. Culinary Trends: Movements like farm-to-table dining and clean eating have encouraged the intake of less processed foods. Consumers are more concerned with the provenance of their food and favour products that are locally sourced and minimally processed. (Fardet, 2018)

3. Ecological Issues: Sustainability has emerged as a crucial determinant affecting consumer decisions. A significant number of consumers are apprehensive about the ecological consequences of food production, encompassing concerns around packaging waste and the carbon emissions linked to the transportation of highly processed meals. Minimally processed foods frequently originate from sustainable sources, attracting environmentally aware consumers. (Perera & Smith, 2013)
4. Convenience: Although minimally processed meals are frequently regarded as healthier, they also provide convenience. Pre-washed salad greens and frozen fruits preserve their nutritional integrity while expediting meal preparation. This equilibrium between health and convenience appeals to busy consumers. (Fardet, 2018)
5. The marketing and labelling of food products profoundly influence consumer perceptions and demand. Products designated as "natural," "organic," or "non-GMO" sometimes command elevated pricing and appeal to health-conscious consumers. Effective marketing methods that emphasize the advantages of minimally processed foods might further augment demand. (Fardet, 2018)

11.6 The Impact of Consumer Behaviour on Market Trends

As consumer preferences shift towards minimally processed foods, businesses must adapt to meet this demand:

- Product Development: Food manufacturers are increasingly emphasizing the development of products that are consistent with consumer preferences for minimal processing and healthful ingredients. This encompasses the repurposing of current products to minimize additives or the introduction of new lines that prioritize whole ingredients. (Zink, 1997)
- Grocery stores are responding by increasing their selection of minimally processed items, whole cereals, and fresh produce. To accommodate this expanding market segment, retailers may establish designated sections for organic or natural products. (Zink, 1997)
- Pricing Strategies: Businesses can establish competitive prices for minimally processed goods while maintaining profitability by comprehending demand dynamics. Economies of scale may also result in reduced prices over time as demand increases, thereby increasing the accessibility of these products to a broader audience. (Admin, 2024)

Advantages of Minimally Processed Foods: The rising consumer demand for minimally processed foods indicates an enhanced awareness of health and nutrition. These minimally processed foods provide several advantages that enhance general well-being. This discourse will examine the principal advantages of minimally processed foods, substantiated by pertinent references. (Admin, 2024 b)

11.7 Nutritional Value of Minimally Processed Foods

When compared to ultra-processed foods, less processed foods have a higher nutritional content. These foods preserve important nutrients, fiber, and beneficial plant chemicals that are frequently lost or reduced during intensive processing. The nutritional benefits of less processed foods include:

1. **Higher Nutrient Density**: Fresh fruits, vegetables, whole grains, and legumes are high in important vitamins, minerals, and antioxidants that are necessary for good health. Fruits and vegetables, for example, have high levels of vitamins C and A, potassium, and dietary fibre. (Kalyan, n.d.)
2. **Conservation of Natural Fibers:** Dietary fiber is essential for gastrointestinal health and has demonstrated efficacy in regulating blood glucose levels and reducing cholesterol. Minimally processed foods such as whole grains, fruits, and vegetables are superior providers of fiber, which is frequently deficient in ultra-processed diets. (Kalyan, n.d.)
3. **Decreased Caloric Density:** Minimally processed foods typically possess a lower calorie density than ultra-processed alternatives. This indicates that they offer a lower caloric density, facilitating weight management by enabling individuals to consume larger servings without excessive caloric consumption. (Kalyan, n.d.)
4. **Antioxidants and Phytochemicals:** Minimally processed foods are rich in antioxidants and phytochemicals, which assist in mitigating oxidative stress and inflammation inside the body. These drugs are associated with a diminished risk of chronic illnesses, including cardiovascular disease and cancer. (Kalyan, n.d.)
5. **Contribution to Nutritional Intake:** Analyses of the National Health and Nutrition Examination Survey (NHANES) data from 2003 to 2008 indicate that processed foods contribute a substantial share of critical nutrients to the American diet. Processed foods contributed to the "nutrients to encourage" as outlined in the Dietary Guidelines for Americans:
    - 55% of dietary fiber
    - 48% of calcium
    - 43% potassium

- 34% of vitamin D content
- 64% of iron
- 65% of folate
- 46% of vitamin B12

Nonetheless, processed foods also contributed significantly to the "constituents to limit":

- 57% of energy
- 52% of saturated fatty acids
- 75% of supplementary sugars
- 57% of sodium

6. **Significance of Selecting Nutrient-Dense Foods**: Diets are more likely to adhere to food guideline recommendations when nutrient-dense foods, regardless of processing, are chosen. By prioritizing minimally processed foods and restricting ultra-processed options, individuals can benefit from modern food technology while providing their bodies with essential nutrients. In conclusion, minimally processed foods deliver greater nutritional value than ultra-processed alternatives. These meals greatly contribute to fulfilling nutrient requirements and enhancing overall health by preserving critical vitamins, minerals, fiber, and advantageous plant chemicals. Integrating minimally processed foods into a balanced diet is essential for enhancing nutrient intake and mitigating the risk of chronic diseases. (Kalyan, n.d.)

11.8 Advantages of Minimally Processed Food in Nutrition

1. **Enhanced Nutrient Density**: Minimally processed foods generally possess more nutritional density compared to ultra-processed meals. Fresh fruits, vegetables, whole grains, and legumes preserve vital vitamins, minerals, and antioxidants that are frequently diminished through prolonged processing. Fruits and vegetables are abundant in vitamins C and A, potassium, and dietary fiber, essential for health maintenance. (Inguglia et al., 2023) (Drewnowski et al., 2022)
2. **Conservation of Natural Fiber**: Dietary fiber is essential for gastrointestinal health and has demonstrated efficacy in regulating blood glucose levels and reducing cholesterol. Minimally processed foods such as whole grains, fruits, and vegetables are outstanding sources of fiber. Conversely, ultra-processed diets frequently comprise refined carbs devoid of this advantageous vitamin. The intake of fiber-dense, less processed meals can enhance satiety, aiding individuals in more efficiently managing their weight. (Kalyan, n.d.)

3. **Decreased Caloric Density**: Minimally processed foods typically possess a lower calorie density than ultra-processed alternatives. This indicates that they offer fewer calories in relation to their volume, which can facilitate weight management by enabling consumers to consume larger quantities without excessive caloric consumption. A diet abundant in whole foods can enhance satiety while reducing caloric intake. (*Processed Foods and Health*, 2024)
4. **Antioxidants and Phytochemicals:** Minimally processed foods are rich in antioxidants and phytochemicals, which aid in mitigating oxidative stress and inflammation within the body. These compounds are associated with a diminished risk of chronic ailments, including cardiovascular disease and cancer. The antioxidants present in fruits and vegetables significantly contribute to cellular protection against harm. (Kalyan, n.d.)

11.9 Advantages of Minimally Processed Food in Human Health

1. **Reduced Risk of Chronic Diseases:** Studies indicate that diets rich in minimally processed foods correlate with a reduced risk of chronic diseases, including obesity, type 2 diabetes, cardiovascular disease, and specific malignancies. The Mediterranean diet, characterized by complete and minimally processed foods such as fruits, vegetables, whole grains, fish, and healthy fats (e.g., olive oil), is renowned for its health advantages and is associated with enhanced longevity and diminished illness risk. (Inguglia et al., 2023) (Rocillo-Aquino et al., 2021)
2. **Enhanced Gastrointestinal Well-being**: Ingesting minimally processed meals can improve intestinal health owing to their elevated fiber content. Fiber functions as a prebiotic, nourishing advantageous gut flora that are essential for digestion and overall health. An optimal gut microbiota correlates with enhanced immune function and decreased inflammation. (Inguglia et al., 2023)
3. **Enhanced Glycemic Regulation:** Dietary regimens abundant in minimally processed foods can aid in stabilizing blood glucose levels. Individuals with diabetes or those at risk of the condition may achieve better glycemic control by consuming whole grains and fiber-rich fruits, as opposed to diets heavy in refined carbs prevalent in many ultra-processed foods. (Inguglia et al., 2023)

11.10 Utilitarian Advantages of Minimally Processed Food

1. **Convenience without Sacrifice**: Minimally processed foods frequently offer convenience while maintaining nutritional integrity. Pre-cut veggies and frozen fruits enable consumers to swiftly create nutritious meals while preserving vital nutrients. This simplicity is especially advantageous for busy folks aiming to uphold a balanced diet. (Knorr & Augustin, 2020)

2. **Food Safety**: Certain minimal processing techniques improve food safety while minimally affecting nutritional content. For example, pasteurization eradicates detrimental microorganisms while maintaining the food's nutritional integrity. This guarantees that people can access safe food choices without sacrificing health. (Knorr & Augustin, 2020)

11.11 Principal Obstacles in the Production of Minimally Processed Foods

1. **Concerns Regarding Food Safety**: Minimally processed goods, like fresh fruits and vegetables, are vulnerable to microbial contamination during the production process. Prevalent sources of contamination encompass soil, irrigation water, and human handling procedures. Substandard hygiene practices among farm labourers and insufficient sanitation infrastructure might intensify these hazards. The existence of harmful microbes can jeopardize food safety and result in foodborne diseases, hence requiring rigorous monitoring and decontamination procedures. (Dávila-Aviña et al., 2014)
2. **Durability and Conservation:** A fundamental problem in the production of minimally processed foods is preserving their freshness and prolonging shelf life without employing chemical preservatives. Conventional preservation techniques may not be appropriate for all minimally treated items. Innovative processing technologies, including high-pressure processing and ultraviolet light therapy, demonstrate the potential to improve food safety while maintaining nutritional integrity. Nevertheless, these technologies necessitate additional research and validation to confirm their efficacy across many food categories. (Dávila-Aviña et al., 2014)
3. **Regulatory Obstacles:** The implementation of innovative processing technologies frequently encounters regulatory obstacles. Prior to the widespread use of novel processing processes, they must undergo stringent safety assessments by regulatory bodies. This procedure may be protracted and could hinder the implementation of advantageous technology in the food sector. Furthermore, there is an absence of standardized rules worldwide concerning innovative food processing methods, which may hinder international commerce and market entry. (Dávila-Aviña et al., 2014)
4. **Acceptance by Consumers:** The attitudes of consumers regarding minimally processed foods can greatly affect demand. Although numerous customers want healthier alternatives, there frequently exists ambiguity around food manufacturing classifications. Terms such as "natural" or "organic" may cause people to presume that all minimally processed goods are inherently safe or superior. Informing customers about the advantages and safety of minimally processed

foods is essential for promoting acceptance and facilitating informed decisions. (Dávila-Aviña et al., 2014) (Fardet, 2018)

5. **Nutritional Excellence:** Although lightly processed foods often preserve more nutrients than ultra-processed alternatives, certain processing techniques may result in nutrient degradation. Peeling fruits or vegetables might eliminate advantageous phytochemicals and fiber. Producers must reconcile the demand for convenience with the maintenance of nutritional integrity throughout processing. (Dávila-Aviña et al., 2014)
6. **Financial Feasibility:** The economic viability of manufacturing minimally processed foods might be difficult. Increased production expenses linked to upholding quality standards, adopting sophisticated processing technologies, and guaranteeing food safety can influence pricing tactics. Producers must identify methods to render minimally processed foods economically accessible for customers while ensuring profitability. (Kalyan, n.d.)

11.12 Factors for Enhancement

To effectively solve these difficulties, numerous factors must be considered:

1. **Capital Allocation in Research and Development**: Investing in research to investigate novel preservation approaches and enhance current processing methods is essential. This involves examining the effectiveness of innovative technologies in preserving food safety and nutritional integrity while prolonging shelf life.
2. **Augmented Food Safety Protocols**: Establishing stringent food safety measures across the production chain is crucial. This encompasses educating agricultural labourers on hygiene protocols, enhancing sanitary infrastructure, and employing efficient decontamination techniques to mitigate microbiological hazards linked to minimally processed foods.
3. **Consumer Education Programs**: Educational efforts designed to enlighten consumers about the advantages of minimally processed foods and to elucidate misconceptions around food processing classifications can improve acceptance. Implementing transparent labelling that explicitly conveys the employed processing processes can assist consumers in making educated decisions.
4. **Cooperation with Regulatory Authorities**: Producers ought to collaborate with regulatory agencies to expedite the approval process for innovative processing technology. Joint initiatives can foster the creation of unified regulations that encourage innovation while safeguarding consumer safety.

5. **Practices for Sustainability**: Integrating sustainable techniques into the manufacture of minimally processed foods can increase their attractiveness to environmentally aware consumers. This entails procuring components locally when feasible and reducing waste during the production process.
6. **Durability and Conservation**: Preserving the freshness and prolonging the shelf life of minimally processed goods presents a significant challenge in the industry. Conventional preservation methods may be inadequate for all minimally processed items, requiring the investigation of innovative strategies to improve food safety while maintaining nutritional integrity.

11.13 Determinants Influencing Shelf Life

Numerous factors can affect the shelf life of minimally processed foods:

1. Microbial contamination: The existence of pathogenic bacteria can jeopardize food safety and result in deterioration. Effective sanitation and hygiene protocols across the production chain are essential to reduce contamination hazards.
2. Enzymatic activity: Naturally occurring enzymes in foods can induce unfavourable alterations in texture, colour, and flavour during storage. Methods such as blanching or the use of antioxidants may assist in deactivating enzymes.
3. Oxidation: Interaction with oxygen may cause the oxidation of lipids and other food constituents, leading to undesirable tastes and nutrient degradation. Utilizing modified environment packaging or antioxidants can effectively reduce oxidation.
4. Moisture content: Excessive moisture can facilitate microbial proliferation and result in deterioration, whilst inadequate moisture may induce textural alterations. Effective moisture regulation is crucial for preserving quality.

11.14 Obstacles and Prospective Pathways in minimal processing

The manufacturing of minimally processed foods entails numerous obstacles and factors that must be handled to guarantee food safety, quality, and consumer approval. With the increasing demand for these foods, producers, regulators, and consumers must comprehend these problems.

Although innovative preservation technologies exhibit potential, issues remain to be addressed:

- Regulatory approval: Innovative technologies frequently encounter regulatory obstacles prior to widespread implementation. Comprehensive safety assessments and the standardization of rules among nations are essential. (Dávila-Aviña et al., 2014)

- Financial implications and scalability: The adoption of innovative technology can incur significant expenses, and increasing production may pose difficulties. Enhancing processes and minimizing expenses are essential for broad implementation. (Dávila-Aviña et al., 2014)
- Consumer acceptance: It is imperative to educate customers of the advantages and safety of innovative preservation techniques to promote acceptance. Explicit labeling and transparency can foster confidence. (Dávila-Aviña et al., 2014)

Ongoing research suggests that the creation of hybrid preservation systems integrating several technologies may enhance the shelf life of minimally processed foods while preserving their nutritional and sensory attributes.

Microwave-Enhanced Processing: Microwave-assisted processing is a novel food technology that use microwave energy to improve the efficacy of food preservation and preparation. This technique utilizes microwave radiation to heat food products, considerably decreasing processing times relative to traditional methods. Microwave-Assisted Thermal Sterilization (MATS) integrates pressurized hot water and microwave energy to sterilize food, thereby prolonging shelf life and maintaining nutritional integrity. This method is especially advantageous for ready-to-eat meals, sauces, and liquids, as it reduces nutritional degradation and enhances food safety. The popularity of microwave-assisted processing is increasing in the food sector because of its efficiency and effectiveness in preserving food quality.

Comparison of energy efficiency between microwave-assisted processing and traditional methods: Microwave-assisted processing has numerous benefits compared to conventional approaches regarding energy efficiency. In contrast to traditional heating methods that depend on external heat sources and often incur considerable energy waste, microwave processing heats food volumetrically, facilitating more efficient energy utilization. This technique immediately stimulates water molecules in the meal, resulting in expedited heating durations and diminished total energy usage. (Weaver et al., 2014)

Comparison of Energy Efficiency

- **Volumetric Heating:** Microwave processing elevates the temperature of food internally, hence minimizing thermal loss and substantially decreasing processing duration. Conventional techniques frequently necessitate extended heating durations to get equivalent interior temperatures, resulting in increased energy usage.

- **Decreased Processing Duration:** Research indicates that microwave-assisted methods can significantly reduce cooking or sterilization times—from hours to minutes—resulting in energy conservation. Microwave-Assisted Thermal Sterilization (MATS) integrates microwave radiation with pressured hot water to sanitize food goods more effectively than traditional retort techniques.
- **Reduced Operating Temperatures:** Microwave processing frequently functions at lower temperatures than conventional technologies, hence improving energy efficiency. Microwaves rapidly attain elevated temperatures without extended exposure, thereby preserving the nutritional integrity of meals and minimizing energy use.

The diminished energy usage linked to microwave processing may result in a reduced carbon footprint for food production. This is especially pertinent in a time when sustainability is gaining prominence in food production.

12. NOVEL DEHYDRATION METHODS

Drying is a prevalent technique for food preservation that prolongs shelf life by eliminating moisture, therefore preventing microbial proliferation and enzymatic processes. Nevertheless, conventional drying techniques frequently result in nutrient depletion and degradation of food quality. (VikumerFD, 2022)

Advanced drying methods, including freeze-drying and vacuum drying, efficiently tackle these issues.

Lyophilization or Freeze Drying: Freeze drying, or lyophilization, entails freezing the food product and subsequently decreasing the surrounding pressure to facilitate the sublimation of ice directly into vapor, bypassing the liquid phase. This procedure efficiently eliminates moisture while maintaining the food's inherent structure.

Advantages: (i) Nutrient Conservation: Freeze drying preserves a significant proportion of vitamins, minerals, and antioxidants in comparison to conventional drying techniques. Research indicates that freeze-dried fruits can preserve up to 90% of their nutrients. (ii) The procedure preserves the food's inherent flavours and fragrances, enhancing its appeal to consumers. (iii) Lightweight and Compact: Freeze-dried items possess low weight and extended shelf life, rendering them suitable for camping, emergency provisions, and space expeditions.(iv) Rehydration Quality: Foods rehydrate rapidly and efficiently, achieving a texture comparable to that of fresh products. (VikumerFD, 2022)

Vacuum Drying: Vacuum drying entails decreasing the pressure within the drying chamber, hence reducing the boiling point of water. This facilitates moisture evaporation at reduced temperatures relative to traditional drying techniques. (VikumerFD, 2022)

Advantages: (i)Quality Maintenance: Vacuum drying, by functioning at reduced temperatures, mitigates the thermal deterioration of delicate nutrients and tastes, hence maintaining the food's overall quality. (ii)Accelerated Drying Times: The diminished pressure expedites the drying process, which is especially beneficial for heat-sensitive goods like as fruits and vegetables, (iii)Minimized Oxidation: The vacuum setting restricts oxygen contact, so diminishing oxidative processes that may result in deterioration or nutrient depletion.(iv)Uniform Drying: Vacuum drying ensures more constant moisture extraction across the product, resulting in uniform quality. (VikumerFD, 2022)

13. ENERGY EFFICIENCY AND WATER CONSERVATION

As consumer demand for sustainable practices escalates, these innovations will be pivotal in cultivating a more sustainable future for food production. Sustainability in food processing is becoming increasingly vital as the business confronts issues pertaining to food waste, energy consumption, and water utilization. Advanced technologies are being implemented to reduce waste and utilize by-products while improving energy efficiency and preserving resources. This discourse will examine sophisticated processing methods aimed at waste minimization and valorization, alongside energy efficiency and water conservation. (Weaver et al., 2014)

Sustainable food processing necessitates the reduction of environmental effects via enhanced energy efficiency and water conservation.

1. **Resistive Heating:** Ohmic heating is a novel technology that transmits electric current through food to produce interior heat. **Energy Efficiency:** This approach diminishes energy usage relative to traditional heating systems by delivering consistent warmth and reducing heat loss. Ohmic heating facilitates expedited processing durations while maintaining the nutritional integrity of the meal. (Chacha et al., 2021) Quality Preservation: Operating at reduced temperatures enhances the retention of flavour, colour, and texture more effectively than conventional thermal procedures, yielding superior products.
2. **Membrane Filtration**: Membrane filtration uses semi-permeable membranes to efficiently separate and concentrate liquids. **Resource Efficiency:** Membrane filtering improves resource efficiency by facilitating the recovery of valuable

constituents from liquid streams while reducing waste production. (Chacha et al., 2021)

3. **Water Conservation:** This method markedly decreases water consumption in processing by recycling water internally within the system. It can be utilized in diverse applications, including juice clarity and whey protein concentration. (*Sustainable Food Processing*, n.d.)

FINAL ASSESSMENT

Cutting-edge technology in food processing is crucial for enhancing sustainability initiatives within the sector. By emphasizing waste reduction and valorization through extraction and fermentation technologies, as well as enhancing energy efficiency and water conservation through techniques such as ohmic heating and membrane filtering, the food sector can markedly diminish its environmental impact. As consumer demand for sustainable methods escalates, these improvements will be pivotal in cultivating a more sustainable future for food production.

CONCLUSION

In conclusion, the food processing sector is undergoing significant transformation driven by the urgent need for quality enhancement and sustainability. As one of the world's leading food producers, India has immense potential to improve its food processing industry by focusing on innovative technologies and practices that prioritize food safety, nutritional integrity, and environmental sustainability. The analysis reveals that capital allocation in research and development is crucial for exploring novel preservation methods and enhancing existing processing techniques, which can lead to improved food quality and extended shelf life. Furthermore, the rising costs of energy and the pressing need for efficient resource management necessitate the adoption of advanced technologies that minimize waste and optimize energy usage.

The exploration of non-thermal preservation methods presents a promising avenue for achieving sustainability in food processing, as these techniques can significantly reduce energy and water consumption while maintaining the nutritional quality of food products. However, the implementation of these innovative technologies is not without challenges. Regulatory hurdles, financial implications, and the need for consumer education regarding the safety and benefits of new preservation methods must be addressed to facilitate widespread adoption. Additionally, the industry must navigate the complexities of maintaining food quality and safety in minimally processed foods, which often require innovative strategies to extend shelf life without relying on chemical preservatives.

As consumer demand for sustainable practices continues to escalate, the food processing industry must prioritize the integration of sustainable techniques into its operations. This includes sourcing ingredients locally, reducing waste during production, and implementing energy-efficient practices. By focusing on these areas, the sector can not only enhance food quality but also contribute positively to environmental sustainability, ultimately benefiting both producers and consumers alike. The future of food processing lies in the ability to balance quality enhancement with sustainable practices, ensuring that the industry can meet the evolving demands of consumers while safeguarding the planet for future generations.

REFERENCES

Adisurya, I P K., & Sari, A R. (2022, January 1). Effects of High-Pressure Processing on Milk, Meat, Fruit, and Vegetable: A Review. Atlantis Press. DOI: 10.2991/absr.k.220305.046

Alamu, E O., & Mooya, A. (2017, January 1). Food Processing Technologies and Value Addition for Improved Food Safety and Security. Elsevier BV, 201-210. DOI: 10.1016/B978-0-12-810521-4.00010-4

Alizadeh, A M., Masoomian, M., Shakooie, M., Khajavi, M Z., & Farhoodi, M. (2020, September 14). Trends and applications of intelligent packaging in dairy products: a review. Taylor & Francis, 62(2), 383-397. DOI: 10.1080/10408398.2020.1817847

Augusto, P E D. (2020, March 20). Challenges, trends and opportunities in food processing. Elsevier BV, 35, 72-78. DOI: 10.1016/j.cofs.2020.03.005

Boekel, V., Martinus, V. F., Pellegrini, N., Stanton, C., Scholz, G., Lalljie, S., Somoza, V., Knorr, D., Jasti, P. R., & Eisenbrand, G. (2010). A review on the beneficial aspects of food processing. *Molecular Nutrition & Food Research*, *54*(9), 1215–1247. DOI: 10.1002/mnfr.200900608 PMID: 20725924

Bokulich, N. A., & Bamforth, C. W. (2013, May 22). The Microbiology of Malting and Brewing. *Microbiology and Molecular Biology Reviews : MMBR*, *77*(2), 157–172. DOI: 10.1128/MMBR.00060-12 PMID: 23699253

Chacha, J. S., Zhang, L., Ofoedu, C. E., Suleiman, R., Dotto, J. M., Roobab, U., Agunbiade, A. O., Duguma, H. T., Mkojera, B. T., Hossaini, S. M., Rasaq, W. A., Шорсткий, И., Okpala, C. O. R., Korzeniowska, M., & Guiné, R. P. F. (2021, June 20). Revisiting Non-Thermal Food Processing and Preservation Methods—Action Mechanisms, Pros and Cons: A Technological Update (2016–2021). *Foods*, *10*(6), 1430–1430. DOI: 10.3390/foods10061430 PMID: 34203089

Coldea, T E. (2019, September 11). Introductory Chapter: A General Overview on Latest Trends in Food Engineering. IntechOpen. DOI: 10.5772/intechopen.86081

Cotacallapa-Sucapuca, M., Vega, E. N., Maieves, H. A., Berrios, J. D. J., Morales, P., Fernández-Ruíz, V., & Cámara, M. (2021, May 15). Extrusion Process as an Alternative to Improve Pulses Products Consumption. A Review. *Foods*, *10*(5), 1096–1096. DOI: 10.3390/foods10051096 PMID: 34063375

Dávila-Aviña, J. E., Solís-Soto, L. Y., Rojas-Verde, G., & Salas, N. A. (2014). Sustainability and Challenges of Minimally Processed Foods. In *Food engineering series* (pp. 279–295). DOI: 10.1007/978-3-319-10677-9_12

Dinkçi, N., Akdeniz, V., & Akalın, A S. (2019, January 1). Survival of probiotics in functional foods during shelf life. Elsevier BV, 201-233. DOI: 10.1016/B978-0-12-817190-5.00006-9

Dodero, A., Escher, A., Bertucci, S., Castellano, M., & Lova, P. (2021, April 15). Intelligent Packaging for Real-Time Monitoring of Food-Quality: Current and Future Developments. *Applied Sciences (Basel, Switzerland)*, *11*(8), 3532–3532. DOI: 10.3390/app11083532

Drago, E., Campardelli, R., Pettinato, M., & Perego, P. (2020, November 7). Innovations in Smart Packaging Concepts for Food: An Extensive Review. *Foods*, *9*(11), 1628–1628. DOI: 10.3390/foods9111628 PMID: 33171881

Drewnowski, A., Detzel, P., & Klassen-Wigger, P. (2022, April 28). Perspective: Achieving Sustainable Healthy Diets Through Formulation and Processing of Foods. Elsevier BV, 6(6), nzac089-nzac089

Enríquez-Castro, C M., Pérez-Nafarrate, M., & Rodríguez, J E G. (2021, June 7). Innovation in Food Products Using Ozone Technology: Impact on Quality Assurance. IntechOpen. DOI: 10.5772/intechopen.96681

Evelyn, E., Chairul, C., Komalasari, K., Pebrianti, E., & Vazirani, W. (2021, October 1). Use of Pulsed Electric Field for the Inactivation of Eupenicillium Javanicum Ascospores in Pineapple Juice. *Journal of Physics: Conference Series*, *2049*(1), 012020–012020. DOI: 10.1088/1742-6596/2049/1/012020

Fadhel, Y. B., Maherani, B., Aragones, M., & Lacroix, M. (2019, November 1). Antimicrobial Properties of Encapsulated Antimicrobial Natural Plant Products for Ready-to-Eat Carrots. *Foods*, *8*(11), 535–535. DOI: 10.3390/foods8110535 PMID: 31683824

Fardet, A. (2018a). Characterization of the Degree of Food Processing in Relation With Its Health Potential and Effects. *Advances in Food and Nutrition Research*, *85*, 79–129. DOI: 10.1016/bs.afnr.2018.02.002 PMID: 29860978

Gao, T., Tian, Y., Zhu, Z., & Sun, D. (2020, May 1). Modelling, responses and applications of time-temperature indicators (TTIs) in monitoring fresh food quality. Elsevier BV, 99, 311-322. DOI: 10.1016/j.tifs.2020.02.019

. Gioia, L C., Ganancio, J R., & Steel, C J. (2017, September 6). Food Additives and Processing Aids used in Breadmaking. DOI: 10.5772/intechopen.70087

Gupta, P. (2023, February 9). Role of oxygen absorbers in food as packaging material, their characterization and applications. Springer Science+Business Media, 61(2), 242-252. DOI: 10.1007/s13197-023-05681-8

Huang, L., & Hwang, C. A. (2012). *In-package pasteurisation of ready-to-eat meat and poultry products Advances in Meat, Poultry and Seafood Packaging*. Woodhead Publishing Limited.

Irradiation, F. What You Need to Know. (2022, January 9). https://www.fda.gov/food/buy-store-serve-safe-food/food-irradiation-what-you-need-know

Jambrak, A., Donsì, F., Paniwnyk, L., & Đjekić, I. (2019, February 20). Impact of Novel Nonthermal Processing on Food Quality: Sustainability, Modelling, and Negative Aspects. *Journal of Food Quality*, *1-2*, 1–2. Advance online publication. DOI: 10.1155/2019/2171375

Jideani, A I O., Billard, H., Mchau, G R., Udoro, E O., & Onipe, O O. (2017, September 13). Processing and Preservation of Fresh-Cut Fruit and Vegetable Products. DOI: 10.5772/intechopen.69763

Kalyan, A. R. (n.d.). *"Ultra-Processed vs. Minimally Processed & Whole Foods."*https://continentalhospitals.com/blog/how-ultra-processed-foods-differ-from-minimally-processed-and-whole-foods/

Katti, K S., Priyadarshini, S., Leena, M M., Moses, J., & Anandharamakrishnan, C. (2019, November 1). Intelligent packaging: Trends and applications in food systems. Elsevier BV, 93, 145-157. DOI: 10.1016/j.tifs.2019.09.008

Knorr, D., & Augustin, M. (2020). Food processing needs, advantages and misconceptions. *Trends in Food Science & Technology*, *108*, 103–110. DOI: 10.1016/j.tifs.2020.11.026

Lee, D S. (2021, February 2). Gas Indicators and Sensors Used for MAP., 215-246. DOI: 10.1002/9781119530916.ch7

Margasahayam, A., & Balraj, Y. (2018, February 20). Properties of food ingredients during processing in a domestic mixer grinder and subsequent storage: A review. Wiley, 41(4). DOI: 10.1111/jfpe.12677

Minimally processed food | TABLE Debates. (n.d.). https://tabledebates.org/glossary/minimally-processed-food

Muller, P. A., & Schmid, M. (2019, January 7). Intelligent Packaging in the Food Sector: A Brief Overview. *Foods*, *8*(1), 16–16. DOI: 10.3390/foods8010016 PMID: 30621006

Mustafa, F., & Andreescu, S. (2018, October 16). Chemical and Biological Sensors for Food-Quality Monitoring and Smart Packaging. *Foods*, *7*(10), 168–168. DOI: 10.3390/foods7100168 PMID: 30332833

Paudyal, R., & Karatzas, K. A. (2016). Stress adaptation of Listeria monocytogenes in acidic ready to-eat products Food Hygiene and Toxicology in Ready-to-Eat Foods ed P Kotzekidou (London: Elsevier Inc.) chapter 10 pp 167-182

Perera, C. O., & Smith, B. (2013). Technology of Processing of Horticultural Crops. In *Elsevier eBooks* (pp. 259–315). DOI: 10.1016/B978-0-12-385881-8.00011-2

Perera, K. Y., Jaiswal, A. K., & Jaiswal, S. (2023, June 20). Biopolymer-Based Sustainable Food Packaging Materials: Challenges, Solutions, and Applications. *Foods*, *12*(12), 2422–2422. DOI: 10.3390/foods12122422 PMID: 37372632

Pham, T. T., Nguyen, L. L. P., Sao, Đ., & Baranyai, L. (2023, March 2). Application of Edible Coating in Extension of Fruit Shelf Life [Review]. *Multidisciplinary Digital Publishing Institute*, *5*(1), 520–536. DOI: 10.3390/agriengineering5010034

Processed Foods and Health. (2024, May 9). The Nutrition Source. https://nutritionsource.hsph.harvard.edu/processed-foods/

. Reshmy, R., Philip, E., Madhavan, A., Sindhu, R., Pugazhendhi, A., Binod, P., Sirohi, R., Awasthi, M K., Tarafdar, A., & Pandey, A. (2021, August 1). Advanced biomaterials for sustainable applications in the food industry: Updates and challenges. Elsevier BV, 283, 117071-117071. DOI: 10.1016/j.envpol.2021.117071

Ricci, A., Bertani, G., Maoloni, A., Bernini, V., Levante, A., Neviani, E., & Lazzi, C. (2021, May 14). Antimicrobial Activity of Fermented Vegetable Byproduct Extracts for Food Applications. *Foods*, *10*(5), 1092–1092. DOI: 10.3390/foods10051092 PMID: 34069051

Señorans, F. J., Ibáñez, E., & Cifuentes, A. (2003). New trends in food processing. *Critical Reviews in Food Science and Nutrition*, *43*(5), 507–526. DOI: 10.1080/10408690390246341 PMID: 14653493

Señoráns, F J., Ibáñez, E., & Cifuentes, A. (2003, September 1). New Trends in Food Processing. Taylor & Francis, 43(5), 507-526. DOI: 10.1080/10408690390246341

Sindhu, R., Binod, P., Ummalyma, S. B., Abraham, A., Mathew, A. K., Madhavan, A., Rebello, S., & Pandey, A. (2018, January 1). Applications of Microbial Enzymes in Food Industry. *Food Technology and Biotechnology*, *56*(1). Advance online publication. DOI: 10.17113/ftb.56.01.18.5491

Singh, M., Rama, E N., Kataria, J., Leone, C., & Thippareddi, H. (2020, July 28). Emerging Meat Processing Technologies for Microbiological Safety of Meat and Meat Products., 4(2)

Souto, L. R. F., Caliari, M., Júnior, M. S. S., Fiorda, F. A., & Garcia, M. C. (2016, June 1). Utilization of residue from cassava starch processing for production of fermentable sugar by enzymatic hydrolysis. Sociedade Brasileira de Ciência e Tecnologia de Alimentos, 37(1), 19-24. https://doi.org/DOI: 10.1590/1678-457x.0023

Sulatri, N. L., Yogeswara, I. B. A., & Nursini, N. W. (2017, June 30). Efektifitas sinar ultraviolet terhadap cemaran bakteri patogen pada makanan cair sonde untuk pasien immune-compremissed. *Diponegoro University*, *5*(2), 112–118. DOI: 10.14710/jgi.5.2.112-118

Sustainable Food Processing. (n.d.). Wiley.com. https://www.wiley.com/en-us/Sustainable+Food+Processing-p-9780470672235

. Ţibulcă, D., & Nagy, M. (2021, November 10). Sustainable and Healthy Food Ingredients: Characterization and Application in Functional Products. IntechOpen. DOI: 10.5772/intechopen.100165

. Tsironi, T., Koutinas, A., Mandala, I., & Stoforos, N G. (2021, June 1). Current and new Green Deal solutions for sustainable food processing. Elsevier BV, 21, 100244-100244. DOI: 10.1016/j.coesh.2021.100244

Types Of Food Processing Equipment. (2021, December 10). https://www .plantautomation-technology.com/articles/types-of-food-processing-equipment

Vikumer, F. D. (2022, September 28). *Differences Between Vacuum Drying and Lyophilization Freeze Drying? - Vikumer Freeze Dry*. Vikumer Freeze Dry. https://vikumer.com/differences-between-vacuum-drying-and-lyophilization-freeze-drying/#:~:text=The%20main%20difference%20between%20them,vapor%20phase%20called%20vacuum%20drying

Wang, L., Wu, Z., & Cao, C. (2019, November 13). Technologies and Fabrication of Intelligent Packaging for Perishable Products. *Applied Sciences (Basel, Switzerland)*, *9*(22), 4858–4858. DOI: 10.3390/app9224858

Weaver, C. M., Dwyer, J., Fulgoni, V. L.III, King, J. C., Leveille, G. A., MacDonald, R. S., Ordovas, J., & Schnakenberg, D. (2014). Processed foods: Contributions to nutrition. *The American Journal of Clinical Nutrition*, *99*(6), 1525–1542. DOI: 10.3945/ajcn.114.089284 PMID: 24760975

Zink, D. (1997). The Impact of Consumer Demands and Trends on Food Processing. *Emerging Infectious Diseases*, *3*(4), 467–469. DOI: 10.3201/eid0304.970408 PMID: 9366598

. Zuo, J., Feng, J., Gameiro, M G., Tian, Y., Liang, J., Wang, Y., Ding, J., & He, Q. (2022, December 1). RFID-based sensing in smart packaging for food applications: A review. Elsevier BV, 6, 100198-100198. DOI: 10.1016/j.fufo.2022.100198

Chapter 15
Intellectual Property and Energy Efficiency in Food Safety:
Legal Protections for Innovations in India

Amrita Singh
https://orcid.org/0000-0002-8394-671X
K.R. Mangalam University, Gurugram, India

Sarita Rana
https://orcid.org/0000-0002-5573-4517
Acropolis Faculty of Management and Research, Indore, India

Priyamvada Tiwari
https://orcid.org/0009-0000-5855-2471
Medicaps University, Indore, India

Kriti Kaushik
https://orcid.org/0000-0003-0898-0173
G.D. Goenka University, Gurugram, India

ABSTRACT

This study examines the relationship between intellectual property rights and energy efficiency within the context of food safety regulations in India, with a focus on innovations that enhance food safety standards. Ensuring public health in a rapidly developing country like India is contingent upon robust food safety standards. Robust intellectual property rights are necessary to both bolster and incentivize innovation, as several breakthroughs necessitate substantial investments in research

DOI: 10.4018/979-8-3693-8342-1.ch015

and development. The authors examine the significant concerns and potential advantages within the existing legal and regulatory structure through the analysis of case studies with newly established food safety systems that have achieved both compliance with safety standards and improvements in energy efficiency. The paper emphasizes the importance of a well-coordinated legal and regulatory framework to encourage innovation in food safety and energy efficiency in India. It attempts to contribute to the ongoing discussion on sustainable development by focusing on these crucial intersections.

INTELLECTUAL PROPERTY RIGHTS IN FOOD SAFETY INNOVATIONS

Intellectual property rights are crucial in promoting food security and Agro-biodiversity management. They provide incentives to private sector actors to develop seedlings with specific characteristics that enhance food security and Agro-biodiversity management. However, IPRs were underdeveloped in the agricultural sector for a long time due to the unrestricted exchange of germplasm and knowledge in various countries and the distinction between agriculture and other fields of technology.

Agriculture has followed two main phases of IPR introduction: first, industrialized countries obtained plant breeder's rights through the patent model, and second, the widespread spread of agro-biotechnology through genetic engineering. Today, every WTO member must adhere to specific baseline criteria provided by the Agreement on Trade-Related Aspects of Intellectual Property Rights, including intellectual property protection for plant varieties and micro-organism patentability.

IPRs can be implemented to promote food security in underdeveloped nations by providing legal protection for private sector participation in Agro-biotechnology. Agro-biotechnology can lead to higher yields, pest resistance, and faster growth using enhanced efficiency in inputs like fertilizers, pesticides, and water.[1] IPRs can increase nutritional content in plants, such as pro-vitamin A rice.[2]

Other arguments for IPR deployment in underdeveloped nations include boosting foreign direct investment, facilitating technology transfer and R&D by international businesses, and providing local players with incentives to be more creative. IPRs play a significant role in enhancing food security and promoting Agro-biodiversity management.[3]

Objectives

1 To examine how intellectual property rights, such as trade secrets and patents; affect and improve energy efficiency in relation to India's food safety laws.

2 To look at how improvements in food processing, preservation, and packaging affect energy efficiency and public health, and to comprehend how strong intellectual property rights are essential to supporting these kinds of breakthroughs.
3 To evaluate the Food Safety and Standards Authority of India's capacity to put energy-saving initiatives into effect and uphold them.
4 To investigate important issues and possible advantages related to India's current legal system that regulates food safety and intellectual property rights, particularly with reference to energy efficiency.
5 In order to support sustainable development in the food industry, we propose to amend intellectual property laws, encourage cooperation between regulatory bodies and industry stakeholders, and increase public awareness of the integration of energy efficiency into food safety protocols.

Patents

Patents plays a vital role in safeguarding breakthroughs across several fields, such as food safety, by granting exclusive rights to methods, technology, and substances. Patentable procedures refer to unique methods of preserving food, such as novel techniques for pasteurisation or inventive processes for sterilisation. Patentable machines encompass energy-efficient refrigeration technology, sophisticated food processing equipment, and specialised automated inspection systems designed to ensure food safety. In addition, patentable compositions of matter encompass novel food additives that improve preservation while maintaining safety, packaging materials that possess the capability to biodegrade, and formulations that prolong the shelf life of food goods.

In order to be eligible for a patent, an innovation must satisfy certain essential criteria. First and foremost, the invention must demonstrate novelty, indicating that it is unique and has not been previously made known to the public, and so is not part of the existing global knowledge (prior art). It should incorporate an innovative breakthrough that would not be immediately evident to a specialist in the domain. Ultimately, the invention must demonstrate its efficacy by delivering a distinct and discernible advantage. Progress in food safety has the capacity to improve the safety, effectiveness, and environmental friendliness of food processing and storage processes.

Patents generally have a lifespan of 20 years, starting from the date they are filed. During this duration, the individual who possesses the patent enjoys sole entitlements to the invention, granting them the authority to prohibit anyone from manufacturing, using, vending, or importing the patented invention without authorization. This exclusivity enables inventors to safeguard their intellectual property, get a competitive advantage, and perhaps generate revenue from their discoveries through licencing or direct commercialization. Therefore, this legislative measure encourages ongoing investment in research and development in the field of food safety, leading to progress that positively impacts public health and sustainability.

Cases related to IPRs in food safety innovations in India involve interpretations and applications of IP laws, especially patents, in the context of food safety and related technologies. *The Monsanto Technology LLC v. Nuziveedu Seeds Ltd.*[4] case revolved around a dispute regarding the ownership of patents for genetically modified cotton seeds, specifically Bt cotton. Monsanto, an American company specialising in agrochemicals and agricultural biotechnology, possessed these patents. Nuziveedu Seeds Ltd., an Indian company, obtained a licence to utilise these seeds but questioned the validity of Monsanto's patents. The lawsuit primarily concerned the legality of Monsanto's patent on Bt cotton seeds under Indian law, and whether Nuziveedu Seeds Ltd. had infringed upon this patent. The Supreme Court of India has affirmed the patent of Monsanto on genetically modified seeds, validating its legal status. However, the Court stressed that Monsanto must ensure that its pricing and licencing conditions comply with Indian regulations, ensuring that the terms are fair and in line with local laws. In the case of *Bayer Corporation v. Union of India*[5], Bayer Corporation, a worldwide pharmaceutical and life sciences company, held the patent for the cancer drug Nexavar. Natco Pharma, an Indian manufacturer of generic pharmaceuticals, has submitted an application for a compulsory licence in order to manufacture a more affordable generic variant of the medication. The issue pertained to Section 84 of the Indian Patents Act, 1970, which empowers the government to provide approval for the utilisation of a patented innovation without the patent holder's consent, under specific conditions. The Intellectual Property Appellate Board has upheld the compulsory licence granted to Natco Pharma. The verdict underscored the significance of public interest, affirming that the accessibility of cost-effective therapy is more important than Bayer's patent rights. This ruling highlighted the balance between protecting patents and ensuring that life-saving drugs are accessible to the public at reasonable prices.

The 2013 case *Novartis AG v. Union of India*[6] concerned a patent application for the cancer drug Glivec that Novartis, a Swiss pharmaceutical company, had filed. The Indian Patent Office denied the patent application on the grounds that the drug did not meet the standards for patentability, as stated in Section 3(d) of the Indian Patents Act. Section 3(d) aims to end "evergreening," a practice in which companies get patents for new versions of popular medications without significantly enhancing their effectiveness. The primary issues in the case were whether Section 3(d) was lawful and if Glivec complied with the conditions necessary to qualify for patent protection under it. The Supreme Court of India turned down Novartis's suit, upholding Section 3(d) and highlighting the need to end evergreening to maintain access to fairly priced drugs. This landmark decision ensured the public's continuous access to essential pharmaceuticals by reiterating India's commitment to finding a balance between patent rights and public health concerns. In the 2008 case *F. Hoffmann-La Roche Ltd. v. Cipla Ltd.*[7], Roche, a multinational healthcare company headquartered in Switzerland, was successful in obtaining a patent for the drug Erlotinib, which treats lung cancer. After Cipla, an Indian pharmaceutical company, began producing and distributing a generic version of the drug, Roche filed a patent infringement action. The primary issue of disagreement was whether Cipla's actions infringed upon Roche's patent rights in India. Ultimately, the Delhi High Court ruled in Cipla's favour, enabling them to continue manufacturing and selling generic Erlotinib. This case illustrated the difficulties in upholding pharmaceutical patents in India and the commitment of the country to prioritising public health over stringent patent rights.

Merck Sharp and Dohme was the patent holder of the diabetes drug Sitagliptin in the 2015 case *Merck Sharp and Dohme Corp. v. Glenmark Pharmaceuticals.*[8] A generic version of the drug was introduced by Glenmark Pharmaceuticals, which said that MSD's patent was not infringed. The major issue of disagreement was whether Glenmark's generic version violated any patents. The Delhi High Court ruled in MSD's favour, directing Glenmark to comply with an injunction that prevented the sale of the generic version. By bringing attention to the enforcement of patent rights against generic manufacturers, this litigation strengthened the protection of patented pharmaceutical advances in India. In the *PepsiCo India Holdings Pvt. Ltd. v. Hindustan Coca Cola Pvt. Ltd*[9] litigation, PepsiCo claimed that Coca-Cola's new product, Maaza Aam Panna, infringed upon their registered design for a similar beverage. The primary issue facing the food and beverage sector was the infringement of intellectual property rights and the protection of registered designs. The Delhi High Court ruled in favour of PepsiCo, concluding that Coca-Cola's product packaging infringed against PepsiCo's registered design. In the case of *Monsanto Technology LLC v. Controller of Patents*[10], Monsanto filed a patent application based on properties of genetically modified (GM) cotton. The Indian Patent Office rejected the application based on Section 3 (j) of the Indian Patents Act, which forbids patenting plants and animals. The primary issue of disagreement in this case was whether genetically modified organisms may be patentable in India. The Intellectual Property Appellate Board, which underlined the necessity of adhering to Indian patent law, which limits the patentability of biological material, backed the ruling. This choice will have a significant influence on biotechnology-related advancements in food safety.

In the case of *Cipla Ltd. v. Union of India*[11], Cipla challenged the Indian Patent Office's decision to grant a patent to Novartis for the drug Vildagliptin, which is used to treat diabetes. Cipla argued that the patent did not meet the requirements of novelty and inventive step. The core issue in this case was the validity of Novartis's patent for Vildagliptin. The Delhi High Court ruled in favour of Cipla, revoking

Novartis's patent. The court determined that the invention lacked an inventive step and was obvious. This case highlights the strict standards for patentability in India. Another case of *Dr. Reddy's Laboratories Ltd. v. AstraZeneca AB*[12]*(2012)*, Dr. Reddy's Laboratories aimed to launch a generic version of AstraZeneca's drug Seroquel XR (quetiapine fumarate). AstraZeneca held a patent for the extended-release formulation of the drug. The central issue was the alleged patent infringement by Dr. Reddy's Laboratories. The Delhi High Court ruled in favour of AstraZeneca, issuing an injunction to prevent Dr. Reddy's Laboratories from launching the generic version.

Patent Policy Implications for Food Safety Innovations

Food safety developments in India depend much on the legal context around intellectual property rights and the rigorous patent policies of that country. Using rigorous criteria, Indian courts assess novelty, non-observance, and utility; so, food safety innovations must be robust and exactly unique from existing technology to ensure patent protection. Cases such as *Monsanto Technology LLC v. Controller of Patents* underline the challenges of patenting biotechnological innovations, particularly those including genetically modified organisms. Food safety technology containing genetically modified organisms has to manage various legal challenges including adherence to Section 3(j) of the Indian Patents Act, which prevents patenting of biological material.

Design protection is another crucial problem that underlines the need of design patents in the food and beverage industry by means of the *PepsiCo v. Coca Cola* case. Original packaging and product design will help you differentiate from the competition and market success. Furthermore, Indian courts occasionally consider public interest, especially in issues involving possibly food safety technologies and fundamental products like drugs. These cases collectively provide a comprehensive understanding of the Indian legal scenario concerning intellectual property rights, which can guide innovators in food safety and other sectors in protecting and enforcing their IP rights effectively.

Trade Secrets

Trade secrets cover a wide range of private business knowledge providing a competitive advantage. This might involve technology, manufacturing techniques, recipes, and approaches inside the scope of food safety. Recipes employing original formulas and ingredient lists could call for odd pairings of food products or preservatives. Manufacturing processes could include specific food preparation, preservation, and packaging techniques aimed to increase efficiency or safety. Techniques could require for novel ideas for food safety testing, contamination avoidance, or quality control. Proprietary technology can demand innovative approaches to reduce waste, energy-efficient methods for food production, or algorithms or software monitoring food safety.

Information must meet particular standards if it is to be regarded as a trade secret. First of all, it has to stay secret, known just to a limited group inside the company and not generally known or readily visible to outsiders. Second, considering its confidentiality, it has to have a financial or competitive advantages. Last but not least, the owner has to act wisely to retain its secrets, enforcing confidence agreements with personnel, contractors, and business partners, limiting access only to need-to-know, and employing both physical and digital security measures to guard the data.

Protection of trade secrets can last probably always. Unlike patents, official registration procedures or a stated term for protection are not followed. Trade secrets maintain their protection as long as the information is private and maintains a competitive edge. Either the secret is disclosed or independently discovered, therefore ending the protection.

Tata Motors has created a special production technique for their Nano automobile in the case *Tata Motors Ltd. v. State of West Bengal*[13]. Problems safeguarding its trade secrets appeared when the business chose to move its production site. The main focus was whether moving Tata Motors' private information and any later exposure of it amounted to trade secret breach. The court underlined the need of keeping trade secrets and intellectual property, therefore supporting the idea that businesses should vigorously defend their personal data. This choice underlined the great significance of companies using strong policies to safeguard their trade secrets, particularly in significant events like migration. In the 2006 case *American Express Bank Ltd. v. Priya Puri*[14], an ex-employee of American Express Bank, Priya Puri was accused of revealing private client information to a rival company. The violation of trade secrets and confidence agreements were the main concern. Declaring that staff members have a fiduciary obligation to guard private information even after leaving the business, the Delhi High Court decided in favour of American Express Bank. This case emphasises Indian trade secret protection and the enforcement of confidentiality agreements.

In the 2010 case *Bombay Dyeing and Manufacturing Co. Ltd. v. Mehar Karan Singh*[15], Bombay Dyeing asserted that Mehar Karan Singh, an ex-employee, had traded secrets to a competitor company. Protecting trade secrets and sensitive corporate information after employment became the main focus. Before the court, Bombay Dyeing won their case underlining the legal responsibility to protect trade secrets and the need of non-disclosure agreements. To stop more information publication, the court granted an injunction.

Trade Secret Implications for Food Safety Innovations

The above-mentioned rulings provide a strong legal precedent for the protection of trade secrets in India, thus the implications for food safety are fairly crucial. Using this precedence, food safety agencies can guard their own information. These illustrations highlight the significance of including confidentially agreements and robust non-disclosure clauses in place with employees, partners, and other third parties to protect trade secrets. Moreover, Indian courts have shown a propensity to grant injunctions and other legal remedies to stop the trade secret exploitation, therefore providing companies with a method of intellectual property protection. The choices also highlight employees' legal responsibilities and their fiduciary commitments, which discourage trade secret theft by implying breaking confidence agreements.

Former workers of the defendant company were accused in *John Richard Brady and Ors. v. Chemical Process Equipments P. Ltd*[16]. of exploiting trade secrets and confidential information stolen from their company to start their own competitor company. The primary issues were the trade secret theft and these former employees breaching their fiduciary duty. The Delhi High Court issued an order against the former employees prohibiting from exploiting the trade secrets and confidential information they came across while working there. Maintaining fair competition depends on protection of trade secrets, the court noted.

In the case of *Ambience India Pvt. Ltd. v. Shri Naveen Jain*[17] (2005), Ambience India Pvt. Ltd. accused its former employee, Shri Naveen Jain, of using its trade secrets and confidential business information in a competing business. The issue before the court was whether the former employee had breached confidentiality agreements and misappropriated trade secrets. The Bombay High Court found that Shri Naveen Jain had indeed breached the confidentiality agreement and used the company's trade secrets for his own benefit. Consequently, the court issued an injunction to prevent further misuse of the confidential information. Signing a bottling deal with Gujarat Bottling Co. Ltd. (GBC), *Coca-Cola Company and Others v. Gujarat Bottling Co. Ltd.*[18] included stipulations protecting Coca-Cola's trade secrets and intellectual property. Later on, GBC began negotiating with PepsiCo, which raised concerns regarding possible Coca-Cola trade secret leak. The main concern was whether enforceability of contractual obligations protecting intellectual information and trade secrets could be achieved. The Indian Supreme Court kept an order against GBC, therefore prohibiting any agreements between it and PepsiCo. The court stressed the significance of honouring contractual obligations to protect intellectual property and trade secrets. Heinz India Pvt. LTD. asserted in *Heinz India Pvt. LTD. v. National Industrial Corporation Ltd.*[19] pilfers trade secrets pertaining to the formulation and production of its food products. Protection of intellectual property and trade secrets in the food industry constituted a current difficulty. The

Delhi High Court underlined the need of safeguarding trade secrets and intellectual knowledge when it decided on an injunction in favour of Heinz India to keep a competitive advantage in the market.

In 2003, *FMC Corporation v. Natco Pharma Lt., FMC Corporation*[20] claimed Natco Pharma Lt. had pilfers trade secrets linked to the production of a pesticide. Dominant all else was protection of trade secrets in the chemical and pharmaceutical sectors. Emphasising the need of safeguarding trade secrets to maintain fair competition and innovation, the Delhi High Court decided in favour of FMC Corporation. The court issued an order against Natco Pharma not using private information any more.

Trademarks

Trademarks in the food sector address many components needed for brand recognition, including brand names, logos, slogans, and certificates. As "SafeFresh" for a range of food preservatives does, brand names help to especially distinguish specific food products or lines of food safety-related products. Logos are graphic symbols or designs connected to a food product or firm, such a distinctive logo reflecting high criteria of food safety. Slogans—such as "Freshness You Can Trust"—are catchphrases or taglines meant to express the message or guarantee of the brand in marketing. Certifications show that a commodity satisfies specific criteria, including eco-labels indicating compliance to food safety and environmental regulations. Unless it has become distinctive by use, a trademark must be unique, able to tell one company's products or services apart from those of another, and neither generic nor strictly descriptive. The mark also has to be applied in business therefore it has to be used or intended for usage on goods or services traded or transported. Trademarks have endless lives as long as they are regularly renewed and applied in business. First serving ten years, trademarks registered in India thereafter can be renewed endlessly every ten years.

Claiming trademark infringement, publisher of the daily "Eenadu," Ushodaya Enterprises sought to ban T.V. Venugopal from using the word "Eenadu" for his product in the case of *T.V. Venugopal v. Ushodaya Enterprises Lt.*[21]. The defence of a well-known trademark and the halt of another company using it dominated the primary issues. Acknowledging "Eenadu" as a well-known trademark, the Indian Supreme Court declared in favour of Ushodaya Enterprises, therefore rendering an injunction against T.V. Venugopal. The situation underscored the significance of protecting well-known trademarks.

Dabur asserted trademark infringement in the 2008 lawsuit *Dabur India Lt. v. K.R. Industries*[22] by suing K.R. Industries for using the trademark "DABUR" for their products. The biggest problems concerned possible customer misperception and registered illegal trademark use. The Delhi High Court decided in Dabur India

Limited's favour, therefore rendering an injunction against K.R. Industries and so preventing the use of the "DABUR" brand. This case underlined the significance of protecting registered trademarks against infringement.

In the case of *Infosys Technologies Ltd. v. Park Infosys*[23], Infosys Technologies Ltd. sought to prevent Park Infosys from using the name "Infosys," claiming it was a well-known trademark and that the use by Park Infosys would cause confusion. The issue centered on the protection of well-known trademarks and the prevention of misleading use by others. The Delhi High Court ruled in favour of Infosys Technologies Ltd., recognizing "Infosys" as a well-known trademark and granting an injunction against Park Infosys. This case reinforced the protection of well-known trademarks.

Marico Limited alleged in *Marico Limited v. Agro Tech Foods Limited*[24] that Agro Tech Foods Limited's use of similar packaging and trade dress for its product was misleading and constituted passing off. The matter at hand was the prevention of passing off and the safeguarding of trade attire. The Bombay High Court ruled in favour of Marico Limited, concluding that the similarly designed packaging could potentially confuse and mislead consumers. Agro Tech Foods Limited was awarded an injunction by the court to prohibit the use of misleading packaging. This case emphasises the significance of safeguarding trade apparel in addition to trademarks.

Trade Secret Implications for Food Safety Innovations

Companies in the food safety sector can utilize trademarks to safeguard their brand names, logos, and slogans. This protection ensures that their products are easily identifiable and associated with quality and safety. As such a well-known example is the trademarked logo of **Nestlé** and branding are closely tied to quality and safety in food products, such as baby formula and dairy products. By trademarking their brand name and logo, Nestlé protects its reputation and ensures consumers can easily recognize its products on store shelves.

Certification marks indicate that a product complies with specific food safety standards, helping consumers make informed choices and enhancing the credibility of the product. The **USDA Organic** certification mark is a prominent example in the food safety sector. Products bearing this mark signify that they meet strict organic farming standards set by the United States Department of Agriculture. This certification helps consumers identify and trust organic products, knowing they comply with specific safety and quality standards.

Strong trademark protection prevents unauthorized use and ensures consumers are not misled by similar or deceptive branding from competitors. The case of **PepsiCo** vs. competitors is an illustrative example. PepsiCo has successfully protected its trademarks against brands that attempt to create packaging or branding similar to its products, such as the distinctive blue colour of its soda cans. This protection helps prevent confusion in the marketplace, ensuring that consumers are not misled into purchasing a competitor's product, believing it to be PepsiCo's.

As long as trademarks are actively used and renewed, they can provide long-term protection, which supports sustained brand recognition and trust. The **Coca-Cola** trademark is a prime example of indefinite protection. Coca-Cola has maintained its brand for over a century by continuously using and renewing its trademark. The iconic script logo and brand name are synonymous with soft drinks worldwide, demonstrating how effective trademark protection can ensure lasting brand recognition and consumer trust in food safety standards.

Copyrights

Copyrights safeguard original works of authorship that are permanently affixed to a tangible medium. This encompasses written manuals, instructional materials, guides, and handbooks that are applicable to energy-efficient practices, food safety protocols, and compliance procedures in the context of energy efficiency and food safety. It also encompasses training programs, which are educational resources that include curricula, videos, and presentations that are intended to educate employees or stakeholders on energy-efficient methods and food safety standards. Furthermore, copyrights encompass software that is designed to ensure regulatory compliance, manage energy consumption, or monitor product safety. In order to be eligible for copyright protection, a work must be independently created by the author and exhibit a minimum level of creativity. It is also necessary to preserve it in a physical or digital form that is sufficiently permanent to be perceived, reproduced, or otherwise communicated for a period exceeding a temporary period. In general, copyright protection is valid for the author's lifetime plus 70 years. The copyright for works created for hire, which are either created by an employee within the scope of their employment or commissioned under specific circumstances, is valid for 95 years from the date of publication or 120 years from the date of creation, whichever is shorter.

In the case of *Eastern Book Company v. D.B. Modak*[25], Eastern Book Company published a law report called Supreme Court Cases, which included headnotes and other editorial matter. D.B. Modak and others were accused of reproducing substantial parts of these headnotes without permission. The issue was whether the editorial matter in SCC was entitled to copyright protection. The Supreme Court of India held that the headnotes and other editorial matter were original literary works deserving of copyright protection, emphasizing the requirement of originality and the investment of skill, labour, and judgment in creating such works.

In the case *University of Oxford & Ors v. Rameshwari Photocopy Services & Anr.*[26], the University of Oxford and other publishers sued Rameshwari Photocopy Services for the production and distribution of photocopies of substantial portions of their copyrighted textbooks. The primary concern was whether the act of photocopying textbooks for educational purposes was a violation of copyright. The Delhi High Court ruled in favour of Rameshwari Photocopy Services, asserting that the photocopying in question was permissible under the fair use exception outlined in Section 52(1)(i) of the Indian Copyright Act.

In the case of *India TV v. Yashraj Films Pvt. Ltd.*[27], India TV used clips from 'Yashraj Films' movies in its television programs without obtaining permission, resulting in a copyright infringement suit. The issue was the unauthorized use of copyrighted film clips in a television broadcast. The Delhi High Court ruled in favour of Yashraj Films, holding that India TV's use of the clips amounted to copyright infringement. In the 2011 case of *Super Cassettes Industries Ltd. v. Myspace Inc.*[28], a significant player in the Indian music industry, filed a lawsuit against Myspace for hosting user-uploaded videos that contained its copyrighted music. The issue at hand was the liability of online platforms for hosting user-generated content that violates copyright. The Delhi High Court ruled in favour of Super Cassettes Industries, asserting that Myspace was obligated to remove infringing content upon notification.

Amar Nath Sehgal, a renowned sculptor, filed a lawsuit against the Union of India in the case of *Amar Nath Sehgal v. Union of India*[29]. He claimed that the Union of India had illegally stored and removed his bronze mural, thereby violating his moral rights as the creator. The matter at hand was the safeguarding of moral rights in accordance with the Indian Copyright Act. The Delhi High Court ruled in favour of Amar Nath Sehgal, emphasising the importance of safeguarding moral rights, such as the right to integrity and the right to attribution. The mural was ordered to be restored by the court, and the artist was awarded compensation.

Copyright Implications for Food Safety Innovations

Copyrights for educational and training materials, such as instructional guides, manuals, and other content, can be obtained by companies in the food safety sector. These organisations are guaranteed exclusive rights to use, reproduce, and distribute their materials as a result of copyright protection. This exclusivity is essential for the preservation of the quality and integrity of training programs, as it prohibits the unauthorised reproduction or distribution of those materials, which could potentially diminish the training's effectiveness. For instance, a company that develops an exhaustive training manual on food safety protocols may copyright the manual to guarantee that only authorised personnel have access to it, thereby preserving the quality of the information that is imparted.

Proprietary software and applications that are designed to enhance energy efficiency and food safety may be safeguarded by copyrights. This is essential for protecting intellectual property from unauthorised use, replication, or distribution. For example, a company that creates a software application to monitor temperature and humidity in food storage facilities can obtain copyright protection for its code and user interface. The software's original developer is granted a competitive advantage in the market by this protection, which prevents competitors from copying or misusing it.

Moral rights are a critical component of copyright that guarantees that authors and creators of copyrighted works are appropriately acknowledged and that their works are not subjected to derogatory treatment. In the food safety sector, this implies that the original creators must be appropriately acknowledged when training materials or software applications are employed. Furthermore, these creators have the right to object to any alterations or uses of their work that could compromise the integrity of the work or their reputation. For instance, the original creator may assert their moral rights to prevent the misuse of a training video by a company that alters it in a manner that misrepresents the content.

The fair use exception in copyright law permits the limited use of copyrighted materials for educational purposes without obtaining permission from the copyright holder. This equilibrium is indispensable in the food safety sector, where compliance and safety necessitate access to precise information and training. For example, during a workshop, a teacher may employ excerpts from a copyrighted food safety training manual to elucidate critical concepts. This restricted utilisation would qualify as fair use, which permits educational advantages while simultaneously honouring the copyright holder's rights.

Companies and individuals in the food safety sector can secure their creative works by comprehending and effectively employing copyright protection. This safeguard not only fosters innovation but also assists in the preservation of a competitive edge in an industry that is becoming increasingly regulated and demanding. Copyrights guarantee the security of intellectual properties, including educational materials and software, thereby promoting an environment that is conducive to the development of creativity and safety.

Energy Efficiency in Food Safety

In order to mitigate the ecological footprint of food production and storage, it is imperative to prioritise energy efficiency in food safety. This endeavour is significantly influenced by the implementation of innovative technologies. For example, smart refrigeration systems that are equipped with sensors can monitor and modify temperatures in real-time, thereby ensuring optimal conditions and reducing energy consumption. An illustration is the utilisation of variable-speed compressors in refrigerators, which regulate their speed in accordance with the chilling demand, resulting in substantial energy savings. Energy-efficient culinary equipment, such as induction cooktops, heats food directly through magnetic induction, rather than using gas or traditional electric elements. This results in faster cooking times and reduced energy consumption. Advanced packaging technologies, such as vacuum-sealed packaging, can extend the shelf life of perishable products, thereby reducing the energy required for storage and the necessity for frequent refrigeration.

Energy efficiency in food safety involves sustainable practices and innovative technologies that promote environmental sustainability and reduce energy consumption. Smart refrigeration systems, such as IoT-enabled refrigeration units, use advanced automation and sensors to optimize temperature control, resulting in reduced energy consumption, lower operational costs, extended expiration life of perishable products, and reduced food waste.

Cultural appliances, such as fryers, convection ovens, and induction cooktops, also focus on energy conservation, offering cost reductions and reduced greenhouse gas emissions. Packaging innovations, such as vacuum packaging, intelligent packaging with sensors, and biodegradable materials, optimize food preservation while reducing energy consumption during production and storage.

Sustainable food safety practices include optimizing supply chain logistics, implementing energy-saving protocols, and utilizing renewable energy sources. Examples include electric or hybrid delivery vehicles, route optimization software, and renewable energy technologies like biogas generators, wind turbines, and solar panels. Implementing energy-saving protocols, such as energy audits, employee training, and regular equipment maintenance, improves energy efficiency, reduces operational costs, and enhances sustainability.

The case studies of Campbell Soup Company, Unilever, and Nestlé demonstrate the impact of these initiatives. Nestlé's energy efficiency initiatives, including manufacturing process optimization and renewable energy investments, have resulted in significant cost savings and reduced energy consumption and carbon emissions. Unilever's Sustainable Living Plan has led to substantial reductions in greenhouse gas emissions and improved energy efficiency in its production facilities.

The food industry's sustainable practices prioritize overall efficacy while simultaneously reducing energy consumption and emissions.

Sustainable practices in the production, refining, and distribution of food can also significantly reduce energy consumption. For instance, the optimisation of supply chain logistics can result in a reduction in petroleum consumption by reducing transportation distances and times. Companies such as Walmart have employed strategies to enhance the efficiency of truck loading and route planning, which has led to significant fuel savings. One notable example is the utilisation of solar-powered cold storage units in rural areas, which aids in the preservation of produce without the need for utility electricity. The implementation of energy-saving protocols, such as the utilisation of energy-efficient lighting and equipment in food processing facilities, can further reduce the overall energy consumption. This practice is exemplified by the utilisation of LED lighting in food manufacturing facilities, which consumes substantially less energy than traditional lighting. Collectively, these technologies and practices contribute to a more energy-efficient and sustainable agricultural industry. Authors have tried to achieve the objectives through following survey:

Figure 1. Pie chart

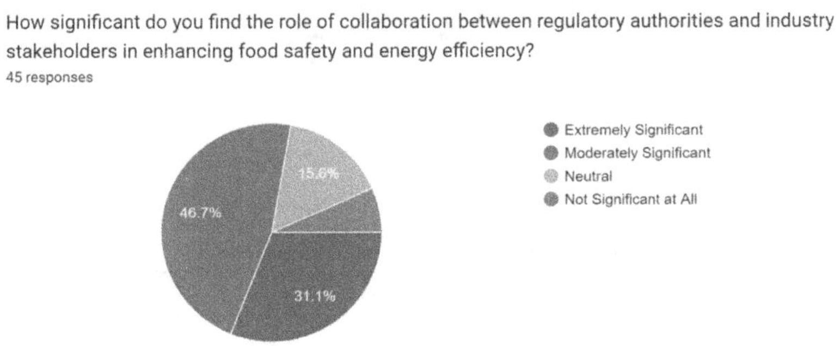

Figure 2. Correlation Analysis

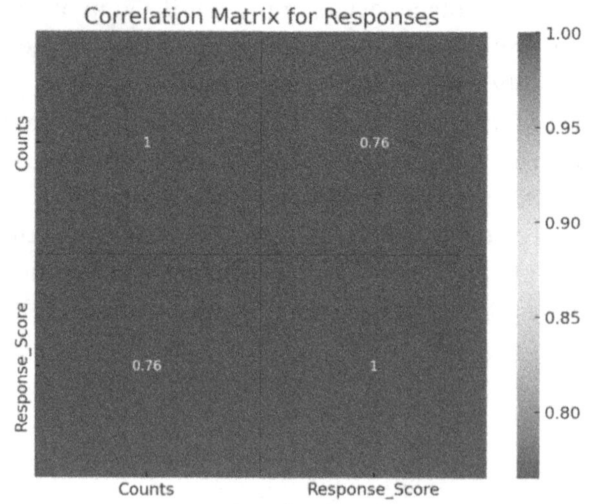

Table 1. The correlation matrix shows the following values:

	Counts	Response_Score
Counts	1.000	0.765
Response Score	0.765	1.000

Counts and Response Score: The correlation coefficient between the counts of responses and the assigned response scores is approximately 0.765. This indicates a strong positive correlation between the frequency of responses and their significance scores.

Strong Positive Correlation (0.765): This means that as the perceived significance of collaboration (from "Not Significant at All" to "Extremely Significant") increases, the number of responses also tends to increase. This suggests that more people view the collaboration between regulatory authorities and industry stakeholders as significant in enhancing food safety and energy efficiency.

INTERPRETATION

The strong positive correlation (0.765) suggests that there is a significant relationship between the perceived importance of collaboration and the number of responses for each category. This indicates that most respondents believe in the significant role of collaboration in enhancing food safety and energy efficiency.

Figure 3. Pie chart

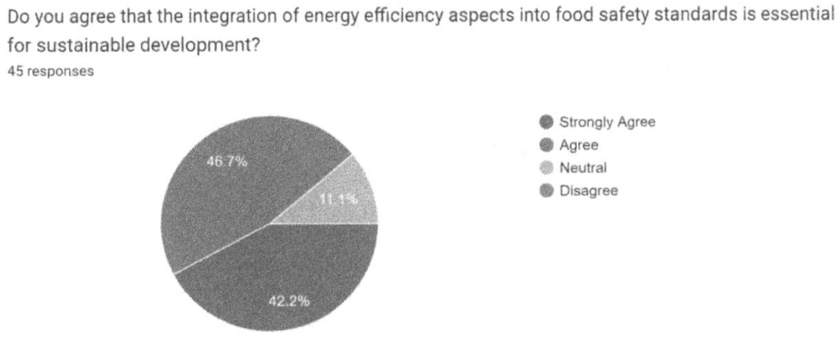

Correlation Analysis:

- **High Agreement:** The majority of respondents either "Strongly Agree" or "Agree" that integrating energy efficiency into food safety standards is essential for sustainable development. Together, these two categories account for **88.9%** of the responses.
- **Neutral:** A smaller portion of respondents, **11.1%**, is neutral, indicating that they may not have strong opinions or feel uncertain about the significance of energy efficiency in food safety standards.
- **Disagreement:** No respondents disagreed with the statement, suggesting a broad consensus on the importance of this integration.

Interpretation:

- **Positive Correlation:** There is a strong positive correlation between the respondents' belief in the importance of integrating energy efficiency and their support for it as part of sustainable development. With a significant majority either strongly agreeing or agreeing, it indicates a shared understanding of the crucial role energy efficiency plays in achieving sustainability goals within the context of food safety standards.
- **Consensus:** The absence of any disagreement and minimal neutrality suggests that the concept of integrating energy efficiency into food

safety standards is widely accepted and considered essential by the respondents.

Figure 4. Pie chart

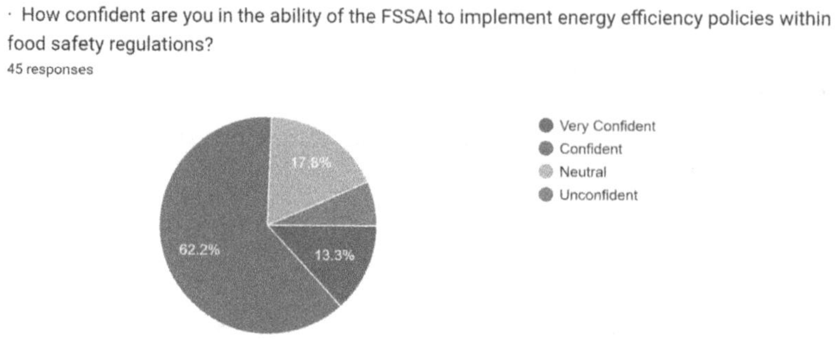

Data Extraction

The chart presents the confidence levels of respondents in FSSAI's ability to implement energy efficiency policies within food safety regulations.

Assumed Data Interpretation

Let's assign numerical values to these levels to calculate the correlation

Correlation Table

Table 2. Correlation table

confidence Level	Percentage	Numerical Value
Very Confident	13.3%	4
Confident	62.2%	3
Neutral	17.8%	2
Unconfident	6.7%	1

Interpretation:

- **Majority Confidence:** A significant percentage of respondents are confident (62.2%) or very confident (13.3%) in FSSAI's ability to implement these policies. This indicates a generally positive perception of FSSAI's capability.
- **Neutral to Low Confidence (Negative Perception):** A smaller portion of respondents is either neutral (17.8%) or unconfident (6.7%), indicating some reservations or doubts about FSSAI's effectiveness.

Observations:

- The confidence levels generally skew towards the positive, suggesting that most respondents believe in FSSAI's ability to implement energy efficiency policies effectively.
- A correlation between confidence in FSSAI and the perceived importance of integrating energy efficiency (from the previous chart) could be inferred. Those who believe in the importance of energy efficiency may also have higher confidence in FSSAI's ability to implement such measures, though this isn't quantitatively assessed here due to a lack of paired data.

Figure 5. Pie chart

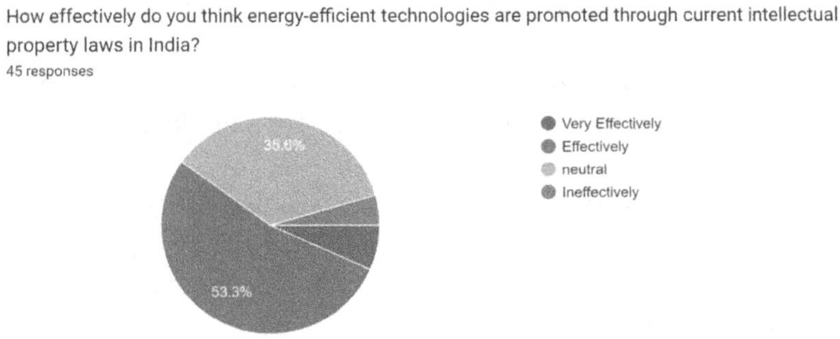

Regression Analysis:

Regression analysis typically involves numerical data, but in this case, we only have categorical data (effectiveness levels). We can still conduct an ordinal regression analysis by assigning numerical values to each category to estimate the relationship between perceived effectiveness and some other variables (if they were available). Here's how we might assign values:

Table 3.

Effectiveness Level	Percentage	Numerical Value
Very Effectively	6.7%	4
Effectively	53.3%	3
Neutral	35.6%	2
Ineffectively	4.4%	1

Interpretation of Regression Analysis:

- **Positive Perception:** The majority of respondents believe that energy-efficient technologies are effectively promoted (53.3% for "Effectively" and 6.7% for "Very Effectively"). This suggests a general positive perception regarding intellectual property laws in India concerning energy efficiency.
- **Neutral and Negative Perception:** A significant portion of respondents (35.6%) is neutral, which may indicate uncertainty or a belief that the promotion of energy-efficient technologies could be improved. Only 4.4% of respondents feel that the laws are ineffective.

Insights:

- **Trend Towards Effectiveness:** The data suggests a trend towards perceiving the promotion of energy-efficient technologies as effective under current intellectual property laws.
- **Potential Area of Improvement:** The relatively high percentage of neutral responses indicates that while the laws are perceived as generally effective, there may still be room for improvement or a need for greater awareness.

LEGAL PROTECTIONS AND REGULATORY FRAMEWORK

In India, legal protections and regulatory frameworks are essential for ensuring food safety and promoting energy efficiency in the food industry. The following aspects highlight how these protections and frameworks are applied within the Indian context:

Food Safety Standards: In India, the Food Safety and Standards Authority of India (FSSAI) is the primary body responsible for regulating and supervising food safety. FSSAI sets guidelines and standards to ensure that food safety innovations comply with health and safety requirements. The Food Safety and Standards Act, 2006, empowers FSSAI to establish scientific standards for food articles and to regulate their manufacture, storage, distribution, sale, and import to ensure the availability of safe and wholesome food for human consumption. For instance, FSSAI's guidelines for the use of food additives and preservatives ensure that any new food safety technology aligns with health standards.

Energy Efficiency Standards: The Bureau of Energy Efficiency under the Ministry of Power is the key agency responsible for setting energy efficiency standards in India. The Energy Conservation Act, 2001, provides a legal framework for promoting energy efficiency and conservation. BEE develops norms and standards for energy-efficient practices and technologies, including those related to the food industry. For example, BEE's standards for energy-efficient refrigeration and air conditioning systems help ensure that food storage technologies minimize energy consumption while maintaining safety standards.

Certification Programs: In India, certification programs play a crucial role in promoting energy-efficient products and technologies. The BEE's Star Rating program is similar to the Energy Star program in the U.S. This program certifies and labels appliances and equipment that meet specific energy efficiency criteria, providing a market advantage for compliant innovations. For example, energy-efficient refrigeration units and kitchen appliances that receive a high star rating are preferred in the market, encouraging manufacturers to adopt energy-saving technologies.

By aligning food safety innovations with regulatory standards, ensuring energy efficiency in food-related technologies, and leveraging certification programs, India fosters a regulatory environment that supports both public health and environmental sustainability.

CHALLENGES AND CONSIDERATIONS

Challenges and considerations in balancing innovation and compliance in the food safety sector involve ensuring that energy-efficient solutions meet intellectual property (IP) protection requirements and regulatory standards. This balance is crucial for successful market adoption and global competitiveness.

Innovators must ensure their energy-efficient food safety solutions comply with regulatory standards while protecting their IP. For example, a company developing a new energy-efficient refrigeration system must adhere to the standards set by the Bureau of Energy Efficiency (BEE) in India or the U.S. Department of Energy (DOE) in the United States. These standards ensure that the product is safe and efficient. At the same time, securing patents for the innovative technology protects the company's investment and prevents others from copying their design. For example, an Indian company, Godrej Appliances, developed an energy-efficient refrigerator that complies with BEE standards. To protect their innovation, they secured patents for their unique cooling technology. Similarly, a U.S. company like Whirlpool must ensure compliance with DOE standards while patenting their energy-efficient technologies.

Successfully bringing an innovation to market requires securing IP rights, gaining consumer trust, and clear the regulatory landscape. Consumers need to trust that the new technology is safe and reliable, which often means meeting stringent regulatory standards. Companies also need to market their innovations effectively to gain consumer acceptance. In India, the introduction of the BEE star rating system helped consumers recognize and trust energy-efficient appliances. A company like LG Electronics, which markets its energy-efficient refrigerators and air conditioners with high BEE star ratings, finds it easier to gain consumer trust. In the U.S., the Energy Star certification plays a similar role. Companies like Samsung, which markets Energy Star-rated appliances, leverage this certification to build consumer trust and drive market adoption.

International IP protection and compliance with global food safety and energy efficiency standards can be complex. Innovators need to ensure their technologies meet various international regulations and standards, which may differ significantly between countries. An Indian company exporting its energy-efficient food processing equipment to the European Union must comply with the EU's energy efficiency and food safety regulations, such as those set by the European Food Safety Authority (EFSA). The company must also secure patents in the EU to protect its IP. Similarly, a U.S. company exporting to India must navigate FSSAI regulations and ensure their products meet BEE standards while securing IP rights in India.

Changes in the Past Five Years

India implemented its national intellectual property rights policy in 2016, with portfolios including Patents, Designs, Trademarks, Copyright, Geographical Indications, and Semiconductor Integrated Circuits Layout Design under the jurisdiction of the Department for Promotion of Industry and Internal Trade (DPIIT). The Cell for IPR Promotion and Management is responsible for executing the policy, promoting commercialization, increasing IP awareness, and expediting processes. Over the past five years, the Indian government has implemented various initiatives to strengthen its intellectual property regime, such as modernizing intellectual property offices, expanding workforce, utilizing information technology, and establishing the Maharashtra IP Crime Unit (MIPCU) to prevent digital piracy.

The United States also cultivates partnerships with Indian customs, police, judiciary officials, and industry representatives to develop strategies for strengthening India's critical enforcement ecosystem. India has laws regulating intellectual property rights and enforcement procedures, but the legislative process is often protracted and uncertain, leading to industries experiencing uncertainty and confounding their strategic IP enforcement decisions. The Commercial Courts Act, which was enacted in 2015 and amended in 2018, has only been beneficial to a limited number of tribunals, and the Designs Act of 2000 permits registration of industrial designs. The Intellectual Property Rights (Imported Goods) Enforcement Rules, 2007, were amended in June 2018 to exclude patents from customs protection.

India lacks data protection provisions for the registration of new agricultural products, such as pesticides, and has taken steps to combat film piracy through the Cinematograph (Amendment) Act, 2023. However, this raises concerns for right holders who need to extend exceptions to infringement from the Copyright Act to film recording or exhibition activities.

The Department Related Parliamentary Standing Committee on Commerce presented the 161st Review of the Intellectual Property Rights Regime in India Report in June 2021, emphasizing the importance of compliance with international agreements, rules, and conventions. The report also highlighted the urgent need to invest in India's patent ecosystem, recommending increased autonomy for the IP Office and addressing procedural issues in patent application and trademark registration procedures. In August 2023, the President granted his assent to the Digital Personal Data Protection Act, 2023, which addresses IP in the context of "certain legitimate uses." The Act primarily addresses IP in the context of "certain legitimate uses," such as processing personal data for employment purposes or safeguarding employers from loss or liability.

The Biological Diversity (Amendment) Act, 2023, would lead to significant changes to the Act, introducing limitations to the application of intellectual property rights to biological resources and decriminalizing cognisable and non-bailable offences. The Jan Vishwas (Amendment of Provisions) Act, 2023, would amend 42 Acts to improve trust-based governance and decriminalize offences.

Obstacles and Regulatory Environment

The commercialisation of intellectual property in India faces regulatory obstacles, including unpredictable and inconsistent patent law interpretation. The Indian government has revised the Patents and Trade Marks Rules in 2016 and 2017 to promote start-up activity and streamline examinations. However, concerns remain about the consistency of patentability decisions. The Draft Patents (Amendment) Rules, 2021, aim to streamline patent procedures and extend charge reductions to all recognized educational institutions. The Copyright (Amendment) Rules, 2021, implement modifications to copyright societies and address trademark opposition backlogs. Despite these positive developments, there are ongoing concerns about inordinate delays in trademark opposition proceedings.

Pharmaceuticals and Medical Devices

The lack of coordination between the Central Drugs Standard Control Organisation and DPIIT, along with the absence of patent linkages and an effective notification system, hinders the commercialization and enforcement of intellectual property rights in India. This lack of coordination encourages abusive infringing activities and complicates the effective IPR regime for the pharmaceutical and medical device industry. The Ministry of Health and Family Welfare abolished the requirement for companies to disclose whether a new drug is registered under a patent, which is seen as regressive and in direct opposition to India's National IPR Policy. The Trade Margin Rationalisation formula for price monitoring is also a concern. The absence of explicit guidance regarding the implementation of Section 3(d) of the Indian Patents Act limits the scope of patent-eligible subject matter, hindering innovation and benefiting Indian patients.

Technology, Brand, and Intellectual Property Licensing

In 2018, the Indian government initiated an inter-ministerial committee to explore the possibility of reintroducing royalty limits for technology collaborations and brand licensing, potentially affecting all sectors of the economy. India's copyright laws are generally adequate, but an amendment to the Copyright Act to include internet and digital broadcasters for individual works has raised concerns among creative, entertainment, and music sectors. The extensive granting of licenses under Chapter VI of the Indian Copyright Act and broad exceptions for specific uses have raised concerns about copyright protection in India. India's commitment to adhering to the World Intellectual Property Organisation Internet Treaties has not been implemented in accordance with international best practices. Form 27, which requires annual submission of commercial activities for patents, may be impeding innovation.

Enforcement Environment

The Intellectual Property Appellate Board was abolished in 2021, transferring authority to High Courts for patent, trademark, and copyright appeals. This has led to uncertainty in IP cases and copyright royalty rates in India. In July 2021, the Delhi High Court established a specialized Intellectual Property Division with three judges. The Indian government is encouraged to encourage High Courts to establish their own IPDs. India has made progress in combating piracy, but lack of a centralised IP enforcement agency, inability to coordinate actions, and weak enforcement by courts and police pose challenges.

Optimisation of Supply Chain Logistics

The optimisation of supply chain logistics involves the enhancement of the efficiency of food supply chains to decrease energy consumption and emissions. This can be achieved by optimising cargo, enhancing route planning, and implementing more energy-efficient transportation methods. Optimising supply chain logistics offers a variety of benefits, including a diminished carbon footprint, improved delivery efficacy, reduced petroleum consumption, and reduced transportation expenses. Such as, organisations may employ electric or hybrid delivery vehicles to mitigate emissions, consolidate shipments to minimise the number of necessary voyages, and implement route optimisation software to determine the most efficient delivery routes. Not only does this approach encourage environmental sustainability, but it also reduces expenses.

Utilisation of Renewable Energy Sources

The integration of renewable energy technologies into food production and processing facilities is another critical sustainable practice. This includes the utilisation of solar panels, wind turbines, and biogas generators to power operations. The utilisation of renewable energy sources offers a variety of benefits, including the potential for cost reductions in energy production, reduced greenhouse gas emissions, and a reduced reliance on fossil fuels. Like, the utilisation of solar-powered irrigation systems can enhance the efficiency of water in agriculture, while wind energy can be employed to power food processing facilities. Biogas systems are capable of converting agricultural refuse into energy, thereby contributing to waste management and providing a sustainable energy source for operations' operations.

Implementation of Energy-Saving Protocols

In order to reduce energy consumption during the production and refining of food, it is necessary to implement energy-saving protocols and implement strategies and best practices. This includes the execution of energy audits, employee training, and routine equipment maintenance. These practices offer a variety of benefits, including improved sustainability, reduced operational costs, and increased energy efficiency. Such as, energy audits enable the identification of facility inefficiencies, which in turn enables targeted improvements. By implementing energy management systems, organisations can effectively monitor and regulate their energy consumption. The training of employees on energy conservation practices can cultivate a culture of sustainability. By adopting these sustainable practices, the food industry can significantly enhance its operational efficiency and make a direct contribution to environmental sustainability.

Implications for Food Safety

Food safety is substantially affected by using sustainable practices and energy-efficient technologies in the food sector. First of all, these technologies assure food safety by helping to preserve ideal storage conditions, therefore lowering spoilage. Advanced refrigeration systems, for instance, can maintain goods at the right temperature, therefore reducing the possibility of bacterial development and foodborne diseases. Second, lower running costs resulting from reduced energy use help companies and consumers alike. Like, a grocery shop that replaces its lighting and cooling equipment with energy-efficient models can drastically reduce its running costs, savings that could be shared with patrons. They are helping environmental sustainability by greatly lowering the carbon footprint of food manufacture and storage is adopting sustainable techniques. By running food processing operations on solar panels or other renewable energy sources, one not only lowers emissions but also advances long-term sustainability. In fact, businesses adopting these ideas are more suited to follow environmental rules and guidelines, thereby preventing any fines and strengthening their image. Food companies who follow energy-efficient policies, for instance, can find it simpler to satisfy the strict criteria established by authorities, therefore enhancing their market position. Adopting sustainable practices and energy-efficient technologies would help the food sector improve food safety, lower environmental impact, and get cost savings, thereby strengthening the resilience of the food system. Consequences for Indian Innovations

Innovators in food safety and energy efficiency depend critically on following legal guidelines and obtaining market access. Companies in India have to follow BEE criteria and FSSAI rules in order to legally promote their goods. Severe penalties, product recalls, and legal action might all follow from non-compliance. For instance, a food company that disregards FSSAI recommendations for food safety runs the risk of paying fines and having to pull its goods off the market, therefore compromising its financial stability and brand. Building customer confidence and enabling market adoption depend on certification in great part. Agencies such as BIS offer certifications guaranteeing customers of the energy economy, quality, and safety of goods. This certification can increase consumer trust and propel market success. Such as, people are more inclined to trust and buy an energy-efficient appliance with BIS certification, therefore increasing the company's sales and profile. Furthermore, included in many regulatory systems are incentives including tax breaks, subsidies, and acknowledgement of companies implementing technology and energy-efficient policies. These rewards inspire creativity and financial commitment in environmentally friendly methods. For instance, a company that runs its operations on solar energy could get tax advantages or subsidies, hence increasing the appeal and financial viability of the venture. Compliance guarantees legal market access and competitiveness as well as helps to achieve sustainable development targets. Authers tried to analyse some more implications for food safety:

Figure 6. Pie chart

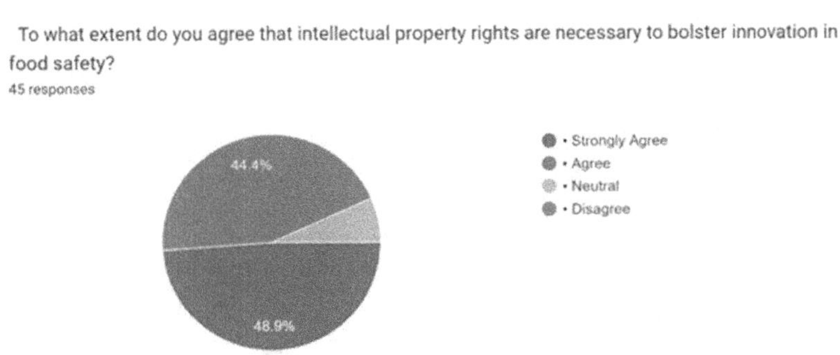

Figure 7. Pie chart

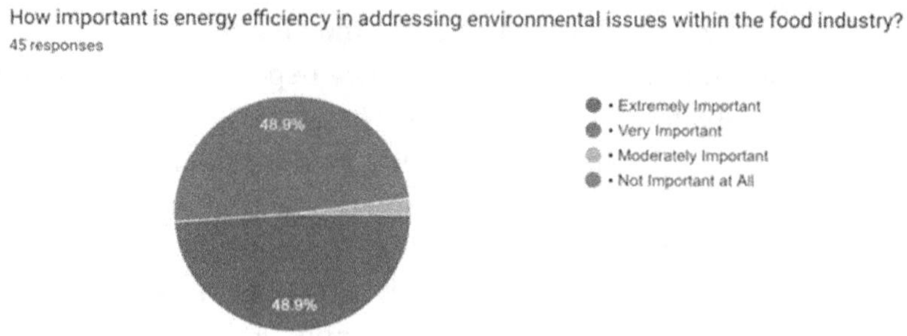

Figure 8. Pie chart

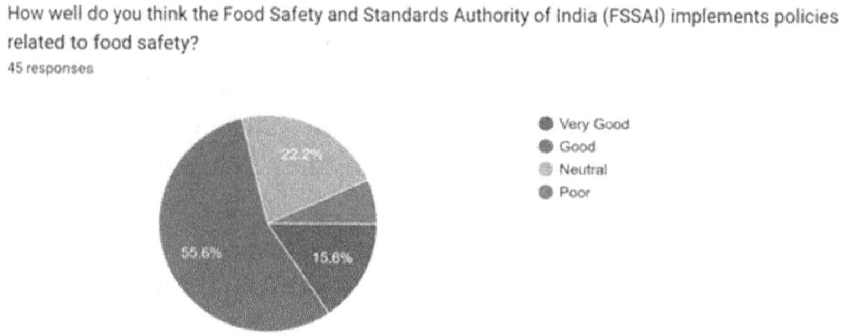

Figure 9. Pie chart

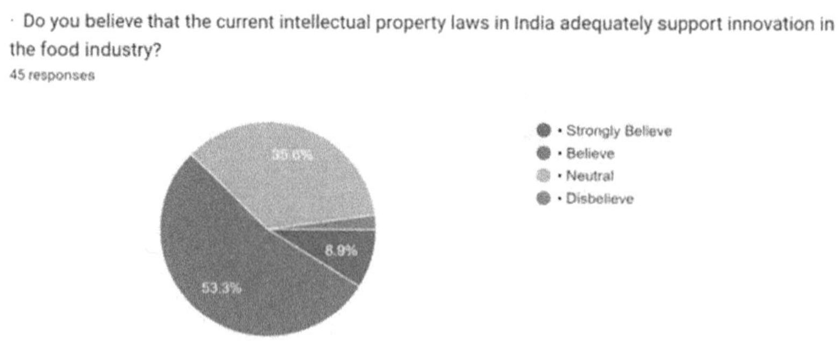

Figure 10. Pie chart

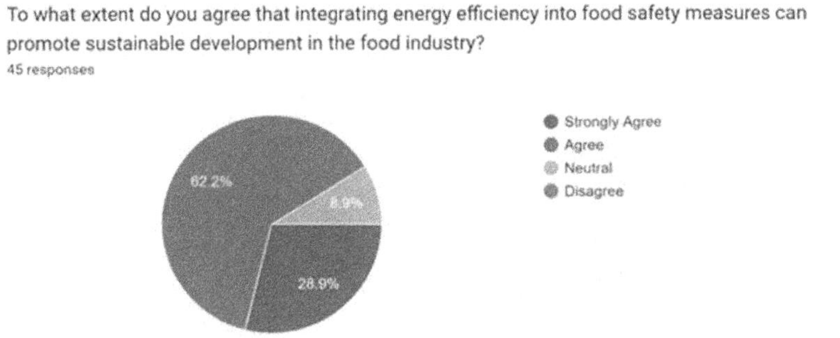

Figure 11. Pie chart

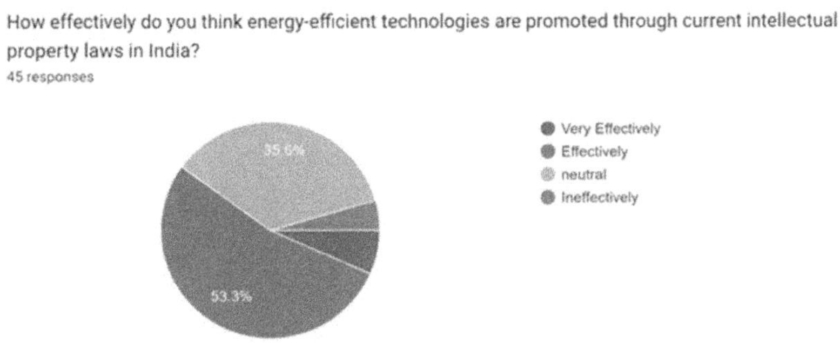

Figure 12. Pie chart

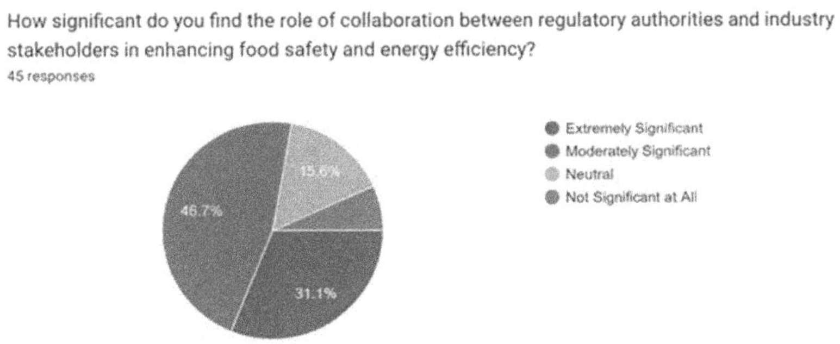

Figure 13. Pie chart

Do you agree that the integration of energy efficiency aspects into food safety standards is essential for sustainable development?

45 responses

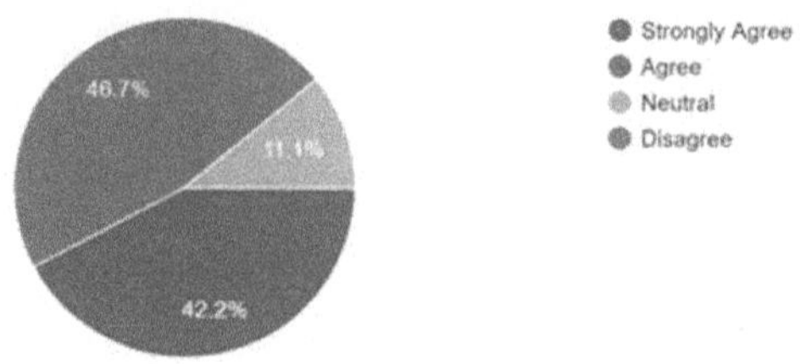

Figure 14. Pie chart

· How confident are you in the ability of the FSSAI to implement energy efficiency policies within food safety regulations?

45 responses

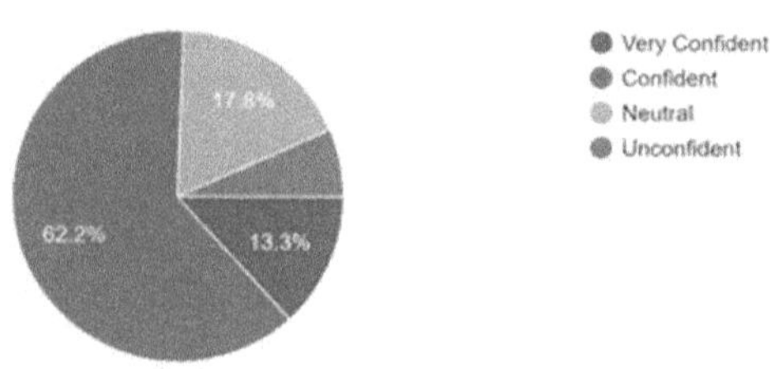

Figure 15. Pie chart

To what extent do you agree that significant concerns exist within the current legal and regulatory structure related to food safety and energy efficiency?

45 responses

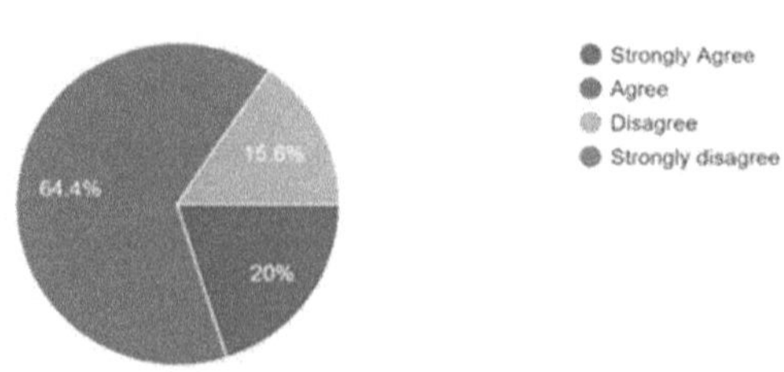

Figure 16. Pie chart

How beneficial do you think the proposed amendments to intellectual property laws would be in promoting energy-efficient technologies in the food industry?
45 responses

Figure 17. Pie chart

How important is the awareness of benefits of integrating energy efficiency into food safety measures among industry stakeholders?
45 responses

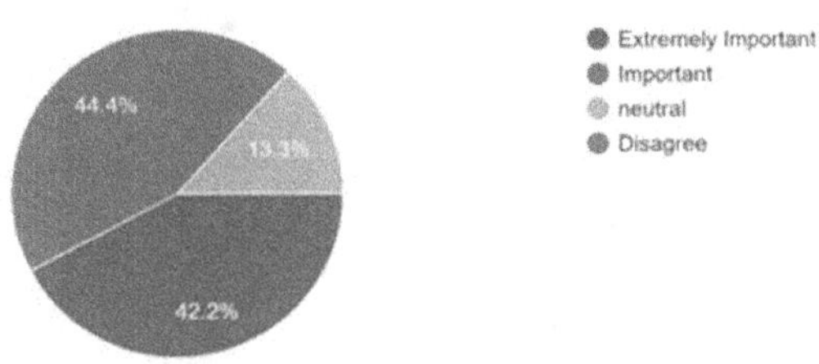

Figure 18. Pie chart

To what extent do you agree that advancements in food processing, preservation, and packaging are crucial for achieving both food safety and energy efficiency?
45 responses

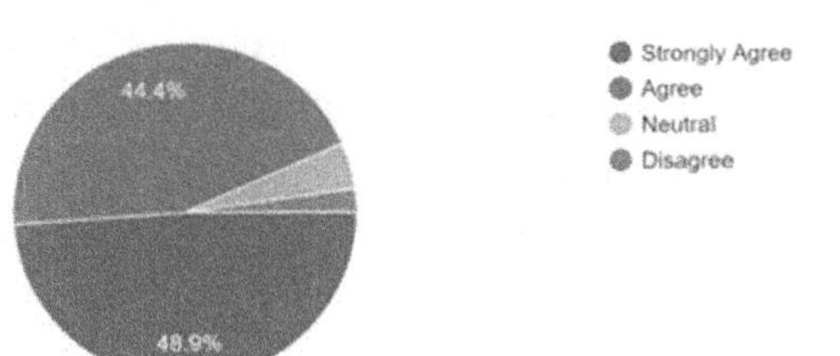

Figure 19. Pie chart

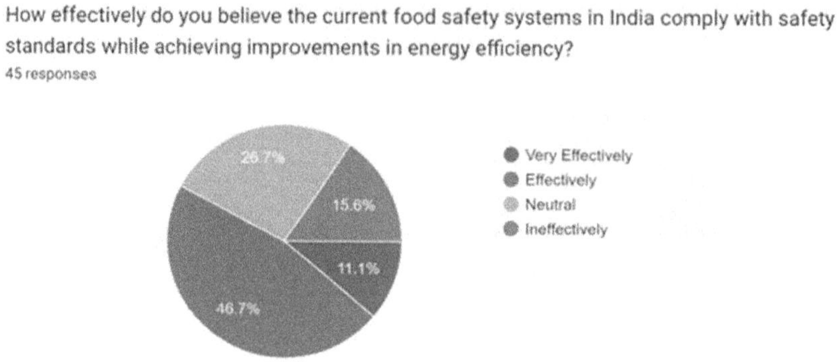

Figure 20. Pie chart

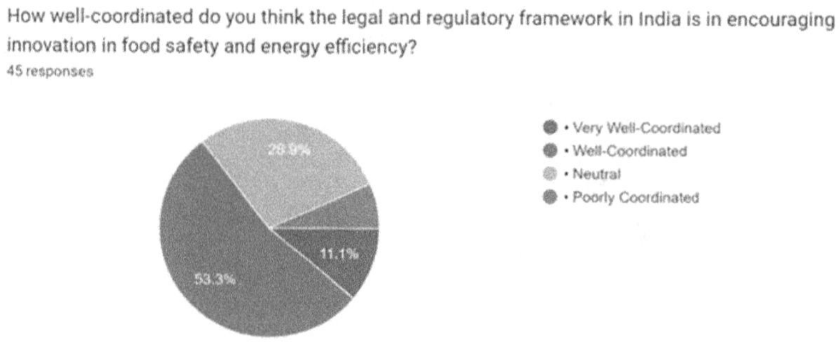

CONCLUSIONS AND SUGGESTIONS

India's strong intellectual property (IP) safeguards, such as patents, trademarks, and copyrights, are crucial for food safety and energy-efficient technology. These safeguards ensure businesses and creators can create innovative ideas, leading to improved food safety, quality, and sustainability. Energy-efficient technologies like advanced refrigeration systems, sustainable packaging, and renewable energy solutions help reduce spoilage, minimize waste, and ensure optimal storage conditions. Patents provide exclusive rights for innovators, promoting research and development.

Several strategic actions are required to improve energy economy in the agricultural sector by means of intellectual property rights and legal systems. First of all, it is absolutely vital to raise knowledge among companies and inventors on the value of intellectual property protections. Programs, seminars, and partnerships with IP offices help one to accomplish this. Clear policies and tools will enable inventors to better grasp the patent and other intellectual property rights acquisition process, hence promoting energy-efficient technologies.

Another vital action is improving the grant procedures and patent inspection. By means of training for patent examiners and implementing sophisticated technologies for patent searches and assessments, one can improve the efficiency of these procedures and thereby lower the time required for issuing patents. Encouragement of cooperation between the patent office and industry specialists will help to guarantee correct evaluation of creative energy-efficient technology in food safety.

Encouragement of group research and growth is absolutely vital. Encouragement of alliances between government agencies, business, and academia will propel creativity in food safety technology with an eye towards energy economy. Through resource pooling, information sharing, and fast development of fresh ideas, such cooperative endeavours can Furthermore crucial is supporting public-private partnerships and joint ventures aiming at commercialising food industry energy-efficient technology.

Encouragement of the acceptance of energy-efficient technologies depends critically on financial incentives and support. Providing firms engaging in energy-efficient food safety technologies grants, subsidies, and tax benefits would help to reduce cost obstacles and inspire more companies to follow sustainable practices. Crucially also is establishing funding programs and innovation centres that give startups and small companies developing IP-protected energy-efficient technology technical and financial help.

Protection of innovators from infringement and piracy depends on strengthened enforcement of IP rights. These covers simplifying legal procedures and stepping up fines for IP infringement. Important is also improving cooperation between industry players and IP enforcement authorities to properly monitor and solve IP-related problems.

Support of Indian entrepreneurs in reaching worldwide markets depends on encouraging international cooperation. Participating in international IP treaties and accords helps one promote international cooperation and harmonisation of IP rules. Encouragement of cross-border alliances that enable the sharing of best practices in energy-efficient food safety innovations and technology helps also. India can establish an environment that promotes innovation in food safety and energy-efficient technologies by enhancing intellectual property rights and offering the required support and incentives. This would not only improve the competitiveness of Indian companies but also help to guarantee a safer and more effective food chain and help to meet objectives for sustainable development.

REFERENCES

Ahuja, V. K. (2017). *Law relating to intellectual property rights*. LexisNexis.

Anant, T. V., & Reddy, M. C. V. (2015). *Sustainable food systems: Intellectual property and energy efficiency in the food sector*. Springer.

Babu, K. G., & Karan, R. K. (2018). *The role of intellectual property in promoting energy efficiency in agriculture and food processing*. Cambridge University Press.

Babu, M. S. R. (2012). *The law of intellectual property in India*. LexisNexis.

Designs Act. (2020). *2000 (with the latest amendment)*. Government of India.

Geographical Indications of Goods (Registration and Protection) Act, 1999. (2019). Government of India.

Indian Council of Agricultural Research. (2017). *Intellectual property and food security: An Indian perspective*. ICAR.

Patil, S. J., & Bhat, A. V. (2016). *Intellectual property rights and food safety: A study of Indian legislation*. Kluwer Law International.

Protection of Plant Varieties and Farmers' Rights Act, 2001. (2021). Government of India.

Quist, D., & Chapela, I. H. (2001). Transgenic DNA introgressed into traditional maize landraces in Oaxaca, Mexico. *Nature*, *414*(6863), 541–543. DOI: 10.1038/35107068 PMID: 11734853

Sharma, K. K. (2019). *Innovation and intellectual property in India: A handbook for practitioners*. Universal Law Publishing.

Singh, R. N., & Shukla, A. K. (2015). *Intellectual property rights in India: A handbook*. Bharat Law House.

The Biological Diversity Act, 2002 (with the latest amendment). (2021). Government of India.

The Copyright Act, 1957 (with the latest amendment). (2012). Government of India.

The Patents Act, 1970 (with the latest amendment). (2020). Government of India.

The Semiconductor Integrated Circuits Layout-Design Act, 2000. (2017). Government of India.

The Trade Marks Act, 1999 (with the latest amendment). (2019). Government of India.

World Intellectual Property Organization. (2018). *Sustainable development goals and intellectual property: The role of IP in promoting innovation for sustainable food systems*. WIPO.

ENDNOTES

1 Sachin Chaturvedi, 'Agricultural Biotechnology and New Trends in IPRs Regime – Challenges before Developing Countries', 37 *Economic & Political Weekly* 1212 (30 March 2002).

2 See, e.g., R. David Kryder et al., The Intellectual and Technical Property Components of pro-Vitamin A Rice (Golden Rice™): A Preliminary Freedom-to-Operate Review (Ithaca, NY: ISAAA, Brief No. 20, 2000). On Vitamin A deficiency, see, e.g., WHO, Micronutrient Deficiencies – Combating Vitamin A Deficiency, available at https://www.who.int/nut/vad.htm.

3 *See, e.g.*, concerning India, Anitha Ramanna, 'Policy Implications of India's Patent Reforms Patent Applications in the Post-1995 Era', XXXVII *Economic & Political Weekly* 2065 (2001).

4 AIR 2019 SC 559, 2019 (3) SCC 381

5 [162(2009) DLT 371]

6 [2013] 13 S.C.R. 148

7 148(2008) DLT 598, MIPR 2008 (2) 35

8 FAO (OS) 190/2013

9 2003 (27) PTC 305 (DEL)

10 AIR 2019 SC 559

11 AIRONLINE 2012 SC 363

12 Civil Action No. 11-2317 (JAP)

13 2012 LawSuit (Cal) 530

14 (2006) IIILLJ 540 DEL

15 (2010 (112) BomLR 375)

16 AIR 1987 DELHI 372

17 2005 (81) DRJ 538

18 AIR 1995 SC 2372

19 AIR1963P&H239

20 2003 (27) PTC 355 (Del)

21 (2011) 4 SCC 85

22 AIR 2008 SC 3123

23 137 (2007) DLT 349

24 CS(OS) No. 1590/2009

25 AIR 2008 SUPREME COURT 809

26 (2016) 16 DRJ (SN) 678

27 AIR 2013 (NOC) 315 (DEL.)

28 FAO(OS) 540/2011, C.M. APPL.20174/2011, 13919 & 17996/2015

29 2005 (30) PTC 253 (Del)

Chapter 16
A Global Study on the Challenges in Implementation of Policy Framework for the Use of Energy-Efficient Technologies for Food Security

Mansi Trivedi
https://orcid.org/0000-0001-5959-8324
Amity Law School, Amity University, Noida, India

Surbhi Girdhar
Shri Vaishnav Institute of Forensic Science, India

Shantanu Bindewari
https://orcid.org/0000-0002-7048-5221
IILM University, Greater Noida, India

Rajat Singh
Indian Institute of Cyber Security and Forensics, India

Rituja Upadhyay
Independent Researcher, India

DOI: 10.4018/979-8-3693-8342-1.ch016

ABSTRACT

Food security has been a major worry globally with a large and diversified population. The legal framework governing food security seeks to alleviate poverty, malnutrition, and fair access to food. Despite strong legal provisions, efficient execution of these regulatory procedures presents numerous obstacles. The study digs at the international legal framework governing food security as well as the issues experienced in enforcing the regulatory procedures established to protect the right to food through various International Covenants.

I INTRODUCTION

Food security, in global Arena is a multifaceted issue influenced by a variety of social, economic, environmental, and political considerations. Since independence (Li & Song, 2022). On the other hand, India has achieved great progress in food production, but issues with distribution, availability, and nutritional quality remain. Some important factors of food security are:

1.1 Agricultural Production

Worldwide, it had an enormous increase in food grain output during the Green Revolution in the 1960s and 1970s (Wang et al., 2024). This resulted in increased wheat and rice production, particularly in the Indian northern states of Punjab, Haryana, and Uttar Pradesh. India is one of the world's leading producers of rice, wheat, sugarcane, cotton, and other fruits and vegetables. However, productivity remains lower than worldwide standards due to tiny landholdings, outmoded farming techniques, and insufficient irrigation.

1.2 Nutritional Security

While countries are self-sufficient in food grains, hunger remains a major issue. There are high rates of stunting, wasting, and undernutrition, especially among children and women. Even with adequate calorie consumption, a lack of key vitamins and minerals remains a serious problem. Anaemia (iron deficiency) is common, particularly among mothers and children. There is a rising emphasis on encouraging the growth and consumption of millets, pulses, and other nutrient-dense foods in order to increase nutritional diversity.

1.3 Issues related to Food Security

The expanding population, puts enormous strain on food production and distribution networks. Agricultural production is under pressure from more unpredictable weather patterns, such as droughts, floods, and high temperatures. Water scarcity and decreased soil fertility also contribute to food poverty. There are various factors that affect the existence of disproportionate in the community, i.e. education, income, access to land, small agricultural farmers, tribes and labourers who are landless.

1.4 Contemporary Advancements

During Covid -19 The government had launched the Pradhan Matri Garib Kalyan Yojana which provided additional food grains to individuals. This commitment of the government highlighted the food insecurity that was created at the time of emergencies. The government has taken steps to resurrect traditional crops which are highly nutritious and climate resistant. These steps are aligned to establish nutritional security.

1.5 Sustainability

The initiative to utilize the Sustainable farming techniques has become important. Techniques like reducing pesticides, efficient management of water flow and organic farming are encouraged to take a step forward. To increase production and resilience for the farmers, precise farming and improved seed quality along with digitalization are a way towards sustainability and attainment of food security in India as compared to other countries. India has put enormous improvement to ensure the availability of food. Even so, there are barriers of accessibility of healthy food. Malnutrition being one of the factors affecting the nutritional level of the people, the distribution of food needs to be checked in a systematized manner. To attain a long-term sustainability of food security the critical difficulties need to be addressed.

1.5.1 Importance of a Robust Legal Framework for Food Security

For the development of the nation, food security is as mandatory as having legal obligations towards society. The individual must be able to have enough nutritious food so that the daily dietary requirement can be sufficiently met. For an healthy and active life, food security is a must. There can be seen different approaches and ideas towards the availability of resources so that nutrition can be emphasized strictly. This approach is inclusive of equal distribution and accessibility of agriculture production as an aspect of ensuring food security. A strict and stringent legal framework is necessary so there can be accountability to achieve the goals and policies given by the government and institutions. To fight hunger issues and starvation, agricultural methods must be sustainable.

a. Food as a Fundamental Right:

One of the key reasons why a legal framework is required for food security is the recognition of food as a fundamental human right. This recognition establishes a legal obligation on governments to ensure that all citizens have access to food. The UDHR i.e Universal Declaration of Human Rights under article 25 and the International Covenant on Economic, Social, and Cultural Rights under article 11 emphasize the right to an adequate standard of living, which includes food. Many countries have included these articles into their constitutions, effectively transforming international commitments into national obligations. A strong legal framework guarantees that the right to food is enforced. Without legislative assurances, food security can be left to political will, which frequently results in unequal access to resources and services.

b. Ensuring Equitable Distribution:

A comprehensive legal framework contributes to equitable food distribution, particularly in nations where discrepancies in wealth and resources can lead to uneven access to food. Legal systems can address the underlying disparities that hinder marginalised populations from receiving food. This is especially important in countries where disadvantaged people, such as the poor, women, children, and tribal tribes, face social, economic, or geographical impediments to accessing food. The legal backing assures that the program targets needy communities while also providing a framework for addressing inefficiencies such as corruption, supply diversion, and exclusion errors. Legal requirements ensure that regulations such as minimum support prices (MSP) for farmers are enforced, avoiding exploitation and ensuring that food producers are paid fairly for their produce, thus contributing to food security by protecting farmers' livelihoods.

c. Addressing Malnutrition and Nutritional Security:

Food security is more than just ensuring that people have enough food to consume; it also includes ensuring that they have access to nutritious food. A strong legislative framework can help to implement policies and initiatives aimed at combating malnutrition, which is still a major problem around the world, especially in poor nations. Legal mandates can ensure the delivery of different and nutritious foods, rather than merely (Blankenship & Blancato, 2022) calorie-rich staples, thereby addressing concerns including stunting, wasting, and micronutrient deficiencies.

Legal frameworks (Clapp et al., 2022). also support activities such as school feeding programs and the distribution of fortified foods to vulnerable groups including pregnant women and young children. Many countries have legislation requiring the fortification of basic foods with important vitamins and minerals. Such legal measures are critical in guaranteeing long-term nutritional security because they establish standards for food quality and safety, ensuring that the food available is both sufficient in quantity and rich in nutrients.

d. Resilience Against Economic Shocks and Emergencies:

Economic shocks, natural disasters, and global crises all pose a significant threat to food security (Shafiee et al., 2022). The COVID-19 pandemic, for example, showed the vulnerability of global food systems, resulting in disruptions in food supply networks and spikes in hunger. A strong legal structure can serve as a safety net during such crises, allowing governments to respond quickly and efficiently. This includes developing emergency food reserves, social safety nets, and legally mandated measures to ensure food supply continuity. Legal frameworks can also help to make food systems more robust to climate change. With climate change having an increasing impact on agricultural productivity, legal rules can encourage sustainable farming techniques, efficient water management, and biodiversity conservation. Legislation, for example, can encourage farmers to practice climate-smart agriculture, which not only contributes to food security but also protects the environment.

e. Strengthening Institutional Capacity and Governance:

Another important feature of a strong legal framework for food security is its ability to boost institutional capacity. Effective governance is essential for implementing food security programs and ensuring that resources are utilized efficiently. A solid legislative framework establishes clear norms and obligations for all players, including government agencies, civil society, and the corporate sector, eliminating bureaucratic inefficiencies and corruption.

It also ensures collaboration at all levels of government—central, state, and local—resulting in more effective service delivery. For example, legal regulations can ensure that local governments are empowered to oversee food distribution at the grassroots level, thereby increasing accountability and decreasing leakages in systems such as PDS.

f. Promoting Sustainable Agriculture:

Sustainable agriculture is critical to long-term food security, and regulatory frameworks play an important role in encouraging sustainable practices. Governments may ensure that agricultural production does not deplete critical resources such as soil, water, and biodiversity by passing laws that restrict their usage. Legal measures can also encourage research and development of sustainable farming technologies, as well as incentivize farmers to use environmentally friendly methods. For example, legislation governing the use of pesticides, fertilizers, and genetically modified organisms (GMOs) (Smyth, 2020). help to protect ecosystems while also assuring safe and sustainable food production. Furthermore, land tenure regulations that safeguard the rights of small farmers, particularly women and indigenous people, are crucial for ensuring that individuals who rely on agriculture for a living have secure access to land.

The value of a strong legal framework for food security cannot be emphasised. Such a framework protects the right to food, promotes equitable distribution, combats hunger, and strengthens resilience to crisis. Ensuring food security for everyone necessitates both political commitment and legal accountability, paving the road for hunger eradication and improved global well-being.

g. Predicaments of Regulatory Mechanism

Regulatory mechanisms are critical instruments for governments and institutions to assure compliance with laws, regulations, and standards that promote fairness, safety, and the public good. These mechanisms cover a wide range of industries, including finance, the environment, public health, agriculture, and food safety. However, successfully applying regulatory procedures presents a number of obstacles. These obstacles are caused by institutional inefficiencies (Nordin et al., 2013), a lack of necessary resources, political and corporate influence, rapid technology innovation, and socioeconomic considerations. Understanding these challenges is critical for developing more effective regulatory frameworks that can serve the public interest while taking into account society's changing nature.

h. Institutional Capacity and Inefficiency:

One of the most major obstacles to establishing regulatory systems is institutional inefficiency. Regulatory authorities frequently lack the organizational capability, financial resources, and technological knowledge required to enforce legislation effectively. This issue is especially acute in underdeveloped nations, where regulatory bodies may be underfunded or suffer from bureaucratic inefficiencies. In many circumstances, enforcement agencies lack adequate staffing, training, and infrastructure, limiting their ability to monitor compliance and respond to infractions. Regulatory agencies in industries such as healthcare, banking, and food safety are responsible for managing complex systems with many stakeholders. The sheer size of these operations can overwhelm institutions, resulting in delays, insufficient monitoring, and enforcement gaps. For example, in the food safety industry, regulatory organizations may struggle to monitor the full supply chain, from farm to fork, due to a lack of manpower or antiquated tracking technology.

i. Corruption and Political Influence:

Corruption is another key issue that undermines the effectiveness of regulatory mechanisms. Political figures or corporate interests can exert undue influence over regulatory authorities, resulting in regulations being manipulated for personal or financial gain. In some situations, large firms may "capture" regulatory agencies, leveraging their influence to craft policies that benefit them at the expense of the public good. This tendency, known as regulatory capture, impairs regulations' effectiveness by making enforcement slack or selectively implemented. For example, in the financial industry, big banks or international firms may advocate for looser restrictions, allowing them to operate with less scrutiny. This can lead to systemic dangers, as shown in the 2008 global financial crisis, which was exacerbated by deregulation and insufficient oversight. Similarly, in the environmental sector, corporations involved in resource extraction or industrial production may advocate for less environmental regulations, resulting in environmental degradation and public health hazards. Political involvement may potentially skew the regulatory process. Elected politicians may encourage regulatory organizations to ease enforcement in exchange for political contributions or to safeguard their voter base.

j. Rapid Technological Advancements:

Regulatory organisations have considerable challenges due to the rapid rate of technology progress. Artificial intelligence (AI), blockchain, cryptocurrency, and biotechnology are evolving faster than regulatory frameworks can keep up. Regulatory agencies are frequently left playing catch-up, attempting to create systems for successfully overseeing industries and activities that did not exist a few years ago. In the case of AI, for example, technological advancements have surpassed legislative solutions. Governments are wrestling with issues of data privacy, ethical considerations, accountability, and transparency in AI-driven decision-making. Meanwhile, industries such as fintech (financial technology) and biotechnology confront comparable issues, with the rate of innovation outpacing regulatory authorities' ability to adopt and enforce rules. This mismatch between technical advances and regulatory frameworks generates supervision gaps, allowing industries to operate in mostly unregulated or under-regulated environments, thus posing societal dangers.

k. Globalization and Cross-Border Issues:

Globalisation has made regulatory enforcement more complicated, as supply chains, financial transactions, and environmental repercussions frequently cross national borders. This interdependence necessitates regulatory procedures that can coordinate across numerous jurisdictions, which is challenging to establish given the wide range of legal systems, regulatory standards, and enforcement capacities between countries. For example, in environmental control, contaminants produced in one country may have an impact on ecosystems in another, necessitating international collaboration to successfully manage. Similarly, multinational firms operating in multiple nations may take advantage of regulatory disparities, using looser rules in one area to save money while dodging tighter restrictions in another. Financial regulation faces similar issues, especially with the rise of cryptocurrencies and globalized money flows. The decentralized structure of digital currencies such as Bitcoin makes it difficult for any single country to adequately control them, as transactions can occur across borders with no monitoring. This hinders efforts to combat criminal activity including money laundering, tax evasion, and terrorist financing.

l. Socio-Economic Disparities:

Socio-economic inequalities complicate the application of regulatory processes, as regulations frequently disproportionately affect different parts of society. In many situations, regulations are droughted with the interests of the wealthiest or most powerful organizations in mind, leaving marginalized populations to suffer the brunt of the bad repercussions. For example, environmental restrictions intended at decreasing pollution may disproportionately affect low-income (Barlow et al., 2020) communities that rely on the businesses targeted by these regulations for employment. In response, these communities may oppose regulatory measures because they see them as a danger to their livelihoods. This opposition may impede the adoption of critical regulations necessary to protect human health and the environment. In industries such as healthcare, regulatory processes intended to ensure access to important services may fail to reach marginalized people due to systemic impediments such as poverty, a lack of education, or inadequate infrastructure. These gaps frequently compound existing inequalities, since wealthier individuals and firms have the resources to navigate complex regulatory systems, leaving poorer communities open to exploitation or exclusion.

m. Resistance from Industry and Public Perception:

Implementing regulatory mechanisms frequently faces opposition from industry actors who see rules as a danger to their profits. Businesses, particularly those in industries such as manufacturing, banking, and energy, may perceive laws as restrictive, increasing operational costs and restricting competition. This resistance might take the form of lobbying, legal challenges or noncompliance, making it difficult for regulatory agencies to successfully enforce rules (Barlow et al., 2020). In rare situations, industries may initiate public relations campaigns to persuade the public that regulatory efforts are excessive or unneeded. This strategy has been widely used in industries such as tobacco, oil, and coal, where restrictions are viewed as a threat to business models based on behaviors that are hazardous to public health or the environment. Some people may perceive stringent building codes or environmental rules as hindering economic progress or infringing on property rights, leading calls for regulatory rollbacks.

n. Monitoring and Enforcement Challenges:

Effective regulation necessitates strong monitoring and enforcement measures, which regulatory agencies frequently fail to provide. Ensuring compliance necessitates regular inspections, audits, and data collecting, all of which require significant resources and experience. In many circumstances, regulatory authorities lack the capacity to monitor every actor within their jurisdiction, resulting in selective enforcement or reliance on industry self-reporting, (Barlow et al., 2020) which can lead to violations being underreported. Furthermore, enforcement frequently relies on the existence of clear and enforceable sanctions for noncompliance. If punishments are ineffective, inconsistent, or improperly implemented, they may fail to deter offences.

o. Lack of Public Awareness and Engagement:

Another issue in adopting regulatory procedures is a lack of public awareness and participation. Regulations are frequently developed and enforced without adequate involvement from the communities and individuals they are supposed to protect. This might lead to a lack of public support or understanding, complicating enforcement efforts. For example, public health restrictions such as vaccine requirements or quarantine measures during a pandemic may meet opposition if people do not completely grasp the reasoning behind them.

1.5.2 Historical Background

i. Evolution of food security policies in India

India's efforts to secure food security for its vast and growing population are deeply embedded in its sociopolitical and economic history. Key historical events, agricultural productivity issues, political developments, and socioeconomic transitions have all influenced the evolution of the country's food security strategies.

The following is a list of significant milestones that led to the development and expansion of Indian food security policies.

ii. Colonial Period and Food Insecurity:

India suffered from acute food insecurity during British colonial rule. Famines were prevalent due to exploitative policies, inefficient agricultural management, and a lack of government action. The colonial administration prioritized revenue collection over Indian welfare, which contributed to rural poverty. The Great Bengal Famine and the Madras Famine were among the most devastating famines of the colonial period. These famines were caused not just by natural causes, but also by British policy of food exportation, a lack of relief measures, and a complete failure to address local food shortages. As a result, the colonial period left India with a vulnerable agricultural sector and a legacy of food poverty that endured even after independence.

iii. Post-Independence Period (1947-1960s):

Following its independence in 1947, India experienced acute food shortages, which were compounded by population increase, partition-related disruptions, and low agricultural production. The country was largely dependent on food imports, especially under the Public Law, which saw the United States supply India with surplus wheat. However, relying on imports was not a long-term option, and there was an urgent need to increase domestic food production and self-sufficiency.

iv. Green Revolution (1960s-1980s):

In the 1960s, India was once again on the verge of a food crisis due to frequent droughts and low agricultural output. The Green Revolution (mid-1960s) marked a watershed moment in Indian agriculture, with a series of programs aimed at raising agricultural production with high-yielding varieties of seeds, chemical fertilizers, and irrigation infrastructure. Key features of the Green Revolution are HYV Seeds, Increased Irrigation and Chemical Inputs. The Green Revolution was primarily centered on wheat and rice, mainly in Punjab, Haryana, and western Uttar Pradesh. This project dramatically increased food grain output, allowing India to become self-sufficient in food grains by the late 1970s. By the early 1980s, India was producing enough food grains to meet domestic demand and had started exporting. However, the Green Revolution resulted in regional imbalances (favoring northern states), increasing dependency on chemical inputs, and environmental damage, such as water depletion and deterioration of soil health. Despite these hurdles, the Green Revolution established the basis for India's food security.

v. Economic Liberalization and Revamped Food Security (1990s):

India's economic liberalisation in 1991 resulted in considerable improvements, but the reforms also had consequences for food security. Essential commodity prices rose in the early 1990s, leading to increased poverty and food insecurity for vulnerable groups. In response, the Indian government implemented many measures to bolster food security procedures. One of the most significant reforms was the implementation of the Targeted Public Distribution System (TPDS) in 1997. Unlike the previous universal PDS, TPDS concentrated on delivering subsidised food grains to households below the poverty line (BPL), whereas households above the poverty line (APL) received food grains at a higher cost. This adjustment was made to improve the efficiency of food distribution and prevent leaks.

vi. National Food Security Mission (2007):

To address concerns about regional inequities and declining productivity in specific sectors, the Indian government established the National Food Security Mission (NFSM) in 2007. This program attempted to boost rice, wheat, and pulse production in underproductive regions. The NFSM aimed to boost agricultural productivity through better seeds, soil health management, and water-use efficiency. The NFSM was part of the larger Eleventh Five-Year Plan (2007-2012), which focused on agricultural expansion as a key driver of food security. The initiative also aimed to address malnutrition issues, as many homes, despite having enough food grains, lacked access to a diverse and nutritious diet.

vii. The National Food Security Act (NFSA), 2013:

he National Food Security Act (NFSA) was passed in 2013, marking a watershed moment in India's food security path. The NFSA represented a change from a welfare-based approach to a rights-based framework for food security. The act aims to give legal entitlements to subsidized food grains for around two-thirds of India's population.The NFSA represents a comprehensive strategy to food security that addresses not just hunger but also malnutrition and improves nutritional outcomes for vulnerable groups.

viii. COVID-19 and Food Security Response:

The COVID-19 pandemic in 2020 intensified food insecurity, especially among migrant workers, day laborers, and the urban poor (Simola et al., 2022). In response, the Indian government implemented the Pradhan Mantri Garib Kalyan Anna Yojana (PMGKAY), which supplied free food grains to approximately 800 million people for several months throughout the pandemic. This emergency step demonstrated the significance of strong food security systems during times of crisis.

India's food security policies have evolved over decades, from crisis-driven responses during famines to a complete legal framework aimed at addressing both food availability and nutrition security. The country has achieved great strides towards food security, notably through the Green Revolution, the PDS, and the NFSA. However, difficulties persist, such as geographical inequities, malnutrition, and the need for more efficient food distribution networks. As India expands, its food security policy must adapt to new issues such as climate change, population expansion, and economic inequality.

1.5.3 Key Milestones in the Development of the Legal Framework

There are a number of milestones that have shaped Global legal framework for food security. Food security policy has progressed from crisis-driven responses to a comprehensive legal framework that ensures fair access to food. These demonstrate the country's efforts to secure access to sufficient, safe, and nutritious food for its population, particularly the disadvantaged groups. Key advancements, such as the PDS, TPDS, NFSM, and the historic NFSA, have revolutionized the country's approach to food security by addressing both availability and access challenges. These policies demonstrate India's determination to eliminate hunger, improve nutrition, and promote food security for all, particularly the most vulnerable members of society. Here are some important milestones arising from legal framework:

The Public Distribution System was founded around 1960 as a mechanism for food security so that the food grains can be distributed in subsidized rates through the fair price shops. The purpose of this system was to cater for the basic needs of people.

In 1997 there came into existence the targeted public distribution system wherein the government focused on people with less income and distributing them food grains at lower prices focusing on above poverty line households but with less income. The Antyodaya Anna Yojana was launched in 2000 with a view to provide subsidies to the poor and most needy individuals. The mid-day Meal system was implemented within the entire country in government aided schools to provide meal to children. This government initiative focused on increasing the presence of children and curb malnutrition in children. With the advent of National Food Security Mission in 2007, there was an initiative to increase the productivity of wheat, rice and pulses to that the scarcity of food should not be there in the country. This was to develop the agricultural industry at large. The government initiated a successful initiative by the government of One Nation One Ration to allow access to food from any fair price shop and boost the mobility of the migrant worker.

1.5.4 Major Food Security Programs and Their Impact

India, with its enormous and diversified population, faces significant challenges in guaranteeing food security for all its residents. The major food security programs have had a significant influence on reducing hunger, increasing access to affordable food, and combating malnutrition. Over the decades, the Indian administration has come up with various aiming at combating hunger, malnutrition, and guaranteeing equitable access to food. These major food security programs have had a significant impact on India's agricultural landscape while also addressing the pressing issue of hunger. While there are still issues, such as distribution inefficiencies and geographical discrepancies, these programs have played an important role in reducing food insecurity and encouraging socioeconomic development throughout the country.

Access to food: The public distribution system has played an important role in increasing food access for vulnerable communities, particularly in rural and economically challenged areas.

Reduction in Hunger: By making food grains more affordable, the PDS has helped to alleviate hunger and starvation in many parts of India.

Poverty Alleviation: TPDS has helped to reduce poverty and improve living conditions in both rural and urban regions by delivering affordable food to low-income people.

Improved Child Nutrition: The MDMS has made a substantial contribution to boosting the nutritional intake of pupils, particularly those from economically disadvantaged homes.

Increased School Attendance: By providing free meals, the initiative has increased school enrolment and attendance, particularly among girls and children from underserved communities.

Improved Agricultural Practices: The mission encouraged sustainable farming methods, which resulted in improved crop production in previously underperforming areas (Klassen & Murphy, 2020).

Improved Nutritional Security: The NFSA increased the nutritional security of children, pregnant women, and breastfeeding mothers by combining the Midday Meal Scheme with Integrated Child Development Services (ICDS).

II LEGAL FRAMEWORK FOR FOOD SECURITY IN INDIA AND INTERNATIONAL ARENA: A COMPARATIVE STUDY

Constitutional Provisions Related to Food Security: The constitution of India does not directly include right to food as a fundamental right but through Article 21, it provides for wider interpretation to include right to food as a part of right to life and personal liberty. The directive principles of state policy under article 39 (a) provide that the state can make any law for the access to food and subsistence of food security in the country. Also, article 47 gives the direction to the state that it is the responsibility of the state to provide legislation for elevating the standard of food and health of people. The Supreme Court in a landmark judgment of PUCL v. Union of India in 2001, decided in the case that there shall be legal obligation to protect the right to food under Article 21 of the Constitution.

The National Food Security Act, 2013 was enacted with an aim to provide food security framework as a legal obligation. The objective of the act is to attain not only food security but also nutrition security with access to food. The act tries to overcome the hunger issues and malnutrition insecurity in the country. The act provides for grievance redressal mechanism for dispute settlement mechanism. It also provides for the security to pregnant women and children to cater their nutritional needs.

Other Relevant Legislations like the Essential Commodities Act enacted in 1955 allows the government to set supply limits, control prices, and prevent stockpiling and black marketing of important commodities. The Public Distribution System (Control) Order, 2001 governs the distribution of food grains and necessary goods at subsidized rates to recipients. This directive provides criteria for identifying beneficiaries, managing fair pricing shops, distributing food grains, and ensuring transparency in the PDS. It emphasizes the obligation of state governments to ensure that eligible households are properly identified and that fair pricing shops run smoothly to eliminate leakages, corruption, and inefficiencies in food grain distribution. The Food Safety and Standards Act (FSSA) of 2006 focuses on the safety and quality of food. It is not directly linked with food security but there are various programs like mid day meal which looks forward for the nutritional concerns of the state. The Food Safety and Standards Authority of India is a body that monitors food quality, oversees the supply chain, and enforces food safety standards to ensure safe and healthy food for the population.

Together, these legal regulatory frameworks provide a strong system aimed at ensuring that India's enormous population gets adequate and inexpensive food, while also addressing food quality, safety, and distribution issues.

2.1 Regulatory Mechanisms for Food Security

2.1.1 Central and State-Level Regulatory Bodies

The central government bears major responsibility for developing food-security policies. The Ministry of Consumer Affairs is responsible for overseeing and implementing different food security initiatives. It oversees buying food grains, managing storage, and distributing supplies to the states. It provide for legal framework that assures subsidized food grains for roughly two-thirds of India's population. States have an important role in enforcing central policies at the local level. They oversee identifying qualified beneficiaries, distributing food grains via fair price shops (FPS), and maintaining local storage facilities. States also contribute by launching their own food security initiatives, such as increased food subsidies or nutritional programs, to supplement central efforts.

2.1.2 Implementation of Mechanisms at Various Administrative Levels

The emphasis is on formulation of policies, coordination, and proper acquisition of food grain. This includes allocating food grains to states and union territories depending on population size and food security requirements. States oversee the operational components of distribution, including ensuring that the PDS functions efficiently. Each state defines the population coverage for food security programs and supervises the allocation to fair-priced stores. States also keep track of stock levels and oversee intrastate trafficking of food grains. The officials are in charge of overseeing the execution of food security programs. They monitor fair-priced stores, review beneficiary lists, and guarantee that food grains reach their proper users. Local organizations, such as village panchayats or urban ward committees, are frequently involved in beneficiary identification and grievance resolution. This decentralized method tries to handle local food security issues more efficiently.

2.1.3 Role of the Food Corporation of India (FCI)

It is crucial to ensuring national food security. Founded in 1965 under the Food Corporation law to achieve goals such as price support, food distribution, and buffer stock maintenance. The primary functions of the FCI include:

a. **Procurement**: FCI purchases food grains at Minimum Support Prices (MSP) from farmers around the country. This not only secures a minimum wage for farmers, but it also ensures that the central government has enough supplies to distribute.
b. **Storage**: FCI manages enormous networks of storage facilities around the country, including warehouses and silos. These facilities hold food grains until they are transferred to state governments.
c. **Distribution**: The FCI provides food grains to state governments for distribution under the PDS. It also keeps buffer inventories on hand to guarantee that food supplies remain stable amid shortages or emergencies.

d. **Market Intervention**: FCI is important in market intervention in stabilizing food prices by releasing supplies during times of scarcity and preventing excessive price swings in open markets.

2.2 Monitoring and Evaluation Systems

2.2.1 Central and State Audits: The national and state governments conduct frequent audits of food security systems. These audits evaluate the financial and operational efficiency of the food procurement, storage, and distribution networks. They also ensure NFSA compliance.

2.2.2 Digital Monitoring: To promote transparency and reduce leakage in the PDS, many states have implemented computerized monitoring systems. Beneficiaries are identified using digitized ration cards that are connected to Aadhaar, and transactions at fair pricing stores are tracked using electronic point-of-sale (ePOS) machines. These systems allow for real-time surveillance, reducing instances of fraud and corruption.

2.2.3 Grievance Redressal Mechanisms: To promote accountability, several states have established grievance redressal processes for food security recipients. Citizens can report difficulties such as non-delivery of food grains or bad quality supplies to helplines, online portals, or local food inspectors. This feedback is critical to enhancing the efficacy of food security programs.

2.2.4 Impact Assessments: Independent bodies and research organisations also perform periodic evaluations of India's food security mechanisms. These studies evaluate food security initiatives in terms of beneficiary satisfaction, coverage, and effectiveness in combating malnutrition and hunger.

India is working to handle the massive challenge of guaranteeing food security for its people through various regulatory mechanisms, with each administrative level playing an important part in the successful implementation of food security policies.

III CHALLENGES IN IMPLEMENTING REGULATORY MECHANISMS

3.1 Administrative Challenges

a. **Bureaucratic Hurdles**: India's multitiered administrative structure, which includes central, state, and municipal governments, frequently causes delays in decision-making and implementation. The overlapping roles of several departments cause delays in approvals and inefficient cooperation, hampering attempts to secure food security.

b. **Corruption and Inefficiency**: Corruption at all levels of food distribution and regulatory systems remains a major impediment. Inefficient processes, rent-seeking behavior, and a lack of transparency in public distribution systems (PDS) reduce the reach and efficacy of food security initiatives. Funds designated for assistance schemes are frequently misused, resulting in poor outcomes for those most in need.

3.2 Technological Challenges

a. **Lack of Infrastructure**: Despite technological advancements, rural India continues to face serious infrastructural shortfalls, such as inadequate storage facilities, poor road connectivity, and limited market access. This causes post-harvest losses and hampers the efficient implementation of food security efforts.

b. **Issues in Digitalization and Data Management**: The effort to digitalize food security measures, such as direct benefit payments and digitized ration cards, has encountered obstacles. Exclusion errors are caused by issues such as limited internet access in rural areas, poor data collecting, and a lack of knowledge among recipients. The lack of accurate data also impedes real-time monitoring and decision-making.

3.3 Socio-economic Challenges

a. **Poverty and Inequality**: With a high proportion of the population living below the poverty line, food insecurity is inextricably linked to income (Breeman et al., 2017) inequality. While food subsidy programs such as the PDS seek to address this issue, they are frequently inefficient and fail to reach the most disadvantaged. Additionally, marginalized people, such as scheduled castes, tribes, and landless laborers, suffer additional barriers to accessing food security programs.

b. **Accessibility and Distribution Issues**: Distribution networks are uneven, and there is frequently an imbalance between areas of food glut and scarcity. Rural areas, particularly those in distant or conflict-prone areas (PUCL v. Union of India, 2001), suffer significant barriers to accessing food due to inadequate transportation infrastructure and distribution channels.

3.4 Political Challenges

a. **Policy Inconsistencies**: Changes in government policy affecting agricultural, food subsidies, and distribution systems impair long-term food security programs. Populist policies, such as farm debt exemptions or price controls (Wang et al., 2024), may bring temporary respite but frequently harm long-term solutions. Inconsistency in policy enforcement between states produces implementation gaps.
b. **Political Interference**: Political influence in the operation of food distribution networks frequently results in favoritism and exclusion of marginalized groups. Political objectives frequently impact resource allocation decisions, beneficiary identification, and welfare program implementation, resulting in inefficiency and inequitable access (PUCL v. Union of India, 2001).

These concerns underline the necessity for a multifaceted, integrated strategy to addressing food security issues in India. Streamlining administrative processes, investing in technology and infrastructure, tackling socioeconomic gaps, and minimizing political influence are all critical to ensuring that regulatory mechanisms work well (Santo & Moragues-Faus, 2019).

IV SUCCESSFUL MEASURES

Globally, for food safety, there has been adopted successful measures to address food security issues, with a focus on increasing agricultural output, food distribution, and social safety nets. Here are some notable examples of successful implementation.

1. Subsidized food grains provide inexpensive access to food for millions, particularly in rural areas.
2. The Legislation has improved food security in India by giving them food grains at subsidized prices. The law encourages women to issue themselves a ration card being the eldest in the house.
3. Mid-day meals have increased school attendance, enrolment, and nutritional status for students across the country. By boosting child nutrition, the program addresses both food security and education, with over 120 million children benefiting each year.
4. The Child Development Services program, which into existence in 1975 provides child under the age of six proper food, education, and primary healthcare, not only to child but the mother as well (Martin-Shields & Stojetz, 2019). The

program provides supplemental nutrition, immunizations, health screenings, and education. It focusses on the most vulnerable populations, particularly those in rural and tribal communities, with the goal of reducing child malnutrition.

5. The MGNREGA Act guarantees work for rural households for at least 100 days. Though primarily an employment program, it indirectly increases food security. By giving income to rural people, they may afford vital commodities and reduce their vulnerability to food shortages. The program also promotes infrastructure development, such as irrigation systems and land development, which serve to boost agricultural production.
6. In 2018, the Indian government began the Zero Hunger pilot project in select areas to improve nutritional outcomes and food security by integrating agriculture, nutrition, and health. The initiative focusses on biofortification (the process of boosting crops' nutritional content) as well as improving nutritional indicators in children and women.
7. The National Agriculture Market is an electronic trading system that connects existing APMC mandis across India, creating a unified national market for agricultural goods. It enables farmers to earn higher pricing for their produce by minimizing intermediaries and enhancing market access. Improved agricultural market efficiency boosts revenues and promotes food security by giving farmers the tools they need to produce more.
8. The Mega Food Parks program intends to develop modern food processing facilities throughout the supply chain, from farm to market. These parks prevent food waste by upgrading storage and processing facilities. They encourage local farmers to negotiate higher prices for their products, eliminate post-harvest losses, and boost the rural economy.

India's food security programs prioritize assuring access to inexpensive food, increasing nutrition, and boosting agricultural output. Successful programs, such the PDS, NFSA, ICDS, and NREGA, have helped vulnerable communities improve their food security significantly.

4.1 Instances of Regulatory Failure

There have been instances where regulatory failings harmed the system's effectiveness. Here are some notable examples of these failures:

One major difficulty with PDS is the leakage of subsidized food grains meant for the poor. Reports of huge quantities being syphoned off into the open market are prevalent. A large portion of food grains never reach its intended recipients, undermining food security efforts.

The NFSA's targeting mechanism can either omit or include eligible individuals, leading to inaccuracies. Poor beneficiary identification means that some of the most disadvantaged groups do not receive benefits, while relatively well-off persons may receive food subsidies unnecessarily. For example, in many states, due to outdated or erroneous statistics, a considerable number of the qualifying population has been denied food entitlements, particularly during situations such as the COVID-19 epidemic.

Although the Mid-Day Meal Scheme has been mainly effective, there have been instances of corruption and inadequate quality control, resulting in substandard or hazardous food being supplied to kids. Poor hygienic procedures in food preparation have resulted in food poisoning occurrences, endangering children's health and safety. Corruption, particularly in procurement, has resulted in the waste of public money and lowered the quality of food delivered.

The Food Corporation of India (FCI), in charge of acquisition, storage, and distribution of food grains, has been criticised for mismanagement, notably in warehousing. Poor storage facilities cause food grains to rot and waste, limiting the effectiveness of procurement and food security strategies. The difficulty to effectively store grains has resulted in periodic shortages and price inflation in some areas.

The government's effort to transition from in-kind food distributions to cash payments via the Direct Benefit Transfer (DBT) system has encountered various hurdles. Many localities lack DBT infrastructure, such as bank branches and digital payment systems, especially in rural and tribal communities. Payment delays, technical difficulties, and beneficiaries' lack of financial knowledge have all contributed to decreased adoption rates and food insecurity in the interim.

NREGA aims to improve food security through employment, but there have been reports of mismanagement, wage delays, and corruption. Delays in wage payments frequently leave rural households without income for extended periods of time, limiting their capacity to buy food. Corruption in the system results in salaries being syphoned off or bogus job cards being issued, preventing genuine recipients from finding work.

The ICDS program, which provides nutrition and health services to children and mothers, has had issues with coverage and service quality. In many cases, the Anganwadi centers that provide ICDS services are understaffed, underfunded, or poorly managed. As a result, many children and pregnant women do not receive adequate nourishment, worsening malnutrition among vulnerable groups.

The COVID-19 lockout caused interruptions in the supply chain for food grains and delayed distribution of benefits under food security programs such as PDS and PMGKAY. Migrant workers and daily wage laborers struggled to obtain food owing to distribution system interruptions, despite the government's efforts to offer free food grains. Inadequate migrant databases and the inability to link ration cards between states exacerbated the situation.

To improve distribution efficiency, some states have decentralized food grain procurement. However, implementation has been hampered by bureaucratic delays and inefficiency. These inefficiencies induce delays in the procurement and delivery of food grains, resulting in food shortages during critical times. In other cases, state governments have struggled to coordinate with the national government, resulting in food shortages. For example, in states such as Uttar Pradesh and West Bengal, decentralized procurement plans have been delayed due to poor coordination among various levels of government, resulting in recurrent food shortages in rural areas.

While India's food security framework has had some notable triumphs, regulatory failures such as leaks, corruption, inadequate storage infrastructure, and bureaucratic slowness have hampered its efficacy. These problems underline the importance of stronger governance, transparency, and technological integration (FAO, n.d.) in improving the reliability and accessibility of food security programs across the country.

4.2 Lessons Learned From Both Successes and Failures

The food regulatory mechanisms in India have taught us some significant lessons, both from their achievements and failings. These lessons can help shape future policy decisions to create more effective and inclusive food security systems.

4.2.1 Lessons from Success

I. Targeted subsidies can significantly increase food access: When properly targeted, food subsidies can directly enhance food security for poor groups. To ensure that the correct persons receive benefits, beneficiary lists must be updated on a regular basis.
II. Integrating Health and Nutrition with Food Security initiatives: Linking food distribution to broader social services such as healthcare and education improves the effectiveness of food security initiatives. Holistic approaches that cover nutrition, health, and education can be more effective in combating food insecurity.
III. Community-Based and Decentralized Models Can Be Effective: Decentralized food systems that incorporate local communities and promote local food crops (such as millets) can boost resistance to climate change and assure the long-term viability of food security initiatives.

IV. Technology Can Improve Transparency and Efficiency: Using technology (such as digital identity, e-governance, and data management systems) can help eliminate leaks, increase transparency, and expand the reach of food security initiatives.
V. Legal Frameworks and Right-Based Approaches Strengthen food security: Rights-based initiatives that are backed by legislation regard food security as a fundamental right, fostering accountability and assuring long-term commitment from the government.

4.3.2 Lessons from Failures

I. Corruption Undermine Policy Effectiveness: Strong monitoring methods, increased transparency, and anti-corruption measures are essential for preventing leaks in food distribution programs. Independent audits and stricter enforcement of regulations are required.
II. Targeting errors can lead to exclusion. To ensure that the most vulnerable groups do not miss out on food security initiatives, databases must be updated on a regular basis and targeting mechanisms improved (such as using real-time data and surveys).
III. Inadequate Infrastructure Causes Food Waste: Investment in modern food storage, supply chain infrastructure, and cold storage is required to reduce post-harvest losses and increase overall food distribution system efficiency.
IV. Implementation gaps lead to inefficiencies. Stronger implementation procedures, improved cooperation between national and state governments, and more effective local governance are crucial to ensure that food security programs run smoothly. A clearer communication and accountability system is also required.
V. Direct Benefit Transfers Require Infrastructure Support: Before moving to cash transfer models like DBT, suitable infrastructure must be in place, such as financial inclusion (bank accounts) and digital literacy among beneficiaries. Phased implementation and localized testing of new systems helps prevent exclusion and promote a smoother transition.
VI. Quality control is critical in food programs. Stringent monitoring, quality inspections, and hygiene standards must be implemented in food programs, particularly those involving vulnerable populations such as children. Employees involved in food preparation should also receive proper training and capacity building.

VII. Social and culture Considerations: Food security strategies must be customized to local culture and culinary preferences. Community involvement in decision-making processes can boost the acceptability and success of food security programs.

VIII. Emergency Preparedness is essential: A more robust food security system must include emergency preparedness measures, such as mobile ration cards, to ensure that disadvantaged populations have access to food during crises. Flexible and adaptive policy frameworks are required to deal with unexpected catastrophes such as pandemics or natural disasters.

IX. The success of India's food security initiatives demonstrates the value of well-targeted subsidies, technological integration, community-based models, and rights-based strategies. However, flaws in corruption control, inadequate infrastructure, exclusion errors, and implementation inefficiencies provide important lessons for future improvement. Moving forward, India must prioritize better governance, stronger infrastructure, and inclusive measures to achieve long-term and equitable food security for all.

V INTERNATIONAL LEGAL FRAMEWORK

Various countries have established legal frameworks to address food security, focusing on ensuring access to affordable, nutritious food for all citizens. Here is a brief overview of food security legal frameworks in different countries:

Brazil initiated the Zero Hunger Program using the legal framework of the National Food and Nutrition Security Law (LOSAN), which was passed in 2006 and established food security as a constitutional right. The National Council on Food and Nutrition Security (CONSEA) was established, with members from the government and civil society. The focus is on policies such as Bolsa Família (conditional cash transfer program) and food acquisition programs, which purchase products from small-scale farmers to serve public institutions like schools and hospitals. Brazil has significantly decreased hunger and malnutrition, setting a global example of integrated food security policies.

South Africa's Constitution (1996) clearly recognizes the right to food in Section 27, which ensures access to adequate food and water. South Africa's Integrated Food Security Strategy (IFSS) seeks to guarantee that vulnerable communities have access to food. Government programs such as the National School Nutrition Program (NSNP) offer free lunches to schoolchildren. Despite legal safeguards, the country continues to suffer food insecurity challenges caused by poverty, inequality, and food distribution inefficiencies.

In the United States of America, the Supplemental Nutrition Assistance (Roberts et al., 2021) Program, is governed by the Food and Nutrition Act of 2008 enacted by U.S. It offers financial assistance to purchase food via an E-Benefits Transfer card. The program is supported by many contributors and organizations (U.S. Government Accountability Office, 2019; Academy of Nutrition and Dietetics, 2021).

In the European Union, the Common Agricultural Policy was founded in 60's, was a set of policies that oversee food production, assuring food security and agricultural sustainability. The program provides financial assistance to farmers and enforces food safety requirements to ensure food supply. The policy includes food aid programs such as the funds for aid to most deprived association which provides disadvantaged communities with food and other necessities.

Canada provides the Right to Food and Social Safety Nets. While there are no explicit food security legislation in Canada, food security is handled through various social welfare measures, including the Canada Child Benefit (CCB) and provincial assistance programs. The Canada Food Guide promotes healthy eating (Berkowitz et al., 2018), while programs such as Nutrition North Canada (NNC) subsidize food in remote northern communities. Food banks and school lunch programs support these initiatives to alleviate food insecurity. Food insecurity remains prevalent among Indigenous peoples and low-income households due to institutional imbalances.

The global experience with legislative frameworks for food security emphasises the significance of integrating food access with broader social welfare systems, maintaining transparency, encouraging local agriculture, and incorporating community-based approaches. Countries that make food security a legal right have more effective and responsible anti-hunger and malnutrition programs. Even with legislative frameworks in place, many people experience difficulties due to economic disparities, infrastructure deficiencies, and governance issues.

5.1 Application of Global Strategies and International Conventions in Indian context

India has actively participated in a number of international treaties and agreements relating to food security, hunger eradication, and nutrition (U.S. Department of Agriculture, 2021). India has signed and approved the following significant international food security conventions and agreements:

5.1.1 **Universal Declaration on the Eradication of Hunger and Malnutrition (1974):**

This proclamation, implemented during the Food Conference held in 1974, emphasizes that everyone has right to be secured against hunger and undernourishment. It emphasizes the international community's responsibility to take the necessary steps to secure food security for everyone. India has endorsed and approved this declaration, committed to worldwide efforts to alleviate deprivation of food and malnutrition as right of human being.

5.1.2 **International Covenant on Economic, Social and Cultural Rights (ICESCR) (1966):**

Article 11 of the ICESCR recognizes everyone's right to a sufficient standard of living, which includes food. It also emphasizes the right to be without hunger. India adopted the ICESCR in 1979, aligning its national policies to provide residents' access to adequate food, shelter, and healthcare.

5.1.3 **World Food Summit (1996)– Rome Declaration on World Food Security:**

The Rome Declaration, adopted at the Food and Agriculture Organization's (FAO) World Food Summit, reaffirmed governments' commitment to attaining universal food security. It intends to cut the number of undernourished persons in half by 2015. India endorsed the declaration, pledging its support for global efforts to combat hunger and malnutrition through sustainable agriculture and poverty alleviation initiatives.

5.1.4 **Sustainable Development Goals (SDGs) (2015):**

The SDGs, particularly Goal 2: Zero Hunger, aim to eradicate hunger, ensure food security, improve nutrition, and promote sustainable agriculture by 2030. India is a signatory to the SDGs and has made significant efforts to achieve them through initiatives such as the National Food Security Act (2013) and agricultural reforms targeted at increasing food production and distribution.

5.1.5 International Treaty on Plant Genetic Resources for Food and Agriculture (ITPGRFA) (2001):

This convention seeks to conserve and sustainably use plant genetic resources for food and agriculture, while also assuring food security through the exchange of genetic materials and technologies. India ratified the ITPGRFA in 2002, so contributing to global efforts to conserve biodiversity and promote sustainable agriculture.

5.1.6 World Trade Organization (WTO) Agreement on Agriculture (AoA) (1995):

The WTO Agreement on Agriculture aims to develop more equitable agricultural trade practices, safeguarding food security by lowering trade barriers and prohibiting subsidies that distort food prices. India is a signatory to the WTO Agreement on Agriculture, and it has negotiated to safeguard its food security interests by pressing for public stockholding programs and other measures that maintain steady access to food for its people.

5.1.7 Sendai Framework for Disaster Risk Reduction (2015):

The Sendai Framework focusses on disaster risk reduction, including strengthening the resilience of agricultural systems and food supply networks in the face of natural catastrophes, climate change, and other crises. India signed the Sendai Framework and has taken initiatives to strengthen its food systems through disaster risk management strategies such as enhanced irrigation practices and drought-resistant crops.

India's ratification and involvement in several international accords demonstrate its commitment to global food security (van der Voort, 2015) and combating poverty and malnutrition. These frameworks also have an impact on India's domestic policy, notably in terms of achieving equitable, long-term food security that is robust to challenges such as climate change, poverty, and inequality (Smyth, 2020).

VI POLICY RECOMMENDATIONS

6.1 Strengthening the Existing Legal Framework

To promote food security, the current legal framework should be strengthened through comprehensive laws that protect food rights and ensure equitable distribution. Governments must focus on connecting national laws with international obligations like the FAO's Right to Food. Emerging concerns such as climate change and market volatility should be addressed through regular modifications and upgrades. Legal reforms should also strive to enhance land tenure, promote sustainable agriculture practices, and enforce anti-food waste laws.

6.2 Enhancing Administrative Efficiency and Transparency

The successful execution of food security programs is dependent on reduced administrative processes and improved governance (FAO, n.d.). Governments should implement measures to decrease bureaucratic delays and corruption in food distribution networks. This involves transparent procurement methods, timely resource allocation, and rigorous supply chain monitoring. Public participation should be encouraged to ensure accountability, and local governments should be given the resources and decision-making authority they need to handle food security issues on the ground.

6.3 Leveraging Technology for Better Implementation

Technology is crucial to furthering food security projects. Policymakers should promote the use of digital systems for real-time tracking of food production, delivery, and storage. Investing in agricultural advances like precision farming, data analytics, and mobile applications for farmers can boost efficiency and productivity. Furthermore, blockchain technology can be used to trace food supply chains, increasing transparency, lowering losses, and eliminating fraud. Governments should also make internet infrastructure accessible to rural and marginalised areas.

6.4 Ensuring Socio-Economic Inclusivity

Food security policy must prioritise the needs of vulnerable groups, such as low-income families, women, and marginalised communities. Food coupons, subsidies, and cash transfers should be established to help these communities cope financially. It is critical to have inclusive agricultural policies that ensure equal access to resources, credit, and training for small-scale farmers. Addressing fundamental disparities in the food chain can help ensure that all people benefit from food security initiatives.

6.5 Promoting Political Commitment and Consistency

Sustained political will is required for food security programs to be effective over time. Governments should demonstrate a strong commitment by prioritising food security in their national development plans. Consistency in policy execution across political cycles is required to sustain momentum and avoid disruption.

Furthermore, international engagement and regional partnerships can promote a cooperative approach to addressing global food security concerns. Leaders must also maintain constant engagement with stakeholders, including civil society, the commercial sector, and international organisations, to guarantee policy alignment and adaptation. These recommendations, if implemented holistically, will help create a robust and sustainable system to achieve food security for all.

VII CONCLUSION

International legal framework for food security, principally based on the National Food Security Act (2013) in India and globally according to various food safety provisions by International Lawis broad in scope and goal (Santo & Moragues-Faus, 2019). It tries to provide adequate food at reasonable prices to its large population, particularly the disadvantaged. However, while the framework is well-designed, implementing its regulatory processes presents major obstacles.

These difficulties originate from a variety of issues, including bureaucratic inefficiency, corruption, limited infrastructure, and the complications of coordinating between the federal and state governments. The Public Distribution System (PDS), for example, suffers from leakage, inefficiency, and exclusion errors, restricting its ability to give food to those in most need. Furthermore, while the digitalization and identification procedures have been progressive, they have not adequately addressed the concerns of marginalized communities, such as migrant workers and those who lack valid papers.

Furthermore, the interaction of many legislative rules, such as those governing agriculture, trade, and public health, results in a complicated regulatory framework that frequently impedes rather than promotes good administration. External factors such as climate change, shifting agricultural productivity, and global market dynamics can all worsen food insecurity and complicate regulatory compliance, putting the legal structure under strain.

To summarize, while legal framework for food security is sound, the effectiveness of its regulatory procedures is dependent on overcoming severe challenges. Strengthening governance, increasing transparency, addressing systemic inefficiencies, and encouraging more collaboration among stakeholders are critical elements towards achieving food security goals and ensuring that no citizen is left behind.

REFERENCES

Academy of Nutrition and Dietetics. (2021). Academy of Nutrition and Dietetics: Increased SNAP funding will improve nation's health.https://www.prnewswire.com/news-releases/academy-of-nutrition-and-dietetics-increased-snap-funding-will-improve-nations-health-301358285.html

Barlow, P., Loopstra, R., Tarasuk, V., & Reeves, A. (2020). Liberal trade policy and food insecurity across the income distribution: An observational analysis in 132 countries, 2014-17. *The Lancet. Global Health, 8*(8), e1090–e1097. DOI: 10.1016/S2214-109X(20)30263-1 PMID: 32710865

Berkowitz, S. A., Seligman, H. K., Meigs, J. B., & Basu, S. (2018). Food insecurity, healthcare utilization, and high cost: A longitudinal cohort study. *The American Journal of Managed Care, 24*(9), 399–404. PMID: 30222918

Blankenship, J., & Blancato, R. B. (2022). Nutrition Security at the Intersection of Health Equity and Quality Care. *Journal of the Academy of Nutrition and Dietetics, 122*(10, 10S), S12–S19. DOI: 10.1016/j.jand.2022.06.017 PMID: 36122954

Breeman, G., Giest, S., & Rimkutė, D. (2017). Food Security and the Sustainability of GMOs in the United States and the European Union. In *Advances in Food Security and Sustainability* (pp. 165–193). Elsevier., DOI: 10.1016/bs.af2s.2017.09.005

Clapp, J., Moseley, W. G., Burlingame, B., & Termine, P. (2022). Viewpoint: The case for a six-dimensional food security framework. *Food Policy, 106*, 102164. DOI: 10.1016/j.foodpol.2021.102164

FAO. (n.d.). *Legal processes, the right to food.* Food and Agriculture Organization of the United Nations. (n.d.). https://www.fao.org/right-to-food/areas-of-work/legal-processes/en/

Hoffmann, V., Moser, C., & Saak, A. (2019). Food safety in low and middle-income countries: The evidence through an economic lens. In *World Development, 123* (Vol. 104611). Elsevier BV., DOI: 10.1016/j.worlddev.2019.104611

Klassen, S., & Murphy, S. (2020). Equity as both a means and an end: Lessons for resilient food systems from COVID-19. In *World Development, 136, 105104*. Elsevier BV., DOI: 10.1016/j.worlddev.2020.105104

Li, J., & Song, W. (2022). Food Security Review Based on Bibliometrics from 1991 to 2021. *Foods, 11*(23), 3915. DOI: 10.3390/foods11233915 PMID: 36496722

Martin-Shields, C. P., & Stojetz, W. (2019). Food security and conflict: Empirical challenges and future opportunities for research and policy making on food security and conflict. In *World Development. 119, 150–164*. Elsevier BV., DOI: 10.1016/j.worlddev.2018.07.011

Nordin, S. M., Boyle, M., & Kemmer, T. M. (2013). Academy of Nutrition and Dietetics. Position of the academy of nutrition and dietetics: nutrition security in developing nations: sustainable food, water, and health. *Journal of the Academy of Nutrition and Dietetics*, *113*(4), 581–595. DOI: 10.1016/j.jand.2013.01.025 PMID: 23522578

PUCL v. Union of India. (2001). AIR 2001 SC 568.

Roberts, S. B., Silver, R. E., Das, S. K., Fielding, R. A., Gilhooly, C. H., Jacques, P. F., Kelly, J. M., Mason, J. B., McKeown, N. M., Reardon, M. A., Rowan, S., Saltzman, E., Shukitt-Hale, B., Smith, C. E., Taylor, A. A., Wu, D., Zhang, F. F., Panetta, K., & Booth, S. (2021, July). Healthy aging—nutrition matters: Start early and screen often. *Advances in Nutrition*, *12*(4), 1438–1448. DOI: 10.1093/advances/nmab032 PMID: 33838032

Santo, R., & Moragues-Faus, A. (2019). Towards a trans-local food governance: Exploring the transformative capacity of food policy assemblages in the US and UK. In *Geoforum, 98, 75–87*. Elsevier BV., DOI: 10.1016/j.geoforum.2018.10.002

Sellberg, M. M., Norström, A. V., Peterson, G. D., & Gordon, L. J. (2020). Using local initiatives to envision sustainable and resilient food systems in the Stockholm city-region. In *Global Food Security, 24, 100334*. Elsevier BV., DOI: 10.1016/j.gfs.2019.100334

Shafiee, M., Keshavarz, P., Lane, G., Pahwa, P., Szafron, M., Jennings, D., & Vatanparast, H. (2022). Food Security Status of Indigenous Peoples in Canada According to the 4 Pillars of Food Security: A Scoping Review. In *Advances in Nutrition, 13(6), 2537–2558*. Elsevier BV., DOI: 10.1093/advances/nmac081

Simola, A., Boysen, O., Ferrari, E., Nechifor, V., & Boulanger, P. (2022). Economic integration and food security – The case of the AfCFTA. In *Global Food Security, 35, 100651*. Elsevier BV., DOI: 10.1016/j.gfs.2022.100651

Smyth, S. J. (2020). Regulatory barriers to improving global food security. In *Global Food Security, 26, 100440*. Elsevier BV., DOI: 10.1016/j.gfs.2020.100440

U.S. Department of Agriculture. (2021). USDA releases study on hurdles to healthy eating in SNAP.https://www.usda.gov/media/press-releases/2021/06/23/usda-releases-study-hurdles-healthy-eating-snap

U.S. Government Accountability Office. (2019). Nutrition assistance programs: Agencies could do more to help address the nutritional needs of older adults.https://www.gao.gov/products/gao-20-18

van der Voort, H. (2015). Co-Regulatory Failure in the Food Industry: Explaining Regulatory Failure by Means of Two Contrasting Interpretations of Governance. *European Journal of Risk Regulation*, *6*(4), 502–511. DOI: 10.1017/S1867299X00005067

Wang, S., Wu, H., Li, J., Xiao, Q., & Li, J. (2024). Assessment of the Effect of the Main Grain-Producing Areas Policy on China's Food Security. *Foods*, *13*(5), 654. DOI: 10.3390/foods13050654 PMID: 38472767

Chapter 17
Future Horizons:
Leveraging Energy-Efficient Technologies for a Sustainable Food Industry

Dipali Saxena
https://orcid.org/0000-0002-8177-319X
Shri Vaishnav Institute of Home Science, Shri Vaishnav Vidyapeeth Vishwavidyalya, India

Shweta Keswani
Shri Vaishnav Institute of Home Science, Shri Vaishnav Vidyapeeth Vishwavidyalaya, India

Manisha Trivedi
Shri Vaishnav Institute of Home Science, Shri Vaishnav Vidyapeeth Vishwavidyalaya, India

ABSTRACT

This chapter examines the role of energy-efficient technologies in advancing a sustainable food industry by improving food safety, quality, and security. As global food systems confront challenges such as climate change, resource scarcity, and rising demand, the integration of energy-saving solutions is vital for enhancing the efficiency and resilience of food production, processing, and distribution. Looking to the future, the chapter explores emerging innovations such as artificial intelligence, machine learning, and the Internet of Things (IoT), which offer new opportunities to optimize food systems for energy efficiency. The chapter concludes with recommendations for policy development and global collaboration to foster the adoption of energy-efficient technologies across the food sector.

DOI: 10.4018/979-8-3693-8342-1.ch017

1. INTRODUCTION

The global food industry is facing an unprecedented set of challenges, including the impacts of climate change, increasing resource scarcity, growing food demand, and the need to ensure food security and safety. As the world's population continues to expand, the pressure on food systems to meet these demands in a sustainable manner has never been greater. Energy consumption in the food industry, from production and processing to transportation and storage, represents a significant portion of the global energy use, contributing to both economic and environmental challenges.

In response, the adoption of energy-efficient technologies is emerging as one of the most promising solutions to address these challenges. By improving the efficiency of food systems, reducing waste, and optimizing resource use, energy-efficient technologies have the potential to create a more sustainable and resilient food industry. This chapter aims to explore key energy-efficient technologies, their applications, and their role in advancing a sustainable food system. Moreover, it examines the future potential of emerging innovations, such as artificial intelligence (AI), machine learning (ML), and the Internet of Things (IoT), as well as circular economy models that promote resource reuse.

2. ENERGY-EFFICIENT TECHNOLOGIES IN FOOD PRODUCTION AND PROCESSING

2.1 Energy-Efficient Refrigeration Systems

Refrigeration is one of the most energy-intensive processes in the food industry. It is vital for preserving perishable foods and ensuring food safety, but traditional refrigeration systems often use inefficient technologies and contribute significantly to global greenhouse gas emissions. However, advancements in refrigeration technologies have led to the development of energy-efficient systems that utilize natural refrigerants, energy recovery mechanisms, and smart control systems to reduce energy consumption.

2.1.1 Natural Refrigerants

Refrigerants such as ammonia, carbon dioxide (CO2), and hydrocarbons offer significantly lower global warming potentials (GWP) compared to synthetic refrigerants like hydrofluorocarbons (HFCs). Ammonia and CO2 are widely used in large-scale food storage and distribution, providing efficient and environmentally-friendly cooling solutions (Mohammed et al., 2021).

2.1.2 Energy Recovery Systems

Many modern refrigeration systems incorporate energy recovery features, such as heat exchangers, which capture excess heat from the cooling process and reuse it for other operations like heating water or air. This integrated approach improves energy efficiency and reduces the need for additional energy inputs (Zhou et al., 2023).

2.1.3 Smart Controls and Sensors

Advanced refrigeration systems integrate smart sensors and control systems that monitor temperature, humidity, and system performance in real-time. These systems can automatically adjust operational parameters to ensure that refrigeration units run at optimal efficiency, reducing energy waste and maintaining food quality (Patel et al., 2023).

2.2 Renewable Energy Integration

The integration of renewable energy sources into food production and processing systems offers a sustainable alternative to traditional fossil fuel-based energy. By utilizing solar, wind, and biomass energy, food manufacturers can reduce their carbon footprint and lower energy costs.

2.2.1 Solar Energy

Solar photovoltaic (PV) systems can be used to generate electricity for food production facilities. Rooftop solar installations are particularly beneficial for food processing plants, where large energy demands for refrigeration, cooking, and drying processes can be met by solar power. Additionally, solar thermal systems can provide heat for processes like pasteurization and drying (Chen et al., 2022).

2.2.2 Biomass Energy

Biomass energy, derived from organic waste products from food production, such as agricultural residues, food scraps, and other organic materials, can be used to generate heat and power. Biomass-based combined heat and power (CHP) systems are increasingly used in food manufacturing to reduce reliance on conventional energy sources and utilize waste materials more efficiently (Jia et al., 2022).

2.2.3 Wind Energy

In regions where wind energy is abundant, food processing plants can integrate wind turbines to generate electricity. Wind energy can complement solar and biomass systems, providing a reliable and renewable energy source for 24/7 operations (Chen et al., 2022).

3. EMERGING INNOVATIONS IN ENERGY EFFICIENCY

3.1 Artificial Intelligence and Machine Learning

Artificial intelligence (AI) and machine learning (ML) technologies are rapidly transforming the way energy efficiency is achieved in the food industry. By analyzing vast amounts of data from production lines, AI can optimize processes in real-time, identify inefficiencies, and suggest improvements. AI algorithms can continuously analyze energy consumption patterns and optimize parameters such as temperature, speed, and pressure to minimize energy use. For example, in baking, AI can adjust oven temperatures and baking times based on product load, thereby reducing energy consumption (Ghosh et al., 2023). AI-powered predictive maintenance tools can identify equipment faults before they cause failures, reducing downtime and ensuring that machinery operates at peak efficiency. In food processing, this can lead to substantial energy savings by preventing energy waste due to malfunctioning or inefficient equipment (Singh & Kumar, 2022).

3.2 Internet of Things (IoT)

The Internet of Things (IoT) is another groundbreaking technology with the potential to improve energy efficiency in the food industry. IoT devices, such as smart sensors and connected equipment, allow food producers to monitor and control energy use in real time.

IoT sensors can track various environmental factors, such as temperature, humidity, and machine performance, and transmit this data to central control systems. These systems can automatically adjust settings based on real-time data, reducing energy use when demand is low (Liu et al., 2021). IoT-based Energy Management Systems can provide insights into energy consumption trends, helping food manufacturers make data-driven decisions on energy optimization. EMS can also facilitate predictive analytics, helping companies anticipate peak energy demands and take preventive measures to reduce overall consumption (Ghosh et al., 2023).

3.3 Sustainable Packaging and Waste Reduction

Sustainable packaging plays a crucial role in reducing the overall environmental impact of food production and distribution. Energy-efficient packaging solutions reduce the amount of energy needed for production, transportation, and disposal.

Innovative packaging materials, such as biodegradable plastics, recyclable materials, and edible packaging, are gaining traction in the food industry. These materials not only reduce waste but also require less energy to produce compared to traditional plastic and metal packaging (Tan & Dodd, 2023). Minimizing packaging size not only reduces material waste but also cuts transportation energy costs. By adopting smaller, more efficient packaging designs, companies can optimize storage and reduce the overall carbon footprint associated with packaging and distribution (Brennan et al., 2022).

Circular economy principles are being adopted in the food industry, promoting the recycling of packaging materials and reducing the need for new resources. Companies are increasingly focused on reusing materials such as glass, cardboard, and plastics, which not only reduces waste but also saves energy in the packaging process (Patel et al., 2023).

3.4 Circular Economy Models in the Food Industry

A circular economy model focuses on the continual use of resources by promoting the reuse, recycling, and regeneration of materials. This model has significant potential for the food industry, which traditionally follows a linear model of production, consumption, and disposal.

Circular models encourage the recovery of energy and materials from food waste, which can then be reintegrated into the production process. For example, organic waste from food production can be converted into bioenergy, which can power the plant and reduce dependence on external energy sources (Jia et al., 2022). Circular economy principles also involve designing food products and packaging for longer lifespans, reducing the need for frequent replacements and minimizing energy consumption in production and transportation (Tan & Dodd, 2023).

4. POLICY DEVELOPMENT AND GLOBAL COLLABORATION

To foster the widespread adoption of energy-efficient technologies in the food industry, it is essential for governments, international organizations, and the private sector to collaborate. Policies that incentivize energy efficiency, promote innovation, and encourage sustainability can drive positive change.

Governments can provide financial incentives, such as subsidies and tax credits, to businesses that invest in energy-efficient technologies. These policies can help reduce the initial investment costs, making energy-efficient solutions more accessible to small and medium-sized food businesses (ISO, 2023). Collaboration between countries, industries, and academic institutions is crucial to accelerate the development and adoption of innovative energy-efficient technologies. Shared research on best practices, technical standards, and policy frameworks can help create a more unified and effective approach to sustainability in the food industry (Zhou et al., 2023).

5. CONCLUSION

As global food systems face increasing pressure to meet growing demands while reducing environmental impacts, energy-efficient technologies are playing an essential role in shaping a more sustainable and resilient food industry. From energy-efficient refrigeration and renewable energy integration to emerging innovations like AI, IoT, and circular economy models, the food industry has a range of opportunities to improve energy efficiency and sustainability. By adopting these technologies and fostering global collaboration, the food industry can contribute to a more sustainable future while ensuring food safety, quality, and security for generations to come.

REFERENCES

Brennan, J. G., O'Callaghan, P., & Ghosh, A. (2022). Innovations in sustainable packaging for food safety and energy efficiency. *Journal of Food Science*, *88*(3), 1157–1172.

Chen, J., Li, X., & Zhang, W. (2022). Integrating renewable energy systems into food production facilities. *Renewable Energy*, *135*, 349–360.

Ghosh, A., Srivastava, A., & Bansal, S. (2023). Harnessing AI and IoT for energy optimization in the food industry. *Energy Efficiency*, *16*(4), 267–284.

Jia, X., Wu, Q., & Wang, R. (2022). Biomass energy applications in food processing. *Renewable & Sustainable Energy Reviews*, *146*, 111062.

Liu, Y., Xu, D., & Zhang, F. (2021). The role of smart sensors and IoT in reducing energy consumption in the food industry. *Sensors (Basel)*, *21*(8), 2716. PMID: 33921508

Mohammed, M. S., Yousif, F., & Rehman, S. (2021). Energy-efficient refrigeration systems in the food sector: A review of technologies and applications. *Journal of Refrigeration Science*, *54*(2), 85–99.

Patel, S., Kumar, N., & Ahuja, A. (2023). Building energy management systems for energy efficiency in food production facilities. *Energy Efficiency*, *16*(1), 1–19.

Singh, R., & Kumar, P. (2022). Energy-efficient motors and drives in the food industry: Trends and challenges. *Industrial Electronics*, *68*(11), 9200–9214.

Tan, D., & Dodd, T. (2023). Circular economy models for sustainable packaging in the food industry. *Sustainable Development*, *31*(2), 115–128.

Zhou, X., Wu, Q., & Chen, Z. (2023). Advancements in energy recovery technologies for food production systems. *Food Science and Technology (Campinas)*, *56*(1), 45–59.

Compilation of References

Aayush, K., McClements, D. J., Sharma, S., Sharma, R., Singh, G. P., Sharma, K., & Oberoi, K. (2022). Innovations in the development and application of edible coatings for fresh and minimally processed Apple. *Food Control*, *141*, 109188. DOI: 10.1016/j.foodcont.2022.109188

Abbruzzese, M., Infante, D., & Smirnova, J. (2024). The Diffusion of Renewable Energy Production in European Countries: The Role of Incentives. *Economics of Innovation and New Technology*, (January), 1–24.

Abdel-Karim, R., Reda, Y., & Abdel-Fattah, A. (2020). Nanostructured materials-based nanosensors. *Journal of the Electrochemical Society*, *167*(3), 37554. DOI: 10.1149/1945-7111/ab67aa

Abeliotis, K., Zachos, F., & Lasaridi, K. (2014). A database tool for raising awareness on the life cycle environmental impacts of household packaging waste. *Journal of Computational Environmental Sciences*, *2014*, 1–6. DOI: 10.1155/2014/809102

Abera, Y. 2023. "Renewable Energy Integration with Sustainable Materials and Technologies." DOI: 10.22541/au.170013430.05838314/v1

Aboagye, I. A., Valappil, G., Dutta, B., Imbeault-Tétreault, H., Ominski, K. H., & Marcos, R. C. (2024). Cordeiro, Roland Kröbel, Sarah J. Pogue, and Tim A. McAllister. "An assessment of the environmental sustainability of beef production in Canada.". *Canadian Journal of Animal Science*, *104*(2), 221–240. DOI: 10.1139/cjas-2023-0077

Academy of Nutrition and Dietetics. (2021). Academy of Nutrition and Dietetics: Increased SNAP funding will improve nation's health.https://www.prnewswire.com/news-releases/academy-of-nutrition-and-dietetics-increased-snap-funding-will-improve-nations-health-301358285.html

Adam, T., & Gopinath, S. C. B. (2022). Nanosensors: Recent perspectives on attainments and future promise of downstream applications. *Process Biochemistry, 117*, 153–173. DOI: 10.1016/j.procbio.2022.03.024

Adamu, H., Bello, U., Yuguda, A. U., Tafida, U. I., Jalam, A. M., Sabo, A., & Qamar, M. (2023). Production processes, techno-economic and policy challenges of bioenergy production from fruit and vegetable wastes. *Renewable & Sustainable Energy Reviews, 186*, 113686. DOI: 10.1016/j.rser.2023.113686

Adisurya, I P K., & Sari, A R. (2022, January 1). Effects of High-Pressure Processing on Milk, Meat, Fruit, and Vegetable: A Review. Atlantis Press. DOI: 10.2991/absr.k.220305.046

Afif, K., Rebolledo, C., & Roy, J. (2021). Drivers, barriers and performance outcomes of sustainable packaging: A systematic literature review. *British Food Journal, 124*(3), 915–935. DOI: 10.1108/BFJ-02-2021-0150

Aghbashlo, M., & Taherzadeh, M. J. (2019). The role of renewable energy in food processing: Potential and opportunities for integration with EMS. *Renewable & Sustainable Energy Reviews, 103*, 311–324. DOI: 10.1016/j.rser.2018.12.046

Agoud, S., Benchrifa, M., Mabrouki, J., Moufti, A., & Slaoui, M. (2024). Study of Energy Production by the Anaerobic Digestion System of Waste Management. In *Technical and Technological Solutions Towards a Sustainable Society and Circular Economy* (pp. 67–76). Springer Nature Switzerland. DOI: 10.1007/978-3-031-56292-1_5

Ahmad, M. S., & Siddiqui, M. W. (2015). *Postharvest quality assurance of fruits*. Springer. DOI: 10.1007/978-3-319-21197-8

Ahmed, C. B., & Rouina, B. B. (2019) Tunisia: Water Resources' Management and Sustainability of Agriculture. Encyclopedia of Environmental Health page: 170-184. https://doi.org/DOI: 10.1016/B978-0-12-409548-9.11854-8

Ahmed, H. A., Yu-Xin, T., & Chang, Y. (2020). Optimal control of environmental conditions affecting lettuce plant growth in a controlled environment with artificial lighting: A review. *South African Journal of Botany, 130*, 75–89. DOI: 10.1016/j.sajb.2019.12.018

Ahmed, I. O., Abdelkader, A., Farrell, C., Rooney, D., & Morgan, K. (2019). Reusing, recycling and up-cycling of biomass: A review of practical and kinetic modelling approaches. *Fuel Processing Technology, 192*, 179–202. DOI: 10.1016/j.fuproc.2019.04.026

Ahuja, V. K. (2017). *Law relating to intellectual property rights*. LexisNexis.

Ajose, D. J., Abolarinwa, T. O., Oluwarinde, B. O., Montso, P. K., Fayemi, O. E., Aremu, A. O., & Ateba, C. N. (2022). Application of plant-derived nanoparticles (PDNP) in food-producing animals as a bio-control agent against antimicrobial-resistant pathogens. *Biomedicines*, *10*(10), 2426. DOI: 10.3390/biomedicines10102426 PMID: 36289688

Akshay Patil, R., Bhavana, A., & Roopa Patil, B. (2024). Cultured Meat: The Upcoming Meat Production having Sustainable Benefits over Conventional Meat Production: A Review. *Agricultural Reviews (Karnal)*, *45*(1).

Al Mamun, M., Wahab, Y. A., Hossain, M. A. M., Hashem, A., & Johan, M. R. (2021). Electrochemical biosensors with aptamer recognition layer for the diagnosis of pathogenic bacteria: Barriers to commercialization and remediation. *Trends in Analytical Chemistry*, *145*, 116458. DOI: 10.1016/j.trac.2021.116458

Al-Alam, J. (2017). *Polluants organiques: analyse, application au {\guillemotleft} biomonitoring{\guillemotright} environnemental et introduction des biopesticides (algues marines) comme alternative*. Université de Strasbourg; Université Libanaise.

Alamu, E O., & Mooya, A. (2017, January 1). Food Processing Technologies and Value Addition for Improved Food Safety and Security. Elsevier BV, 201-210. DOI: 10.1016/B978-0-12-810521-4.00010-4

Alegbeleye, O., Odeyemi, O. A., Strateva, M., & Stratev, D. (2022). Microbial spoilage of vegetables, fruits and cereals. *Applied Food Research*, *2*(1), 100122. DOI: 10.1016/j.afres.2022.100122

Alharbi, S. S., Al Mamun, M., Boubaker, S., & Rizvi, S. K. A. (2023). Green finance and renewable energy: A worldwide evidence. *Energy Economics*, *118*, 106499.

Ali, S., Dash, N., & Pradhan, A. (2012). Role of renewable energy on agriculture. *International Journal of Engineering Sciences Emerging Technologies*, *4*(1), 51–57.

Alizadeh, A M., Masoomian, M., Shakooie, M., Khajavi, M Z., & Farhoodi, M. (2020, September 14). Trends and applications of intelligent packaging in dairy products: a review. Taylor & Francis, 62(2), 383-397. DOI: 10.1080/10408398.2020.1817847

Allahverdiyev, E. (2023). Use of renewable energy resources in agriculture. *Voprosy Istorii (Moscow, Russia)*, *2*(2), 250–255.

Allison, A., Lorencatto, F., Michie, S., & Miodownik, M. (2021). Barriers and enablers to buying biodegradable and compostable plastic packaging. *Sustainability (Basel)*, *13*(3), 1463. DOI: 10.3390/su13031463

Almeida, R. M., et al. (2022). *The use of solar energy in dairy processing: A case study of energy savings*. Journal of Clean Energy Technologies.

Al-Salem, S. M., Lettieri, P., & Baeyens, J. (2009). Recycling and recovery routes of plastic solid waste (PSW): A review. *Waste Management (New York, N.Y.)*, *29*(10), 2625–2643. DOI: 10.1016/j.wasman.2009.06.004 PMID: 19577459

Alwahab, A., Maharani, P. T., Nur K, W. O. M., Ahmad, L. O., Alimin, A., & Zaeni, A. (2024). Optimization of Biogas Liquid Waste from Livestock Manure as a Source of Renewable Energy through Microbial Fuel Cell (MFC) Technology. *Advance Sustainable Science, Engineering and Technology*, *6*(2), 0240202. DOI: 10.26877/asset.v6i2.17931

Amit, S. K., Uddin, M. M., Rahman, R., Islam, S. M. R., & Khan, M. S. (2017). A review on mechanisms and commercial aspects of food preservation and processing. *Agriculture \&. Food Security*, *6*, 1–22.

Anant, T. V., & Reddy, M. C. V. (2015). *Sustainable food systems: Intellectual property and energy efficiency in the food sector*. Springer.

Anderson, S. W., Daly, J. D., & Johnson, M. F. (1999). Why firms seek ISO 9000 certification: Regulatory compliance or competitive advantage? *Production and Operations Management*, *8*(1), 28–43. DOI: 10.1111/j.1937-5956.1999.tb00059.x

Andrés-Bello, A., Barreto-Palacios, V. I. V. I. A. N., García-Segovia, P., Mir-Bel, J., & Martinez-Monzo, J. (2013). Effect of pH on color and texture of food products. *Food Engineering Reviews*, *5*(3), 158–170. DOI: 10.1007/s12393-013-9067-2

Ani, O. I., Aniokete, T., & Agbo, A. (2024). Assessing Potential of Biogas: Harnessing Sustainable Energy from Biomass for Renewable Solutions. Al-Rafidain Journal of Engineering Sciences, 330-349.

Anquez, E., Raab, K., Cechella, F., & Wagner, R. (2022). Consumers' perception of sustainable packaging in the food industry. *Revista Direitos Culturais*, *17*(41), 251–265. DOI: 10.20912/rdc.v17i41.728

An, X., Liu, P., Meng, Q., Su, C., & Zhao, S. (2014). Research on life cycle environmental property of biomass biodegradable packaging material. *Applied Mechanics and Materials*, *670-671*, 1620–1623. . DOI: 10.4028/www.scientific.net/AMM .670-671.1620

Aqilah, N. M. N., Rovina, K., Felicia, W. X. L., & Vonnie, J. M. (2023). A review on the potential bioactive components in fruits and vegetable wastes as value-added products in the food industry. *Molecules (Basel, Switzerland)*, *28*(6), 2631. DOI: 10.3390/molecules28062631 PMID: 36985603

Ariwaodo, C. A. (2022). Handling strategies and facilities for horticultural crops. *OAlib*, *9*(5), 1–29. DOI: 10.4236/oalib.1108577

Armghan Khalid, M., Niaz, B., Saeed, F., Afzaal, M., Islam, F., Hussain, M., Mahwish, , Muhammad Salman Khalid, H., Siddeeg, A., & Al-Farga, A. (2022). Edible coatings for enhancing safety and quality attributes of fresh produce: A comprehensive review. *International Journal of Food Properties*, *25*(1), 1817–1847. DOI: 10.1080/10942912.2022.2107005

Arora, N. K. (2019). Impact of climate change on agriculture production and its sustainable solutions. *Environmental Sustainability*, *2*(2), 95–96. DOI: 10.1007/s42398-019-00078-w

Arora, N. K., & Mishra, I. (2022). Current scenario and future directions for sustainable development goal 2: A roadmap to zero hunger. *Environmental Sustainability*, *5*(2), 129–133. DOI: 10.1007/s42398-022-00235-8 PMID: 37521584

Arumugam, T., Sona, C. L., & Maheswari, M. U. (2021). Fruits and vegetables as Superfoods: Scope and demand. *Journal of Pharmaceutical Innovation*, *10*, 119–129.

Ashok, A., Rejeesh, C., & Renjith, R. (2016). Biodegradable polymers for sustainable packaging applications: A review. *IJBB*, *1*, 11.

Astuti, R., & Hidayati, L. (2023). How might blockchain technology be used in the food supply chain? A systematic literature review. Cogent Business &amp. *Cogent Business & Management*, *2*(2), 2246739. Advance online publication. DOI: 10.1080/23311975.2023.2246739

Augusto, P E D. (2020, March 20). Challenges, trends and opportunities in food processing. Elsevier BV, 35, 72-78. DOI: 10.1016/j.cofs.2020.03.005

Avramescu, S. M., Butean, C., Popa, C. V., Ortan, A., Moraru, I., & Temocico, G. (2020). Edible and functionalized films/coatings—Performances and perspectives. *Coatings*, *10*(7), 687. DOI: 10.3390/coatings10070687

Awad, S., Zhou, Y., Katsou, E., & Fan, M. (2020). Polymer matrix systems used for date palm composite reinforcement. *Date Palm Fiber Composites: Processing, Properties and Applications*, 119–159.

Awasthi, P., Joshi, D., Chataut, G., Bogati, S., & Bhandari, R. (2022). Biochemical Changes Due To Storable Packing Materials In Litchi (Litchi Chinensis Sonn) Fruit. [RFNA]. *Reviews in Food and Agriculture*, *3*(2), 80–86. DOI: 10.26480/rfna.02.2022.80.86

Aydoğan, B., & Vardar, G. (2020). Evaluating the role of renewable energy, economic growth and agriculture on CO2 emission in E7 countries. *International Journal of Sustainable Energy*, *39*(4), 335–348. DOI: 10.1080/14786451.2019.1686380

Ayres, R., Turton, H., & Casten, T. (2007). Energy efficiency, sustainability and economic growth. *Energy*, *32*(5), 634–648. DOI: 10.1016/j.energy.2006.06.005

Azis, R., Priyanti, N. A., & Wahyuni, S. (2024). RadityaPratama, Muhammad Rizza Pahlevi, and PraptopYuliadi. "Development of a Super Hybrid Duck Cultivation and breeding business using energy efficient hatching machines in Blitar Regency.". *International Journal of Asia Pacific Community Service*, *1*(1), 24–30. DOI: 10.71131/jmm26n60

B. Xiang, X. Zhang (2023) Advancements in the development of field precooling of fruits and vegetables with/without phase change materials. Journal of Energy Storage 73(B): 109007. https://doi.org/DOI: 10.1016/j.est.2023.109007

Babatunde, O., Denwigwe, I., Adedoja, O., Babatunde, D., & Gbadamosi, S. (2019). Harnessing Renewable Energy for Sustainable Agricultural Applications. *International Journal of Energy Economics and Policy*, *9*(5), 308–315. DOI: 10.32479/ijeep.7775

Babu, K. G., & Karan, R. K. (2018). *The role of intellectual property in promoting energy efficiency in agriculture and food processing*. Cambridge University Press.

Babu, M. S. R. (2012). *The law of intellectual property in India*. LexisNexis.

Babu, P. J. (2022). Nanotechnology mediated intelligent and improved food packaging. *International Nano Letters*, *12*(1), 1–14. DOI: 10.1007/s40089-021-00348-8

Badawi, A. K., Hassan, R., Farouk, M., Bakhoum, E. S., & Salama, R. S. (2024). Optimizing the coagulation/flocculation process for the treatment of slaughterhouse and meat processing wastewater: Experimental studies and pilot-scale proposal. *International Journal of Environmental Science and Technology*, *21*(13), 1–16. DOI: 10.1007/s13762-024-05591-y

Bai, J., & Plotto, A. (2012). Coatings for fresh fruits and vegetables. *Edible Coatings and Films to Improved Food Quality. Second Edition. Taylor \& Francis Group, United States*, 185–242.

Bajwa, U., & Sandhu, K. S. (2014). Effect of handling and processing on pesticide residues in food-a review. *Journal of Food Science and Technology*, *51*(2), 201–220. DOI: 10.1007/s13197-011-0499-5 PMID: 24493878

Bannister, K. E. (2024). Key Performance Indicators. In Practical Lubrication for Industrial Facilities (pp. 717–721). River Publishers. https://doi.org/DOI: 10.1201/9781032632360-43

Barker, M., & Williams, P. (2020). Implementing energy management systems in the food industry: Benefits, challenges, and best practices. *Energy Efficiency, 13*(3), 555–574. DOI: 10.1007/s12053-020-09728-3

Barlow, C.Y., Morgan, D.C., (2013). Polymer film packaging for food: an environmental assessment. Resour. Conserv. Recycl. 78, 74e80.

Barlow, P., Loopstra, R., Tarasuk, V., & Reeves, A. (2020). Liberal trade policy and food insecurity across the income distribution: An observational analysis in 132 countries, 2014-17. *The Lancet. Global Health, 8*(8), e1090–e1097. DOI: 10.1016/S2214-109X(20)30263-1 PMID: 32710865

Barone, , Matheus, J. R. V., de Souza, T. S. P., Moreira, R. F. A., & Fai, A. E. C. (2021). Green-based active packaging: Opportunities beyond COVID-19, food applications, and perspectives in circular economy—A brief review. *Comprehensive Reviews in Food Science and Food Safety, 20*(5), 4881–4905. Advance online publication. DOI: 10.1111/1541-4337.12812 PMID: 34355490

Basavegowda, N., & Baek, K.-H. (2021). Synergistic antioxidant and antibacterial advantages of essential oils for food packaging applications. *Biomolecules, 11*(9), 1267. DOI: 10.3390/biom11091267 PMID: 34572479

Basumatary, I. B., Mukherjee, A., Katiyar, V., & Kumar, S. (2022). Biopolymer-based nanocomposite films and coatings: Recent advances in shelf-life improvement of fruits and vegetables. *Critical Reviews in Food Science and Nutrition, 62*(7), 1912–1935. DOI: 10.1080/10408398.2020.1848789 PMID: 33249872

Basu, S., & Rao, Y. V. (2020). Environmental effects and management strategies of the herbicides. *International Journal of Bio-Resource and Stress Management, 11*(6), 518–535. DOI: 10.23910/1.2020.2069d

Batista, L., Bourlakis, M., Smart, P., Maull, R., (2018). In search of a circular supply chain archetypeea content-analysis-based literature review. Prod. Plan. Control 29 (6), 438e451.

Begum, A., Alam, S. N., & Jalal Uddin, M. (2017). Management of pesticides: Purposes, uses, and concerns. *Pesticide Residue in Foods: Sources, Management, and Control*, 53–86.

Behera, U. K., & France, J. (2016). Integrated farming systems and the livelihood security of small and marginal farmers in India and other developing countries. *Advances in Agronomy*, *138*, 235–282. DOI: 10.1016/bs.agron.2016.04.001

Belanche, A.. (2019). The role of alternative feed ingredients in the sustainability of livestock systems. *Animal Feed Science and Technology*, *252*, 38–47.

Benavente, R. M., Perdigones, A., Baptista, F., García, J. L., & Mazarrón, F. R. (2024). Influence of hot water consumption patterns on the viability of low-temperature solar thermal systems in the food industry. *Results in Engineering*, *22*, 102140. DOI: 10.1016/j.rineng.2024.102140

Berkowitz, S. A., Seligman, H. K., Meigs, J. B., & Basu, S. (2018). Food insecurity, healthcare utilization, and high cost: A longitudinal cohort study. *The American Journal of Managed Care*, *24*(9), 399–404. PMID: 30222918

Bertoluci, , Leroy, Y., & Olsson, A. (2014). Exploring the environmental impacts of olive packaging solutions for the European food market. *Journal of Cleaner Production*, *64*, 234–243. Advance online publication. DOI: 10.1016/j.jclepro.2013.09.029

Bhatia, L., Jha, H., Sarkar, T., & Sarangi, P. K. (2023). Food Waste Utilization for Reducing Carbon Footprints towards Sustainable and Cleaner Environment: A Review. *International Journal of Environmental Research and Public Health*, *20*(3), 2318. DOI: 10.3390/ijerph20032318 PMID: 36767685

Bhatnagar, P., Gururani, P., Bisht, B., Kumar, V., Kumar, N., Joshi, R., & Vlaskin, M. S. (2022). Impact of irradiation on physico-chemical and nutritional properties of fruits and vegetables: A mini review. *Heliyon*, *8*(10), e10918. DOI: 10.1016/j.heliyon.2022.e10918 PMID: 36247116

Bhat, R., Kamaruddin, N. S., Min-Tze, L., & Karim, A. A. (2018). Advanced technologies for meat processing. *Food Engineering Series*, *12*, 85–108. DOI: 10.1007/978-1-4614-3315-5_8

Bhavsar, D., Limbasia, B., Mori, Y., Aglodiya, M. I., & Shah, M. (2023). A comprehensive and systematic study in smart drip and sprinkler irrigation systems. *Smart Agricultural Technology*, *5*, 100303. DOI: 10.1016/j.atech.2023.100303

Bhunia, S., Karmakar, S., Bhattacharjee, S., Roy, K., Kanthal, S., Pramanick, M., Baishya, A., & Mandal, B. (2021). Optimization of energy consumption using data envelopment analysis (DEA) in rice-wheat-green gram cropping system under conservation tillage practices. *Energy*, *236*, 121499. DOI: 10.1016/j.energy.2021.121499

Bhuvaneswari, P., Steffi, P. F., Thirumalaiyammal, B., & Mishel, P. F. (2023). 9 Potential antimicrobial uses agents of natural and their applications as bio-preservatives. *Antimicrobials in Food Science and Technology*, 154.

Bielenberg, J.. (2021). Improving energy efficiency in dairy plants: The role of automation and high-efficiency motors. *Energy Reports*.

Binasova, V., Micieta, B. and Kubinec, L., (2015). *Information and Communication Technology. Enabled Energy Efficiency*. International Scientific Book; 2015.

Bisht, B., Bhatnagar, P., Gururani, P., Kumar, V., Tomar, M. S., Sinhmar, R., Rathi, N., & Kumar, S. (2021). Food irradiation: Effect of ionizing and non-ionizing radiations on preservation of fruits and vegetables--a review. *Trends in Food Science \& Technology, 114*, 372–385.

Bisht, A., & Singh, S. P. (2024). Postharvest Losses and Management of Horticultural Produce: A Review. *Journal of Scientific Research and Reports*, *30*(3), 305–320. DOI: 10.9734/jsrr/2024/v30i31881

Bizymis, A.-P., & Tzia, C. (2022). Edible films and coatings: Properties for the selection of the components, evolution through composites and nanomaterials, and safety issues. *Critical Reviews in Food Science and Nutrition*, *62*(31), 8777–8792. DOI: 10.1080/10408398.2021.1934652 PMID: 34098828

Blackburn, J., & Scudder, G. (2009). Supply chain strategies for perishable products: The case of fresh produce. *Production and Operations Management*, *18*(2), 129–137. DOI: 10.1111/j.1937-5956.2009.01016.x

Blankenship, J., & Blancato, R. B. (2022). Nutrition Security at the Intersection of Health Equity and Quality Care. *Journal of the Academy of Nutrition and Dietetics*, *122*(10, 10S), S12–S19. DOI: 10.1016/j.jand.2022.06.017 PMID: 36122954

Bock, M., & Meyerding, S. G. H. (2023). Consumer perception of food product packaging materials sustainability versus life cycle assessment results: The case of processed tomatoes—a quantitative study in Germany. *Sustainability (Basel)*, *15*(23), 16370. DOI: 10.3390/su152316370

Bodbodak, S., & Moshfeghifar, M. (2016). Advances in modified atmosphere packaging of fruits and vegetables. In *Eco-friendly technology for postharvest produce quality* (pp. 127–183). Academic Press.

Boekel, V., Martinus, V. F., Pellegrini, N., Stanton, C., Scholz, G., Lalljie, S., Somoza, V., Knorr, D., Jasti, P. R., & Eisenbrand, G. (2010). A review on the beneficial aspects of food processing. *Molecular Nutrition & Food Research*, *54*(9), 1215–1247. DOI: 10.1002/mnfr.200900608 PMID: 20725924

Bokulich, N. A., & Bamforth, C. W. (2013, May 22). The Microbiology of Malting and Brewing. *Microbiology and Molecular Biology Reviews : MMBR*, *77*(2), 157–172. DOI: 10.1128/MMBR.00060-12 PMID: 23699253

Boukhatem, M. N., & Setzer, W. N. (2020). Aromatic herbs, medicinal plant-derived essential oils, and phytochemical extracts as potential therapies for coronaviruses: Future perspectives. *Plants*, *9*(6), 800. DOI: 10.3390/plants9060800 PMID: 32604842

Boukouvalas, C., Kekes, T., Oikonomopoulou, V., & Krokida, M. (2024). Life Cycle Assessment of Energy Production from Solid Waste Valorization and Wastewater Purification: A Case Study of Meat Processing Industry. *Energies*, *17*(2), 487. DOI: 10.3390/en17020487

Bozchalui, M. C., Canizares, C. A., & Bhattacharya, K. (2015). Optimal energy management of greenhouses in smart grids. *IEEE Transactions on Smart Grid*, *6*(2), 827–835. DOI: 10.1109/TSG.2014.2372812

Boz, Z., Korhonen, V., & Sand, C. (2020). Consumer considerations for the implementation of sustainable packaging: A review. *Sustainability (Basel)*, *12*(6), 2192. DOI: 10.3390/su12062192

Braun, V., & Clarke, V. (2013). Successful qualitative research: A practical guide for beginners. *Sage (Atlanta, Ga.)*.

Breeman, G., Giest, S., & Rimkutė, D. (2017). Food Security and the Sustainability of GMOs in the United States and the European Union. In *Advances in Food Security and Sustainability* (pp. 165–193). Elsevier., DOI: 10.1016/bs.af2s.2017.09.005

Brennan, J. G., O'Callaghan, P., & Ghosh, A. (2022). Advancements in food drying technologies: Energy efficiency and sustainability. *Journal of Food Engineering*, *323*(10), 24–36.

Brennan, J. G., O'Callaghan, P., & Ghosh, A. (2022). Innovations in sustainable packaging for food safety and energy efficiency. *Journal of Food Science*, *88*(3), 1157–1172.

Briam, R., Walker, M., & Masanet, E. (2015). A comparison of product-based energy in tensity metrics for cheese and whey processing. *Journal of Food Engineering*, *151*, 25–33. DOI: 10.1016/j.jfoodeng.2014.11.011

Brito, I. P. C., & Silva, E. K. (2024). Pulsed electric field technology in vegetable and fruit juice processing: A review. *Food Research International*, *184*, 114207. DOI: 10.1016/j.foodres.2024.114207 PMID: 38609209

Buchanan, R. A., & Wright, S. P. (2019). Leveraging IoT technologies for energy optimization in the food industry: A case study of intelligent temperature control systems. *Energy Reports*, *5*, 563–570. DOI: 10.1016/j.egyr.2019.07.004

Bulak, P., Pietrycha, K., Kaczor, M., Pawłowska, M., & Bieganowski, A. (2023). Post-breeding waste from Zophobas morio for biogas and energy generation via anaerobic methane fermentation. *Fuel*, *351*, 128844. DOI: 10.1016/j.fuel.2023.128844

Bulantekin, Ö. and Alp, D. (2023). Perspective chapter: development of food packaging films from microorganism-generated polyhydroxyalkanoates.. DOI: 10.5772/intechopen.108802

Bundschuh, J., Chen, G., & Mushtaq, S. (2014). Towards a sustainable energy technologies based agriculture, in Sustain. *Energy Solution Agricultural ROUT LEDGEinassociation with GSEResearch*, *3*(15), 3–15.

Bvenura, C., & Sivakumar, D. (2017). The role of wild fruits and vegetables in delivering a balanced and healthy diet. *Food Research International*, *99*, 15–30. DOI: 10.1016/j.foodres.2017.06.046 PMID: 28784471

Calín-Sánchez, Á., Lipan, L., Cano-Lamadrid, M., Kharaghani, A., Masztalerz, K., Carbonell-Barrachina, Á. A., & Figiel, A. (2020). Comparison of traditional and novel drying techniques and its effect on quality of fruits, vegetables and aromatic herbs. *Foods*, *9*(9), 1261.

Camargo, P. V. M., Gladness, E. K., Clech, M., Milutinovic, S., Bustamate, E. L., & Chau, T. T. H. (2022). Group H. *Youth Actions and Innovations towards Climate Change Adaptation and Mitigation to Promote Sustainable Agriculture in Their Communities.*, 109.

Cammarelle, A., Viscecchia, R., & Bimbo, F. (2021). Intention to purchase milk packaged in biodegradable packaging: Evidence from italian consumers. *Foods*, *10*(9), 2068. DOI: 10.3390/foods10092068 PMID: 34574179

Capehart, B. L., Turner, W. C., & Kennedy, W. J. (2008). *Guide to Energy Management*. The Fairmont Press, Inc.

Capper, J. L.. (2014). The environmental sustainability of dairy production in the United States: A synthesis of greenhouse gas emissions, water use, and land use efficiency. *Journal of Dairy Science*.

Caputo, V., Sun, J., Staples, A. J., & Taylor, H. (2024). Market outlook for meat alternatives: Challenges, opportunities, and new developments. *Trends in Food Science & Technology*, *148*, 104474. DOI: 10.1016/j.tifs.2024.104474

Carruthers, G., & Vanclay, F. (2007). Enhancing the social content of environmental management systems in Australian agriculture. *International Journal of Agricultural Resources, Governance and Ecology, 3*(3), 326. DOI: 10.1504/IJARGE.2007.012840

Casnici, C. V. C., & Schulzmann, D. (2024). Future Feast: Mapping the Cultivated Meat Sector. In *Cultivated Meat: Technologies, Commercialization and Challenges* (pp. 407–423). Springer Nature Switzerland. DOI: 10.1007/978-3-031-55968-6_20

Casson, A., Ferrazzi, G., Guidetti, R., Bellettini, C., Narote, A. D., Rollini, M., Piccardo, A., Volturo, E., Cosentino, M., & Limbo, S. (2024). Wholesale fruit and vegetable market in Milan: Turning food surpluses into environmental gains. *Journal of Cleaner Production, 462*, 142625. DOI: 10.1016/j.jclepro.2024.142625

Castellanos, R., & Balaguer, J. (2017). Modeling energy efficiency in food production through machine learning and data analytics. *Energy Procedia, 105*, 1704–1711. DOI: 10.1016/j.egypro.2017.03.680

Castell-Perez, M. E., & Moreira, R. G. (2021). Irradiation and consumers acceptance. *Innovative Food Processing Technologies*, 122.

Cattaneo, A., Sánchez, M. V., Torero, M., & Vos, R. (2021). Reducing food loss and waste: Five challenges for policy and research. *Food Policy, 98*, 101974. DOI: 10.1016/j.foodpol.2020.101974 PMID: 33012956

Chaabouni, T., & Khemakhem, M. (2017). Energy management strategy in cloud computing: A perspective study. *The Journal of Supercomputing, 12*(12), 6569–6597. DOI: 10.1007/s11227-017-2154-z

Chacha, J. S., Zhang, L., Ofoedu, C. E., Suleiman, R., Dotto, J. M., Roobab, U., Agunbiade, A. O., Duguma, H. T., Mkojera, B. T., Hossaini, S. M., Rasaq, W. A., Шорсткий, И., Okpala, C. O. R., Korzeniowska, M., & Guiné, R. P. F. (2021, June 20). Revisiting Non-Thermal Food Processing and Preservation Methods—Action Mechanisms, Pros and Cons: A Technological Update (2016–2021). *Foods, 10*(6), 1430–1430. DOI: 10.3390/foods10061430 PMID: 34203089

Chandrasekaran, S. K., & Rajasekaran, V. A. (2024). Energy-efficient cluster head using modified fuzzy logic with WOA and path selection using enhanced CSO in IoT-enabled smart agriculture systems. *The Journal of Supercomputing, 80*(8), 11149–11190. DOI: 10.1007/s11227-023-05780-5

Chatterjee, B., & Mazumder, D. (2024). Valorization of fruit and vegetable waste in a novel three-stage hybrid anaerobic digester for enhanced biogas production: Performance study and microbial community analysis. *Biochemical Engineering Journal, 209*, 109403. DOI: 10.1016/j.bej.2024.109403

Chatti, W., & Majeed, M. T. (2024). Meat production, technological advances, and environmental protection: Evidence from a dynamic panel data model. *Environment, Development and Sustainability, 26*(12), 1–26. DOI: 10.1007/s10668-023-04449-6

Chel, A., & Kaushik, G. (2011). Renewable energy for sustainable agriculture. *Agronomy for Sustainable Development, 31*, 91–118.

Chemat, F., Rombaut, N., Meullemiestre, A., Turk, M., Perino, S., Fabiano-Tixier, A. S., & Abert-Vian, M. (2017). Review of green food processing techniques. Preservation, transformation, and extraction. *Innovative Food Science & Emerging Technologies, 41*, 357–377. DOI: 10.1016/j.ifset.2017.04.016

Chen, C., Sujanto, R., Tseng, M., Chiu, A., & Lim, M. (2021). How is the sustainable consumption intention model in food industry under preference uncertainties? the consumer willingness to pay on recycled packaging material. *Sustainability (Basel), 13*(21), 11578. DOI: 10.3390/su132111578

Cheng, Y.. (2020). Biogas production from dairy farm waste: Efficiency, challenges, and future trends. *Waste Management & Research.*

Chen, H., Wang, J., Cheng, Y., Wang, C., Liu, H., Bian, H., Pan, Y., Sun, J., & Han, W. (2019). Application of protein-based films and coatings for food packaging: A review. *Polymers, 11*(12), 2039. DOI: 10.3390/polym11122039 PMID: 31835317

Chen, J., Li, X., & Zhang, W. (2022). Integrating renewable energy sources into the food industry: A review of technologies and applications. *Renewable & Sustainable Energy Reviews, 150*, 111488.

Chen, J., Li, X., & Zhang, W. (2022). Integrating renewable energy systems into food production facilities. *Renewable Energy, 135*, 349–360.

Chew, C. L., Tan, B. A., Low, J. Y. S., Mohd Hakimi, N. I. N., Kua, S. F., & Lim, C. M. (2021). Exogenous ethylene application on postharvest oil palm fruit bunches improves crude palm oil quality. *Food Science \& Nutrition, 9*(10), 5335–5343.

Chiellini, E. (2008): Enviromentally compatible food packaging. Woodhead publishing limited, Cambridge England, 8–10

Chirilli, C., Molino, M., & Torri, L. (2022). Consumers' awareness, behavior and expectations for food packaging environmental sustainability: Influence of socio-demographic characteristics. *Foods, 11*(16), 2388. DOI: 10.3390/foods11162388 PMID: 36010389

Cho, I.-H., Radadia, A. D., Farrokhzad, K., Ximenes, E., Bae, E., Singh, A. K., Oliver, H., Ladisch, M., Bhunia, A., Applegate, B., Mauer, L., Bashir, R., & Irudayaraj, J. (2014). Nano/Micro and Spectroscopic Approaches to Food Pathogen Detection. *Annual Review of Analytical Chemistry (Palo Alto, Calif.)*, *1*(1), 65–88. DOI: 10.1146/annurev-anchem-071213-020249 PMID: 24896312

Choi, B., Yoo, S., & Park, S. (2018). Carbon footprint of packaging films made from ldpe, pla, and pla/pbat blends in south korea. *Sustainability (Basel)*, *10*(7), 2369. DOI: 10.3390/su10072369

Choi, J., Byun, S.-W., Lim, T., Lee, H. I., Shin, H. S., Kim, G. W., Kwon, J.-K., & Kang, B.-C. (2023). Artificial Intelligence-Based Plant Breeding using Genotype and Phenotype Data: Methods and Future Work. *2023 IEEE International Conference on Big Data (BigData)*, 3458–3463. https://doi.org/DOI: 10.1109/BigData59044.2023.10386874

Chopra, M. A., & Mishra, R. (n.d.). *ROLE OF FOOD PROCESSING INDUSTRY IN FOOD AND NUTRITIONAL SECURITY IN INDIA.*

Choudhary, M., & Singh, P. (2022). Renewable energy applications in dairy processing: A review. *Journal of Dairy Science*, *105*(4), 1541–1550. DOI: 10.3168/jds.2021-20984

Chowdhury, T., & Kabir, G. (2023). Life cycle assessment (lca) of package deliveries: Sustainable decision-making for the academic institutions. *Green and Low-Carbon Economy*, *2*(1), 1–13. DOI: 10.47852/bonviewGLCE3202824

Christiaensen, L., Rutledge, Z., & Taylor, J. E. (2021). The future of work in agri-food. *Food Policy*, *99*, 101963. DOI: 10.1016/j.foodpol.2020.101963 PMID: 33071436

Chua, K. J., Chou, S. K., Yang, W. M., & Yan, J. (2013). Achieving better energy-efficient air conditioning–a review of technologies and strategies. *Applied Energy*, *104*, 87–104. DOI: 10.1016/j.apenergy.2012.10.037

Cimpan, C., Maul, A., Jansen, M., Pretz, T., & Wenzel, H. (2015). Central sorting and recovery of MSW recyclable materials: A review of technological state-of-the-art, cases, practice and implications for materials recycling. *Journal of Environmental Management*, *156*, 181–199. DOI: 10.1016/j.jenvman.2015.03.025 PMID: 25845999

Civancik-Uslu, D., Puig, R., Voigt, S., Walter, D., & Fullana–i–Palmer, P. (2019). Improving the production chain with lca and eco-design: Application to cosmetic packaging. *Resources, Conservation and Recycling*, *151*, 104475. DOI: 10.1016/j.resconrec.2019.104475

Clairand, J, M., Marco, B, L., Guillermo, E, E. and Pantaleo, A, M., (2020). *Review of Energy Efficiency Technologies in the Food Industry: Trends, Barriers, and Opportunities*. IEEE Access journals, 4, 1-15.

Clapp, J., Moseley, W. G., Burlingame, B., & Termine, P. (2022). Viewpoint: The case for a six-dimensional food security framework. *Food Policy*, *106*, 102164. DOI: 10.1016/j.foodpol.2021.102164

Clement, O., Akinyele, O., Oladimeji, S., & Oladipo, O. (2018). Renewable Energy Usage for Agricultural Practices: A review. 2018 International Conference of The Nigerian Institution of Agricultural Engineers, Lagos, Nigeria, September, 2018. EIP-AGRI. (2019). Enhancing production and use of renewable energy on the farm. Brussels, Belgium.

Cloete, L., Picot-Allain, C., Ramasawmy, B., Neetoo, H., Ramful-Baboolall, D., & Emmambux, M. N. (2023). Drivers and barriers for commercial uptake of edible coatings for fresh fruits and vegetables industry-A review. *Food Reviews International*, *39*(6), 3481–3514. DOI: 10.1080/87559129.2021.2012795

Coelho, P. M., Corona, B., ten Klooster, R., & Worrell, E. (2020). Sustainability of reusable packaging–Current situation and trends. In *Resources, Conservation and Recycling: X* (Vol. 6). Elsevier B.V. DOI: 10.1016/j.rcrx.2020.100037

Coldea, T E. (2019, September 11). Introductory Chapter: A General Overview on Latest Trends in Food Engineering. IntechOpen. DOI: 10.5772/intechopen.86081

Coma, V., (2008). Bioactive packaging technologies for extended shelf life of meat-based products. Meat Science. 78 (1 2), 90 103

Conrad, Z., & Blackstone, N. T. (2021). Identifying the links between consumer food waste, nutrition, and environmental sustainability: A narrative review. *Nutrition Reviews*, *79*(3), 301–314. DOI: 10.1093/nutrit/nuaa035 PMID: 32585005

Corato, U. D., & Cancellara, F. A. (2019). Measures, technologies, and incentives for cleaning the minimally processed fruits and vegetables supply chain in the Italian food industry. *Journal of Cleaner Production*, *237*, 117735. DOI: 10.1016/j.jclepro.2019.117735

Corigliano, O., & Algieri, A. (2024). A comprehensive investigation on energy consumptions, impacts, and challenges of the food industry. *Energy Conversion and Management*, *X*, 100661. DOI: 10.1016/j.ecmx.2024.100661

Cotacallapa-Sucapuca, M., Vega, E. N., Maieves, H. A., Berrios, J. D. J., Morales, P., Fernández-Ruíz, V., & Cámara, M. (2021, May 15). Extrusion Process as an Alternative to Improve Pulses Products Consumption. A Review. *Foods, 10*(5), 1096–1096. DOI: 10.3390/foods10051096 PMID: 34063375

Cottle, D. J.. (2011). *Livestock and greenhouse gas emissions: Facts and figures*. University of Queensland.

Coudard, A., Corbin, E., de Koning, J., Tukker, A., & Mogollon, J. M. (2021). Global water and energy losses from consumer avoidable food waste. *Journal of Cleaner Production, 326*, 129342. DOI: 10.1016/j.jclepro.2021.129342

Cox, S. (2016). *"Financial Incentives to Enable Clean Energy Deployment: Policy Overview and Good Practices." Office of Scientific and Technical Information*. OSTI., DOI: 10.2172/1239887

Creswell, J. W., & Creswell, J. D. (2018). *Research design: Qualitative, quantitative, and mixed methods approaches*. Sage publications.

Cui, Y.. (2020). Barriers and opportunities for implementing sustainable dairy farming practices: A global perspective. *Environmental Sustainability*.

Darbari, A., & Nayar, H. S. (2020). *Indian spice exports and sanitary and phytosanitary measures related challenges: a study of the legal and policy framework*.

Davidovic´, A., & Savic´, A. (2010). Microbial production of biodegradable polymer. [in Croatian]. *Tehnologica Acta, 3*, 3–13.

Dávila-Aviña, J. E., Solís-Soto, L. Y., Rojas-Verde, G., & Salas, N. A. (2014). Sustainability and Challenges of Minimally Processed Foods. In *Food engineering series* (pp. 279–295). DOI: 10.1007/978-3-319-10677-9_12

De Benedetto, L., & Klemeš, J. (2009). The Environmental Performance Strategy Map: An integrated LCA approach to support the strategic decision-making process. *Journal of Cleaner Production, 10*(10), 900–906. DOI: 10.1016/j.jclepro.2009.02.012

De Boer, A., Krul, L., Fehr, M., Geurts, L., Kramer, N., Urbieta, M. T., & Hepburn, P. A. (2020). Animal-free strategies in food safety and nutrition: What are we waiting for? Part I: Food safety. *Trends in Food Science & Technology, 106*, 469–484. DOI: 10.1016/j.tifs.2020.10.034

de Siqueira Oliveira, L., Eça, K. S., de Aquino, A. C., & da Silva, L. M. R. (2020). Modified and controlled atmosphere packaging. In *Fresh-cut fruits and vegetables* (pp. 151–164). Elsevier. DOI: 10.1016/B978-0-12-816184-5.00007-0 de Sousa, R., Bragança, L., da Silva, M. V., & Oliveira, R. S. (2024). Challenges and Solutions for Sustainable Food Systems: The Potential of Home Hydroponics. *Sustainability (Basel)*, *16*(2), 817. DOI: 10.3390/su16020817

Debebe, S. (2022). Post-harvest losses of crops and its determinants in Ethiopia: Tobit model analysis. *Agriculture \&. Food Security*, *11*(1), 1–8. PMID: 35529169

Degerli, B., Nazir, S., Sorgüven, E., Hitzmann, B., & Ozilgen, M. (2015). As sessment of the energy and exergy efficiencies of farm to fork grain cultivation and bread making processes in Turkey and Germany. *Energy*, *93*, 421–434. DOI: 10.1016/j.energy.2015.08.095

Dena, A., Honrado, L., Lin, P., & Dotong, E. (2023). Promoting the usage of eco-friendly tertiary packaging: A market research on the perceived behavior of filipino consumers based on sustainability factors. *American Journal of Social Development and Entrepreneurship*, *2*(1), 26–31. DOI: 10.54536/ajsde.v2i1.1223

Department for Business, Energy and Industrial Strategy. (2018). Digest of United Kingdom Energy Statistics (DUKES) July 2018. London, UK. https://assets .publishing.service. gov.uk/government/uploads/system/uploads/attachment_data/file/736148/ DUKES_2018.pdf, Accessed date: 16 January 2025.

Derkus, B. (2016). Applying the miniaturization technologies for biosensor design. *Biosensors & Bioelectronics*, *79*, 901–913. DOI: 10.1016/j.bios.2016.01.033 PMID: 26800206

Deshmukh, R. K., & Gaikwad, K. K. (2022). Natural antimicrobial and antioxidant compounds for active food packaging applications. *Biomass Conversion and Biorefinery*, •••, 1–22.

Designs Act. (2020). *2000 (with the latest amendment)*. Government of India.

Dewi, L. 2024. "Exploring Stakeholder Perspectives on Supply Chain Sustainability in Renewable Energy Industries." *Research Square*. DOI: 10.21203/rs.3.rs-4301842/v1

Díaz-Montes, E., & Castro-Muñoz, R. (2021). Edible films and coatings as food-quality preservers: An overview. *Foods*, *10*(2), 249.

Dinkçi, N., Akdeniz, V., & Akalın, A S. (2019, January 1). Survival of probiotics in functional foods during shelf life. Elsevier BV, 201-233. DOI: 10.1016/B978-0-12-817190-5.00006-9

Dladla, S. (2020). Effects of different packaging materials and storage conditions on the quality attributes of tomato fruits across the south african supply chain. (Doctoral dissertation, University of KwaZulu-Natal Pietermaritzburg).

Dodero, A., Escher, A., Bertucci, S., Castellano, M., & Lova, P. (2021, April 15). Intelligent Packaging for Real-Time Monitoring of Food-Quality: Current and Future Developments. *Applied Sciences (Basel, Switzerland)*, *11*(8), 3532–3532. DOI: 10.3390/app11083532

Dong, J., Gruda, N., Li, X., Cai, Z., Zhang, L., & Duan, Z. (2022). Global vegetable supply towards sustainable food production and a healthy diet. *Journal of Cleaner Production*, *369*, 133212. DOI: 10.1016/j.jclepro.2022.133212

Drago, E., Campardelli, R., Pettinato, M., & Perego, P. (2020). Innovations in smart packaging concepts for food: An extensive review. *Foods*, *9*(11), 1628. DOI: 10.3390/foods9111628 PMID: 33171881

Drewnowski, A., Detzel, P., & Klassen-Wigger, P. (2022, April 28). Perspective: Achieving Sustainable Healthy Diets Through Formulation and Processing of Foods. Elsevier BV, 6(6), nzac089-nzac089

Driesen, D. M. (2002). Sustainable Development and Air Quality: The Need to Replace Basic Technologies with Cleaner Alternatives. *Envtl. L. Rep. News & Analysis*, *32*, 10277.

Driscoll, R. H., & Rahman, M. S. (2020). Types of Packaging Materials Used for Foods. In *Handbook of Food Preservation* (pp. 905–922). CRC Press. DOI: 10.1201/9780429091483-59

Dufresne, A., & Charlebois, S. (2021). Leveraging big data for energy efficiency and sustainability in food production. *Trends in Food Science & Technology*, *107*, 30–40. DOI: 10.1016/j.tifs.2020.11.010

Dukare, A. S., Singh, R. K., Jangra, R. K., & Bhushan, B. (2022). Non-fungicides-based promising technologies for managing post-production penicillium induced spoilage in horticultural commodities: A comprehensive review. *Food Reviews International*, *38*(3), 227–267. DOI: 10.1080/87559129.2020.1727497

Dumitran, G. E., Vuta, L. I., Negrusa, E., & Birdici, A. (2024). Reducing greenhouse gas emissions in Romanian agriculture using renewable energy sources. *Journal of Cleaner Production*, *467*, 142918. DOI: 10.1016/j.jclepro.2024.142918

Du, Y., Wang, H., Yuan, S., Yu, H., Xie, Y., Guo, Y., Cheng, Y., & Yao, W. (2024). Dielectric barrier discharge plasma pretreatment: A cleaner new way to improve energy efficiency and quality of wolfberry drying. *Journal of Cleaner Production*, *450*, 141951. DOI: 10.1016/j.jclepro.2024.141951

Dybikowska, A., & Graczyk, M. (2019). Technological Eco-Innovations Related to Res – Opportunities and Barriers. *Nongye Jingji Wenti, 1*(358), 120–139. DOI: 10.30858/zer/104699

Dziadowiec, D., Matykiewicz, D., Szostak, M., & Andrzejewski, J. (2023). Overview of the cast polyolefin film extrusion technology for multi-layer packaging applications. *Materials (Basel), 16*(3), 1071. DOI: 10.3390/ma16031071 PMID: 36770077

Ebrahimi, A., Zabihzadeh Khajavi, M., Ahmadi, S., Mortazavian, A. M., Abdolshahi, A., Rafiee, S., & Farhoodi, M. (2022). Novel strategies to control ethylene in fruit and vegetables for extending their shelf life: A review. *International Journal of Environmental Science and Technology, 19*(5), 4599–4610. DOI: 10.1007/s13762-021-03485-x

El Jaouhari, A., Arif, J., Jawab, F., Samadhiya, A., & Kumar, A. (2024). Unfolding the role of metaverse in agri-food supply chain security: Current scenario and future perspectives. International Journal of Food Science &amp. *International Journal of Food Science & Technology, 5*(5), 3451–3460. DOI: 10.1111/ijfs.16972

Elik, A., Yanik, D. K., Istanbullu, Y., Guzelsoy, N. A., Yavuz, A., & Gogus, F. (2019). Strategies to reduce post-harvest losses for fruits and vegetables. *Strategies (La Jolla, Calif.), 5*(3), 29–39.

Energy Information Administration. EIA (2017). International Energy Outlook 2017. https://www.eia.gov/outlooks/archive/ieo17/pdf/exec_summ.pdf, Accessed date 23 January 2015 2018.

Engler, N., & Krarti, M. (2021). Review of energy efficiency in controlled environment agriculture. *Renewable & Sustainable Energy Reviews, 141*, 110786. DOI: 10.1016/j.rser.2021.110786

Enríquez-Castro, C M., Pérez-Nafarrate, M., & Rodríguez, J E G. (2021, June 7). Innovation in Food Products Using Ozone Technology: Impact on Quality Assurance. IntechOpen. DOI: 10.5772/intechopen.96681

Erekath, S., Seidlitz, H., Schreiner, M., & Dreyer, C. (2024). Food for future: Exploring cutting-edge technology and practices in vertical farm. *Sustainable Cities and Society, 106*, 105357. DOI: 10.1016/j.scs.2024.105357

Erkan, M., & Yıldırım, I. (2017). Postharvest quality and safety of fresh-cut vegetables. Minimally processed refrigerated fruits and vegetables, 271-326.

Escursell, S., Llorach-Massana, P., & Roncero, M. (2021). Sustainability in e-commerce packaging: A review. *Journal of Cleaner Production, 280*, 124314. DOI: 10.1016/j.jclepro.2020.124314 PMID: 32989345

Ettadili, F. E., Aghris, S., Laghrib, F., Farahi, A., Saqrane, S., Bakasse, M., Lahrich, S., & El Mhammedi, M. A. (2022). Recent advances in the nanoparticles synthesis using plant extract: Applications and future recommendations. *Journal of Molecular Structure*, *1248*, 131538. DOI: 10.1016/j.molstruc.2021.131538

European Commission. data– Eurostat, Energy. balancesheets [Online]. (2017) Available: http://ec.europa.eu/eurostat/documents/3217494/8113778/KS-EN 17-001- EN-N.pdf/99cc20f1-cb11-4886-80f9-43ce0ab7823c

Evelyn, E., Chairul, C., Komalasari, K., Pebrianti, E., & Vazirani, W. (2021, October 1). Use of Pulsed Electric Field for the Inactivation of Eupenicillium Javanicum Ascospores in Pineapple Juice. *Journal of Physics: Conference Series*, *2049*(1), 012020–012020. DOI: 10.1088/1742-6596/2049/1/012020

Fabbriani, L., & Calili, R. (2018). Proposal of energy efficiency policies for food and beverage industry in Brazil. *Journal of Renewable and Sustainable Energy*, *10*(6), 065903. DOI: 10.1063/1.5002088

Facchini, F., Silvestri, B., Digiesi, S., & Lucchese, A. (2022). Agri-food loss and waste management: Win-win strategies for edible discarded fruits and vegetables sustainable reuse. *Innovative Food Science \& Emerging Technologies*, 103235.

Fadhel Khelifa, R., & Jelassi, K. (2016). An energy monitoring and management system based on key performance indicators. 2016 IEEE 21st International Conference on Emerging Technologies and Factory Automation (ETFA), 1–6. https://doi.org/DOI: 10.1109/ETFA.2016.7733542

Fadhel, Y. B., Maherani, B., Aragones, M., & Lacroix, M. (2019, November 1). Antimicrobial Properties of Encapsulated Antimicrobial Natural Plant Products for Ready-to-Eat Carrots. *Foods*, *8*(11), 535–535. DOI: 10.3390/foods8110535 PMID: 31683824

FAO (Food and Agriculture Organization of the United Nations). (2019). FAO's Work on Climate Change, United Nations Climate Change Conference 2019.. Available online: https://www.fao.org/3/ca7126en/CA7126EN.pdf (accessed on 25 January 2023).

FAO (Food and Agriculture Organization of the United Nations). (2023). *Livestock*. Retrieved from FAO – Livestock.

FAO. (2013). *Tackling Climate Change Through Livestock: A Global Assessment of Emissions and Mitigation Opportunities*. Food and Agriculture Organization of the United Nations.

FAO. (2017). The future of food and agriculture– Trends and challenges. Rome, Italy: Food and Agriculture Organization of the United Nations. http://www.fao.org/3/a-i6583e. pdf, Accessed date: 20 January 20.

FAO. (2020). [Food and Agriculture Organization of the United Nations.]. *The State of Food and Agriculture*, •••, 2020.

FAO. (2023). *Energy-smart food for people and climate: The dairy sector perspective*. FAO Publications.

FAO. (n.d.). *Legal processes, the right to food*. Food and Agriculture Organization of the United Nations. (n.d.). https://www.fao.org/right-to-food/areas-of-work/legal-processes/en/

Fardet, A. (2018a). Characterization of the Degree of Food Processing in Relation With Its Health Potential and Effects. *Advances in Food and Nutrition Research*, *85*, 79–129. DOI: 10.1016/bs.afnr.2018.02.002 PMID: 29860978

Farias, A., Lane, H., Killingsworth, J., Warden, J. M., & Wais, S. (2023). Exploring the interplay of pro-environmental attitudes, dietary choices, and packaging preferences: A virtual reality restaurant scenario study. *Challenges*, *15*(1), 1. DOI: 10.3390/challe15010001

Fauziana, D. R., Suwarsinah, H. K., & Prasetio, E. A. (2023). What factors impact the adoption of postharvest loss-reduction technologies in mangosteen supply chain? *Journal of Open Innovation*, *9*(3), 100102. DOI: 10.1016/j.joitmc.2023.100102

Fazlollahtabar, H. (2024). Sustainable Renewable Energy Supply Chain with Current Technological Adaptation: Macro Energy Progress Policy in Iran. *Environmental Progress & Sustainable Energy*, *43*(June), e14433. Advance online publication. DOI: 10.1002/ep.14433

FDF. (2018). Facts and Stats. Stats at a glance. London, UK: Food and Drink Federation. www.fdf.org.uk/statsataglance.aspx, Accessed date: 15 January 2025.

Fellows, P. (2017). *Food Processing Technology: Principles and Practice*. Woodhead Publishing.

Feng, X., Li, S., Sun, Z., Yuan, H., Li, R., Yu, N., Zhang, Y., & Chen, X. (2024). The preservation effect of chitosan-hawthorn leaf extract coating on strawberries. *Journal of Food Protection*, *87*(4), 100244. DOI: 10.1016/j.jfp.2024.100244 PMID: 38378071

Ferdousi, J., Hossain, M. I., Saha, S. R., Rob, M., Pramanik, T. A. S., Islam, M. R., & Nath, D. D. (2024). POSTHARVEST PHYSIOLOGY OF FRUITS AND VEGETABLES AND THEIR MANAGEMENT TECHNOLOGY: A REVIEW. *The Journal of Animal and Plant Sciences*, *34*(2), 291–303. DOI: 10.36899/JAPS.2024.2.0717

Fernqvist, F., Olsson, A., & Spendrup, S. (2015). What's in it for me? food packaging and consumer responses, a focus group study. *British Food Journal*, *117*(3), 1122–1135. DOI: 10.1108/BFJ-08-2013-0224

Ferreira, F. J., Baoming, G., & de Almeida, A. T. (2015). Reliability and operation of high-efficiency induction motors. In *2015 IEEE/IAS 51st Industrial and Commercial Power Systems Technical Conference (I andCPS)* (pp. 1-13). IEEE. DOI: 10.1109/ICPS.2015.7266412

Ferreira, A. R. V., Alves, V. D., & Coelhoso, I. M. (2016). Polysaccharide-based membranes in food packaging applications. *Membranes*, *6*(2), 22. DOI: 10.3390/membranes6020022 PMID: 27089372

Field, A. (2018). Discovering statistics using IBM SPSS statistics. *Sage (Atlanta, Ga.)*.

Fischer, J. R., Finnell, J., & Lavoie, B. (2006). *Renewable Energy in Agriculture: Back to the Future?* Business Choices. The Magazine of Food, Farm, and Resources Issues.

Floros, J. D., Newsome, R., Fisher, W., Barbosa-Cánovas, G. V., Chen, H., Dunne, C. P., & Ziegler, G. R. (2010). Feeding the world today and tomorrow: the importance of food science and technology: an IFT scientific review. *Comprehensive Reviews in Food Science and Food Safety*, *9*(5), 572–599. DOI: 10.1111/j.1541-4337.2010.00127.x PMID: 33467827

Food and Agriculture Organization of the United Nations. (2021). *The State of Food Security and Nutrition in the World 2021: Transforming food systems for food security, improved nutrition and affordable healthy diets for all*. Rome, FAO. https://www.fao.org/publications/sofi/2021/en/

Foster, C., Green, K., Bleda, M., Dewick, P., Evans, B., Flynn, A., & Mylan, J. (2014). Environmental impacts of food production and consumption: A report to the Department of Environment, Food and Rural Affairs. Manchester Business School. DEFRA, London.

Gac, A., H. Chambaut, C. Perrot, A. Lorilloux, C. Mosnier, and J. B. Dollé. "28. Environmental footprint of milk and meat from the French cattle sector: improvements since 1990 and future trends until 2035." (2035).

Gadekar, K., & Kshirsagar, S. (2020). Energy management practices in the food processing industry: Challenges and opportunities. *Sustainable Energy Technologies and Assessments*, *39*, 100677. DOI: 10.1016/j.seta.2020.100677

Galanakis, C. M. (2012). Recovery of high added-value components from food wastes: Conventional, emerging technologies and commercialized applications. *Trends in Food Science & Technology*, *26*(2), 68–87. DOI: 10.1016/j.tifs.2012.03.003

Galanakis, C. M. (2024). The future of food. *Foods*, *13*(4), 506. DOI: 10.3390/foods13040506 PMID: 38397483

Gao, T., Tian, Y., Zhu, Z., & Sun, D. (2020, May 1). Modelling, responses and applications of time-temperature indicators (TTIs) in monitoring fresh food quality. Elsevier BV, 99, 311-322. DOI: 10.1016/j.tifs.2020.02.019

Garcia, A., Alzate, J. M., & Barrera, J. (2012). Regulatory Design and Incentives for Renewable Energy. *Journal of Regulatory Economics*, *41*(3), 315–336. DOI: 10.1007/s11149-012-9188-1

Gardas, B. B., Raut, R. D., & Narkhede, B. (2017). Modeling causal factors of post-harvesting losses in vegetable and fruit supply chain: An Indian perspective. *Renewable & Sustainable Energy Reviews*, *80*, 1355–1371. DOI: 10.1016/j.rser.2017.05.259

Gardas, B. B., Raut, R. D., & Narkhede, B. (2018). Evaluating critical causal factors for post-harvest losses (PHL) in the fruit and vegetables supply chain in India using the DEMATEL approach. *Journal of Cleaner Production*, *199*, 47–61. DOI: 10.1016/j.jclepro.2018.07.153

Gardas, B. B., Raut, R. D., & Narkhede, B. (2019). Identifying critical success factors to facilitate reusable plastic packaging towards sustainable supply chain management. *Journal of Environmental Management*, *236*, 81–92. DOI: 10.1016/j.jenvman.2019.01.113 PMID: 30716694

Garnett, T. (2011). Where are the best opportunities for reducing greenhouse gas emissions in the food system (including the food chain)? *Food Policy*, *36*(S1), S23–S32. DOI: 10.1016/j.foodpol.2010.10.010

Garza-Juárez, A., Pérez-Carrillo, E., Arredondo-Espinoza, E. U., Islas, J. F., Benítez-Chao, D. F., & Escamilla-García, E. (2023). Nutraceuticals and Their Contribution to Preventing Noncommunicable Diseases. *Foods*, *12*(17), 3262. DOI: 10.3390/foods12173262 PMID: 37685194

Gaspar, M. C., & Braga, M. E. M. (2023). Edible films and coatings based on agrifood residues: A new trend in the food packaging research. *Current Opinion in Food Science*, *50*, 101006. DOI: 10.1016/j.cofs.2023.101006

Gaspar, R. F., Juliano, J. V., Natanauan, J., Zapanta, D. J., Santos, A., & Gevaña, S. (2023). Low-Cost Solar Powered Automated Modular Aquaponic System. In *TENCON 2023 - 2023 IEEE Region 10 Conference (TENCON)*. IEEE., DOI: 10.1109/TENCON58879.2023.10322522

Gatiso, T. T., Kulik, L., Bachmann, M., Bonn, A., Bösch, L., Freytag, A., Heurich, M., Wesche, K., Winter, M., Ordaz-Németh, I., Sop, T., & Kühl, H. S. (2022). Sustainable Protected Areas: Synergies between Biodiversity Conservation and Socioeconomic Development. *People and Nature*, *4*(4), 893–903. DOI: 10.1002/pan3.10326

Gatt, I., & Refalo, P. (2021). Life cycle assessment of recyclable, reusable and dematerialised plastic cosmetic packages. *IOP Conference Series. Materials Science and Engineering*, *1196*(1), 012022. DOI: 10.1088/1757-899X/1196/1/012022

Gebresenbet, G., & Bosona, T. (2012). Logistics practices in Ethiopia. *International Journal of Logistics Systems and Management*, *12*(2), 133–154.

Gembicki, J. (2016). Energy efficiency in the agricultural and food industry illustrated with the example of the feed production plant. In *E3S Web of Conferences* (Vol. 10, p. 00138). EDP Sciences.

Gentle, A. R., Aguilar, J. L. C., & Smith, G. B. (2011). Optimized cool roofs: Integrating albedo and thermal emittance with R-value. *Solar Energy Materials and Solar Cells*, *95*(12), 3207–3215. DOI: 10.1016/j.solmat.2011.07.018

Geographical Indications of Goods (Registration and Protection) Act, 1999. (2019). Government of India.

Geueke, B., Groh, K., & Muncke, J. (2018). Food packaging in the circular economy: Overview of chemical safety aspects for commonly used materials. *Journal of Cleaner Production*, *193*, 491–505. DOI: 10.1016/j.jclepro.2018.05.005

Ghisellini, P., Ncube, A., Rotolo, G., Vassillo, C., Kaiser, S., Passaro, R., & Ulgiati, S. (2023). Evaluating Environmental and Energy Performance Indicators of Food Systems, within Circular Economy and "Farm to Fork". *Energies*, *4*(4), 1671. DOI: 10.3390/en16041671

Ghosh, A., Srivastava, A., & Bansal, S. (2023). Energy optimization and automation in food manufacturing processes. *International Journal of Energy Research*, *47*(6), 245–263.

Ghosh, A., Srivastava, A., & Bansal, S. (2023). Harnessing AI and IoT for energy optimization in the food industry. *Energy Efficiency*, *16*(4), 267–284.

Giannakourou, M. C., & Tsironi, T. N. (2021). Application of processing and packaging hurdles for fresh-cut fruits and vegetables preservation. *Foods*, *10*(4), 830. DOI: 10.3390/foods10040830 PMID: 33920447

Gielen, D., & Taylor, P. (2009). Indicators for industrial energy efficiency in India. *Energy*, *8*(8), 962–969. DOI: 10.1016/j.energy.2008.11.008

Gillani, S. A., Abbasi, R., Martinez, P., & Ahmad, R. (2022). Review on Energy Efficient Artificial Illumination in Aquaponics. *Cleaner and Circular Bioeconomy*, *2*(100015), 100015. DOI: 10.1016/j.clcb.2022.100015

Giménez, B., Zaritzky, N., & Graiver, N. (2024). Ozone treatment of meat and meat products: A review. *Frontiers in Food Science and Technology*, *4*, 1351801. DOI: 10.3389/frfst.2024.1351801

Gong, M., & Wall, G. (2001). On exergy and sustainable development—Part 2: Indicators and methods. Exergy. *International Journal (Toronto, Ont.)*, *4*(4), 217–233. DOI: 10.1016/S1164-0235(01)00030-9

González-González, A., et al. (2020). *Energy consumption optimization in dairy*

González-González, A.. (2019). Energy use in the dairy industry: Overview and opportunities for improvements. *Dairy Science & Technology*.

Gonzalez, M. A., Hernandez, L., Toro, C. R., Ruiz, H. A., Valdes, J. A., Herrera, R., & Aguilar, C. N. (2017). Rhizopus oryzae–Ancient microbial resource with importance in modern food industry. *International Journal of Food Microbiology*, *257*, 110–127. DOI: 10.1016/j.ijfoodmicro.2017.06.012 PMID: 28651077

Gorjian, S., Fakhraei, O., Gorjian, A., Sharafkhani, A., & Aziznejad, A. (2022). Sustainable Food and Agriculture: Employment of Renewable Energy Technologies. *Current Robotics Reports*, *3*(3), 153–163. DOI: 10.1007/s43154-022-00080-x PMID: 35600253

Gorjian, S., Hashjin, T., & Ghobadian, B. (2011). Solar Powered Greenhouses. *10th International Conference on Sustainable Energy Technologies on Sustainable Energy Technologies*

Gorjian, S., Kamrani, F., Ebadi, H., Samanta, S., & Savoldi, L. (2022). Applications of Renewable Energy Sources in Agriculture from a Complementarity Perspective. In *Complementarity of Variable Renewable Energy Sources* (pp. 615–647). Elsevier. DOI: 10.1016/B978-0-323-85527-3.00019-4

Gorris, L. G. M., & Peppelenbos, H. W. (2020). Modified-atmosphere packaging of produce. In *Handbook of food preservation* (pp. 349–362). CRC Press. DOI: 10.1201/9780429091483-26

Gouda, M. H. B., & Duarte-Sierra, A. (2024). An Overview of Low-Cost Approaches for the Postharvest Storage of Fruits and Vegetables for Smallholders, Retailers, and Consumers. *Horticulturae*, *10*(8), 803. DOI: 10.3390/horticulturae10080803

Gould, G. W. (1996). Methods for preservation and extension of shelf life. *International Journal of Food Microbiology*, *33*(1), 51–64. DOI: 10.1016/0168-1605(96)01133-6 PMID: 8913809

Gradziuk, P., Klepacki, B., & Stolarski, M. J. (Eds.). (2022). *Energy Supply within Sustainable Agricultural Production: Challenges, Policies and Mechanisms*. Mdpi AG. DOI: 10.3390/books978-3-0365-5176-0

Granato, G., Fischer, A., & Trijp, J. (2022). The price of sustainability: How consumers trade-off conventional packaging benefits against sustainability. *Journal of Cleaner Production*, *365*, 132739. DOI: 10.1016/j.jclepro.2022.132739

Greco, C., Campiotti, A., De Rossi, P., Febo, P., & Giagnacovo, G. (2020). Energy consumption and improvement of energy efficiency for the European agricultural-food system. *Rivista di Studi sulla Sostenibilità*, *1*(1), 91–103. DOI: 10.3280/RISS2020-001006

Grgic, I. (2024). *Unfair Trading Practices in Contract Farming-Ensuring Commercial Justice and Contract Enforcement*. University of Pittsburgh.

Grigore, M. E. (2017). Methods of recycling, properties and applications of recycled thermoplastic polymers. *Recycling*, *2*(4), 24. DOI: 10.3390/recycling2040024

Grondahl, M., Gustafsson, A., & Gatenholm, P. (2006). Gas-phase surface flourination of arabinoxlyan films. *Macromolecules*, *39*(7), 2718–2712. DOI: 10.1021/ma052066q

Guillard, V., Gaucel, S., Fornaciari, C., Angellier-Coussy, H., Buche, P., & Gontard, N. (2018). The next generation of sustainable food packaging to preserve our environment in a circular economy context. *Frontiers in Nutrition*, *5*, 121. Advance online publication. DOI: 10.3389/fnut.2018.00121 PMID: 30564581

Güler, O., & Uyar, A. (2018). Dynamic optimization of energy consumption in the food industry: A review of strategies and solutions. *Energy Efficiency*, *11*(3), 729–743. DOI: 10.1007/s12053-017-9577-0

Gunawan, A., Amalia, F. A., Ramadhan, M., & Bansah, P. F. (2024). Exploring the reasons of indonesian young adult consumers toward sustainably packaged food & beverages product. *Journal of Marketing Innovation (Jmi)*, *4*(1). Advance online publication. DOI: 10.35313/jmi.v4i1.106

Guo, Y., Han, M., Chen, L., Zeng, X., Wang, P., Xu, X., Feng, X., & Lu, X. (2024). Pulsed electric field: A novel processing technology for meat quality enhancing. *Food Bioscience*, *58*, 103645. DOI: 10.1016/j.fbio.2024.103645

Gupta, P. (2023, February 9). Role of oxygen absorbers in food as packaging material, their characterization and applications. Springer Science+Business Media, 61(2), 242-252. DOI: 10.1007/s13197-023-05681-8

Gupta, A., & Dey, P. K. (2017). Energy management in food industry: A review. *Journal of Food Engineering*, *193*, 47–58. DOI: 10.1016/j.jfoodeng.2016.09.015

Gupta, P., Toksha, B., & Rahaman, M. (2022). A review on biodegradable packaging films from vegetative and food waste. *Chemical Record (New York, N.Y.)*, *22*(7), e202100326. DOI: 10.1002/tcr.202100326 PMID: 35253984

Gupta, S. P., & Bhardwaj, A. (2024). Technology of Sensors. In *Advances in Environmental Engineering and Green Technologies* (pp. 1–9). IGI Global., DOI: 10.4018/979-8-3693-1930-7.ch001

Gustavo, J.Jr, Pereira, G., Bond, A., Viegas, C., & Borchardt, M. (2018). Drivers, opportunities and barriers for a retailer in the pursuit of more sustainable packaging redesign. *Journal of Cleaner Production*, *187*, 18–28. DOI: 10.1016/j.jclepro.2018.03.197

Gutiérrez, J., Villa-Medina, J. F., Nieto-Garibay, A., & Porta-Gándara, M. A. (2013). Automated irrigation system using a wireless sensor network and GPRS module. *IEEE Transactions on Instrumentation and Measurement*, *63*(1), 166–176. DOI: 10.1109/TIM.2013.2276487

Guven, T. B., Degirmenci, H., Gunerhan, A., & Altuntas, O. (2024). HursitDegirmenci, Ali Gunerhan, and Onder Altuntas. "Energy efficiency and sustainability: Implementing circular economy principles for cabin waste management in aviation.". *Energy for Sustainable Development*, *81*, 101515. DOI: 10.1016/j.esd.2024.101515

Guzdemir, O. (2024). Sustainable Food Packaging Solutions: Polysaccharide-Based Films and Coatings. In *Sustainable Engineering: Concepts and Practices* (pp. 73–95). Springer.

Haas, C.. (2017). Energy consumption and greenhouse gas emissions in milk powder production. *Renewable & Sustainable Energy Reviews*.

Haddad, S. M.. (2020). The Internet of Things in Food Safety: Research Progress and Future Directions. *Comprehensive Reviews in Food Science and Food Safety, 19*(3), 1192–1209.

Hahladakis, J. N., Iacovidou, E., & Gerassimidou, S. (2024). Plastic waste in a circular economy. In *Environmental Materials and Waste* (pp. 99–134). Elsevier. DOI: 10.1016/B978-0-443-22069-2.00003-6

Haji, M., & Himpel, F. (2024). Building Resilience in Food Security: Sustainable Strategies Post-COVID-19. *Sustainability (Basel), 16*(3), 995. DOI: 10.3390/su16030995

Halpin, M. (2013). Advancing standards in energy efficiency. *IEEE Industry Applications Magazine, 19*(4), 78–79. DOI: 10.1109/MIAS.2013.2255426

Hamid, L. and Samy, I. (2022). Fabricating natural biocomposites for food packaging.. DOI: 10.5772/intechopen.100907

Han, J. H. (2014). A review of food packaging technologies and innovations. Innovations in food packaging, 3-12.

Han, J., Dong, P., & Benjamin, W. B. (2024). Holman, Huixuan Yang, Xue Chen, Lixian Zhu, Xin Luo, Yanwei Mao, and Yimin Zhang. "Processing interventions for enhanced microbiological safety of beef carcasses and beef products: A review.". *Critical Reviews in Food Science and Nutrition, 64*(8), 2105–2129. DOI: 10.1080/10408398.2022.2121258 PMID: 36148812

Han, J., Ruiz-García, L., Qian, J., & Yang, X. (2018). Food packaging: A comprehensive review and future trends. *Comprehensive Reviews in Food Science and Food Safety, 17*(4), 860–877. DOI: 10.1111/1541-4337.12343 PMID: 33350114

Hasan, I., Habib, M., & Mohamed, Z. (2023). Blockchain Database and IoT: A Technology driven Agri-Food Supply Chain. *International Supply Chain Technology Journal, 3*(3), 40–45. DOI: 10.20545/isctj.v09.i03.01

Haskell, M. J.. (2020). Animal welfare in dairy production: An overview of recent research and practices. *Animal Welfare (South Mimms, England).*

Herbes, C., Beuthner, C., & Ramme, I. (2020). How green is your packaging—A comparative international study of cues consumers use to recognize environmentally friendly packaging. *International Journal of Consumer Studies, 44*(3), 258–271. DOI: 10.1111/ijcs.12560

Hernandez-Escobedo, Q., Muñoz-Rodríguez, D., Vargas-Casillas, A., José, M. J. L., Aparicio-Martínez, P., Martínez-Jiménez, M. P., & Perea-Moreno, A.-J. (2022). Renewable Energies in the Agricultural Sector: A Perspective Analysis of the Last Three Years. *Energies*, *16*(1), 345. DOI: 10.3390/en16010345

Herrero, M.. (2021). Livestock and climate change: A review of the potential impacts of climate change on dairy production systems. *Environmental Research Letters*.

Himanshu, D. D. S., & Bhavani, M. (2023). *Polysaccharide, protein, lipid-based coatings and their impact on fruit crops*.

Ho An, C., Jong Ri, H., Han, T., Kim, S., & Song Ju, U. (2022). Feasibility of winter cultivation of fruit vegetables in a solar greenhouse in temperate zone; experimental and numerical study. *Solar Energy*, *233*, 18–30. DOI: 10.1016/j.solener.2022.01.024

Hoffmann, V., Moser, C., & Saak, A. (2019). Food safety in low and middle-income countries: The evidence through an economic lens. In *World Development, 123* (Vol. 104611). Elsevier BV., DOI: 10.1016/j.worlddev.2019.104611

Holka, M., Kowalska, J., & Jakubowska, M. (2022). Reducing Carbon Footprint of Agriculture—Can Organic Farming Help to Mitigate Climate Change? *Agriculture*, *12*(9), 1383. DOI: 10.3390/agriculture12091383

Honcharuk, I. (2023). European experience of biogas and biomethane production from waste according to the circular economy principle in agriculture. Economy, finances, management: topical issues of science and practical activity, 4(66), 7-19.

Hossain, R., Ghinangju, B. S., Biswal, S., Schandl, H., & Sahajwalla, V. (2024). Current technological options for recycling packaging waste: Challenges and opportunities in India. *Wiley Interdisciplinary Reviews. Energy and Environment*, *13*(1), e500. DOI: 10.1002/wene.500

Hristov, A. N.. (2022). Mitigation of methane emissions from ruminants through dietary strategies: A review. *Journal of Animal Science*.

Huang, L., & Hwang, C. A. (2012). *In-package pasteurisation of ready-to-eat meat and poultry products Advances in Meat, Poultry and Seafood Packaging*. Woodhead Publishing Limited.

Hu, G., & You, F. (2022). Renewable Energy-Powered Semi-Closed Greenhouse for Sustainable Crop Production Using Model Predictive Control and Machine Learning for Energy Management. *Renewable & Sustainable Energy Reviews*, *168*(112790), 112790. DOI: 10.1016/j.rser.2022.112790

Hughes, M. L., & Garcia, L. A. (2019). Advances in energy-efficient technologies for food safety, quality, and sustainability. *Comprehensive Reviews in Food Science and Food Safety*, *18*(2), 445–463. DOI: 10.1111/1541-4337.12456

Hui, L., Gui-rong, W., Jian-ping, W., & Peiyong, D. (2017, May). Monitoring platform of energy management system for smart community. In 2017 29th Chinese Control And Decision Conference (CCDC) (pp. 1832-1836). IEEE. DOI: 10.1109/CCDC.2017.7978814

Hyder, M. J., Khan, M. J., Khan, M. A., & Saeed, S. (2023). The Design and Development of a Solar Dehydrator for Fruits. *Engineering Proceedings*, *2023*(45), 48.

Ibatulin, I., Nedaskivsky, V., Golembivsky, S., Artimonova, I., & Prysiazhnyuk, N. (2024). Problems and prospects of the functioning of agricultural chains in the meat breeding of Ukraine. *Ekonomichnyyanaliz*, *34*(2), 9–19. DOI: 10.35774/econa2024.02.009

Indian Council of Agricultural Research. (2017). *Intellectual property and food security: An Indian perspective*. ICAR.

Indiarto, R., Pratama, A. W., Sari, T. I., & Theodora, H. C. (2020). Food irradiation technology: A review of the uses and their capabilities. *International Journal of Engineering Trends and Technology*, *68*(12), 91–98. DOI: 10.14445/22315381/IJETT-V68I12P216

Iñiguez-Moreno, M., Ragazzo-Sánchez, J. A., & Calderón-Santoyo, M. (2021). An extensive review of natural polymers used as coatings for postharvest shelf-life extension: Trends and challenges. *Polymers*, *13*(19), 3271. DOI: 10.3390/polym13193271 PMID: 34641086

International Energy Agency (IEA). (2021). Tracking Industry 2021: Focus on dairy sector energy use.

International Organization for Standardization, "ISO 50001. (2011). Swedish Standards Institute.

IRENA. (2016). *Solar pumping for irrigation: Improving livelihoods and sustainability*. The International Renewable Energy Agency.

Irradiation, F. What You Need to Know. (2022, January 9). https://www.fda.gov/food/buy-store-serve-safe-food/food-irradiation-what-you-need-know

Ishangulyyev, R., Kim, S., & Lee, S. H. (2019). Understanding food loss and waste—Why are we losing and wasting food? *Foods*, *8*(8), 297. DOI: 10.3390/foods8080297 PMID: 31362396

ISO. (2023). *ISO 50001:2018 - Energy management systems*. International Organization for Standardization.

Ivonkovic, A., Zeljko, K., & Talic, S., & Lasic, M. (2017). Biodegradable packaging in the food industry. *J. Food Saf. Food Qual*, *68*(2), 26–38.

Jackson-Davis, A., White, S., Kassama, L. S., Coleman, S., Shaw, A., Mendonca, A., Cooper, B., Thomas-Popo, E., Gordon, K., & London, L. (2023). A review of regulatory standards and advances in essential oils as antimicrobials in foods. *Journal of Food Protection*, *86*(2), 100025. DOI: 10.1016/j.jfp.2022.100025 PMID: 36916569

Jadhav, N., Premalatha, S., & Poongodi, P. (2024). Enhancing Crop Growth Efficiency through IoT-enabled Smart Farming System. EAI Endorsed Transactions on Internet of Things, 10.

Jafari, A., & Kheradmand, A. (2021). Data-driven energy optimization in the food industry: Insights from industrial case studies. *Journal of Cleaner Production*, *285*, 125417. DOI: 10.1016/j.jclepro.2020.125417

Jafarzadeh, S., Nafchi, A. M., Salehabadi, A., Oladzad-Abbasabadi, N., & Jafari, S. M. (2021). Application of bio-nanocomposite films and edible coatings for extending the shelf life of fresh fruits and vegetables. *Advances in Colloid and Interface Science*, *291*, 102405. DOI: 10.1016/j.cis.2021.102405 PMID: 33819726

Jagtap, S., Rahimifard, S., & Duong, L. N. K. (2019a). Real-time data collection to improve energy efficiency: A case study of food manufacturer. *Journal of Food Processing and Preservation*, *8*. Advance online publication. DOI: 10.1111/jfpp.14338

Jahangiri, F., Mohanty, A. K., & Misra, M. (2024). Sustainable biodegradable coatings for food packaging: Challenges and opportunities. *Green Chemistry*, *26*(9), 4934–4974. DOI: 10.1039/D3GC02647G

Jain, S., Saxena, S., Minz, V., Behera, S. D., Harini, K., Mishra, S., & Nidhi, N. (2023). Post Harvest Handling of Fruit Crops. *International Journal of Environment and Climate Change*, *13*(11), 1990–1999. DOI: 10.9734/ijecc/2023/v13i113357

Jaiswal, S., Rawat, G., Khadse, C., & Sharma, S. (2024). A Smart Hydroponics System for Sustainable Agriculture. In *Data Science for Agricultural Innovation and Productivity* (pp. 25–47). BENTHAM SCIENCE PUBLISHERS. DOI: 10.2174/9789815196177124010006

Jambrak, A., Donsì, F., Paniwnyk, L., & Đjekić, I. (2019, February 20). Impact of Novel Nonthermal Processing on Food Quality: Sustainability, Modelling, and Negative Aspects. *Journal of Food Quality*, *1-2*, 1–2. Advance online publication. DOI: 10.1155/2019/2171375

James, S. J., & James, C. (2010). The food cold-chain and climate change. *Food Research International, 43*(7), 1944–1956. DOI: 10.1016/j.foodres.2010.02.001

Jarman, A., Thompson, J., McGuire, E., Reid, M., Rubsam, S., Becker, K., & Mitcham, E. (2023). Postharvest technologies for small-scale farmers in low-and middle-income countries: A call to action. *Postharvest Biology and Technology, 206*, 112491. DOI: 10.1016/j.postharvbio.2023.112491

Javaid, M., Haleem, A., Singh, R. P., Rab, S., & Suman, R. (2021). Exploring the potential of nanosensors: A brief overview. *Sensors International, 2*, 100130. DOI: 10.1016/j.sintl.2021.100130

Jedlicka, W. (2008). *Packaging sustainability: tools, systems and strategies for innovative package design*. John Wiley and Sons.

Jeya Jeevahan, J., Chandrasekaran, M., Venkatesan, S. P., Sriram, V., Britto Joseph, G., Mageshwaran, G., & Durairaj, R. B. (2020). Scaling up difficulties and commercial aspects of edible films for food packaging: A review. *Trends in Food Science & Technology, 100*, 210–222. DOI: 10.1016/j.tifs.2020.04.014

Jha, S., Akula, B., Enyioma, H., Novak, M., Amin, V., & Liang, H. (2024). Biodegradable Biobased Polymers: A Review of the State of the Art, Challenges, and Future Directions. *Polymers, 16*(16), 2262. DOI: 10.3390/polym16162262 PMID: 39204482

Jhaveri, R. H., Revathi, A., Ramana, K., Raut, R., & Dhanaraj, R. K. (2022). A Review on Machine Learning Strategies for Real-World Engineering Applications. *Mobile Information Systems, 2022*(1), 1833507. DOI: 10.1155/2022/1833507

Jia, X., Wu, Q., & Wang, R. (2022). Biomass energy applications in food processing. *Renewable & Sustainable Energy Reviews, 146*, 111062.

Jideani, A I O., Billard, H., Mchau, G R., Udoro, E O., & Onipe, O O. (2017, September 13). Processing and Preservation of Fresh-Cut Fruit and Vegetable Products. DOI: 10.5772/intechopen.69763

Ji, L., Nie, Q., Zhou, Y., Ning, J., Wang, S., Zhou, C., Luo, Y., Yang, R., & Zhang, J. (2024). Advances and applications in water retention technology for meat and meat products: Status and future research directions. *International Journal of Food Science & Technology, 59*(5), 2823–2836. DOI: 10.1111/ijfs.16980

John, A., Isaac, W.-A., Daley, O., & Kahramano\uglu, \.Ibrahim. (2024). Biological and Environmental Factors Affecting Postharvest Quality of Fruits and Vegetables. In *Postharvest Physiology and Handling of Horticultural Crops* (pp. 36–64). CRC Press.

John, A. T., Murugappan, K., Nisbet, D. R., & Tricoli, A. (2021). An outlook of recent advances in chemiresistive sensor-based electronic nose systems for food quality and environmental monitoring. *Sensors (Basel)*, *21*(7), 2271. DOI: 10.3390/s21072271 PMID: 33804960

Jones, S., & Mark Gillott, T. (2021). A multi-energy system optimisation software for advanced process control using hypernetworks and a micro-service architecture. *Energy Reports*, *7*(4), 167–175. DOI: 10.1016/j.egyr.2021.08.159

Jose, A., Pareek, S., & Radhakrishnan, E. K. (2020). Advances in edible fruit coating materials. *Advances in Agri-Food Biotechnology*, 391–408.

Joshi, A., Gupta, A. K., Semwal, S., Deoli, N., Rather, M. A., Naik, B., Kumar, V., Jha, A. K., Rustagi, S., & Preet, M. S. (2024). Sustainable solutions for food security: Evaluating pre-treatment technologies in the growing fruits and vegetables industry of India. *Sustainable Chemistry and Pharmacy*, *39*, 101580. DOI: 10.1016/j.scp.2024.101580

Jouhara, H., Khordehgah, N., Almahmoud, S., Delpech, B., Chauhan, A., & Tassou, S. A. (2018). Waste heat recovery technologies and applications. *Thermal Science and Engineering Progress*, *6*, 268–289. DOI: 10.1016/j.tsep.2018.04.017

Joy, G.-F. F., Sunday, O. Z., Remilekun, A., & Olusegun, A. D. (2022). Waste-to-wealth; nutritional potential of five selected fruit peels and their health benefits: A review. *African Journal of Food Science*, *16*(7), 172–183. DOI: 10.5897/AJFS2021.2138

Junaid, M. D., & Gokce, A. F. (2024). Global agricultural losses and their causes. *Bulletin of Biological and Allied Sciences Research*, *2024*(1), 66. DOI: 10.54112/bbasr.v2024i1.66

Jursová, S., Burchart-Korol, D., & Pustějovská, P. (2019). Carbon footprint and water footprint of electric vehicles and batteries charging in view of various sources of power supply in the czech republic. *Environments (Basel, Switzerland)*, *6*(3), 38. DOI: 10.3390/environments6030038

Kadi, R. H. (2023). Development of zinc oxide nanoparticles as safe coating for the shelf life extension of grapes (Vitisvinifera L., Red Globe) fruits. *Materials Express*, *13*(1), 182–188. DOI: 10.1166/mex.2023.2334

Kalyan, A. R. (n.d.). *"Ultra-Processed vs. Minimally Processed & Whole Foods."*https://continentalhospitals.com/blog/how-ultra-processed-foods-differ-from-minimally-processed-and-whole-foods/

Kaminski, J., & Leduc, G. (2010). *Energy efficiency improvement options for the EU food industry.* POLITYKA ENERGETYCZNA Tom13 Zeszyt 1 PL ISSN:1429-6675.

Kanchiralla, F. M., Jalo, N., Thollander, P., Andersson, M., & Johnsson, S. (2020). Energy end-use categorization and performance indicators for energy management in the engineering industry. *Energies*, *13*(2), 369. DOI: 10.3390/en13020369

Kang, M., Lee, J. E., Sohn, S., Kim, J., Cho, S., Park, J., Kang, N., & Park, K. S. (2024). Estimation of cucumber net primary production using environmental and control information in a smart multi-span plastic greenhouse. *Computers and Electronics in Agriculture*, *219*, 108819. DOI: 10.1016/j.compag.2024.108819

Kariyanna, B., & Sowjanya, M. (2024). Unravelling the use of artificial intelligence in management of insect pests. *Smart Agricultural Technology*, *8*, 100517. DOI: 10.1016/j.atech.2024.100517

Karkou, E., Teo, C. J., Savvakis, N., Poinapen, J., & Arampatzis, G. "Industrial Circular Water Use Practices Through the Application of a Conceptual Water Efficiency Framework in the Process Industry." *Available atSSRN* 4790097. DOI: 10.1016/j.jenvman.2024.122596

Kar, S. K., Pradhan, N. C., Swain, S. S., & Kar, S. (2023). Applicability of Various Renewable Energy Sources to Agricultural Applications. In *Handbook of Energy Management in Agriculture* (pp. 1–19). Springer Nature Singapore. DOI: 10.1007/978-981-19-7736-7_22-1

Kasera, R. K., & Acharjee, T. (2024). A Comprehensive IoT edge based smart irrigation system for tomato cultivation. *Internet of Things : Engineering Cyber Physical Human Systems*, *28*, 101356. DOI: 10.1016/j.iot.2024.101356

Kashyap, P. K., Kumar, S., Jaiswal, A., Prasad, M., & Gandomi, A. H. (2021). Towards precision agriculture: IoT-enabled intelligent irrigation systems using deep learning neural network. *IEEE Sensors Journal*, *21*(16), 17479–17491. DOI: 10.1109/JSEN.2021.3069266

Kathpalia, A., Arora, S., & Sharma, J. G. (2022). Post-Harvest Pest Management of Perishables. *RE:view*.

Katti, K S., Priyadarshini, S., Leena, M M., Moses, J., & Anandharamakrishnan, C. (2019, November 1). Intelligent packaging: Trends and applications in food systems. Elsevier BV, 93, 145-157. DOI: 10.1016/j.tifs.2019.09.008

Kaur, R., & Bhagat, D. (2020). Application of Internet of Things (IoT) in energy management systems for the food industry. *Energy Reports*, *6*, 1245–1256. DOI: 10.1016/j.egyr.2020.05.008

Kaushani, K., Rathnasinghe, N., Katuwawila, N., Jayasinghe, R., Nilmini, A., & Priyadarshana, G. (2022). Trends in smart packaging technologies for sustainable monitoring of food quality and safety. International Journal of Research and Innovation in Applied Science, 07(07), 07-30. DOI: 10.51584/IJRIAS.2022.7702

Kazancoglu, Y., Ozbiltekin-Pala, M., Sezer, M. D., Luthra, S., & Kumar, A. (2022). Resilient reverse logistics with blockchain technology in sustainable food supply chain management during COVID-19. *Business Strategy and the Environment*, *4*(4), 2327–2340. DOI: 10.1002/bse.3251

Kebour, O., & Kebour, D. (2023). A practical approach for Energy Management, in the Agro-food sector industry in Algeria. *2023 International Conference on Advances in Electronics, Control and Communication Systems (ICAECCS)*, 1–6. https://doi.org/DOI: 10.1109/ICAECCS56710.2023.10105084

Kelliher, M., Bogueva, D., & Marinova, D. (2024). Meta-Analysis and Ranking of the Most Effective Methane Reduction Strategies for Australia's Beef and Dairy Sector. *Climate (Basel)*, *12*(4), 50. DOI: 10.3390/cli12040050

Khan, M. S., & Abbas, A. (2021). Exploring AI-based energy optimization in food processing industries: A systematic review. *Energy and Building*, *245*, 111027. DOI: 10.1016/j.enbuild.2021.111027

Khan, O. A., Zaidi, S., Islam, R. U., Naseem, S., & Junaid, P. M. (2023). Enhanced shelf-life of peach fruit in alginate based edible coating loaded with TiO2 nanoparticles. *Progress in Organic Coatings*, *182*, 107688. DOI: 10.1016/j.porgcoat.2023.107688

Khayatnezhad, M., Fataei, E., & Imani, A. (2023). Retracted: Integrated Modeling of Food–water–energy Nexus for Maximizing Water Productivity. *Water Science and Technology: Water Supply*, *23*(3), 1362–1374. DOI: 10.2166/ws.2023.038

Kheiralipour, K., Rafiee, S., Karimi, M., Nadimi, M., & Paliwal, J. (2024). The environmental impacts of commercial poultry production systems using life cycle assessment: A review. *World's Poultry Science Journal*, *80*(1), 33–54. DOI: 10.1080/00439339.2023.2250326

Kim, Y.-J., Cha, J. Y., Kim, T.-K., Lee, J. H., Jung, S., & Choi, Y.-S. (2024). The Effect of Irradiation on Meat Products. *Food Science of Animal Resources*, *44*(4), 779–789. DOI: 10.5851/kosfa.2024.e35 PMID: 38974724

Kinney, C., Dehghani-Sanij, A., Mahbaz, S., Dusseault, M. B., Nathwani, J. S., & Fraser, R. A. (2019). Geothermal Energy for Sustainable Food Production in Canada's Remote Northern Communities. *Energies*, *12*(21), 4058. DOI: 10.3390/en12214058

Kirci, M., Isaksson, O., & Seifert, R. (2022). Managing perishability in the fruit and vegetable supply chains. *Sustainability (Basel), 14*(9), 5378. DOI: 10.3390/su14095378

Klassen, S., & Murphy, S. (2020). Equity as both a means and an end: Lessons for resilient food systems from COVID-19. In *World Development, 136, 105104*. Elsevier BV., DOI: 10.1016/j.worlddev.2020.105104

Kling, M. M., & Hough, I. J. (2010). *The American Carbon Footprint: Un derstanding your food's impact on climate change, Bright*. Planet, Inc.

Knorr, D., & Augustin, M. (2020). Food processing needs, advantages and misconceptions. *Trends in Food Science & Technology, 108*, 103–110. DOI: 10.1016/j.tifs.2020.11.026

Koch, S., & Müller, S. (2019). The role of energy management systems in achieving sustainability targets in food production. *Sustainability, 11*(11), 3050. DOI: 10.3390/su11113050

Kocira, A., Kozłowicz, K., Panasiewicz, K., Staniak, M., Szpunar-Krok, E., & Hortyńska, P. (2021). Polysaccharides as edible films and coatings: Characteristics and influence on fruit and vegetable quality—A review. *Agronomy (Basel), 11*(5), 813. DOI: 10.3390/agronomy11050813

Koeijer, B., Gelhard, C., & Klooster, R. (2019). Sustainability priorities across the strategic and operational level in packaging development. *Packaging Technology & Science, 32*(12), 618–629. DOI: 10.1002/pts.2480

Kok, C. L., Bagus Pradnya Kusuma, I. M., Yan Koh, Y., Tang, H., & Boon Lim, A. (2024). Smart Aquaponics: An Automated Water Quality Management System for Sustainable Urban Agriculture. *Electronics (Basel), 13*(5), 820. DOI: 10.3390/electronics13050820

Koulouris, A., Misailidis, N., & Petrides, D. (2021). Applications of process and digital twin models for production simulation and scheduling in the manufacturing of food ingredients and products. *Food and Bioproducts Processing, 126*, 317–333. DOI: 10.1016/j.fbp.2021.01.016

Kovacic, D., Loncaric, Z., Jovic, J., Samac, D., Popovic, B., & Tisma, M. (2022). Digestate management and processing practices: A review. *Applied Sciences (Basel, Switzerland), 12*(18), 9216. DOI: 10.3390/app12189216

Kozik, N. (2020). Sustainable packaging as a tool for global sustainable development. In *SHS web of conferences* (Vol. 74, p. 04012). EDP Sciences. DOI: 10.1051/shsconf/20207404012

Kraśniewska, K., Galus, S., & Gniewosz, M. (2020). Biopolymers-based materials containing silver nanoparticles as active packaging for food applications—A review. *International Journal of Molecular Sciences*, *21*(3), 698. DOI: 10.3390/ijms21030698 PMID: 31973105

Krithika, V., & Subramani, C. (2018). A comprehensive review on choice of hybrid vehicles and power converters, control strategies for hybrid electric vehicles. *International Journal of Energy Research*, *42*(5), 1789–1812. DOI: 10.1002/er.3952

Kumar, N., Devgan, K., Kumar, A., Kaur, P., & Mahajan, P. (2024). Active and Passive Modified Atmosphere Packaging: Recent Advances. *Nonthermal Food Engineering Operations*, 225–259.

Kumar, S. N., & Chakabarti, B. (2019). nergy and carbon footprint of food industry, in Energy footprints food Text. *Section. Springer*, pp. 19 44.

Kumari, M., Kumar, M., Solankey, S. S., & Tomar, S. (2021). Effect of Green House Gases on Vegetable Production. *Advances in Research on Vegetable Production Under a Changing Climate*, *1*, 211–219. DOI: 10.1007/978-3-030-63497-1_11

Kumar, P., & Kalita, P. (2017). Reducing postharvest losses during storage of grain crops to strengthen food security in developing countries. *Foods*, *6*(1), 8. DOI: 10.3390/foods6010008 PMID: 28231087

Kumar, R. (2016). Health effects of morpholine based coating for fruits and vegetables. *International Journal of Medical Research & Health Sciences*, *5*(9), 32–38.

Kumar, R., Chandra, S., Singh, B., Kumar, R., & Kumar, A. A. (2018). Zero energy cool chamber for food commodities: Need of eco-friendly storage facility for farmers: A review. *Journal of Pharmacognosy and Phytochemistry*, *7*(5), 2293–2301.

Kumar, R., & Kapur, S. (2016). Morpholine: A glazing agent for fruits and vegetables coating/waxing. *Int. J. Sci. Technol. Eng*, *2*(11), 694–697.

Kumar, R., & Sharma, A. (2024). Review of pet food packaging in the us market: Future direction towards innovation and sustainability. *Annual Research & Review in Biology*, *39*(6), 16–30. DOI: 10.9734/arrb/2024/v39i62085

Kumar, R., Verma, A., Shome, A., Sinha, R., Sinha, S., Jha, P., & Prasad, P. (2021). Impacts of plastic pollution on ecosystem services, sustainable development goals, and need to focus on circular economy and policy interventions. *Sustainability (Basel)*, *13*(17), 9963. DOI: 10.3390/su13179963

Kumar, S.. (2019). Sustainable livestock production: Environmental issues and their management. *Journal of Animal Science and Technology*, *61*, 1–11.

Kumar, S., & Rosen, M. A. (2016). A critical review of photovoltaic systems integrated with storage and possible configurations for grid-connected applications. *Renewable & Sustainable Energy Reviews*, *62*, 1192–1206.

Kumar, V., Sharma, K. V., Kedam, N., Patel, A., Kate, T. R., & Rathnayake, U. (2024). A comprehensive review on smart and sustainable agriculture using IoT technologies. *Smart Agricultural Technology*, *8*, 100487. DOI: 10.1016/j.atech.2024.100487

Ladha-Sabur, A., Bakalis, S., Fryer, P. J., & Lopez-Quiroga, E.Ladha-Sabura. (2019). Review Mapping energy consumption in food manufacturing. *Trends in Food Science & Technology*, *86*, 270–280. DOI: 10.1016/j.tifs.2019.02.034

Lal, R., & Yadav, S. S. (2018). Vulnerability of women to climate change in arid and semi-arid regions: The case of India and South Asia. *Journal of Arid Environments*, *149*, 4–17. DOI: 10.1016/j.jaridenv.2017.08.001

Lambert, S. (2013). *Environmental risk of polymer and their degradation products*. University of York.

Larramendy, M. L., & Soloneski, S. (2022). *Pesticides: Updates on Toxicity, Efficacy and Risk Assessment*. BoD--Books on Demand.

Laubinger, F., Brown, A., Dubois, M., & Börkey, P. (2021). Modulated fees for extended producer responsibility schemes (epr).. DOI: 10.1787/19970900

Laverdure, R. B. (2009). *Wind Systems Technology*. Renewable Energies., DOI: 10.1002/9780470612002.ch5

Lawrence, M. J. (2017). Disposal of plastics. In *A guide to the manufacture, performance, and potential of plastics in agriculture* (pp. 187–195). Elsevier. DOI: 10.1016/B978-0-08-102170-5.00010-5

Lebreton, L., Slat, B., Ferrari, F., Sainte-Rose, B., Aitken, J., Marthouse, R., & Reisser, J. (2018). Evidence that the Great Pacific Garbage Patch is rapidly accumulating plastic. *Scientific Reports*, *8*(1), 1–15. DOI: 10.1038/s41598-018-22939-w PMID: 29568057

Lee, D S. (2021, February 2). Gas Indicators and Sensors Used for MAP., 215-246. DOI: 10.1002/9781119530916.ch7

Lee, J. and Park, K. (2020). Because it is green or beautiful? consumer's perceived value of a unique type of sustainable packaging, store evaluation, and store patronage intentions.. DOI: 10.31274/itaa.12190

Lee, D., & Cheng, C. (2016). Energy savings by energy management systems: A review. *Renewable & Sustainable Energy Reviews*, *56*, 760–777. DOI: 10.1016/j.rser.2015.11.067

Lee, D., Shayan, M., Gwon, J., Picha, D. H., & Wu, Q. (2022). Effectiveness of cellulose and chitosan nanomaterial coatings with essential oil on postharvest strawberry quality. *Carbohydrate Polymers*, *298*, 120101. DOI: 10.1016/j.carbpol.2022.120101 PMID: 36241324

Lei, T., Min, Z., Gao, Q., Song, L., Zhang, X., & Zhang, X. (2022). The Architecture Optimization and Energy Management Technology of Aircraft Power Systems: A Review and Future Trends. *Energies*, *11*(11), 4109. DOI: 10.3390/en15114109

Leite, T., & Vicente, A. (2022). Design of a Smart Microgrid with Small-Scale Hydro Generation: A Practical Case Study. *Revista Facultad de Ingenieria Universidad de Antioquia (Medellín)*, (May). Advance online publication. DOI: 10.17533/udea.redin.20220577

Lemaire, G.. (2021). Regenerative agriculture practices for reducing greenhouse gas emissions in dairy farming. *Agricultural Systems*.

Li, B. (2024). Leading role of natural resources, eco-efficiency assessment, and energy transition in environmental sustainability: A depth of digital transformation. *Resources Policy*, *94*, 105145. DOI: 10.1016/j.resourpol.2024.105145

Li, D., Gan, M., Liu, X., Hu, Q., & Liu, X. (2024). Energy efficiency and driving factors of railway cold chain transportation in China. *Journal of Environmental Planning and Management*, *67*(4), 852–869. DOI: 10.1080/09640568.2022.2133688

Lignou, S., & Oloyede, O. (2021). Consumer acceptability and sensory profile of sustainable paper-based packaging. *Foods*, *10*(5), 990. DOI: 10.3390/foods10050990 PMID: 34062787

Liguori, G., Greco, G., Salsi, G., Garofalo, G., Gaglio, R., Barbera, M., Greco, C., Orlando, S., Fascella, G., & Mammano, M. M. (2024). Effect of the gellan-based edible coating enriched with oregano essential oil on the preservation of the 'Tardivo di Ciaculli'mandarin (Citrus reticulata Blanco cv. Tardivo di Ciaculli). *Frontiers in Sustainable Food Systems*, *8*, 1334030. DOI: 10.3389/fsufs.2024.1334030

Li, H.. (2020). Solar energy utilization in livestock farms: A review. *Renewable & Sustainable Energy Reviews*, *134*, 110124.

Li, J., & Song, W. (2022). Food Security Review Based on Bibliometrics from 1991 to 2021. *Foods*, *11*(23), 3915. DOI: 10.3390/foods11233915 PMID: 36496722

Li, L., Sun, J., Gao, H., Shen, Y., Li, C., Yi, P., He, X., Ling, D., Sheng, J., Li, J., Liu, G., Zheng, F., Xin, M., Li, Z., & Tang, Y. (2017). Effects of polysaccharide-based edible coatings on quality and antioxidant enzyme system of strawberry during cold storage. *International Journal of Polymer Science, 2017*, 1–8. DOI: 10.1155/2017/9746174

Li, M. J., & Tao, W. Q. (2017). Review of methodologies and polices for evaluation of energy efficiency in high energy-consuming industry. *Applied Energy*, *187*, 203–215. DOI: 10.1016/j.apenergy.2016.11.039

Lin, B., & Xie, X. (2015). Factor substitution and rebound effect in China's food industry. *Energy Conversion and Management*, *105*, 20–29. DOI: 10.1016/j.enconman.2015.07.039

Lioutas, E. D., Charatsari, C., De Rosa, M., & La Rocca, G. (2019). Key questions on the use of big data in farming: An activity theory approach. *Precision Agriculture*, *20*, 1105–1120.

Litter, M. I., & Ahmad, A. (2023). The world of nanotechnology. In *Industrial applications of nanoparticles* (pp. 1–15). CRC Press. DOI: 10.1201/9781003183525-1

Liu, X., Liu, J., Chi, Y., & Yang, Y. (2024). A coordinated planning and management framework for transmission and distribution systems with novel bilateral sharing energy storage model and time-phased consumption subsidy strategy. *Journal of Energy Storage*, *95*, 112377. DOI: 10.1016/j.est.2024.112377

Liu, Y., Xu, D., & Zhang, F. (2021). The role of smart sensors and IoT in reducing energy consumption in the food industry. *Sensors (Basel)*, *21*(8), 2716. PMID: 33921508

Li, Y., Erhunmwunsee, F., Liu, M., Yang, K., Zheng, W., & Tian, J. (2022). Antimicrobial mechanisms of spice essential oils and application in food industry. *Food Chemistry*, *382*, 132312. DOI: 10.1016/j.foodchem.2022.132312 PMID: 35158267

Lopez-Gonzalez, J.. (2021). Waste heat recovery in dairy processing plants: A review of opportunities and challenges. *Energy Conversion and Management*.

López-González, R., & Peña-García, A. (2021). Industrial energy efficiency and optimization in food manufacturing: A case study approach. *Journal of Industrial Engineering and Management*, *14*(1), 47–60. DOI: 10.3926/jiem.3336

Lukose, S., & Singh, K. "Sustainability in Food Process Development: A Comprehensive Carbon Footprint Analysis Tool." In *Sustainable Food Systems (Volume I) SFS: Framework, Sustainable Diets, Traditional Food Culture & Food Production*, pp. 55-71. Cham: Springer Nature Switzerland, 2024. DOI: 10.1007/978-3-031-47122-3_4

Lycourghiotis, S. (2022). Trends in renewable energy: An overview. *Global NEST: the International Journal, 24*(3), 505–525.

Maftoon Azad, N. (2006). *Evaluation of edible films and coatings for extending the postharvest shelf life of avocado.*

Magama, P., Chiyanzu, I., & Mulopo, J. (2022). A systematic review of sustainable fruit and vegetable waste recycling alternatives and possibilities for anaerobic biorefinery. *Bioresource Technology Reports, 18*, 101031. DOI: 10.1016/j.biteb.2022.101031

Magnier, L., & Schoormans, J. (2015). Consumer reactions to sustainable packaging: The interplay of visual appearance, verbal claim and environmental concern. *Journal of Environmental Psychology, 44*, 53–62. DOI: 10.1016/j.jenvp.2015.09.005

Mahajan, P. V., Rodrigues, F. A., Motel, A., & Leonhard, A. (2008). Development of a moisture absorber for packaging of fresh mushrooms (Agaricus bisporous). *Postharvest Biology and Technology, 48*(3), 408–414. DOI: 10.1016/j.postharvbio.2007.11.007

Mahmoud, A. R.. (2021). Predictive maintenance in food processing: A review. *Computers and Electronics in Agriculture, 183*, 106052.

Mahmoudi, M., & Parviziomran, I. (2020). Reusable packaging in supply chains: A review of environmental and economic impacts, logistics system designs, and operations management. *International Journal of Production Economics, 228*, 107730. DOI: 10.1016/j.ijpe.2020.107730

Majeed, Y., Khan, M. U., Waseem, M., Zahid, Z., Mahmood, F., Majeed, F., Sultan, M., & Raza, A. (2023). Renewable energy as an alternative source for energy management in agriculture. *Energy Reports, 10*, 344–359. DOI: 10.1016/j.egyr.2023.06.032

Makinde, P., & Obikoya, E. (2024). Implementation of Solar System for Electricity Generation for Rural Farmers: A Review. *World Journal of Advanced Research and Reviews, 22*(3), 458–471. DOI: 10.30574/wjarr.2024.22.3.1705

Mangaraj, S., & Goswami, T. K. (2009). Modified atmosphere packaging of fruits and vegetables for extending shelf-life-A review. *Fresh Produce, 3*(1), 1–31.

Mankhin, B., & Md, A. (2023). Market attractiveness of pineapple and banana agroforestry systems of Madhupur Sal (Shorea robusta) forest: A sustainable way of generating income and conserving forests. *Journal of Agriculture and Food Research, 11*, 100475. DOI: 10.1016/j.jafr.2022.100475

Manzoor, S., Fayaz, U., Dar, A. H., Dash, K. K., Shams, R., Bashir, I., Pandey, V. K., & Abdi, G. (2024). Sustainable Development Goals Through Reducing Food Loss and Food Waste: A Comprehensive Review. *Future Foods : a Dedicated Journal for Sustainability in Food Science, 9*, 100362. DOI: 10.1016/j.fufo.2024.100362

Margasahayam, A., & Balraj, Y. (2018, February 20). Properties of food ingredients during processing in a domestic mixer grinder and subsequent storage: A review. Wiley, 41(4). DOI: 10.1111/jfpe.12677

Marsden, T., & Morley, A. (2014). Current food questions and their scholarly challenges: Creating and framing a sustainable food paradigm. In *Sustainable food systems* (pp. 1–29). Routledge. DOI: 10.4324/9780203083499-9

Marsh, K., & Bugusu, B. (2007). Food packaging—Roles, materials, and environmental issues. *Journal of Food Science, 72*(3), R39–R55. DOI: 10.1111/j.1750-3841.2007.00301.x PMID: 17995809

Martinez, C. I. P., & Poveda, A. C. (2022). Characterization of cooling equipment in the food industry: Case study of the Colombian meat, dairy, and fruit and vegetable sectors. *Environmental Development, 41*, 100693. DOI: 10.1016/j.envdev.2021.100693

Martínez-Hernández, E., & Rodríguez, A. (2020). Energy-saving measures and energy management in food processing: A holistic approach. *Food Research International, 132*, 109040. DOI: 10.1016/j.foodres.2020.109040

Martin-Shields, C. P., & Stojetz, W. (2019). Food security and conflict: Empirical challenges and future opportunities for research and policy making on food security and conflict. In *World Development. 119, 150–164*. Elsevier BV., DOI: 10.1016/j.worlddev.2018.07.011

Martins, V. F. R., Pintado, M. E., Morais, R. M. S. C., & Morais, A. M. M. B. (2024). Recent Highlights in Sustainable Bio-Based Edible Films and Coatings for Fruit and Vegetable Applications. *Foods, 13*(2), 318. DOI: 10.3390/foods13020318 PMID: 38275685

Mason, R. E., White, A., Bucini, G., Anderzén, J., Méndez, V. E., & Merrill, S. C. (2020). The evolving landscape of agroecological research. *Agroecology and Sustainable Food Systems, 4*(4), 551–591. DOI: 10.1080/21683565.2020.1845275

Masriadi, L. F., & Gusmailina, L. M. P. (2024). Addressing the Global Food Security Crisis and Energy Shortages: Innovative Solutions and Policy Interventions for Sustainable Development. *The International Journal of Science in Society*, *6*(1), 748–759. DOI: 10.54783/ijsoc.v6i1.1060

Matloob, A., Ayub, H., Mohsin, M., Ambreen, S., Khan, F. A., Oranab, S., Rahim, M. A., Khalid, W., Nayik, G. A., Ramniwas, S., & Ercisli, S. (2023). A review on edible coatings and films: Advances, composition, production methods, and safety concerns. *ACS Omega*, *8*(32), 28932–28944. DOI: 10.1021/acsomega.3c03459 PMID: 37599927

Mattia, G., Leo, A., & Pratesi, C. (2021). Recognizing the key drivers and industry implications of sustainable packaging design: A mixed-method approach. *Sustainability (Basel)*, *13*(9), 5299. DOI: 10.3390/su13095299

Ma, W., & Li, L. (2024). Trends and prospects in sustainable food packaging materials. *Foods*, *13*(11), 1744. DOI: 10.3390/foods13111744 PMID: 38890972

May, G., Taisch, M., Prabhu, V. V., & Barletta, I. (2013). Energy Related Key Performance Indicators – State of the Art, Gaps and Industrial Needs. In IFIP Advances in Information and Communication Technology (pp. 257–267). Springer Berlin Heidelberg. https://doi.org/DOI: 10.1007/978-3-642-41266-0_32

Mayanja, I. K., & Oluk, I. (2023). Intermediate Technologies: The Key to Eradicating Post-harvest Losses (PHLs) in Sub-Saharan Africa (SSA). *Journal of Advances in Food Science \& Technology*, *10*(4), 64–74.

Mbadlisa, G., & Jokonya, O. (2024). Factors affecting the adoption of blockchain technologies in the food supply chain. *Frontiers in Sustainable Food Systems*, *8*, 1497599. Advance online publication. DOI: 10.3389/fsufs.2024.1497599

McCarthy, B. (2024). Green, guilty, and frugal: Facilitators of energy saving and the choice of energy-efficient appliances in the Australian energy market. *Energy Strategy Reviews*, *53*, 101400. DOI: 10.1016/j.esr.2024.101400

McNabola, A., García, A. M., & Juan, A. R. D. (2022). The Role of Micro-Hydropower Energy Recovery in the Water-Energy-Food Nexus. In *EWaS5*. MDPI., DOI: 10.3390/environsciproc2022021027

McPherson, B., Sharip, M., & Grimmond, T. (2019). The impact on life cycle carbon footprint of converting from disposable to reusable sharps containers in a large us hospital geographically distant from manufacturing and processing facilities. *PeerJ*, *7*, e6204. DOI: 10.7717/peerj.6204 PMID: 30809428

Meherishi, L., Narayana, S. A., & Ranjani, K. S. (2019). Sustainable packaging for supply chain management in the circular economy: A review. *Journal of Cleaner Production, 237*, 117–582. DOI: 10.1016/j.jclepro.2019.07.057

Mehmood, M. U., & Akhtar, P. (2020). A review of energy management systems in food processing and manufacturing: Current trends and future prospects. *Food Control, 110*, 107036. DOI: 10.1016/j.foodcont.2020.107036

Mekonnen, M. M., & Hoekstra, A. Y. (2016). Water footprint assessment of dairy products: A global perspective. *Water Resources and Industry*.

Mennerich, C., Pattison, M., & Morrow, W. (2017). The effect of LED lighting on broiler performance, energy use, and carcass quality. *Poultry Science, 96*(2), 423–431.

Menon, S., & Jain, K. (2024). Blockchain Technology for Transparency in Agri-Food Supply Chain: Use Cases, Limitations, and Future Directions. *IEEE Transactions on Engineering Management, 71*, 106–120. DOI: 10.1109/TEM.2021.3110903

Merino, D., Quilez-Molina, A. I., Perotto, G., Bassani, A., Spigno, G., & Athanassiou, A. (2022). A second life for fruit and vegetable waste: A review on bioplastic films and coatings for potential food protection applications. *Green Chemistry, 24*(12), 4703–4727. DOI: 10.1039/D1GC03904K

Mesterházy, Á., Oláh, J., & Popp, J. (2020). Losses in the grain supply chain: Causes and solutions. *Sustainability (Basel), 12*(6), 2342. DOI: 10.3390/su12062342

Mihalca, V., Kerezsi, A. D., Weber, A., Gruber-Traub, C., Schmucker, J., Vodnar, D. C., Dulf, F. V., Socaci, S. A., Fărca , A., Mure an, C. I., Suharoschi, R., & Pop, O. L. (2021). Protein-based films and coatings for food industry applications. *Polymers, 13*(5), 769. DOI: 10.3390/polym13050769 PMID: 33801341

Miller, D. A., & McGinity, J. W. (2008). Aqueous polymeric film coating. In *Pharmaceutical Dosage Forms-Tablets* (pp. 415–454). CRC Press.

Miller, J. D., & Babcock, B. A. (2019). The Role of Livestock in Sustainable Agricultural Production. *Journal of Animal Science, 97*(5), 1–12. PMID: 30873559

Minimally processed food | TABLE Debates. (n.d.). https://tabledebates.org/glossary/minimally-processed-food

Ministry of new and Renewable energy, Government of India. https://eerem.delhi.gov.in/sites/default/files/eerem/generic_multiple_files/com_solar_energy.pdf. Accessed date 14 January 2025. Identification of Industrial Sectors Promising for Commercialisation of Solar Energy Commercialisation of Solar Energy in Urban and Industrial Areas – ComSolar

Mishenin, Y., Zinovchuk, N., Vysochanska, M., & Botsula, A. (2023). Environmental and economic sustainability and conservation of natural resources in the context of rural territories. *Ukrainian Journal of Applied Economics and Technology*, *8*(4), 305–312. DOI: 10.36887/2415-8453-2023-4-49

Mitani, T. (2018). Recent Progress on Microwave Processing of Biomass for Bioenergy Production. *Journal of the Japan Petroleum Institute*, *61*(2), 113–120. DOI: 10.1627/jpi.61.113

Moeini, A., Pedram, P., Fattahi, E., Cerruti, P., & Santagata, G. (2022). Edible polymers and secondary bioactive compounds for food packaging applications: Antimicrobial, mechanical, and gas barrier properties. *Polymers*, *14*(12), 2395. DOI: 10.3390/polym14122395 PMID: 35745971

Mohamed, S. A. A., El-Sakhawy, M., & El-Sakhawy, M. A.-M. (2020). Polysaccharides, protein and lipid-based natural edible films in food packaging: A review. *Carbohydrate Polymers*, *238*, 116178. DOI: 10.1016/j.carbpol.2020.116178 PMID: 32299560

Mohammed, M. S., Yousif, F., & Rehman, S. (2021). Energy-efficient refrigeration systems in the food sector: A review of technologies and applications. *Journal of Refrigeration Science*, *54*(2), 85–99.

Mohammed, M. S., Yousif, F., & Rehman, S. (2021). Energy-efficient refrigeration systems in the food sector: Recent developments and future perspectives. *Journal of Refrigeration Science*, *54*(2), 85–99.

Mohan, A., Krishnan, R., Arshinder, K., Vandore, J., & Ramanathan, U. (2023). Management of postharvest losses and wastages in the Indian tomato supply chain—A temperature-controlled storage perspective. *Sustainability (Basel)*, *15*(2), 1331. DOI: 10.3390/su15021331

Mohd Basri, M. S., Abdul Karim Shah, N. N., Sulaiman, A., Mohamed Amin Tawakkal, I. S., Mohd Nor, M. Z., Ariffin, S. H., Abdul Ghani, N. H., & Mohd Salleh, F. S. (2021). Progress in the valorization of fruit and vegetable wastes: Active packaging, biocomposites, by-products, and innovative technologies used for bioactive compound extraction. *Polymers*, *13*(20), 3503. DOI: 10.3390/polym13203503 PMID: 34685262

Molina, J., Frías-Celayeta, J., Bolton, D., & Botinestean, C. (2024). Molina, Jan Roland G., Jesús M. Frías-Celayeta, Declan J. Bolton, and Cristina Botinestean. "A Comprehensive Review of Cured Meat Products in the Irish Market: Opportunities for Reformulation and Processing.". *Foods*, *13*(5), 746. DOI: 10.3390/foods13050746 PMID: 38472858

Mondal, K., Bhattacharjee, S. K., Mudenur, C., Ghosh, T., Goud, V. V., & Katiyar, V. (2022). Development of antioxidant-rich edible active films and coatings incorporated with de-oiled ethanolic green algae extract: A candidate for prolonging the shelf life of fresh produce. *RSC Advances, 12*(21), 13295–13313. DOI: 10.1039/D2RA00949H PMID: 35520137

Moradinezhad, F. (2021). Quality improvement and shelf life extension of minimally fresh-cut mango fruit using chemical preservatives. *Journal of Horticulture and Postharvest Research, 4*(Special Issue-Fresh-cut Products), 13–24.

Mora, N.. (2022). Biogas as an energy resource for livestock operations: A review. *Renewable Energy, 186*, 1299–1310.

Morashti, J., An, Y., & Hyunmi, J. (2022). A systematic literature review of sustainable packaging in supply chain management. *Sustainability (Basel), 14*(9), 4921. DOI: 10.3390/su14094921

Morrison, J. H.. (2019). The impact of energy-efficient technologies on meat quality: A comprehensive review. *Foods, 8*(5), 155. PMID: 31067817

Mouzakitis, C.-K., Sereti, V., Matsakidou, A., Kotsiou, K., Biliaderis, C. G., & Lazaridou, A. (2022). Physicochemical properties of zein-based edible films and coatings for extending wheat bread shelf life. *Food Hydrocolloids, 132*, 107856. DOI: 10.1016/j.foodhyd.2022.107856

Muchtarom, A., & Jamaluddin, J. (2024). Seamless Solar-Powered Hydroponics Revolutionizes Urban Farming in Indonesia. *Indonesian Journal of Innovation Studies, 25*(4). Advance online publication. DOI: 10.21070/ijins.v25i4.1192

Mudaliar, K., Sharma, V., Agnihotri, C., Agnihotri, S., Deora, A., & Singh, B. P. (2023). Microbiological impact and control strategies to monitor postharvest losses in fruits and vegetables. In *Postharvest Management of Fresh Produce* (pp. 113–147). Elsevier. DOI: 10.1016/B978-0-323-91132-0.00003-4

Mugiya, D., & Hofisi, C. (2017). Climate change adaptation challenges confronting small-scale farmers. *Environment and Ecology*, (8, Iss. 1), 57–65.

Muharja, M., Khotimah, H., Fitria, D. R., Andika, P., Isnan, K. A., & Badril, A. (2024). Techno-Economic Analysis of Biogas Production from Cow Manure. *Jurnal Teknik Kimia Dan Lingkungan, 8*(1), 1–13. DOI: 10.33795/jtkl.v8i1.3335

Mukherjee, I. (2016). Solar energy: A solution for sustainable rural cold storage in India. *Journal of Energy Storage, 5*, 153–159.

Muller, A., Jawtusch, J., & Gattinger, A. (2011). Mitigating greenhouse gases in agriculture.

Muller, P. A., & Schmid, M. (2019, January 7). Intelligent Packaging in the Food Sector: A Brief Overview. *Foods*, *8*(1), 16–16. DOI: 10.3390/foods8010016 PMID: 30621006

Munawar, A., Ong, Y., Schirhagl, R., Tahir, M. A., Khan, W. S., & Bajwa, S. Z. (2019). Nanosensors for diagnosis with optical, electric and mechanical transducers. *RSC Advances*, *9*(12), 6793–6803. DOI: 10.1039/C8RA10144B PMID: 35518460

Muñoz-Tebar, N., Pérez-Álvarez, J. A., Fernández-López, J., & Viuda-Martos, M. (2023). Chitosan edible films and coatings with added bioactive compounds: Antibacterial and antioxidant properties and their application to food products: A review. *Polymers*, *15*(2), 396. DOI: 10.3390/polym15020396 PMID: 36679276

Mustafa, F., & Andreescu, S. (2018, October 16). Chemical and Biological Sensors for Food-Quality Monitoring and Smart Packaging. *Foods*, *7*(10), 168–168. DOI: 10.3390/foods7100168 PMID: 30332833

Nair, S., Trafiałek, J., & Kolanowski, W. (2023). Edible packaging: A technological update for the sustainable future of the food industry. *Applied Sciences (Basel, Switzerland)*, *13*(14), 8234. DOI: 10.3390/app13148234

Naser, A. Z., Deiab, I., & Darras, B. M. (2021). Poly (lactic acid)(PLA) and polyhydroxyalkanoates (PHAs), green alternatives to petroleum-based plastics: A review. *RSC Advances*, *11*(28), 17151–17196. DOI: 10.1039/D1RA02390J PMID: 35479695

Nasiri, E., Rocha-Meneses, L., Inayat, A., & Kikas, T. (2022). Impact of Policy Instruments in the Implementation of Renewable Sources of Energy in Selected European Countries. *Sustainability (Basel)*, *14*(10), 6314. DOI: 10.3390/su14106314

Nassif, Y.. (2020). Innovations in livestock feeding strategies to enhance energy efficiency in animal production. *Food and Energy Security*, *9*(1), e224.

Nasso, M., Blasi, E., Pezzoli, F., & Cicatiello, C. (2024). Investigating Co-Innovation strategies to prevent food loss in the fruits and vegetables sector. *Journal of Cleaner Production*, *467*, 142984. DOI: 10.1016/j.jclepro.2024.142984

Nayak, P., & Solanki, H. (2021). Pesticides and Indian agriculture—A review. *Int J Res Granthaalayah*, *9*(5), 250–263. DOI: 10.29121/granthaalayah.v9.i5.2021.3930

Nayak, S. L., Sharma, R. R., & Sethi, S. (2021). Traditional/Conventional Cooling Systems. In *Cold Chain Management for the Fresh Produce Industry in the Developing World* (pp. 33–44). CRC Press. DOI: 10.1201/9781003056607-5

Nayeri, S., Dehghanian, Z., Lajayer, B. A., Thomson, A., Astatkie, T., & Price, G. W. (2023). CRISPR/Cas9-Mediated genetically edited ornamental and aromatic plants: A promising technology in phytoremediation of heavy metals. *Journal of Cleaner Production*, *428*, 139512. DOI: 10.1016/j.jclepro.2023.139512

Ncube, L. K., Ude, A. U., Ogunmuyiwa, E. N., Zulkifli, R., & Beas, I. N. (2020). Environmental impact of food packaging materials: A review of contemporary development from conventional plastics to polylactic acid based materials. *Materials (Basel)*, *13*(21), 4994. DOI: 10.3390/ma13214994 PMID: 33171895

Neri, A., Hummel, F., Benalia, S., Zimbalatti, G., Gabauer, W., Mihajlovic, I., & Bernardi, B. (2024). Use of Continuous Stirred Tank Reactors for Anaerobic Co-Digestion of Dairy and Meat Industry By-Products for Biogas Production. *Sustainability (Basel)*, *16*(11), 4346. DOI: 10.3390/su16114346

Ng, A. K., Jing, Z. C., & Mahkeswaran, R. (2023). "Energy Conservation Strategies for Increased Sustainability and Energy Efficiency in Urban Farming." In *2023 IEEE International Conference on Service Operations and Logistics, and Informatics (SOLI)*. IEEE. DOI: 10.1109/SOLI60636.2023.10425538

Nguyen, M. L.. (2021). Technologies for nutrients recovery from livestock waste: A review. *Waste Management (New York, N.Y.)*, *120*, 67–78.

Nguyen, M. V., Arason, S., Gissurarson, M., & Pálsson, P. G. (2015). *Uses of geothermal energy in food and agriculture: Opportunities for developing countries. Food and Agriculture Organization of the United Nations*. FAO.

Nicola, S., Cocetta, G., Ferrante, A., & Ertani, A. (2022). Fresh-cut produce quality: Implications for postharvest. In *Postharvest Handling* (pp. 187–250). Elsevier. DOI: 10.1016/B978-0-12-822845-6.00007-5

Niemira, B. A., & Fan, X. (2014) FRUITS AND VEGETABLES | Advances in Processing Technologies to Preserve and Enhance the Safety of Fresh and Fresh-Cut Fruits and Vegetables. Encyclopedia of Food Microbiology. (Pages) 983-991. https://doi.org/DOI: 10.1016/B978-0-12-384730-0.00428-6

Nilsen-Nygaard, J., Fernández, E., Radusin, T., Rotabakk, B., Sarfraz, J., Sharmin, N., Sivertsvik, M., Sone, I., & Pettersen, M. (2021). Current status of biobased and biodegradable food packaging materials: Impact on food quality and effect of innovative processing technologies. *Comprehensive Reviews in Food Science and Food Safety*, *20*(2), 1333–1380. DOI: 10.1111/1541-4337.12715 PMID: 33547765

Nirmal, N. P., Khanashyam, A. C., Mundanat, A. S., Shah, K., Babu, K. S., Thorakkattu, P., Al-Asmari, F., & Pandiselvam, R. (2023). Valorization of Fruit Waste for Bioactive Compounds and Their Applications in the Food Industry. *Foods*, *12*(3), 556. DOI: 10.3390/foods12030556 PMID: 36766085

Noble, K., Huatuco, L. H., Dyke, A., & Green, J. "The Environmental Impacts of Cultured Meat Production: A Systematic Literature Review." In *International Conference on Sustainable Design and Manufacturing*, pp. 91-101. Springer, Singapore, 2024. DOI: 10.1007/978-981-99-8159-5_8

Nonhebel, S. (2005). Renewable Energy and Food Supply: Will There Be Enough Land? *Renewable & Sustainable Energy Reviews*, *9*(2), 191–201. DOI: 10.1016/j.rser.2004.02.003

Noor, A., Moyle, P. M., Malik, A., Ziora, Z. M., & Pant, K. K. (2024). Transformative upcycling of fruit-vegetable waste for nutraceutical and pharmaceutical breakthroughs and circular economy evolution. *Process Safety and Environmental Protection*, *187*, 1022–1036. DOI: 10.1016/j.psep.2024.05.002

Nordin, S. M., Boyle, M., & Kemmer, T. M. (2013). Academy of Nutrition and Dietetics. Position of the academy of nutrition and dietetics: nutrition security in developing nations: sustainable food, water, and health. *Journal of the Academy of Nutrition and Dietetics*, *113*(4), 581–595. DOI: 10.1016/j.jand.2013.01.025 PMID: 23522578

Norton, V., Waters, C., Oloyede, O., & Lignou, S. (2022). Exploring consumers' understanding and perception of sustainable food packaging in the uk. *Foods*, *11*(21), 3424. DOI: 10.3390/foods11213424 PMID: 36360036

Nsabiyeze, A., Ma, R., Li, J., Zhao, Q., & Zhang, M. (2024). Mitigating greenhouse gas emissions from sheep production system in China: An integrated approach of data envelopment analysis and life cycle assessment. *Resources, Conservation and Recycling*, *207*, 107695. DOI: 10.1016/j.resconrec.2024.107695

Nunes, C., Silva, M., Farinha, D., Sales, H., Pontes, R., & Nunes, J. (2023). Edible Coatings and Future Trends in Active Food Packaging—Fruits' and Traditional Sausages' Shelf Life Increasing. *Foods*, *12*(17), 3308. DOI: 10.3390/foods12173308 PMID: 37685240

Nur Rothy, J. N. (2023). *Effectiveness of master bag system packaging with oxygen scavengers and CO2/N2 modified atmosphere to extend bison meat shelf life*.

Nusri-Un, J., Kabsuk, J., Binsulong, B., & Sommart, K. (2024). JirapornKabsuk, BhoowadolBinsulong, and KritaponSommart. "Effects of Cattle Breeds and Dietary Energy Density on Intake, Growth, Carcass, and Meat Quality under Thai Feedlot Management System.". *Animals (Basel)*, *14*(8), 1186. DOI: 10.3390/ani14081186 PMID: 38672334

Nxumalo, K. A., Fawole, O. A., & Oluwafemi, O. S. (2022). Evaluating the efficacy of gum arabic-zinc oxide nanoparticles composite coating on shelf-life extension of mandarins (cv. kinnow). *Frontiers in Plant Science*, *13*, 953861. DOI: 10.3389/fpls.2022.953861 PMID: 35937367

Ofon, U. A., Ndubuisi-Nnaji, U. U., Udo, I. M., Udofia, E. S., Fatunla, O. K., & Shaibu, S. E. (2024). Biogas Production Potential from Anaerobic Co-Digestion of Food Waste and Animal Manure. *Journal of Chemical Society of Nigeria*, *49*(1). Advance online publication. DOI: 10.46602/jcsn.v49i1.949

Ogedengbe, T. C., Malomo, O. J., & Akanji, N. E. (2022). Post-Harvest Losses and Reduction Techniques in Crop Production: A Review. *International Journal of Agricultural Science, Research and Technology in Extension and Education Systems*, *12*(4), 225–233.

Ogunnupebi, T. A., Oluyori, A. P., Dada, A. O., Oladeji, O. S., Inyinbor, A. A., & Egharevba, G. O. (2020). Promising natural products in crop protection and food preservation: Basis, advances, and future prospects. *International Journal of Agronomy*, *2020*(1), 8840046. DOI: 10.1155/2020/8840046

Oke, O. E., Akosile, O. A., Uyanga, V. A., Oke, F. O., Oni, A. I., Tona, K., & Onagbesan, O. M. (2024). Climate change and broiler production. *Veterinary Medicine and Science*, *10*(3), e1416. DOI: 10.1002/vms3.1416 PMID: 38504607

Olabi, A. G., & Abdelkareem, M. A. (2022). Renewable energy and climate change. *Renewable & Sustainable Energy Reviews*, *158*, 112111.

Oloyede, O., & Lignou, S. (2021). Sustainable paper-based packaging: A consumer's perspective. *Foods*, *10*(5), 1035. DOI: 10.3390/foods10051035 PMID: 34068639

Oluwole, O., Ibidapo, O., Arowosola, T., Raji, F., Zandonadi, R. P., Alasqah, I., Lho, L. H., Han, H., & Raposo, A. (2023). Sustainable transformation agenda for enhanced global food and nutrition security: A narrative review. *Frontiers in Nutrition*, *10*, 10. DOI: 10.3389/fnut.2023.1226538 PMID: 37599683

Omerali, M., & Kaya, T. (2023). Energy Efficiency Optimization Application in Food Production Using IIOT Based Machine Learning. In Applied Innovation and Technology Management (pp. 185–204). Springer International Publishing. http://dx.doi.org/DOI: 10.1007/978-3-031-38387-8_11

Onut, S., Tuzkaya, U. R., & Dogaç, B. (2008). A particle swarm optimization algorithm for the multiple-level warehouse layout design problem. *Computers & Industrial Engineering*, *54*(4), 783–799. DOI: 10.1016/j.cie.2007.10.012

Oriekhoe, O. I., Ashiwaju, B. I., Ihemereze, K. C., Ikwue, U., & Udeh, C. A. (2024). Review of technological advancements in food supply chain management: A comparative study between the US and Africa. *International Journal of Management & Entrepreneurship Research*, *6*(1), 132–149.

Otto, S., Strenger, M., Maier-Nöth, A., & Schmid, M. (2021). Food packaging and sustainability–Consumer perception vs. correlated scientific facts: A review. *Journal of Cleaner Production*, *298*, 126733. DOI: 10.1016/j.jclepro.2021.126733

Ottosson, M., Thollander, P., Schulze, M., & Nehler, H. (2016). Energy management in industry - A systematic review of previous findings and an integrative conceptual framework. *Journal of Cleaner Production*, *112*(3), 692–708.

Oyama, S., Kashihara, H., Kazari, M., Matsuda, Y., Shimizu, M., Yamada, K., & Watanabe, T. (2009). "D105 Development of an Efficient Small-Scale Distributed Power Generation System Using Gasification of Woody Biomass(biomass-2)." *The Proceedings of the International Conference on Power Engineering (ICOPE)* 2009.1 (0): _1–213_ – _1–218_. DOI: 10.1299/jsmeicope.2009.1._1-213_

Ozilgen, M. (2016). Energy utilization and carbon dioxide emission during production of snacks. *Journal of Cleaner Production*, *112*, 2601–2612. DOI: 10.1016/j.jclepro.2015.10.031

Ozilgen, M. (2018). Nutrition and production related energies and exergies of foods. *Renewable & Sustainable Energy Reviews*, *96*(August), 275–295. DOI: 10.1016/j.rser.2018.07.055

Ozturk, B., Aglar, E., Saracoglu, O., Karakaya, O., & Gun, S. (2022). Effects of GA3, CaCl2 and modified atmosphere packaging (MAP) applications on fruit quality of sweet cherry at cold storage. *International Journal of Fruit Science*, *22*(1), 696–710. DOI: 10.1080/15538362.2022.2113597

Palumbo, M., Attolico, G., Capozzi, V., Cozzolino, R., Corvino, A., de Chiara, M. L. V., Pace, B., Pelosi, S., Ricci, I., Romaniello, R., & Cefola, M. (2022). Emerging postharvest technologies to enhance the shelf-life of fruit and vegetables: An overview. *Foods*, *11*(23), 3925. DOI: 10.3390/foods11233925 PMID: 36496732

Panchal, N., Das, K., Prabhakar, P. K., & Ghanghas, N. (2022). Edible films and coatings for fruits and vegetables: Composition, functions, and regulatory aspects. *Edible Food Packaging: Applications, Innovations and Sustainability*, 191–216.

Panda, S. N., Saikia, R., Sagar, , Swamy, G. N., Panotra, N., Yadav, K., Singh, B. V., Rathi, S., Singh, R., & Pandey, S. K. (2024). Solar Energy's Role in Achieving Sustainable Development Goals in Agriculture. *International Journal of Environment and Climate Change*, *14*(5), 10–31. DOI: 10.9734/ijecc/2024/v14i54167

Pang, Q., Qiu, M., Zhang, L., & Chiu, Y. (2024). Congestion effects of energy and its influencing factors: China's transportation sector. *Socio-Economic Planning Sciences*, *92*, 101850. DOI: 10.1016/j.seps.2024.101850

Papadopoulou, C.-A., Papadopoulou, M. P., & Laspidou, C. (2022). Implementing Water-Energy-Land-Food-Climate Nexus Approach to Achieve the Sustainable Development Goals in Greece: Indicators and Policy Recommendations. *Sustainability (Basel)*, *14*(7), 4100. DOI: 10.3390/su14074100

Pape, R. A. (1997). The limits of precision-guided air power. *Security Studies*, *7*(2), 93–114. DOI: 10.1080/09636419708429343

Park, J., & Waqar, Z. (2022). Life cycle assessment of returnable mailers used for apparel electronic commerce: A case study in canada. *Packaging Technology & Science*, *35*(9), 651–662. DOI: 10.1002/pts.2653

Pascall, M. A. (2020). The role and importance of packaging and labeling in assuring food safety, quality \& compliance with regulations I: Packaging basics. In *Food safety and quality systems in developing countries* (pp. 261–283). Elsevier. DOI: 10.1016/B978-0-12-814272-1.00006-1

Pashova, S. (2023). Application of Plant Waxes in Edible Coatings. *Coatings*, *13*(5), 911. DOI: 10.3390/coatings13050911

Pasqualino, J., Meneses, M., & Castells, F. (2011). The carbon footprint and energy consumption of beverage packaging selection and disposal. *Journal of Food Engineering*, *103*(4), 357–365. DOI: 10.1016/j.jfoodeng.2010.11.005

Patel, N., Feofilovs, M., & Romagnoli, F. (2023). Carbon footprint evaluation tool for packaging marketplace. *Environmental and Climate Technologies*, *27*(1), 368–378. DOI: 10.2478/rtuect-2023-0027

Patel, S., Kumar, N., & Ahuja, A. (2023). Building energy management systems for energy efficiency in food production facilities. *Energy Efficiency*, *16*(1), 1–19.

Patil, S. J., & Bhat, A. V. (2016). *Intellectual property rights and food safety: A study of Indian legislation*. Kluwer Law International.

Pattabhiramaiah, M., & Mallikarjunaiah, S. (2023). "Advances in Innovative and Emerging Technologies in Food Processing." In *Advances in Environmental Engineering and Green Technologies*, 1–32. IGI Global. DOI: 10.4018/978-1-6684-9094-5.ch001

Pattison, M. (2019). The potential benefits of LED lighting for poultry production. *World's Poultry Science Journal*, *75*(3), 435–446. DOI: 10.1017/S0043933919000339

Patton, M. Q. (2015). *Qualitative research & evaluation methods: Integrating theory and practice* (4th ed.). Sage publications.

Paucar-Menacho, L. M., Moreno-Rojo, C., & Chuqui-Diestra, S. R. (2024). Emerging Non-Thermal Technologies in the Food Industry: Advances and Potential Applications in Food Processing. *Scientia Agropecuaria*, *15*(1), 65–83. DOI: 10.17268/sci.agropecu.2024.006

Paudyal, R., & Karatzas, K. A. (2016). Stress adaptation of Listeria monocytogenes in acidic ready to-eat products Food Hygiene and Toxicology in Ready-to-Eat Foods ed P Kotzekidou (London: Elsevier Inc.) chapter 10 pp 167-182

Payne, J., McKeown, P., & Jones, M. D. (2019). A circular economy approach to plastic waste. *Polymer Degradation & Stability*, *165*, 170–181. DOI: 10.1016/j.polymdegradstab.2019.05.014

Peerzada Gh, J., Sinclair, B. J., Perinbarajan, G. K., Dutta, R., Shekhawat, R., Saikia, N., Chidambaram, R., & Mossa, A.-T. (2023). An overview on smart and active edible coatings: Safety and regulations. *European Food Research and Technology*, *249*(8), 1–18. DOI: 10.1007/s00217-023-04273-2

Pereira, R., & Silva, F. (2020). Real-time energy monitoring and control systems in food manufacturing: A critical review. *Food and Bioprocess Technology*, *13*(7), 1251–1265. DOI: 10.1007/s11947-020-02463-z

Perera, C. O., & Smith, B. (2013). Technology of Processing of Horticultural Crops. In *Elsevier eBooks* (pp. 259–315). DOI: 10.1016/B978-0-12-385881-8.00011-2

Perera, K., Jaiswal, A., & Jaiswal, S. (2023). Biopolymer-based sustainable food packaging materials: Challenges, solutions, and applications. *Foods*, *12*(12), 2422. DOI: 10.3390/foods12122422 PMID: 37372632

Pérez, A. V., Castillo, G. A. L., Alava, L. A. C., & Chilan, J. C. H. (2016). The regulatory framework for renewable energy sources. Journal of College and University. This is an open access article under the, 2395, 7492.

Pérez-Santaescolástica, C., Munekata, P. E. S., Feng, X., Liu, Y., Bastianello Campagnol, P. C., & Lorenzo, J. M. (2022). Active edible coatings and films with Mediterranean herbs to improve food shelf-life. *Critical Reviews in Food Science and Nutrition*, *62*(9), 2391–2403. DOI: 10.1080/10408398.2020.1853036 PMID: 33251820

Perez-Vazquez, A., Barciela, P., Carpena, M., & Prieto, M. A. (2023). Edible coatings as a natural packaging system to improve fruit and vegetable shelf life and quality. *Foods*, *12*(19), 3570. DOI: 10.3390/foods12193570 PMID: 37835222

Perry, G. A., & Welsh, T. H.Jr. (2024). The importance of developmental programming in the beef industry. *Animal Reproduction Science*, *265*, 107488. DOI: 10.1016/j.anireprosci.2024.107488 PMID: 38701640

Perveen, S., Anwar, M. J., Ismail, T., Hameed, A., Naqvi, S. S., Mahomoodally, M. F., Saeed, F., Imran, A., Hussain, M., Imran, M., Ur Rehman, H., Khursheed, T., Tufail, T., Mehmood, T., Ali, S. W., & Al Jbawi, E. (2023). Utilization of biomaterials to develop the biodegradable food packaging. *International Journal of Food Properties*, *26*(1), 1122–1139. DOI: 10.1080/10942912.2023.2200606

Pešta, J., Šerešová, M., & Kočí, V. (2020). Carbon footprint assessment of construction waste packaging using the package-to-product indicator. *Sustainability (Basel)*, *12*(23), 10094. DOI: 10.3390/su122310094

Pham, T. T., Nguyen, L. L. P., Sao, Đ., & Baranyai, L. (2023, March 2). Application of Edible Coating in Extension of Fruit Shelf Life [Review]. *Multidisciplinary Digital Publishing Institute*, *5*(1), 520–536. DOI: 10.3390/agriengineering5010034

Piližota, V. (2023). Fruits and vegetables (including herbs). In *Food Safety Management* (pp. 235–268). Elsevier. DOI: 10.1016/B978-0-12-820013-1.00039-5

Pillai, A. R. S., Eapen, A. S., Zhang, W., & Roy, S. (2024). Polysaccharide-Based Edible Biopolymer-Based Coatings for Fruit Preservation: A Review. *Foods*, *13*(10), 1529. DOI: 10.3390/foods13101529 PMID: 38790829

Piñeiro, V., Arias, J., Dürr, J., Elverdin, P., Ibáñez, A., Kinengyere, A., Opazo, C. M., Owoo, N., Page, J. R., Prager, S. D., & Torero, M. (2020). A scoping review on incentives for adoption of sustainable agricultural practices and their outcomes. *Nature Sustainability*, *3*(10), 809–820. DOI: 10.1038/s41893-020-00617-y

Piselli, C., Balocco, C., Forastiere, S., Silei, A., Sciurpi, F., & Cotana, F. (2024). Energy efficiency in the commercial sector. Thermodynamics fundamentals for the energy transition. *Energy Reports*, *11*, 4601–4621. DOI: 10.1016/j.egyr.2024.04.033

Porat, R., Lichter, A., Terry, L. A., Harker, R., & Buzby, J. (2018). Postharvest losses of fruit and vegetables during retail and in consumers' homes: Quantifications, causes, and means of prevention. *Postharvest Biology and Technology*, *139*, 135–149. DOI: 10.1016/j.postharvbio.2017.11.019

Porta, R., Sabbah, M., & Di Pierro, P. (2020). Biopolymers as food packaging materials. *International Journal of Molecular Sciences*, *21*(14), 4942. DOI: 10.3390/ijms21144942 PMID: 32668678

Pradella, A. M., & Loures, E. de F. R., Costa, S. E. G. da, & Lima, E. P. de. (2019). Energy Efficiency in the Food Industry: A Systematic Literature Review. Brazilian Archives of Biology and Technology, spe. https://doi.org/DOI: 10.1590/1678-4324-smart-2019190002

Pretty, J. (2008). Agricultural sustainability: Concepts, principles and evidence. *Philosophical Transactions of the Royal Society of London. Series B, Biological Sciences*, *363*(1491), 447–465. DOI: 10.1098/rstb.2007.2163 PMID: 17652074

Priya, K., Thirunavookarasu, N., & Chidanand, D. V. (2023). Recent advances in edible coating of food products and its legislations: A review. *Journal of Agriculture and Food Research*, *12*, 100623. DOI: 10.1016/j.jafr.2023.100623

Processed Foods and Health. (2024, May 9). The Nutrition Source. https://nutritionsource.hsph.harvard.edu/processed-foods/

Protection of Plant Varieties and Farmers' Rights Act, 2001. (2021). Government of India.

PUCL v. Union of India. (2001). AIR 2001 SC 568.

Pudenz, C. C., & Schulz, L. L. (2024). Multi-plant coordination in the US beef packing industry. *American Journal of Agricultural Economics*, *106*(1), 382–415. DOI: 10.1111/ajae.12391

Quist, D., & Chapela, I. H. (2001). Transgenic DNA introgressed into traditional maize landraces in Oaxaca, Mexico. *Nature*, *414*(6863), 541–543. DOI: 10.1038/35107068 PMID: 11734853

Raak, N., Symmank, C., Zahn, S., Aschemann-Witzel, J., & Rohm, H. (2017). Processing-and product-related causes for food waste and implications for the food supply chain. *Waste Management (New York, N.Y.)*, *61*, 461–472. DOI: 10.1016/j.wasman.2016.12.027 PMID: 28038904

Rahman, Z., Johnson, J., Tripathi, S., & Krithika, D. J. (2023). The future of packaging: Minimalism in modern era. *Interantional Journal of Scientific Research in Engineering and Management*, *07*(10), 1–11. DOI: 10.55041/IJSREM26073

Raj, R., Dhiman, S., Shakya, L., & others. (2022). *Nanotechnological interventions in sustainable agriculture.*

Rajput, S., Jadhav, A., Gadge, J., Tilani, D., & Dalgade, V. (2023). Agricultural Food supply chain Traceability using Blockchain. 2023 4th International Conference on Innovative Trends in Information Technology (ICITIIT), 1–6. https://doi.org/DOI: 10.1109/ICITIIT57246.2023.10068564

Raman, R., & Vasmatkar, A. 2024. "IoT-Enabled Aquaponics and Hydroponics for Efficient Indoor Farming Systems with Cloud Computing." In *2024 International Conference on Advances in Data Engineering and Intelligent Computing Systems (ADICS)*. IEEE. DOI: 10.1109/ADICS58448.2024.10533550

Ramjan, M. D., & Ansari, M. T. (2018). Factors affecting of fruits, vegetables and its quality. *Faslnamah-i Giyahan-i Daruyi*, *6*(6), 16–18.

Rana, A. (2022). Potential of renewable energy in water-energy-food nexus in Jordan. *Energy Nexus*, *7*, 100140. DOI: 10.1016/j.nexus.2022.100140

Rana, M., Arshad, R. N., Malek, Z., Roobab, U., Ranjha, M., Jambrak, A., Muhammad, I. Q., Khan, N., & Lorenzo, J. M. (2022). Nonthermal food processing: A step towards a circular economy to meet the sustainable development goals. *Food Chemistry: X*, *16*, 100516. DOI: 10.1016/j.fochx.2022.100516 PMID: 36439936

Rani, P., Dubey, A., Kumar, P., & Kumar, A. (2023). Sustainable Renewable Energy Sources for Food and Dairy Processing. In *Food Process Engineering and Technology* (pp. 65–95). Springer Nature Singapore. DOI: 10.1007/978-981-99-6831-2_4

Rastogi, V. K., & Samyn, P. (2015). Bio-based coatings for paper applications. *Coatings*, *5*(4), 887–930. DOI: 10.3390/coatings5040887

Rasyid, H., & Ningsih, G. M. (2024). Application of Sustainable Agricultural Technology to Increase Productivity and Food Security. *Journal of Wood Science*, *3*(4), 500–506.

Rathor, A., Choudhury, S., Sharma, A., Nautiyal, P., & Shah, G. (2024). Empowering vertical farming through IoT and AI-Driven technologies: A comprehensive review. *Heliyon, 10*(15), e34998. DOI: 10.1016/j.heliyon.2024.e34998 PMID: 39157372

Rathore, S., Schuler, B., & Park, J. (2023). Life cycle assessment of multiple dispensing systems used for cosmetic product packaging. *Packaging Technology & Science, 36*(7), 533–547. DOI: 10.1002/pts.2729

Raymond, L., & Leung, G. C. K. (2021). The relationship between energy prices, economic growth and renewable energy consumption. *Energy Reports, 7*, 712–1719.

Razza, F., Innocenti, F.D., (2012). Bioplastics from renewable resources: the benefits of biodegradability. Asia Pac. J. Chem. Eng. 7, S301eS309

Rebollar, R., Gil, I., Lidón, I., Martín, J., Fernández, M., & Rivera, S. (2017). How material, visual and verbal cues on packaging influence consumer expectations and willingness to buy: The case of crisps (potato chips) in spain. *Food Research International, 99*, 239–246. DOI: 10.1016/j.foodres.2017.05.024 PMID: 28784480

Rempelos, L., Kabourakis, E., & Leifert, C. (2023). Innovative Organic and Regenerative Agricultural Production. *Agronomy (Basel), 13*(5), 1344. DOI: 10.3390/agronomy13051344

Renewable Energy World. (2021). Solar thermal applications in the dairy industry.

Rennie, T. J., & Sunjka, P. S. (2017). Modified atmosphere for storage, transportation, and packaging. In *Novel postharvest treatments of fresh produce* (pp. 433–480). CRC Press. DOI: 10.1201/9781315370149-14

Ren, Z., Zhang, D., & Zhu, G. (2022). Sustainable design strategy of cosmetic packaging in china based on life cycle assessment. *Sustainability (Basel), 14*(13), 8155. DOI: 10.3390/su14138155

Reynolds, S. 2024. "Exploring the Influence of Corporate Social Responsibility on Supply Chain Sustainability in Renewable Energy." *Preprints*. DOI: 10.20944/preprints202405.1888.v1

Rezende, V. T., Araújo Nacimento, R., Ali, S., Roberto Dias Rodrigues, G., Libório Romanelli, T., Noely dos Santos Gonçalves Cyrillo, J., Bonaudo, T., Lescoat, P., & Hauber Gameiro, A. (2024). Understanding nitrogen dynamics in the Brazilian beef industry: A comprehensive decadal analysis. *The Science of the Total Environment, 921*, 171045. DOI: 10.1016/j.scitotenv.2024.171045 PMID: 38402966

Ricci, A., Bertani, G., Maoloni, A., Bernini, V., Levante, A., Neviani, E., & Lazzi, C. (2021, May 14). Antimicrobial Activity of Fermented Vegetable Byproduct Extracts for Food Applications. *Foods, 10*(5), 1092–1092. DOI: 10.3390/foods10051092 PMID: 34069051

Rizan, C., Lillywhite, R., Reed, M., & Bhutta, M. (2021). Minimising carbon and financial costs of steam sterilisation and packaging of reusable surgical instruments. *British Journal of Surgery, 109*(2), 200–210. DOI: 10.1093/bjs/znab406 PMID: 34849606

Rizwan, A., Khan, A. N., Ibrahim, M., Ahmad, R., Iqbal, N., & Kim, D. H. (2024). Optimal environment control and fruits delivery tracking system using blockchain for greenhouse. *Computers and Electronics in Agriculture, 220*, 108889. DOI: 10.1016/j.compag.2024.108889

Roberts, S. B., Silver, R. E., Das, S. K., Fielding, R. A., Gilhooly, C. H., Jacques, P. F., Kelly, J. M., Mason, J. B., McKeown, N. M., Reardon, M. A., Rowan, S., Saltzman, E., Shukitt-Hale, B., Smith, C. E., Taylor, A. A., Wu, D., Zhang, F. F., Panetta, K., & Booth, S. (2021, July). Healthy aging—nutrition matters: Start early and screen often. *Advances in Nutrition, 12*(4), 1438–1448. DOI: 10.1093/advances/nmab032 PMID: 33838032

Robson, J. R. (Ed.). (2023). *Famine: Its Causes, Effects and Management.* Taylor and Francis. DOI: 10.4324/9781003412366

Rodrigues, E. T., & Han, J. H. (2003). Intelligent packaging. In Heldman, D. R. (Ed.), *Encyclopedia of Agricultural and Food Engineering* (pp. 528–535). Marcel Dekker.

Rønning, S. B., Pedersen, M. E., & Bjørnerud, E. (2024). Emerging Food Trends: Cellular Agriculture—novel Food Production Technology. [Elsevier.]. *Food Industries, 4*(0), 233–246. DOI: 10.1016/B978-0-443-15516-1.00011-6

Rosen, M. K., & Meilke, K. (2023). Livestock Production Systems: An Overview. *Agricultural Systems Review, 196*, 107832.

Rosin, A., Drovtar, I., & Kilter, J. (2024). "Solutions and Active Measures for Wind Power Integration." In *Large Scale Grid Integration of Renewable Energy Sources: Solutions and Technologies*, 91–137. Institution of Engineering and Technology. DOI: 10.1049/PBPO222E_ch4

Roy, R., Marakkar, S., Vayalil, M. P., Shahanaz, A., Anil, A. P., Kunnathpeedikayil, S., & Yadav, K. K. (2022). Drug-food interactions in the era of molecular big data, machine intelligence, and personalized health. *Recent Advances in Food, Nutrition & Agriculture*, *13*(1), 27–50. DOI: 10.2174/2212798412666220620104809 PMID: 36173075

Ruppenthal, R. J. (2008). *Fresh Food from Small Spaces: The Square-inch Gardener's Guide to Year-round Growing, Fermenting, and Sprouting*. Chelsea Green Publishing.

Russo, G. L., Langellotti, A. L., Torrieri, E., & Masi, P. (2024). Emerging technologies in seafood processing: An overview of innovations reshaping the aquatic food industry. *Comprehensive Reviews in Food Science and Food Safety*, *23*(1), e13281. DOI: 10.1111/1541-4337.13281 PMID: 38284572

Rutta, E. W. (2022). *Barriers Impeding the Deployment and Uptake of Solar-Powered Cold Storage Technologies for Postharvest Loss Reduction in Tomato Value Chain in Africa: Empirical Evidence from Tanzania*. Queen's University.

Saboia, L. and Oliveira, R. (2021). Replacement of conventional plastic shampoo packaging for biodegradable ones – a survey.. Revista Eletrônica Em Gestão Educação E Tecnologia Ambiental, 25, e15. DOI: 10.5902/2236117065105

Sæter, F., Alvarado, I. O., & Pettersen, I. N. (2020). Reuse principle for primary packaging circularity in the food system. DS 101: *Proceedings of NordDesign 2020*, Lyngby, Denmark, 12th-14th August 2020, 1-12.

Sagar, N. A., Kumar, N., Choudhary, R., Bajpai, V. K., Cao, H., Shukla, S., & Pareek, S. (2022). Prospecting the role of nanotechnology in extending the shelf-life of fresh produce and in developing advanced packaging. *Food Packaging and Shelf Life*, *34*, 100955. DOI: 10.1016/j.fpsl.2022.100955

Saidur, R., Mekhilef, S., Ali, M. B., Safari, A., & Mohammed, H. A. (2012). Applications of variable speed drive (VSD) in electrical motors energy savings. *Renewable & Sustainable Energy Reviews*, *16*(1), 543–550. DOI: 10.1016/j.rser.2011.08.020

Sai, M., Singh, S., Gupta, M., Nimdeo, Y. M., Raushan, R., Deorankar, A. V., Ananda, T. M., Rout, P., Chanotiya, C. S., Pakhale, V. D., & Nannaware, A. (2023). Solar energy: A promising renewable source for meeting energy demand in Indian agriculture applications. *Sustainable Energy Technologies and Assessments*, *55*, 102905. DOI: 10.1016/j.seta.2022.102905

Saitone, T. L., Schaefer, K. A., Scheitrum, D., Arita, S., Breneman, V., Boehm, R. N., & Maples, J. G. (2024). Consolidation and concentration in US meat processing: Updated measures using plant-level data. *Review of Industrial Organization, 64*(1), 35–56. DOI: 10.1007/s11151-023-09923-z

Saklani, P. S. (2019). A review of edible packaging for foods. *International Journal of Current Microbiology and Applied Sciences, 8*(7), 2885–2895. DOI: 10.20546/ijcmas.2019.807.359

Saleh, N., & Yousaf, Z. (2018). Tools and techniques for the optimized synthesis, reproducibility and scale up of desired nanoparticles from plant derived material and their role in pharmaceutical properties. In *Nanoscale Fabrication, Optimization, Scale-Up and Biological Aspects of Pharmaceutical Nanotechnology* (pp. 85–131). Elsevier. DOI: 10.1016/B978-0-12-813629-4.00003-6

Saleh, T. A. (2020). Nanomaterials: Classification, properties, and environmental toxicities. *Environmental Technology \& Innovation, 20*, 101067.

Salem, S. S., & Fouda, A. (2021). Green synthesis of metallic nanoparticles and their prospective biotechnological applications: An overview. *Biological Trace Element Research, 199*(1), 344–370. DOI: 10.1007/s12011-020-02138-3 PMID: 32377944

Salgado, P., Giorgio, L., Musso, Y., & Mauri, A. (2021). Recent developments in smart food packaging focused on biobased and biodegradable polymers. *Frontiers in Sustainable Food Systems, 5*, 630393. Advance online publication. DOI: 10.3389/fsufs.2021.630393

Sandström, V., Kastner, T., Schwarzmueller, F., & Kummu, M. (2024). The Potential to Increase Food System Resilience by Replacing Feed Imports with Domestic Food System Byproducts. [Web Site]. *Environmental Research Letters, 19*(8), 084018. DOI: 10.1088/1748-9326/ad5ab3

Sandu, M. (2015). Food Safety-Component of National Security. *Pub. Sec. Stud., 4*, 15.

Sani, M. A., Zhang, W., Abedini, A., Khezerlou, A., Shariatifar, N., Assadpour, E., Zhang, F., & Jafari, S. M. (2024). Intelligent packaging systems for the quality and safety monitoring of meat products: From lab scale to industrialization. *Food Control, 160*, 110359. DOI: 10.1016/j.foodcont.2024.110359

Santo, R., & Moragues-Faus, A. (2019). Towards a trans-local food governance: Exploring the transformative capacity of food policy assemblages in the US and UK. In *Geoforum, 98, 75–87*. Elsevier BV., DOI: 10.1016/j.geoforum.2018.10.002

Sarkar, M. R., Rashid, M. H., Rahman, A., Kafi, M. A., Hosen, M. I., Rahman, M. S., & Khan, M. N. (2022). Recent advances in nanomaterials based sustainable agriculture: An overview. *Environmental Nanotechnology, Monitoring \& Management, 18*, 100687.

Saxena, A., Sharma, L., & Maity, T. (2020). Enrichment of edible coatings and films with plant extracts or essential oils for the preservation of fruits and vegetables. In *Biopolymer-based formulations* (pp. 859–880). Elsevier. DOI: 10.1016/B978-0-12-816897-4.00034-5

Saxena, P., Caldwell, J., Ramos, A., Flores, D., Shah, D., & Kuo, T. (2022). Food Security, Fruit and Vegetable Intake, and Chronic Conditions among Supplemental Nutrition Assistance Program Education Participants Attending Free Food and Produce Events. *Obesities*, *2*(3), 264–275. DOI: 10.3390/obesities2030021

Schumm, G., Philipp, M., Schlosser, F., Hesselbach, J., Walmsley, T. G., & Atkins, M. J. (2018). Hybrid heating system for increased energy efficiency and f lexible control of low temperature heat. *Energy Efficiency*, *11*(5), 1117–1133. DOI: 10.1007/s12053-017-9584-6

Sedighi, Md., & Zakariapour, M. (2014)... *Sustainable Energy.*, *2*(2), 44–51. DOI: 10.12691/rse-2-2-2

Seglia, D., Krasanova, I., Heidemane, G., Kampuse, S., Dukalska, L., & Kampus, K. (2009). Packaging tehnology influence on the shelf life extention of fresh raspberries. *Acta Horticulturae*, (877), 433–440.

Sellberg, M. M., Norström, A. V., Peterson, G. D., & Gordon, L. J. (2020). Using local initiatives to envision sustainable and resilient food systems in the Stockholm city-region. In *Global Food Security, 24, 100334*. Elsevier BV., DOI: 10.1016/j.gfs.2019.100334

Semkov, K., Mooney, E., Connolly, M., & Adley, C. (2014). Efficiency improvement through waste heat reduction. *Applied Thermal Engineering*, *1*(1), 716–722. DOI: 10.1016/j.applthermaleng.2014.05.030

Señorans, F. J., Ibáñez, E., & Cifuentes, A. (2003). New trends in food processing. *Critical Reviews in Food Science and Nutrition*, *43*(5), 507–526. DOI: 10.1080/10408690390246341 PMID: 14653493

Senturk Parreidt, T., Müller, K., & Schmid, M. (2018). Alginate-based edible films and coatings for food packaging applications. *Foods*, *7*(10), 170. DOI: 10.3390/foods7100170 PMID: 30336642

Šerešová, M., & Kočí, V. (2020). Proposal of package-to-product indicator for carbon footprint assessment with focus on the Czech Republic. *Sustainability (Basel), 12*(7), 3034. DOI: 10.3390/su12073034

Shafiee, M., Keshavarz, P., Lane, G., Pahwa, P., Szafron, M., Jennings, D., & Vatanparast, H. (2022). Food Security Status of Indigenous Peoples in Canada According to the 4 Pillars of Food Security: A Scoping Review. In *Advances in Nutrition, 13(6), 2537–2558*. Elsevier BV., DOI: 10.1093/advances/nmac081

Shahbaz, M. U., Arshad, M., Mukhtar, K., Nabi, B. G., Goksen, G., Starowicz, M., Nawaz, A., Ahmad, I., Walayat, N., Manzoor, M. F., & Aadil, R. M. (2022). Natural plant extracts: An update about novel spraying as an alternative of chemical pesticides to extend the postharvest shelf life of fruits and vegetables. *Molecules (Basel, Switzerland), 27*(16), 5152. DOI: 10.3390/molecules27165152 PMID: 36014396

Sharma, A., Kumar, V., Shahzad, B., Tanveer, M., Sidhu, G. P. S., Handa, N., Kohli, S. K., Yadav, P., Bali, A. S., Parihar, R. D., Dar, O. I., Singh, K., Jasrotia, S., Bakshi, P., Ramakrishnan, M., Kumar, S., Bhardwaj, R., & Thukral, A. K. (2019). Worldwide pesticide usage and its impacts on ecosystem. *SN Applied Sciences, 1*(11), 1–16. DOI: 10.1007/s42452-019-1485-1

Sharma, A., Saxena, R., & Sethi, M. (2018). Solar energy in India: Strategies, policies, perspectives and future potential. *Renewables: Wind, Water, and Solar, 5*(1), 9. DOI: 10.1186/s40807-017-0056-9

Sharma, B., & Kumari, A. (2020). Role of Post-Harvest Physiology in Evolution of Transgenic Crops. In *Plant Breeding-Current and Future Views*. IntechOpen.

Sharma, C., Pathak, P., Yadav, S. P., & Gautam, S. (2024). Potential of emerging "all-natural" edible coatings to prevent post-harvest losses of vegetables and fruits for sustainable agriculture. *Progress in Organic Coatings, 193*, 108537.

Sharma, H. B., Kumar, R. V., Cheela, V. R. S., Ranjan, V. P., Jaglan, A. K., Dubey, B., Goel, S., & Bhattacharya, J. (2020). Challenges, opportunities, and innovations for effective solid waste management during and post COVID-19 pandemic. *Resources, Conservation and Recycling, 162*, 105052. DOI: 10.1016/j.resconrec.2020.105052 PMID: 32834486

Sharma, K. K. (2019). *Innovation and intellectual property in India: A handbook for practitioners*. Universal Law Publishing.

Sharma, P., Sharma, S. R., & Mittal, T. C. (2020). Effects and application of ionizing radiation on fruits and vegetables: A review. *Journal of Agricultural Engineering, 57*(2), 97–126. DOI: 10.52151/jae2020572.1708

Sharma, P., Shehin, V. P., Kaur, N., & Vyas, P. (2019). Application of edible coatings on fresh and minimally processed vegetables: A review. *International Journal of Vegetable Science, 25*(3), 295–314. DOI: 10.1080/19315260.2018.1510863

Sharma, R. R., Raghavendra, H. R., & Sethi, S. (2021). Postharvest Treatments for Horticultural Produce. In *Postharvest Handling and Diseases of Horticultural Produce* (pp. 43–56). CRC Press. DOI: 10.1201/9781003045502-4

Sharma, R., & Ghoshal, G. (2018). Emerging trends in food packaging. *Nutrition & Food Science, 48*(5), 764–779. DOI: 10.1108/NFS-02-2018-0051

Sharma, S., Nakano, K., Kumar, S., & Katiyar, V. (2024). Edible packaging to prolong postharvest shelf-life of fruits and vegetables: A review. *Food Chemistry Advances, 4*, 100711. DOI: 10.1016/j.focha.2024.100711

Sharma, S., Sudhakara, P., Singh, J., Ilyas, R. A., Asyraf, M. R. M., & Razman, M. R. (2021). Critical review of biodegradable and bioactive polymer composites for bone tissue engineering and drug delivery applications. *Polymers, 13*(16), 2623. DOI: 10.3390/polym13162623 PMID: 34451161

Sheth, J., & Parvatiyar, A. (2020). Sustainable marketing: Market-driving, not market-driven. *Journal of Macromarketing, 41*(1), 150–165. DOI: 10.1177/0276146720961836

Shiekh, K. A., Ngiwngam, K., & Tongdeesoontorn, W. (2021). Polysaccharide-based active coatings incorporated with bioactive compounds for reducing postharvest losses of fresh fruits. *Coatings, 12*(1), 8. DOI: 10.3390/coatings12010008

Shih, J.-C., Lin, F.-Y., Hong, M.-D., Lin, H.-R., & Wen, J.-C. 2022. "Application on the Micro-Hydropower Generation Benefits of Agricultural Channels and the Water- Energy-Food Nexus." DOI: 10.5194/egusphere-egu22-2132

Shin, J., & Selke, S. E. M. (2014). Food packaging. In Clark, S., Jung, S., & Lamsal, B. (Eds.), *Food processing: principles and applications* (pp. 249–273). Wiley., DOI: 10.1002/9781118846315.ch11

Shoreh, M. H., Siano, P., Shafie-khah, M., Loia, V., & Catalao, J. P. (2016). A survey of industrial applications of Demand Response. *Electric Power Systems Research, 141*, 31–49. DOI: 10.1016/j.epsr.2016.07.008

Shrouf, F., & Miragliotta, G. (2015). Energy management based on Internet of Things: Practices and framework for adoption in production management. *Journal of Cleaner Production, 100*, 235–246. DOI: 10.1016/j.jclepro.2015.03.055

Sillman, J., Uusitalo, V., Ruuskanen, V., Ojala, L., Kahiluoto, H., Soukka, R., & Ahola, J. (2020). A Life Cycle Environmental Sustainability Analysis of Microbial Protein Production via Power-to-Food Approaches. *The International Journal of Life Cycle Assessment*, *25*(11), 2190–2203. DOI: 10.1007/s11367-020-01771-3

Silvenius, F., Grönman, K., Katajajuuri, J., Soukka, R., Koivupuro, H., & Virtanen, Y. (2013). The role of household food waste in comparing environmental impacts of packaging alternatives. *Packaging Technology & Science*, *27*(4), 277–292. DOI: 10.1002/pts.2032

Simola, A., Boysen, O., Ferrari, E., Nechifor, V., & Boulanger, P. (2022). Economic integration and food security – The case of the AfCFTA. In *Global Food Security, 35, 100651*. Elsevier BV., DOI: 10.1016/j.gfs.2022.100651

Sims, R. E. H. (2014) *Global energy resources, supply and demand, energy security and on-farm energy efficiency, in Sustainable Energy Solutions in Agriculture*. J. Bundschuh and G. Chen, Eds, 1, 1952

Sims, R. E., Bundschuh, J., & Chen, G. (2014). Global energy resources, supply and demand, energy security and on-farm energy efficiency. *Sustainable Energy Solutions in Agriculture*, *1*, 19–52.

Sims, R., Puri, A. F. M., & Bracco, S. (2015). *Opportunities For Agri-Food Chains To Become Energy-Smart. Food and Agriculture Organization of the United Nations*. FAO.

Sindhu, R., Binod, P., Ummalyma, S. B., Abraham, A., Mathew, A. K., Madhavan, A., Rebello, S., & Pandey, A. (2018, January 1). Applications of Microbial Enzymes in Food Industry. *Food Technology and Biotechnology*, *56*(1). Advance online publication. DOI: 10.17113/ftb.56.01.18.5491

Singh, M., Rama, E N., Kataria, J., Leone, C., & Thippareddi, H. (2020, July 28). Emerging Meat Processing Technologies for Microbiological Safety of Meat and Meat Products., 4(2)

Singh, B., & Singh, S. (2018). *Advances in postharvest technologies of vegetable crops*. CRC Press. DOI: 10.1201/9781315161020

Singh, D., Tembhare, M., Dikshit, A., Dangi, M. B., & Kumar, S. (2024). Technical and operational challenges in setting up a decentralized biogas plant: Opportunities and future perspective toward sustainable nation. *Process Safety and Environmental Protection*, *185*, 392–407. DOI: 10.1016/j.psep.2024.02.079

Singh, M., Kaul, A., Pandey, V., & Bimbraw, A. S. (2019). Weed management in vegetable crops to reduce the yield losses. *International Journal of Current Microbiology and Applied Sciences*, *8*(7), 1241–1258. DOI: 10.20546/ijcmas.2019.807.148

Singh, R. N., & Shukla, A. K. (2015). *Intellectual property rights in India: A handbook*. Bharat Law House.

Singh, R. P., & Ramanathan, A. L. (2019). Urbanization, River Yamuna Pollution, and the Impact of Plastic Waste on Water Quality. *Environmental Pollution*, *252*, 799–809.

Singh, R., & Kumar, P. (2022). Energy-efficient motors and drives in the food industry: Trends and challenges. *Industrial Electronics*, *68*(11), 9200–9214.

Singh, S., Dubey, A., Gangwar, V., Kumar, A., Kumar, A., Kumar, M., & Wamiq, M. (2023). Edible coatings for improving the storability of fresh fruits and vegetables: A review. *The Pharma Innovation*, *12*(6), 3992–4002.

Singh, S., Kumar, J., & Rao, P. (2018). Environmental impact assessment framework for product packaging. *Management of Environmental Quality*, *29*(3), 499–515. DOI: 10.1108/MEQ-10-2017-0105

Singh, T. (2022). Economic growth and the state of poverty in India: Sectoral and provincial perspectives. *Economic Change and Restructuring*, *55*(3), 1251–1302. DOI: 10.1007/s10644-021-09345-5

Siracusa, V., Ingrao, C., Lo Giudice, A., Mbohwa, C., & Dalla Rosa, M. (2014). Environmental assessment of a multilayer polymer bag for food packaging and preservation: An LCA approach. *Food Research International*, *62*, 151–161. DOI: 10.1016/j.foodres.2014.02.010

Sjah, T., & Zainuri, Z. (2020). Agricultural supply chain and food security. In *Zero Hunger* (pp. 79–88). Springer. DOI: 10.1007/978-3-319-95675-6_82

Sleem, S. T., Salam, D. A., Ghaddar, N., Abou Ghali, K., Chehab, G., Daghir, N., Doughan, Y., & Haddad, N. (2024). NuhadDaghir, Youssef Doughan, and Nicolas Haddad. "Solar-assisted poultry production in small-scale farms: A case study in the Bekaa semi-arid region, Lebanon.". *Energy, Sustainability and Society*, *14*(1), 8. DOI: 10.1186/s13705-023-00437-w

Smith, J., Firth, J., & Clark, S. (2020). Use of existing databases in clinical epidemiology. In Schmidt, S., & Pedersen, B. K. (Eds.), *Epidemiology: How to do clinical practice research* (pp. 111–130). Routledge.

Smith, L., & Jones, K. (2020). Advances in sustainable dairy farming. *Sustainability, 12*(7), 2985. DOI: 10.3390/su12072985

Smith, P., & Fitton, N. (2019). The role of renewable energy in the global energy transformation. *Energy Strategy Reviews, 24*, 38–50. DOI: 10.1016/j.esr.2019.01.006

Smith, P., & Gregory, P. J. (2013). Climate change and sustainable food production. *The Proceedings of the Nutrition Society, 72*(1), 21–28. DOI: 10.1017/S0029665112002832 PMID: 23146244

Smyth, S. J. (2020). Regulatory barriers to improving global food security. In *Global Food Security, 26, 100440*. Elsevier BV., DOI: 10.1016/j.gfs.2020.100440

Soltani, M., Moradi Kashkooli, F., Souri, M., Rafiei, B., Jabarifar, M., Gharali, K., & Nathwani, J. S. (2021). Environmental, Economic, and Social Impacts of Geothermal Energy Systems. *Renewable & Sustainable Energy Reviews, 140*(110750), 110750. DOI: 10.1016/j.rser.2021.110750

Sonka, S., Lee, H., & Shah, S. (2023). The YieldWise Approach to Post-Harvest Loss Reduction: Creating Market-Driven Supply Chains to Support Sustained Technology Adoption. *Agriculture, 13*(4), 910. DOI: 10.3390/agriculture13040910

Souto, L. R. F., Caliari, M., Júnior, M. S. S., Fiorda, F. A., & Garcia, M. C. (2016, June 1). Utilization of residue from cassava starch processing for production of fermentable sugar by enzymatic hydrolysis. Sociedade Brasileira de Ciência e Tecnologia de Alimentos, 37(1), 19-24. https://doi.org/DOI: 10.1590/1678-457x.0023

Sperberg, R., Kniss, R., & Ruegg, T. (2018). Energy Efficiency in California Cement Plants: Energy Savings Potential Realized by Accessing Utility Energy-Efficiency Program Resources. *IEEE Industry Applications Magazine, 24*(4), 12–16. DOI: 10.1109/MIAS.2017.2740453

Springmann, M., & Freund, F. (2022). Options for reforming agricultural subsidies from health, climate, and economic perspectives. *Nature Communications, 13*(1), 82. DOI: 10.1038/s41467-021-27645-2 PMID: 35013212

Steinfeld, H.. (2006). *Livestock's Long Shadow: Environmental Issues and Options*. FAO.

Stroker, J. J. (2003). What's the real cost of higher efficiency? *IEEE Industry Applications Magazine, 9*(3), 32–37. DOI: 10.1109/MIA.2003.1195680

Sturm, B., Meyers, S., Jones, M. C., & Schmitt, B. (2015). A survey of the energy consumption patterns and potential methods for energy efficiency and renewable energy integration within the food and beverage industries of five European countries. In *2015 ASABE Annual International Meeting* (p. 1). American Society of Agricultural and Biological Engineers.

Sulatri, N. L., Yogeswara, I. B. A., & Nursini, N. W. (2017, June 30). Efektifitas sinar ultraviolet terhadap cemaran bakteri patogen pada makanan cair sonde untuk pasien immune-compremissed. *Diponegoro University*, *5*(2), 112–118. DOI: 10.14710/jgi.5.2.112-118

Suresh, B. V., Shireesha, Y., Kishore, T. S., Dwivedi, G., Haghighi, A. T., & Patro, E. R. (2023). Natural energy materials and storage systems for solar dryers: State of the art. *Solar Energy Materials and Solar Cells*, *255*, 112276. DOI: 10.1016/j.solmat.2023.112276

Suryana, E. A., Mauludyani, A. V. R., Sumedi, S., Ariani, M., & Munarso, S. J. (2024). Reducing post-harvest loss and waste along the rice supply chain and consumption. *BIO Web of Conferences, 119*, 1013.

Sustainable Food Processing. (n.d.). Wiley.com. https://www.wiley.com/en-us/Sustainable+Food+Processing-p-9780470672235

ŚwiĄtek, L.ŚwiĄtek. (2020). A Case Study of Geothermal Resources Use for the Innovative Aquaculture from Perspective of Syntropic Development Concept. *International Journal of Energy Production and Management*, *5*(1), 60–69. DOI: 10.2495/EQ-V5-N1-60-69

Syamaladevi, R. M., Tang, J., Villa-Rojas, R., Sablani, S., Carter, B., & Campbell, G. (2016). Influence of water activity on thermal resistance of microorganisms in low-moisture foods: A review. *Comprehensive Reviews in Food Science and Food Safety*, *15*(2), 353–370. DOI: 10.1111/1541-4337.12190 PMID: 33371598

T. Qi, J. Ji, X. Zhang, L. Liu, X. Xu, K. Ma, Y. Gao (2022) Research progress of cold chain transport technology for storage fruits and vegetables. Journal of Energy Storage 56(B): 105958. https://doi.org/DOI: 10.1016/j.est.2022.105958

Tadesse, K. T. (2024). The Role of Post-Harvest Management in Ensuring Food Security in a Changing World: Review Article. *J Clin Res Case Stud*, *2*(3), 1–14. DOI: 10.61440/JCRCS.2024.v2.41

Tadvi, S., Jain, V., & Thakkar, K. (2014). A Review: Solar Water Heating Systems. DOI: 10.13140/2.1.1910.5281

Tajeddin, B. (2017). Packaging composite materials from renewable resources. In *Handbook of Composites from Renewable Materials* (Vols. 1–8, pp. 525–561). wiley. DOI: 10.1002/9781119441632.ch58

Tan, D., & Dodd, T. (2023). Circular economy models for sustainable packaging in the food industry. *Sustainable Development*, *31*(2), 115–128.

Tapia, M. S., Alzamora, S. M., & Chirife, J. (2020). Effects of water activity (aw) on microbial stability as a hurdle in food preservation. *Water Activity in Foods: Fundamentals and Applications*, 323–355.

Tarroja, B., AghaKouchak, A., & Samuelsen, S. (2016). Quantifying climate change impacts on hydropower generation and implications on electric grid greenhouse gas emissions and operation. *Energy*, *111*, 295–305. DOI: 10.1016/j.energy.2016.05.131

Teague, W. R.. (2016). The role of rotational grazing and agroforestry in promoting soil health and biodiversity in dairy farming. *Land Use Policy*.

Teixeira, R. F., Balbinot Filho, C. A., & Borges, C. D. (2022). Essential oils as natural antimicrobials for application in edible coatings for minimally processed apple and melon: A review on antimicrobial activity and characteristics of food models. *Food Packaging and Shelf Life*, *31*, 100781. DOI: 10.1016/j.fpsl.2021.100781

Teutsch, B., & Kitinoja, L. (2019). *100 under $100: Tools for reducing postharvest losses. The Postharvest Education Foundation*. PEF.

Thaker, M., Zambre, A., & Bhosale, H. (2018). Wind Farms Have Cascading Impacts on Ecosystems across Trophic Levels. *Nature Ecology & Evolution*, *2*(12), 1854–1858. DOI: 10.1038/s41559-018-0707-z PMID: 30397304

Thakur, J. K., Singh, S. K., & Ekanthalu, V. K. (2017). Integrating remote sensing, geographic information systems and global positioning system techniques with hydrological modeling. *Applied Water Science*, *7*(4), 1595–1608. DOI: 10.1007/s13201-016-0384-5

Thanasi Boçe, M., & Hoxha, J. (2024). Blockchain Technology as a Catalyst for Sustainable Development: Exploring Economic, Social, and Environmental Synergies. *Academic Journal of Interdisciplinary Studies*, *13*(2), 151. DOI: 10.36941/ajis-2024-0041

Thangapandi, P., Perception Of Sustainable Chai-Flasks & Its Influence On Gen Z's Buying Behavior Dr. N. Bargavi S. Bharhane Raghaventheran.

The Biological Diversity Act, 2002 (with the latest amendment). (2021). Government of India.

The Copyright Act, 1957 (with the latest amendment). (2012). Government of India.

The Patents Act, 1970 (with the latest amendment). (2020). Government of India.

The Semiconductor Integrated Circuits Layout-Design Act, 2000. (2017). Government of India.

The Trade Marks Act, 1999 (with the latest amendment). (2019). Government of India.

The World Bank. 2022, Climate Change. Available online: https://www.worldbank .org/en/topic/climatechange/overview (accessed on 2 December 2022).

Thevanes, N. (2020). Environmental Management Practices of an Apparel Manufacturer in Sri Lanka: A Critique. South Asian Journal of Social Studies and Economics, 1–10. https://doi.org/DOI: 10.9734/sajsse/2019/v5i330149

Thompson, R. B., Gallardo, M., & Elia, A. (2020). Decision support systems and models for aiding irrigation and nutrient management of vegetable crops. *Agricultural Water Management*, *240*, 106209. DOI: 10.1016/j.agwat.2020.106209

Thompson, T. M., Young, B. R., & Baroutian, S. (2020). Pelagic Sargassum for energy and fertiliser production in the Caribbean: A case study on Barbados. *Renewable & Sustainable Energy Reviews*, *118*, 109564. DOI: 10.1016/j.rser.2019.109564

Thorarinsdottir, R., & Unnthorsson, R. (2018). Direct Use of Geothermal Resources for Circular Food Production. *Proceedings*, *2*(8), 497.

Thornton, P. K. (2010). Livestock production: Recent trends, future prospects. *Philosophical Transactions of the Royal Society of London. Series B, Biological Sciences*, *365*(1554), 2853–2867. DOI: 10.1098/rstb.2010.0134 PMID: 20713389

Thyberg, K. L., & Tonjes, D. J. (2016). Drivers of food waste and their implications for sustainable policy development. *Resources, Conservation and Recycling*, *106*, 110–123. DOI: 10.1016/j.resconrec.2015.11.016

TI, M. S., Prakash, S., & Vikraman, K. (2024). Cooked Food Supply Chain ERP Model.

Tissot, R. N. (2009). Small Hydropower. In Renewable Energies (pp. 227-260). DOI: 10.1002/9780470612002.ch8

Tomoda, B. T., Yassue-Cordeiro, P. H., Ernesto, J. V., Lopes, P. S., Péres, L. O., da Silva, C. F., & de Moraes, M. A. (2020). Characterization of biopolymer membranes and films: Physicochemical, mechanical, barrier, and biological properties. In *Biopolymer membranes and films* (pp. 67–95). Elsevier. DOI: 10.1016/B978-0-12-818134-8.00003-1

Torshizi, M. V., Mighani, A. (2017). The application of solar energy in agricultural systems. 3(2), 234-240.

Touil, A., Ayad, K., El Hamidi, N., & Babounia, A. (2023). Toward greener supply chains: Analysis of the determining factors. *Environment and Ecology*, *14*(2), 114–126. DOI: 10.21511/ee.14(2).2023.09

Trinh, H. A., Truong, H. V. A., Do, T. C., Nguyen, M. H., & Ahn, K. K. (2022). Optimization-based energy management strategies for hybrid construction machinery: A review. *Energy Reports*, *8*, 6035–6057. DOI: 10.1016/j.egyr.2022.04.050

Truong, Q., (2019). Food packaging process material's effect on environment and their control measures.

Tsai, H.-W., & Lee, Y.-C. (2023). Effects of land use change and crop rotation practices on farmland ecosystem service valuation. *Ecological Indicators*, *155*, 110998. DOI: 10.1016/j.ecolind.2023.110998

Tunc, Y., & Dincer, I. (2021). Energy efficiency and sustainability in food industries: Strategies and case studies. *Journal of Cleaner Production*, *277*, 124040. DOI: 10.1016/j.jclepro.2020.124040

Türkmen, B. A. (2020). Renewable Energy Applications for Sustainable Agricultural Systems. *International Journal of Innovative Approaches in Agricultural Research*, *4*(4), 497–504. DOI: 10.29329/ijiaar.2020.320.11

Tuschl, M., & Kurevija, T. (2023). Revitalization Modelling of a Mature Oil Field with Bottom-Type Aquifer into Geothermal Resource—Reservoir Engineering and Techno-Economic Challenges. *Energies*, *16*(18), 6599. DOI: 10.3390/en16186599

Twidell, J. (2021). *Renewable Energy Resources* (4th ed.). Routledge., DOI: 10.4324/9780429452161

Types Of Food Processing Equipment. (2021, December 10). https://www.plantautomation-technology.com/articles/types-of-food-processing-equipment

Tze-Zhang, A., Salem, M., Kamarol, M., Das, H. S., Nazari, M. A., & Prabaharan, N. (2022) A comprehensive study of renewable energy sources: Classifications, challenges and suggestions, Energy Strategy Reviews, Volume 43, 100939, ISSN 2211-467X

Tzia, C., Tasios, L., Spiliotaki, T., Chranioti, C., & Giannou, V. (2016). 16 Edible Coatings and Films to Preserve Quality of Fresh Fruits and Vegetables. *Food Preservation, 531*.

Tzounis, A., Katsoulas, N., Bartzanas, T., & Kittas, C. (2017). Impact of precision agriculture on farming sustainability in Greece. *Renewable & Sustainable Energy Reviews*, *78*, 654–663. DOI: 10.1016/j.rser.2017.05.041

U.S department of energy About the Office of Energy Efficiency and Renewable Energy | Department of Energy (access on 16 January 2025).

U.S. Department of Agriculture. (2021). USDA releases study on hurdles to healthy eating in SNAP.https://www.usda.gov/media/press-releases/2021/06/23/usda-releases-study-hurdles-healthy-eating-snap

U.S. Government Accountability Office. (2019). Nutrition assistance programs: Agencies could do more to help address the nutritional needs of older adults.https://www.gao.gov/products/gao-20-18

UCS. (2008). Renewable Energy and Agriculture: A Natural Fit. Retrieved 20.07.2020 from https://www.ucsusa.org/resources/renewable-energy-and-agriculture?print=t # 27

Umaraw, P., & Verma, A. K. (2024). V. P. Singh, Pavan Kumar, N. Mehta, and Devendra Kumar. "Rabbit meat processing and safety.". In *Veterinary Care of Farm Rabbits: A Complete Practice Guide to Rabbit Medicine and Production* (pp. 317–335). Springer International Publishing. DOI: 10.1007/978-3-031-44542-2_14

Umeohia, U. E., & Olapade, A. A. (2024). Physiological Processes Affecting Post-harvest Quality of Fresh Fruits and Vegetables. *Asian Food Science Journal*, *23*(4), 1–14. DOI: 10.9734/afsj/2024/v23i4706

Ungureanu, C., Tihan, G., Zgârian, R., & Pandelea, G. (2023). Bio-Coatings for Preservation of Fresh Fruits and Vegetables. *Coatings*, *13*(8), 1420. DOI: 10.3390/coatings13081420

UNIDO. (2022). *Energy efficiency guide for dairy processing plants*. UNIDO Publications.

United Nations, Department of Economic and Social Affairs, Population Division. (2019). *World Population Prospects 2019: Highlights* (ST/ESA/SER.A/423). New York: United Nations. https://population.un.org/wpp/Publications/Files/WPP2019_Highlights.pdf

USDA (U.S. Department of Agriculture). (2022). *Livestock and Poultry: World Markets and Trade*. Retrieved from USDA Foreign Agricultural Service.

van der Voort, H. (2015). Co-Regulatory Failure in the Food Industry: Explaining Regulatory Failure by Means of Two Contrasting Interpretations of Governance. *European Journal of Risk Regulation*, *6*(4), 502–511. DOI: 10.1017/S1867299X00005067

Van Gorp, J. C. (2005). Using Key Performance Indicators To Manage Energy Costs. *Strategic Planning for Energy and the Environment*, *2*(2), 9–25. DOI: 10.1080/10485230509509683

Van Henten, E. J. (2020). Greenhouse technology for sustainable production in the 21st century. *Acta Horticulturae*, (1271), 15–26. DOI: 10.17660/ActaHortic.2020.1271.2

Van Milgen, J.. (2008). Precision nutrition in growing pigs: Potential and limitations. *Animal Feed Science and Technology*, *143*, 205–238.

Van Os, E. A. (2019). Geothermal energy in Dutch greenhouse horticulture: A review on economics and system development. *Energies*, *12*(10), 1983. DOI: 10.3390/en12101983

Varun, S. A., & Nautiyal, H. (2015). Environmental impacts of packaging materials. In Environmental Footprints of Packaging (pp. 115-137). Singapore: Springer Singapore.

Varžinskas, V., & Markevičiūtė, Z. (2020). Sustainable food packaging: Materials and waste management solutions. *Environmental Research, Engineering and Management*, *76*(3), 154–164. DOI: 10.5755/j01.erem.76.3.27511

Vásquez Araya, G. (2020). Sistema de Gestión de la Energía para una planta de conservas. *Yulök Revista de Innovación Académica*, *1*(1). Advance online publication. DOI: 10.47633/yulk.v4i1.263

Vazirov, R. A., Yu Sokovnin, S., Krivonogova, A. S., & Isaeva, A. G. (2024). Radiation surface antimicrobial processing of poultry meat and by-products using the nanosecond low-energy electron beam. *Radiation Physics and Chemistry*, *217*, 111528. DOI: 10.1016/j.radphyschem.2024.111528

Vega-Garzon, L. P., Jeimy, A. P. R., Maria, P. G. S., Maria, A. R. G., & Liz, A. P. J. (2022). Use of Geothermal Energy in the Food Industry: A Review. *Revista Ingenierias Univsidad de Medellin*, *21*(40), 67–86. DOI: 10.22395/rium.v21n40a5

Venendaal, R., Jørgensen, U., & Foster, C. A. (1997). European energy crops: A synthesis. *Biomass and Bioenergy*, *13*(3), 147–185. DOI: 10.1016/S0961-9534(97)00029-9

Venkatesh, J., Lee, S. Y., Back, S., Kim, T. G., Kim, G. W., Kim, J. M., Kwon, J. K., & Kang, B. C. (2023). Update on the genetic and molecular regulation of the biosynthetic pathways underlying pepper fruit color and pungency. *Current Plant Biology, 100303*(35-36), 100303. DOI: 10.1016/j.cpb.2023.100303

Victory, I., Adetunji, C. O., Olaniyan, O. T., Dauda, W. P., Adetuyi, B. O., Adetunji, J. B., Inobeme, A., Habib, A. I., Adeyomoye, O., & Adeniyi, M. J. (2023). Zero-energy evaporative cooling technology and the prospective for fresh produce postharvest handling most especially for the preservation of fruits and vegetable. In *Engineering principles, modeling and economics of evaporative coolers* (pp. 95–102). Elsevier. DOI: 10.1016/B978-0-323-90039-3.00018-9

Vieira, I. R. S., de Carvalho, A. P. A., & Conte-Junior, C. A. (2022). Recent advances in biobased and biodegradable polymer nanocomposites, nanoparticles, and natural antioxidants for antibacterial and antioxidant food packaging applications. *Comprehensive Reviews in Food Science and Food Safety, 21*(4), 3673–3716. DOI: 10.1111/1541-4337.12990 PMID: 35713102

Vieira, T. M., Moldão-Martins, M., & Alves, V. D. (2021). Composite coatings of chitosan and alginate emulsions with olive oil to enhance postharvest quality and shelf life of fresh figs (Ficus carica L. cv.'Pingo De Mel'). *Foods, 10*(4), 718. DOI: 10.3390/foods10040718 PMID: 33805309

Vignesh, N. S., Vimali, E., Rajkeerthana, S., Kaleeshwari, R., Kiruthika, M., Ashokkumar, B., & Rajendran, K. (2024). Utilization of woody biomass for biogas production. In *Sustainable Biorefining of Woody Biomass to Biofuels and Biochemicals* (pp. 103–123). Woodhead Publishing.

Vikumer, F. D. (2022, September 28). *Differences Between Vacuum Drying and Lyophilization Freeze Drying? - Vikumer Freeze Dry*. Vikumer Freeze Dry. https://vikumer.com/differences-between-vacuum-drying-and-lyophilization-freeze-drying/#:~:text=The%20main%20difference%20between%20them,vapor%20phase%20called%20vacuum%20drying

Vinitha, V., Preeyanghaa, M., Vinesh, V., Dhanalakshmi, R., Neppolian, B., & Sivamurugan, V. (2021). Two is better than one: Catalytic, sensing and optical applications of doped zinc oxide nanostructures. *Emergent Materials, 4*(5), 1093–1124. DOI: 10.1007/s42247-021-00262-x

Viswanadham, N., & Narahari, Y. (2015). *Performance modeling of automated systems*. PHI Learning Pvt. Ltd.

Vittuari, M., De Menna, F., García-Herrero, L., Pagani, M., Brenes-Peralta, L., & Segrè, A. (2019). Food systems sustainability: The complex challenge of food loss and waste. In *Sustainable Food Supply Chains* (pp. 249–260). Academic Press.

Vivek, K., & Singh, S. S., & RC, P. (2019). A review on postharvest management and advances in the minimal processing of fresh-cut fruits and vegetables. *Journal of Microbiology, Biotechnology and Food Sciences, 8*(5).

von Braun, J., Chichaibelu, B. B., Laborde Debucquet, D., & Torero, M. (2024). *Cost of ending hunger: Consequences of complacency, and financial needs for SDG2 achievement.*

Wackett, L. P. (2008). Polylactic acid (PLA) An annotated selection of World Wide Web sites relevant to the topics in Environmental Microbiology. *Microbial Biotechnology, 1*(5), 432–433. DOI: 10.1111/j.1751-7915.2008.00056.x

Wagner, M., Lask, J., Kiesel, A., Lewandowski, I., Weselek, A., Högy, P., Trommsdorff, M., Schnaiker, M.-A., & Bauerle, A. (2023). Agrivoltaics: The Environmental Impacts of Combining Food Crop Cultivation and Solar Energy Generation. *Agronomy (Basel), 13*(2), 299. DOI: 10.3390/agronomy13020299

Wallace, T. C., Bailey, R. L., Blumberg, J. B., Burton-Freeman, B., Chen, C. O., Crowe-White, K. M., Drewnowski, A., Hooshmand, S., Johnson, E., Lewis, R., Murray, R., Shapses, S. A., & Wang, D. D. (2020). Fruits, vegetables, and health: A comprehensive narrative, umbrella review of the science and recommendations for enhanced public policy to improve intake. *Critical Reviews in Food Science and Nutrition, 60*(13), 2174–2211. DOI: 10.1080/10408398.2019.1632258 PMID: 31267783

Walmsley, T. G., Varbanov, P. S., Su, R., Ong, B., & Lal, N. (2018). Frontiers in process development, integration and intensification for circular life cycles and reduced emissions. *Journal of Cleaner Production, 201*, 178–191. DOI: 10.1016/j.jclepro.2018.08.041

Wang, X., Puri, V. M., & Demirci, A. (2020). Equipment cleaning, sanitation, and maintenance. Food safety engineering, 333-353.

Wang, J., Li, W., Haq, S., & Shahbaz, P. (2023). Adoption of Renewable Energy Technology on Farms for Sustainable and Efficient Production: Exploring the Role of Entrepreneurial Orientation, Farmer Perception and Government Policies. *Sustainability (Basel), 15*(7), 5611. DOI: 10.3390/su15075611

Wang, K., Wang, Y., Sun, Y., Guo, S., & Wu, J. (2016). Green Industrial Internet of Things Architecture: An Energy-Efficient Perspective. *IEEE Communications Magazine*, *54*(11), 48–54. DOI: 10.1109/MCOM.2016.1600399CM

Wang, L. (2008). *Energy Efficiency and Management in Food Processing Facili ties*. CRC Press. DOI: 10.1201/9781420063394

Wang, L., Wang, B., Cen, W., Xu, R., Huang, Y., Zhang, X., Han, Y., & Zhang, Y. (2024). Ecological Impacts of the Expansion of Offshore Wind Farms on Trophic Level Species of Marine Food Chain. *Journal of Environmental Sciences (China)*, *139*(May), 226–244. DOI: 10.1016/j.jes.2023.05.002 PMID: 38105050

Wang, L., Wu, Z., & Cao, C. (2019, November 13). Technologies and Fabrication of Intelligent Packaging for Perishable Products. *Applied Sciences (Basel, Switzerland)*, *9*(22), 4858–4858. DOI: 10.3390/app9224858

Wang, S., Liu, X., Yang, M., Zhang, Y., Xiang, K., & Tang, R. (2015). Review of time temperature indicators as quality monitors in food packaging. *Packaging Technology & Science*, *28*(10), 839–867. DOI: 10.1002/pts.2148

Wang, S., Wei, Z., & Wang, L. (2024). Improving slaughterhouse byproducts utilization via anaerobic digestion, composting, and rendering. *Renewable & Sustainable Energy Reviews*, *189*, 113881. DOI: 10.1016/j.rser.2023.113881

Wang, S., Wu, H., Li, J., Xiao, Q., & Li, J. (2024). Assessment of the Effect of the Main Grain-Producing Areas Policy on China's Food Security. *Foods*, *13*(5), 654. DOI: 10.3390/foods13050654 PMID: 38472767

Wang, W., Wang, Y., Liu, Z., Han, Y., & Wang, C. (2019). Study on application performance of oxidized polyethylene wax in powder coatings. *Progress in Organic Coatings*, *136*, 105294. DOI: 10.1016/j.porgcoat.2019.105294

Wang, Y.. (2022). Utilizing big data in animal agriculture: A review. *Computers and Electronics in Agriculture*, *186*, 106215.

Wani, N. R., Rather, R. A., Farooq, A., Padder, S. A., Baba, T. R., Sharma, S., Mubarak, N. M., Khan, A. H., Singh, P., & Ara, S. (2024). New insights in food security and environmental sustainability through waste food management. *Environmental Science and Pollution Research International*, *31*(12), 17835–17857. DOI: 10.1007/s11356-023-26462-y PMID: 36988800

Warriner, K., Reddy, S. M., Namvar, A., & Neethirajan, S. (2014). Developments in nanoparticles for use in biosensors to assess food safety and quality. Trends in Food Science &amp. *Trends in Food Science & Technology*, *2*(2), 183–199. DOI: 10.1016/j.tifs.2014.07.008

Wasatkar, N., Namrata, P. G. C., & Kolekar, V. K. 2023. "IoT and AI-Based Smart Farm: Optimizing Crop Yield and Sustainability." In *Research Trends in Artificial Intelligence: Internet of Things*, 1–17. BENTHAM SCIENCE PUBLISHERS.

Wattie, Y., Suparno, O., & Raharja, S. (2024, June). Enhancing competitiveness in the Indonesian meat processing industry: A literature review. [). IOP Publishing.]. *IOP Conference Series. Earth and Environmental Science*, *1358*(1), 012042.

Watto, M. A.CSIRO Environment Business Unit,(2024). Solar-Powered Irrigation Systems: Recent Developments and Future Trends. In *Burleigh Dodds Series in Agricultural Science* (pp. 209–232). Burleigh Dodds Science Publishing.

Weaver, C. M., Dwyer, J., Fulgoni, V. L.III, King, J. C., Leveille, G. A., MacDonald, R. S., Ordovas, J., & Schnakenberg, D. (2014). Processed foods: Contributions to nutrition. *The American Journal of Clinical Nutrition*, *99*(6), 1525–1542. DOI: 10.3945/ajcn.114.089284 PMID: 24760975

Weidema, B. P., Pizzol, M., Schmidt, J., & Thoma, G. (2015). Methodology for life cycle assessment of milk production worldwide. *International Dairy Journal*, *31*, S66–S73.

Wei, H., Seidi, F., Zhang, T., Jin, Y., & Xiao, H. (2021). Ethylene scavengers for the preservation of fruits and vegetables: A review. *Food Chemistry*, *337*, 127750. DOI: 10.1016/j.foodchem.2020.127750 PMID: 32861166

Weiland, P. (2006). Biomass Digestion in Agriculture: A Successful Pathway for the Energy Production and Waste Treatment in Germany. Engineering in Life Sciences, 6, 302-309.p https://doi.org/doi:10.1002/elsc.200620128

Wijewardane, N. (2022). Sustainable Postharvest Management: The Key to Attaining Food Security. *78th 78th, 11*, 15.

Williams, H., & Wikström, F. (2011). Environmental impact of packaging and food losses in a life cycle perspective: A comparative analysis of five food items. *Journal of Cleaner Production*, *19*(1), 43–48. DOI: 10.1016/j.jclepro.2010.08.008

Williams, P. M., & Arezoo Dadrasnia, I. (2021). Sustainable nutrient recovery from animal manure: A review of current best practice technology and the potential for freeze concentration. *Journal of Cleaner Production*, *315*, 128106. DOI: 10.1016/j.jclepro.2021.128106

Wojdalski, J., Grochowicz, J., Dróżdż, B., Bartoszewska, K., Zdanowska, P., Kupczyk, A., Ekielski, A., Florczak, I., Hasny, A., & Wójcik, G. (2015). Energy efficiency of a confectionery plant – Case study. *Journal of Food Engineering*, *146*, 182–191. DOI: 10.1016/j.jfoodeng.2014.08.019

Workineh, M., & Enyew, M. (2021). *Review the Extent and Cause of Post Harvest Loss of Fruits and Vegetables in Ethiopia.*

World Intellectual Property Organization. (2018). *Sustainable development goals and intellectual property: The role of IP in promoting innovation for sustainable food systems*. WIPO.

World Resources Institute (WRI). (2022). The role of precision agriculture in reducing dairy sector emissions.

Wunderlich, S. M., & Martinez, N. M. (2018). Conserving natural resources through food loss reduction: Production and consumption stages of the food supply chain. *International Soil and Water Conservation Research*, *6*(4), 331–339. DOI: 10.1016/j.iswcr.2018.06.002

Wünsche, J. F., & Fernqvist, F. (2022). The potential of blockchain technology in the transition towards sustainable food systems. *Sustainability (Basel)*, *14*(13), 7739. DOI: 10.3390/su14137739

Wu, W., Ni, X., Shao, P., & Gao, H. (2021). Novel packaging film for humidity-controlled manipulating of ethylene for shelf-life extension of Agaricus bisporus. *Lebensmittel-Wissenschaft + Technologie*, *145*, 111331. DOI: 10.1016/j.lwt.2021.111331

Wu, Y., Wu, H., & Hu, L. (2023). Recent Advances of Proteins, Polysaccharides and Lipids-Based Edible Films/Coatings for Food Packaging Applications: A Review. *Food Biophysics*, •••, 1–17.

Xiaofeng, X., Xuelai, Z., & Munyalo, J. M. (2017). Simulation Study on Temperature Field and Cold Plate Melting of Cold Storage Refrigerator Car. *Energy Procedia*, *142*, 3394–3400. DOI: 10.1016/j.egypro.2017.12.476

Xue, Y., Qi, Z., Yan, J., Li, D., Zhao, H., & Zheng, H. (2024). Technical Efficiency and Allocative Efficiency of Beef Cattle Fattening in the Content of Digital Economy: An Empirical Study Based on Survey in China. *Agriculture*, *14*(7), 1007. DOI: 10.3390/agriculture14071007

Xu, Y., & Szmerekovsky, J. (2017). System dynamic modeling of energy savings in the US food industry. *Journal of Cleaner Production*, *165*, 13–26. DOI: 10.1016/j.jclepro.2017.07.093

Yadav, A., Kumar, N., Upadhyay, A., Sethi, S., & Singh, A. (2022). Edible coating as postharvest management strategy for shelf-life extension of fresh tomato (Solanum lycopersicum L.): An overview. *Journal of Food Science*, *87*(6), 2256–2290. DOI: 10.1111/1750-3841.16145 PMID: 35502679

Yahia, E. M., Fonseca, J. M., & Kitinoja, L. (2019). Postharvest losses and waste. *Postharvest Technology of Perishable Horticultural Commodities*, 43–69.

Yamprayoon, J., & Noomhorm, A. (2000). Effects of preservation methods on geosmin content and off-flavor in Nile tilapia (Oreochromis niloticus). *Journal of Aquatic Food Product Technology*, *9*(4), 95–107. DOI: 10.1300/J030v09n04_09

Yoe, H. (2024). ICT-based Digital Agriculture. 2024 IEEE/ACIS 22nd International Conference on Software Engineering Research, Management and Applications (SERA), 1–1. https://doi.org/DOI: 10.1109/SERA61261.2024.10685615

Yousuf, B., Sun, Y., & Wu, S. (2022). Lipid and lipid-containing composite edible coatings and films. *Food Reviews International, 38*(sup1), 574–597.

Yousuf, B., Qadri, O. S., & Srivastava, A. K. (2018). Recent developments in shelf-life extension of fresh-cut fruits and vegetables by application of different edible coatings: A review. *Lebensmittel-Wissenschaft + Technologie*, *89*, 198–209. DOI: 10.1016/j.lwt.2017.10.051

Yu, X., Pu, H., & Sun, D. W. (2023). Nano light conversion agents: Recent advances, future prospects and challenges in enhancing plant green cultivation. *Journal of Cleaner Production*, *426*, 138919. DOI: 10.1016/j.jclepro.2023.138919

Zahran, S. (2024). Optimizing Supply Chain Management of Fresh E-Commerce Agri-Consumer Products Using Energy-Efficient Vehicle Routing. *IEEE Transactions on Consumer Electronics*, *70*(1), 1685–1693. DOI: 10.1109/TCE.2024.3370845

Zaidi, N. A., Tahir, M. W., Vellekoop, M. J., & Lang, W. (2017). A gas chromatographic system for the detection of ethylene gas using ambient air as a carrier gas. *Sensors (Basel)*, *17*(10), 2283. DOI: 10.3390/s17102283 PMID: 28991173

Zainalabidin, F. A., Sagrin, M. S., Azmi, W. N. W., & Ghazali, A. S. (2019). Optimum postharvest handling-effect of temperature on quality and shelf life of tropical fruits and vegetables. [JTRSS]. *Journal of Tropical Resources and Sustainable Science*, *7*(1), 23–30. DOI: 10.47253/jtrss.v7i1.505

Zaller, J. G., & Zaller, J. G. (2020). What is the problem? Pesticides in our everyday life. *Daily Poison: Pesticides-an Underestimated Danger*, 1–125.

Zenda, M., & Malan, P. J. (2024). Investigation of Challenges Faced by Small-Scale Sheep Farmers in the Northern Cape (Hantam Karoo), South Africa. [SAJAE]. *South African Journal of Agricultural Extension*, *52*(1), 189–209. DOI: 10.17159/2413-3221/2024/v52n1a14663

Zhang, J. (2023). Energy Management System: The Engine for Sustainable Development and Resource Optimization. Highlights in Science, Engineering and Technology, 618–624. https://doi.org/DOI: 10.54097/cvfd9m83

Zhang, B.. (2020). Consumer awareness and demand for sustainable livestock products. *Sustainability*, *12*(5), 2188.

Zhang, L., Yin, G., Wei, Z., Li, W., Cui, C., Wang, M., Zhao, C., Zhao, H., & Xue, F. (2024). Potential and Pathways of Carbon Emission Reduction in China's Beef Production from the Supply Chain Perspective. *Agriculture*, *14*(7), 1190. DOI: 10.3390/agriculture14071190

Zhang, W. (2018). Global pesticide use: Profile, trend, cost/benefit and more. *Proceedings of the International Academy of Ecology and Environmental Sciences*, *8*(1), 1.

Zhang, W., & Rhim, J.-W. (2022). Titanium dioxide (TiO2) for the manufacture of multifunctional active food packaging films. *Food Packaging and Shelf Life*, *31*, 100806. DOI: 10.1016/j.fpsl.2021.100806

Zhang, X., Ismail, B. B., Cheng, H., Jin, T. Z., Qian, M., Arabi, S. A., Liu, D., & Guo, M. (2021). Emerging chitosan-essential oil films and coatings for food preservation-A review of advances and applications. *Carbohydrate Polymers*, *273*, 118616. DOI: 10.1016/j.carbpol.2021.118616 PMID: 34561014

Zhang, X., Jiang, D., Li, J., Zhao, Q., & Zhang, M. (2024). Carbon emission oriented life cycle assessment and optimization strategy for meat supply chain. *Journal of Cleaner Production*, *439*, 140727. DOI: 10.1016/j.jclepro.2024.140727

Zhang, Z., Zou, S., & Li, P. (2023). Aging of plastics in aquatic environments: Pathways, environmental behavior, ecological impacts, analyses and quantifications. *Environmental Pollution*, •••, 122926. PMID: 37963513

Zhao, J. (2023). Exploring the development trend of packaging design in the new era., 1272-1278. DOI: 10.2991/978-2-38476-092-3_162

Zhao, P., Ndayambaje, J. P., Liu, X., & Xia, X. (2022). Microbial spoilage of fruits: A review on causes and prevention methods. *Food Reviews International, 38*(sup1), 225–246.

Zhao, L., & Chien, S. (2018). Optimization of energy consumption in food processing industries using energy management systems. *Renewable & Sustainable Energy Reviews*, *81*, 296–304. DOI: 10.1016/j.rser.2017.07.040

Zhao, X.. (2020). Energy management in dairy processing: A comprehensive review. *Journal of Dairy Science*.

Zhou, X., Wu, Q., & Chen, Z. (2023). Advancements in energy recovery technologies for food production systems. *Food Science and Technology (Campinas)*, *56*(1), 45–59.

Zhou, X., Wu, Q., & Chen, Z. (2023). Advancements in heat recovery technologies for energy efficiency in food processing. *Journal of Food Science and Technology*, *60*(5), 1327–1339.

Zink, D. (1997). The Impact of Consumer Demands and Trends on Food Processing. *Emerging Infectious Diseases*, *3*(4), 467–469. DOI: 10.3201/eid0304.970408 PMID: 9366598

Zinkernagel, J., Valero, J. F., Seresti, S. Y., & Intrigliolo, D. S. (2020). New technologies and practical approaches to improve irrigation management of open field vegetable crops. *Agricultural Water Management*, *242*, 106404. DOI: 10.1016/j.agwat.2020.106404

Zisopoulos, F. K., Rossier-Miranda, F. J., van der Goot, A. J., & Boom, R. M. (2015). The use of exergetic indicators in the food industry – A review. *Critical Reviews in Food Science and Nutrition*, *1*(1), 197–211. DOI: 10.1080/10408398.2014.975335 PMID: 26066523

Živkovic´, N. (2009). *Polyhydroxy alkanoates, Green Chemistry (seminar)*. Food Technology Zagreb. (in Croatian)

Zuba-Ciszewska, M., & Suchoń, A. (2024). The Role of State Aid in the Achievement of the Energy Efficiency Objective in the Food Industry—The Example of Poland. *Energies*, *17*(12), 2895. DOI: 10.3390/en17122895

About the Contributors

Ahmed Hamad is a renowned food science and nutrition expert with years of experience in teaching, research, and industry. Dr. Ahmed has published numerous papers and books on food hygiene, nutrition, and bioscience and has worked with various government and non-government organizations on food safety and nutrition projects. Ahmed has been teaching and researching food science, Safety, technology, and quality for Twelve (12) years. Ahmed teaches Food Chemistry, Microbiology, Food Technology and processing, Food Safety, and TQM courses in the food industry. Ahmed has many publications (+30) in highly prestigious peer-reviewed journals. He also contributes to many funded scientific projects in food safety, antimicrobial resistance, biofilm formation and control, Meat quality improvement, and environmental pollution. Ahmed participated as a peer reviewer for many prestigious food journals. Ahmed Has a role as a Book Editor for IGI Global, Wiley, Springer and CRC. Ahmed is a member of many professional associations and societies, such as RSPH, IFST, IAFP, AMSA, and ASM. He has also worked as an associate food safety and quality consultant from 2012 till now. He conducts over 2000+ Food safety and quality audits and leads Consulting projects for ISO 22000, ISO 9001, BRC, IFS, SQF, FSSC 22000, and HACCP/HARPC. He also designs and develops SOPs, SSOPs, Food safety policies for the Food Industry, and Food catering.

Pallavi Singh is currently working as an Associate Professor in Faculty of Public Health, Poornima University, Jaipur Rajasthan India. She has received her B. Sc. Health Science (Health Management) Degree from SHIATS, Allahabad, M. Sc. In Nutritional Sciences Degree from University of Allahabad, India and Ph.D. in the area of Public Health Nutrition from SHUATS, Allahabad. She also qualified UGC-NET and worked as a Junior Research Fellow in the DST-UKIERI funded International Project. She published more than 10 papers in Peer Reviewed and Refereed Journals, 1 Patent publication, 3 books with International Publishers and 2 books with National Publishers, 5 book chapters and many articles in the magazines. She has organized more than 10 workshops, summer internships, training programs and national seminars and international conference as coordinator, convener, organizing secretary and chair. She has worked as a session chair, conference steering committee member, editorial board member, and reviewer in international/national Journal/ conferences. Her current research interests are Value addition, Community Nutrition and Public Health Nutrition.

Dipali Saxena, an accomplished Assistant Professor, boasts over 10 years of academic experience in Food Sciences and Technology. She earned her Ph.D. in Food Technology from Jiwaji University, Gwalior, specializing in Nutraceuticals and functional foods. Graduating as the university topper and receiving the MSc. Gold Medal in Food Technology from Jiwaji University, Gwalior, Madhya Pradesh, highlights her academic excellence. Currently, Dr. Saxena serves as the Senior Faculty and Assistant Professor in the Department of Food Nutrition at Shri Vaishnav Institute of Home Science and Shri Vaishnav Institute of Agriculture, part of Shri Vaishnav Vidyapeeth Vishwavidyalaya, Indore. Additionally, she holds positions as a Member of the Board of Studies (BOS), Faculty of Studies (FOS), and various committees including Admission, Cultural, Anti-ragging, ABC credit, and Library Committees at Shri Vaishnav Vidyapeeth Vishwavidyalaya, Indore. Furthermore, Dr. Saxena guides PhD and MSc dissertations at the institute. She has developed Elective and Vocational courses in the Department of Food and Nutrition, aligning with the New Education Policy 2020. Additionally, she has restructured bachelor's and master's degree courses as per the same policy.

A. K. Haghi, Ph.D., is a retired professor and has written, co-written, edited or co-edited more than 1000 publications, including books, book chapters, and papers in refereed journals with over 4000 citations and h-index of 33, according to Google Scholar database. He was appointed as Honorary Research Associate (HRA) of Bioprocesses and Bioproducts Group, Facultad de Ciencias Químicas, Universidad Autónoma de Coahuila, Saltillo, Coahuila, México. He has also been appointed as Honorary Research Associate (HRA) at Chemistry Centre, University of Coimbra, Portugal and held a Senior Research Fellowship (SRF) at the Tbilisi State University. He has supervised several PhD and MSc theses at the University of Guilan and co-supervised international doctoral projects. Professor Haghi holds a BSc in urban and environmental engineering from the University of North Carolina (USA) and holds two MSc, one in mechanical engineering from North Carolina State University (USA) and another one in applied mechanics, acoustics, and materials from the Université de Technologie de Compiègne (France). He was awarded a PhD in engineering sciences at Université de Franche-Comté (France). Professor

Haghi is a frequent speaker to national and international audiences. He is the Founder and former Editor-in-Chief of the International Journal of Chemoinformatics and Chemical Engineering and Polymers Research Journal. He has acted as an editorial board member of many international journals and is a persistent reviewer of leading international journals.

***

Madhumathi A is a B.Tech Food Technology student at K.S. Rangasamy College of Technology, Tiruchengode, with a CGPA of 8.1. She is a motivated individual with a strong interest in quality control, research and development, and new product development in the food industry. Through internships at Christy and Earthy Origins, she gained hands-on experience in Product processing, meat production monitoring, and food safety management systems (FSMS). During her internship at Christy Pvt Ltd, Madhumathi assisted in producing various baking products and improved his skills in processing, quality analysis, and equipment handling. At Earthy Origins, she focused on meat processing compliance, new product development, and temperature control systems to ensure proper storage conditions. She has undertaken significant academic major project on developing gummies using Hemidesmus Indicus root extract, which he presented at an international conference. His personal skills include leadership, time management, and the ability to adapt to industry best practices.

Ruby Celsia Arul Selvaraj is a dedicated educator and researcher specializing in Nanobiotechnology, with a passion for advancing scientific knowledge through teaching, research, and service. She currently serves as an Associate Professor at K.S. Rangasamy College of Technology, Tamil Nadu, where she teaches courses in Food Toxicology, Microbiology, Fermentation Technology, and Nano and Space Food to both undergraduate and postgraduate students. Dr. Ruby is highly committed to student success, employing experiential and activity-based learning approaches to ensure academic excellence. With a Ph.D. in Biotechnology from Anna University, her research focuses on applying nanomaterials in medical implants and animal models. She has extensive experience in coordinating accreditation processes like NAAC and NBA, organizing social outreach programs, and mentoring students. Her administrative roles include serving as Research and Development Coordinator, Internship Coordinator, and Lab Coordinator, highlighting her comprehensive contributions to academic institutions. Dr. Ruby's research has been funded by prestigious grants, including the DST Nanomission Project. She has published scholarly articles in reputed journals and co-authored the book Futuristic Trends in Biotechnology. Her technical expertise spans microscopy, image analysis tools, statistical software, and educational animation tools, reflecting her versatility in both academic and practical applications. With a strong focus on academic integrity, continuous learning, and student engagement, Dr. Ruby is committed to shaping future leaders in biotechnology while contributing to the global scientific community through her research and publications.

E. Chandrakumar is a Assistant Professor in the department of Electrical and Electronics Engineering working at K.S. Rangasamy college of Technology. He has received his M.E degree from Anna University Chennai. He has a teaching experience 2.8 years in total. His research interest includes Electronic Product Design, Embedded System, Power System, Electric Vehicle. He is currently working in research area focusing on Electric Vehicle. He has published more than 04 paper in international journal peer reviewed publications. He has published 06 papers in international conferences. He has 01 patent owned for SAFETECH PEN. He has published 01 book chapter. He has received Startup fund of Rs.7.5 Lakhs from MSME – for the project "SAFETECH PEN". He has additional responsibilities of Associate NCC Officer, KSRCT Idea Lab Tech Guru and IIC Coordinator.

Tatheer Fatima is a research scholar in the Food and Nutrition department at Era University, Lucknow. She is pursuing her PhD, focusing on the hepatoprotective effects of seeds and food products of Bunium persicum. Tatheer has actively contributed to the research field of food and nutrition through her participation in numerous conferences, where she has presented her findings. She has also published papers that emphasize the therapeutic properties of various spices and herbs. Known for her dedication, consistency, and significant contributions, Tatheer remains committed to advancing research in food and nutrition, aiming to uncover new insights and applications in this vital area of study.

Surendra Prakash Gupta is working as Associate Professor in the Department of Biotechnology, SVIS, Shri Vaishnav Vidhyapeeth Viswavidhayalaya, Indore. He has 16 years of academic experience under various capacities in University. He has worked in the area of Drug-DNA interactions, Bioenergy, and Computational Biology. He has qualified GATE in 2006 in Life Science. He has published 05 research papers in peer reviewed Journals, 11 book chapters, 05 conference papers and more than 35 conference abstracts. He has supervised B.Tech and M.Sc Biotechnology students successfully. He is also a member of several reputed professional bodies like, IAENG, Education and Development Research and ASR etc. and Lifetime Member of IARA, and ISTD, Indore Chapter.

Vishal Jain is presently working as a Professor at the Department of Computer Science and Engineering, School of Engineering and Technology, Sharda

University, Greater Noida, India. Before that, he worked for several years as an Associate Professor at Bharati Vidyapeeths Institute of Computer Applications and Management (BVICAM), New Delhi. He has more than 18 years of experience in the academics. He has earned degrees: Ph.D (CSE), M.Tech (CSE), MBA (HR), MCA, MCP, and CCNA. He has more than 1900 research citations with Google Scholar (h-index score 21 and i-10 index 49) and has authored more than 150 research papers in professional journals and conferences. He has authored and edited more than 60 books (most of them are indexed at the Scopus) with various reputed publishers, including Elsevier, Springer, IET, Apple Academic Press, CRC, Taylor and Francis Group, Scrivener, Wiley, Emerald, NOVA Science, River Publishers, IGI-Global and Bentham Science. He has more than 160 Scopus publications. He is the book series editor of 10 book series with the reputed international publishers. His research areas include machine learning, information retrieval, semantic web, ontology engineering, data mining, ad hoc networks, sensor networks and network security. He received a Young Active Member Award for the year 2012–13 from the Computer Society of India, and Best Faculty Award for the year 2017 and Best Researcher Award for the year 2019 from BVICAM, New Delhi.

Kriti Kaushik is currently working as an Assistant Professor of Law at GD Goenka University Gurugram Haryana. She has a teaching experience of 2 years and research experience of 4 years. She has pursued her Masters in Law from NALSAR University of Law, Hyderabad and completed her doctoral study from Manipal University Jaipur in 2024. Her key interest area is in Criminal Law Enforcement and Justice.

Shweta Keswani is an expert in Clinical Nutrition, Community Nutrition with 20 years of experience in National and International organizations. Past Working Organization as Nutrition Consultant/ Program Manager/ Project Officer Nutrition International MP - fortification program manager Medanta Indore - Chief dietician UNICEF MP - Nutrition consultant UNDP - Nutrition consultant Children International - program Manager National Institute of Public Cooperation and Child Development (NIPCCD) Delhi - Project Associate Nestle Delhi - Nutritionist National Certifications for Master Trainer Infant and Young Child Feeding practices Severe Acute Malnutrition Management NIPI/SABLA Master Trainer Diabetes Educator (Hope Foundation) Life Membership Indian Dietetics Association (IDA) Indian Society for training and development (ISTD) Indian association of parenteral and enteral nutrition (IAPEN)

Salman Khalil is a dedicated student pursuing an M.Sc. (Hons.) in Food Science and Technology at the Department of Food Science, MNS-University of Agriculture,

Multan. He is focused on advancing his knowledge in food science, with a keen interest in innovative food processing techniques and nutrition. Salman is committed to contributing to the field through his academic endeavors and research.

Sapna Kundu is currently pursuing her Ph.D from the University Institute of Biotechnology (UIBT), Chandigarh University, Mohali, Punjab. She also has done a 6-month research project on "Preparation of fermented guava juice with lactic acid bacteria and check its antimicrobial activity" from Orbit Biotech, Mohali, Punjab. She has done her post-graduation in Biotechnology from Chaudhary Devi Lal University, Sirsa, Haryana. Till now, she has published 1 review article in a peer-reviewed international journal. She has participated in 3 International conferences and also attended another 2 International conferences. Her area of interest is sustainable packaging, edible films, edible coatings, and microbiology. She is actively engaged in research activities.

Archan Mitra is an Assistant Professor at School of Media Studies (SOMS) at Presidency University, Bangalore. He is the author of two book "Cases for Classroom Media and Entertainment Business" and "Multiverse and Media", he also has other several edited books to his credit. He has done his doctorate from Visva-Bharati Santiniketan, West Bengal in the field of "environmental informatics and communication for sustainability". In addition to that he is a certified Science Communicator and Journalism from Indian Science Communication Society (ISCOS), certified Corporate Trainer with Amity Institute of Training and Development, Certified Social Media Network Analyst. He has a strong interest in environmental communication. He was awarded certificate of merit by PRSI, Kolkata Chapter and Medal of Honor by Journalistic Club of Kolkata. He was working as a research assistant with the World Bank's "Environmental Capacity Building in Southeast Asia" project at IIM Kashipur. He was instrumental in launching the World Bank's Green MBA MOOC, he has also assisted in the research project on Uttarakhand disaster mitigation by ICSSR, the leading research on Uttarakhand disaster.

Fatima Noor is a dedicated research scholar in the Department of Home Science, specializing in the risk factors associated with Non-Alcoholic Fatty Liver Disease (NAFLD). Her academic journey is marked by a commitment to advancing knowledge in her field, leading her to publish numerous articles in esteemed peer-reviewed journals. In addition to her journal publications, Fatima has contributed chapters to various academic books, showcasing her expertise and insights on NAFLD. Fatima has actively participated in the academic community, presenting her research through posters and presentations at both national and international conferences. Her work not only highlights critical health issues but also aims to inform strategies for prevention and management. As a passionate researcher, Fatima continues to explore innovative approaches to enhance understanding of NAFLD, making significant contributions to the field of home science and public health.

Sarita Rana is a distinguished academician with over 14 years of extensive experience in the field of commerce and finance. She holds a PhD in Commerce and an MBA with a specialization in Finance, underscoring her deep knowledge and expertise in these areas. Dr. Rana's commitment to academic excellence and research is evident through her numerous accolades and contributions to the scholarly community.She has guided many students in their minor research project and guided PhD Students as supervisor from DAVV, Indore. She has a good background in research, has published many researcher papers in many respected journals, and has also choreographed and organized international conferences.

Jyoti Rani's doctoral research was carried with thesis title "Cyto-morphological, Biochemical and Molecular Characterization of Indian Germplasm of Catharanthus roseus (L.) G. Don". The study was carried out to evaluate the genetic diversity of Catharanthus roseus germplasm collected on the basis of morphological variations. Cytological, morphological, biochemical and molecular characterization of the 25 accessions was done. Meiotic studies, morphological traits, pollen fertility, pollen size, pollen shape (SEM) were used for cyto-morphological analysis. Biochemical characterization with respect to total phenolic content, total flavonoid content, SDS-PAGE analysis, HPLC, and GC-MS analysis; antioxidant and anti-microbial activities, and Simple Sequence Repeat markers were used for molecular diversity analysis.

Dhanush Anand S is a B.Tech Food Technology student at K.S. Rangasamy College of Technology, Tiruchengode, with a CGPA of 8.7. He is a motivated individual with a strong interest in quality control, research and development, and new product development in the food industry. Through internships at SKA Dairy Foods and Earthy Origins, he gained hands-on experience in milk product processing, meat production monitoring, and food safety management systems (FSMS). During his internship at SKA Dairy Foods, Dhanush assisted in producing various dairy products and improved his skills in milk processing, quality analysis, and equipment handling. At Earthy Origins, he focused on meat processing compliance, new product development, and temperature control systems to ensure proper storage conditions. He has undertaken significant academic projects, including a mini-project on gluten-free product development using dried banana powder and a major project on developing gummies using Hemidesmus Indicus root extract, which he presented at an international conference. His personal skills include leadership, time management, and the ability to adapt to industry best practices.

Kavin S is a B.Tech Food Technology student at K.S. Rangasamy College of Technology, Tiruchengode, with a CGPA of 7.9. He is a motivated individual with a strong interest in quality control, research and development, and new product development in the food industry. Through internships at Aavin and Earthy Origins, he gained hands-on experience in milk product processing, meat production monitoring, and food safety management systems (FSMS). During his internship at aavin, Kavin assisted in producing various dairy products and improved his skills in milk processing, quality analysis, and equipment handling. At Earthy Origins, he focused on meat processing compliance, new product development, and temperature control systems to ensure proper storage conditions. He has undertaken significant academic projects, including a mini-project on gluten-free product development using dried banana powder and a major project on developing gummies using Hemidesmus Indicus root extract, which he presented at an international conference. His personal skills include leadership, time management, and the ability to adapt to industry best practices.

Amrita Singh possesses experience as an Assistant Professor of Law, spanning over eight years. Her expertise encompasses a diverse range of subjects including Criminal Law, Administrative Law, Family Law, Constitutional Law, and Contract Law. Previously, She served as an Assistant Professor at both Shri Vaishnav Vidyapeeth Vishwavidyalaya and Indore Institute of Law, located in Indore, Madhya Pradesh. Ms. Singh has demonstrated her commitment to academic research by presenting over 30 research papers at various National and International Seminars & Conferences. Additionally, her scholarly contributions extend to publications in reputable platforms such as Scopus and the UGC Care List, showcasing the depth of her knowledge and engagement in the field of law.

Ainee Tanwir is a qualified Dietitian and Nutritionist currently working as a Nutrition Insight Architect with the Pikky Tech Private Limited organization. A lifetime member of the Indian Dietetic Association (IDA), she possesses over six years of experience in nutritional therapy, focusing on pediatric nutrient management, sports nutrition, and allergy testing in children. Ainee specializes in women's health issues, including PCOS, diabetes, and postpartum nutrition. She also has expertise in nutrigenomics, exploring how genetic variations influence dietary needs. Additionally, she is a certified Food Safety Lead Auditor with experience in regulatory affairs and new product development.

Priyamvada Tiwari is an accomplished legal scholar and educator with a distinguished academic background. Dr. Tiwari is the Associate Professor and Head of the Department of Law at the Faculty of Law, Medicaps University. With over 12 years of teaching and research experience, she has previously held notable positions, including Vice Principal, Head of the Department, and Academic Coordinator.

Manisha Trivedi holds a Master's degree in Food and Nutrition from DAVV, Indore. Working as an Assistant Professor at Shri Vaishnav Vidyapeeth Vishwavidyalaya. Formerly a Senior Dietician at Babyliss Slimming Clinic, with over 12 years of experience in the field of nutrition and dietetics, where she provided personalized nutritional counseling, developed tailored diet plans, and helped clients improve their health, energy, and fitness levels. Additionally, served as a Guest Faculty at Choithram College of Nursing, contributing to the academic development of students in food and nutrition.

Index

F

G

I

L

M

N

P

R

S

T

V

W

www.ingramcontent.com/pod-product-compliance
Lightning Source LLC
LaVergne TN
LVHW080614300926
849162LV00005B/1151